Elliptic Curves

Elliptic Curves

by

Anthony W. Knapp

Mathematical Notes 40

PRINCETON UNIVERSITY PRESS

PRINCETON, NEW JERSEY
1992

Copyright © 1992 by Princeton University Press
ALL RIGHTS RESERVED

The Princeton Mathematical Notes are edited by
Luis A. Caffarelli, John N. Mather, and Elias M. Stein

Princeton University Press books are printed on acid-free paper, and meet the guidelines for permanence and durability of the Committee on Production Guidelines for Book Longevity of the Council on Library Resources

Printed in the United States of America

Library of Congress Cataloging-in-Publication Data

Knapp, Anthony W.
  Elliptic curves / by Anthony W. Knapp.
  p. cm.—(Mathematical notes ; 40)
  Includes bibliographical references and index.
  ISBN 0-691-08559-5 (PB)
  1. Curves, Elliptic. I. Title. II. Series: Mathematical notes
(Princeton University Press) ; 40.
QA567.2.E44K53 1993   516.3'52—dc20   92-22183

3 5 7 9 10 8 6 4

To Susan

CONTENTS

	List of Figures	x
	List of Tables	x
	Preface	xi
	Standard Notation	xv
I.	Overview	3
II.	Curves in Projective Space	
	1. Projective Space	19
	2. Curves and Tangents	24
	3. Flexes	32
	4. Application to Cubics	40
	5. Bezout's Theorem and Resultants	44
III.	Cubic Curves in Weierstrass Form	
	1. Examples	50
	2. Weierstrass Form, Discriminant, j-invariant	56
	3. Group Law	67
	4. Computations with the Group Law	74
	5. Singular Points	77
IV.	Mordell's Theorem	
	1. Descent	80
	2. Condition for Divisibility by 2	85
	3. $E(\mathbb{Q})/2E(\mathbb{Q})$, Special Case	88
	4. $E(\mathbb{Q})/2E(\mathbb{Q})$, General Case	92
	5. Height and Mordell's Theorem	95
	6. Geometric Formula for Rank	102
	7. Upper Bound on the Rank	107
	8. Construction of Points in $E(\mathbb{Q})$	115
	9. Appendix on Algebraic Number Theory	122
V.	Torsion Subgroup of $E(\mathbb{Q})$	
	1. Overview	130
	2. Reduction Modulo p	134
	3. p-adic Filtration	137
	4. Lutz-Nagell Theorem	144
	5. Construction of Curves with Prescribed Torsion	145
	6. Torsion Groups for Special Curves	148

CONTENTS

VI. Complex Points
1. Overview — 151
2. Elliptic Functions — 152
3. Weierstrass $\wp$ Function — 153
4. Effect on Addition — 162
5. Overview of Inversion Problem — 165
6. Analytic Continuation — 166
7. Riemann Surface of the Integrand — 169
8. An Elliptic Integral — 174
9. Computability of the Correspondence — 183

VII. Dirichlet's Theorem
1. Motivation — 189
2. Dirichlet Series and Euler Products — 192
3. Fourier Analysis on Finite Abelian Groups — 199
4. Proof of Dirichlet's Theorem — 201
5. Analytic Properties of Dirichlet L Functions — 207

VIII. Modular Forms for $SL(2, \mathbf{Z})$
1. Overview — 221
2. Definitions and Examples — 222
3. Geometry of the q Expansion — 227
4. Dimensions of Spaces of Modular Forms — 231
5. L Function of a Cusp Form — 238
6. Petersson Inner Product — 241
7. Hecke Operators — 242
8. Interaction with Petersson Inner Product — 250

IX. Modular Forms for Hecke Subgroups
1. Hecke Subgroups — 256
2. Modular and Cusp Forms — 261
3. Examples of Modular Forms — 265
4. L Function of a Cusp Form — 267
5. Dimensions of Spaces of Cusp Forms — 271
6. Hecke Operators — 273
7. Oldforms and Newforms — 283

X. L Function of an Elliptic Curve
1. Global Minimal Weierstrass Equations — 290
2. Zeta Functions and L Functions — 294
3. Hasse's Theorem — 296

CONTENTS ix

XI.	Eichler-Shimura Theory	
	1. Overview	302
	2. Riemann surface $X_0(N)$	311
	3. Meromorphic Differentials	312
	4. Properties of Compact Riemann Surfaces	316
	5. Hecke Operators on Integral Homology	320
	6. Modular Function $j(\tau)$	333
	7. Varieties and Curves	341
	8. Canonical Model of $X_0(N)$	349
	9. Abstract Elliptic Curves and Isogenies	359
	10. Abelian Varieties and Jacobian Variety	367
	11. Elliptic Curves Constructed from $S_2(\Gamma_0(N))$	374
	12. Match of L Functions	383
XII.	Taniyama-Weil Conjecture	
	1. Relationships among Conjectures	386
	2. Strong Weil Curves and Twists	392
	3. Computations of Equations of Weil Curves	394
	4. Connection with Fermat's Last Theorem	397
	Notes	401
	References	409
	Index of Notation	419
	Index	423

LIST OF FIGURES

1.1	Method for obtaining **Q** solutions of $x^2 + y^2 = 1$	6
1.2	Newton's explanation of the Diophantus method	10
1.3	Chord-tangent composition rule	11
1.4	Group laws relative to different base points O	11
1.5	Negatives relative to the group law	12
1.6	Singular behavior	13
1.7	Graphs of **R** points of the elliptic curve $y^2 = P(x)$	14
3.1	Configuration for $(PP')(QQ') = (PQ)(P'Q')$	69
8.1	Fundamental domain for $SL(2,\mathbf{Z})$	228
8.2	Contour for calculating number of zeros	232
9.1	Fundamental domain for $\Gamma_0(2)$	260

LIST OF TABLES

1.1	Values of $\prod'_{p \leq R} N(p)/p$ for $y^2 = x^3 + ax$	17
3.1	Some elliptic curves with small negative discriminant	64
3.2	Some elliptic curves with small positive discriminant	65
4.1	Image of φ for $y^2 = x^3 - 4x$ with $x \notin \{-2, 0, 2\}$	111
4.2	Image of φ for $y^2 = x^3 - 4x$ with $x \in \{-2, 0, 2\}$	111
4.3	Adjusted image of φ for $y^2 = x^3 - 4x$	111
4.4	Image of φ for $y^2 = x^3 - p^2 x$ with $x \notin \{-p, 0, p\}$	113
4.5	Adjusted image of φ for $y^2 = x^3 - p^2 x$	113
4.6	Some primes $p \equiv 5 \bmod 8$ that are congruent numbers	117
4.7	Effect on $E: y^2 = x^3 + 8x$ of Fermat's descent	121
5.1	Examples of torsion subgroups of $E(\mathbf{Q})$	133
10.1	Effect of an admissible change of variables	291
12.1	Conductors of some elliptic curves	391
12.2	Curves $X_0(N)$ of low genus	395

PREFACE

For introductory purposes, an elliptic curve over the rationals is an equation $y^2 = P(x)$, where P is a monic polynomial of degree three with rational coefficients and with distinct complex roots.

The points on such a curve, together with a point at infinity, form an abelian group under a geometric definition of addition. Namely if we take two points on the curve and connect them by a line, the line will intersect the curve in a third point. The reflection of that third point in the x-axis is taken as the sum of the given points. The identity is the point at infinity. According to Mordell's Theorem, the abelian group of points on the curve with rational coordinates is finitely generated. A theorem of Lutz and Nagell describes the torsion subgroup completely, but the rank of the free abelian part is as yet not fully understood.

This is the essence of the basic theory of rational elliptic curves. The first five of the twelve chapters of this book give an account of this theory, together with many examples and number-theoretic applications. This is beautiful mathematics, of interest to people in many fields. Except for one small part of the proof of Mordell's Theorem, it is elementary, requiring only undergraduate mathematics. Accordingly the presentation avoids most of the machinery of algebraic geometry.

A related theory concerns elliptic curves over the complex numbers, or Riemann surfaces of genus one. This subject requires complex variable theory and is discussed in Chapter VI. It leads naturally to the topic of modular forms, which is the subject of Chapters VIII and IX.

But the book is really about something deeper, the twentieth-century discovery of a remarkable connection between automorphy and arithmetic algebraic geometry. This connection first shows up in the coincidence of L functions that arise from some very special modular forms ("automorphic" L functions) with L functions that arise from number theory ("arithmetic" or "geometric" L functions, also called "motivic"). Chapter VII introduces this theme. The automorphic L functions have manageable analytic properties, while the arithmetic L functions encode subtle number-theoretic information. The fact that the arithmetic L functions are automorphic enables one to bring a great deal of mathematics to bear on extracting the number-theoretic information from the L function.

The prototype for this phenomenon is the Riemann zeta function $\zeta(s)$, which should be considered as an arithmetic L function defined initially for Re $s > 1$. An example of subtle number-theoretic information that $\zeta(s)$ encodes is the Prime Number Theorem, which follows from the nanvanishing of $\zeta(s)$ for Re $s = 1$. In particular, this property of $\zeta(s)$ is a property of points s outside the initial domain of $\zeta(s)$. To get a handle on analytic properties of $\zeta(s)$, one proves that $\zeta(s)$ has an analytic continuation and a functional equation. These properties are completely formal once one establishes a relationship between $\zeta(s)$ and a theta function with known transformation properties. Establishing this relationship is the same as proving that $\zeta(s)$ is an automorphic L function.

The main examples of Chapter VII are the Dirichlet L functions $L(s,\chi)$. These too are arithmetic L functions defined initially for Re $s > 1$. They encode Dirichlet's Theorem on primes in arithmetic progressions, which follows from the nonvanishing of all $L(s,\chi)$ at $s = 1$. As with $\zeta(s)$, the relevant properties of $L(s,\chi)$ are outside the initial domain. Also as with $\zeta(s)$, one gets at the analytic continuation and functional equation of $L(s,\chi)$ by identifying $L(s,\chi)$ with an automorphic L function.

The examples at the level of Chapter VII are fairly easy. Further examples, generalizing the Dirichlet L functions in a natural way, arise in abelian class field theory, are well understood even if not easy, and will not be discussed in this book. The simplest L functions that are not well understood come from elliptic curves. An elliptic curve has a geometric L function $L(s,E)$ initially defined for Re $s > \frac{3}{2}$. An example conjecturally of the subtle information that $L(s,E)$ encodes is the rank of the free abelian group of rational points on the curve. This rank is believed to be the order of vanishing of $L(s,E)$ at $s = 1$. Once again, the relevant property of $L(s,E)$ is outside the initial domain. To address the necessary analytic continuation, one would like to know that $L(s,E)$ is an automorphic L function. Work of Eichler and Shimura provides a clue where to look for such a relationship. Eichler and Shimura gave a construction for passing from certain cusp forms of weight two for Hecke subgroups of the modular group to rational elliptic curves. Under this construction, the L function of the cusp form (which is an automorphic L function) equals the L function of the elliptic curve. The Taniyama-Weil Conjecture expects conversely that every elliptic curve arises from this construction, followed by a relatively simple map between elliptic curves. This conjecture appears to be very deep; a theorem of Frey, Serre, and Ribet says that it implies Fermat's Last Theorem. The final three chapters discuss these matters, the last two take for granted more mathematics than do the earlier chapters.

If the theme were continued beyond the twelve chapters that are here, eventually it would lead to the Langlands program, which brings in representation theory on the automorphic side of this correspondence. As a representation theorist, I come to elliptic curves from the point of view of the Langlands program. Although the book neither uses nor develops any representation theory, elliptic curves do give the simplest case of the program where the correspondence of L functions is not completely understood. Furthermore representation-theoretic methods occasionally yield results about elliptic curves that seem inaccessible by classical methods. From my point of view, they are an appropriate place to begin to study and appreciate the Langlands program. A beginning guide to the literature in this area appears in the section of Notes at the end of the book.

This book grew out of a brilliant series of a half dozen lectures by Don Zagier at the Tata Institute of Fundamental Research in Bombay in January 1988. The book incorporates notes from parts of courses that I gave at SUNY Stony Brook in Spring 1989 and Spring 1990. The organization owes a great deal to Zagier's lectures, and I have reproduced a number of illuminating examples of his. I am indebted to Zagier for offering his series of lectures.

Much of the mathematics here can already be found in other books, even if it has not been assembled in quite this way. Some sections of this book follow sections of other books rather closely. Notable among these other books are Fulton [1989],* Hartshorne [1977], Husemoller [1987], Lang [1976] and [1987], Ogg [1969a], Serre [1973a], Shimura [1971a], Silverman [1986], and Walker [1950]. The expository paper Swinnerton-Dyer and Birch [1975] was also especially helpful. Detailed acknowledgments of these dependences may be found in the section of Notes at the end.

In addition, I would like to thank the following people for help in various ways, some large and some small: H. Farkas, N. Katz, S. Kudla, R. P. Langlands, S. Lichtenbaum, H. Matumoto, C.-H. Sah, V. Schechtman, and R. Stingley. The type-setting was by $\mathcal{A}_\mathcal{M}\mathcal{S}$-TeX, and the Figures were drawn with Mathematica®. Financial support in part was from the National Science Foundation in the form of grants DMS 87-23046 and DMS 91-00367.

A. W. Knapp
January, 1992

---

*A name followed by a bracketed year is an allusion to the list of References at the end of the book. The date is followed by a letter in case of ambiguity.

STANDARD NOTATION

Item	Meaning
$\#S$ or $|S|$	number of elements in S
$\emptyset$	empty set
A^c	A complement
n positive	$n > 0$
$\mathbf{Z}, \mathbf{Q}, \mathbf{R}, \mathbf{C}$	integers, rationals, reals, complex numbers
Re z	real part of z
Im z	imaginary part of z
$O(1)$	bounded term
$o(1)$	term tending to 0
$\doteq$	approximately numerically equal
$\sim$	asymptotic to, with ratio tending to 1
$\mathbf{Z}_m$	integers modulo m
$a \equiv b \bmod m$	m divides $a - b$
$a \mid b$	a divides b
$\mathrm{GCD}(a, b)$	greatest common divisor
1	multiplicative identity
1 or I	identity matrix
$\dim V$	dimension of vector space
V'	dual of vector space
$\mathrm{End}_k(V)$	linear maps of vector space to itself
$GL(n, k)$	general linear group over a field k
$SL(2, \mathbf{R})$, $SL(2, \mathbf{Z})$	group of 2-by-2 matrices of determinant 1
Tr A	trace of A
A^{tr}	transpose of A
$R^\times$	multiplicative group of invertible elements
$\bar{k}$	algebraic closure of k
$[A : B]$	index of B in A, or degree of A in B
$\mathrm{Aut}_k(K)$	automorphism group of K fixing k
$\sum \oplus$	direct sum (for emphasis)
π_1	fundamental group

Elliptic Curves

CHAPTER I

OVERVIEW

Diophantus lived in Alexandria around 250 A.D. He published a series of books, *Arithmetica*, in 13 volumes, which were lost for more than a thousand years. They were found again about 1570, and in the next century Fermat studied a translation. Despite the passage of so much time, much of Diophantus's work had still not been rediscovered. The books are in the style of problems and solutions. The early volumes introduced, apparently for the first time, algebraic notation and equations, as well as negative numbers. Later volumes dealt with number theory. The work of Diophantus is so stunning that equations to be solved in number theory are often called Diophantine equations in his honor.

Basic Problem (affine). We consider the locus $f(x,y) = 0$ with f a nonzero polynomial. To fix the ideas, think of $\mathbb{Q}$ coefficients and of solutions with x and y in $\mathbb{Q}$. Clearing denominators, we can always adjust f so that we have $\mathbb{Z}$ coefficients and are seeking $\mathbb{Q}$ solutions, i.e., solutions with x and y in $\mathbb{Q}$.

Basic Problem (projective). We study the locus $F(x,y,w) = 0$ with F a nonzero homogeneous polynomial of some degree d, and we seek "projective" $\mathbb{Q}$ solutions. This means we identify solutions (x,y,w) and $(\lambda x, \lambda y, \lambda w)$ for $\lambda \neq 0$ and we discard $(0,0,0)$. Rational solutions automatically give us integer solutions, by clearing denominators.

The two problems are related, in that we can pass from each to the other. Two examples will illustrate.

EXAMPLE 1. Fermat equation $x^d + y^d = z^d$. This equation is projective. We can reduce to the affine case by dividing by z^d: $u^d + v^d = 1$. Relatively prime integer solutions of $x^d + y^d = z^d$ (with $z \neq 0$) correspond to rational solutions of $u^d + v^d = 1$ provided that we identify a solution (x,y,z) with its negative $(-x,-y,-z)$. This process of reducing to the affine case loses "affine solutions at ∞," those with $z = 0$, but they are trivial here. **Fermat's Last Theorem** is the conjecture that the equation has no nontrivial solutions for $d > 2$.

3

EXAMPLE 2. The affine equation $y^2 = x^3 + 1$. This becomes a projective equation if we put $y = \frac{v}{w}$ and $x = \frac{u}{w}$. We get $wv^2 = u^3 + w^3$. (Or we simply insert powers of w to make all terms cubic: $wy^2 = x^3 + w^3$.)

Diophantus considered the affine problem in degrees 1, 2, and 3.

Case of degree 1.

An affine line is of the form

$$ax + by + cz = 0 \qquad \text{with } a \text{ and } b \text{ not both } 0.$$

Two such lines are the same if and only if the coefficients of one are a multiple of the coefficients of the other.

A projective line is of the form

$$ax + by + cw = 0 \qquad \text{with } a, b, c \text{ not all } 0.$$

Again, two such lines are the same if and only if the coefficients of one are a multiple of the coefficients of the other. The line with $a = b = 0$ and $c = 1$ is $w = 0$, which is the "line at ∞."

Two facts are clear:
1) $\mathbb{Q}$ solutions exist (projectively).
2) We can parametrize all $\mathbb{Q}$ solutions.

Case of degree 2.

First we consider this case projectively. Then it turns out to be possible to make a linear change of variables so that the equation is

$$ax^2 + by^2 + cw^2 = 0. \tag{1.1}$$

The questions we shall address are:

(I) Do (nonzero) solutions exist?

(II) If solutions exist, how are they parametrized?

Of these, only the second question was considered by Diophantus.

We shall return to the second question shortly. We begin with a discussion of Question I, first giving two examples.

EXAMPLE 1. $x^2 + y^2 + z^2 = 0$ has no $\mathbb{Q}$ solutions since it has no $\mathbb{R}$ solutions.

EXAMPLE 2. $x^2 + y^2 = 3z^2$ has no $\mathbb{Q}$ solutions since it has no solutions modulo 9 in which some variable is not divisible by 3. This condition is necessary since existence of a $\mathbb{Q}$ solution implies existence of a relatively prime $\mathbb{Z}$ solution. To see that the condition fails, we argue as follows: Modulo 3, the equation is $x^2 + y^2 \equiv 0$, which forces $x \equiv y \equiv 0$ mod 3. Then 3^2 divides $x^2 + y^2$ and so 3 divides z. Thus 3 divides all of x, y, and z, a contradiction.

More generally, suppose that a, b, c are relatively prime integers and no square divides a coefficient. Then two necessary conditions for the existence of solutions of (1.1) can be seen to be that

(1) (1.1) must have a nonzero solution in **R**
(2) (1.1) must have a relatively prime solution modulo p^m for each prime p and each integer $m > 0$.

It turns out that (2) is automatic for all m if p is odd and does not divide abc. If $p = 2$ or p divides abc, then the weaker necessary condition (3a) or (3b) below implies (2) for the corresponding p:

(3a) when an odd prime p divides abc, (1.1) must have a relatively prime solution modulo p
(3b) when $p = 2$, (1.1) must have a relatively prime solution modulo 16.

Theorem 1.1 (Legendre). If (1) holds and if (3) holds for $p = 2$ and for all odd primes p dividing abc, then (1.1) has a **Q** solution.

REMARKS: i) (1) and (3) are decidable. So we can decide the existence question, Question I. (And, parenthetically, the hypothesis on $p = 2$ in the theorem is not needed.)

ii) To fix the ideas, think of (1) and (2) as the conditions. The Hasse-Minkowski theorem generalizes the Legendre theorem to more variables: A homogeneous quadratic polynomial in n variables with integer coefficients has a nonzero solution in **Q** if and only if it has a nonzero solution in **R** and a relatively prime solution modulo p^m for each prime and each integer $m > 0$.

iii) The field of p-adic numbers, described below in §V.3, can be used to combine congruence information modulo p^m for all m. With this device, the Hasse-Minkowski theorem says: A homogeneous quadratic polynomial in n variables with integer coefficients has a nonzero solution in **Q** if and only if it has a nonzero solution in **R** and in the p-adic numbers for every prime p.

The idea that solutions over **R** and the p-adics ought to control solutions over **Q** is called the **Hasse Principle**. Unfortunately it is not universally true. As we shall see shortly, it fails even for homogeneous cubics in three variables. Nevertheless, the Hasse Principle is a useful place to begin the study of a class of projective equations over **Q**. We shall see better how the principle operates in Chapters V and VII.

In the language of algebraic number theory, the p-adics and **R** are indexed by "places." A narrow (but still false) form of the Hasse Principle then says: If a projective equation is locally solvable at every place, then it is globally solvable. This terminology is supposed to suggest an analogy with what happens in the theory of several complex variables. In that theory, we consider holomorphic functions and make a ring out of those near a point (technically by passing to germs). The ring that we make is a "local ring" in the sense that it has a unique maximal ideal. The global ring is the ring of all globally holomorphic functions on a given open set.

In number theory, we attempt the same thing. Our global number field is $\mathbb{Q}$. To this we attach "places," analogous to points, by an abstract definition that we shall not pursue. The places work out to be the primes p and also ∞. For each place, there is a "local field" - the p-adics and the reals. The sense in which the p-adics are a "local field" is that the p-adic integers form a "local ring," a ring with a unique maximal ideal. Local solvability is to be solvability in the local field for each place. The hope expressed in the Hasse Principle is that we can obtain global solutions from local ones, because finding local ones is easier.

We turn to a discussion of Question II, the parametrization of solutions. Let us begin with Diophantus's method to solve $x^2 + y^2 = 1$. Take a particular solution $(-1, 0)$. Choose a linear relation it satisfies, say $x = -1 + 2y$.

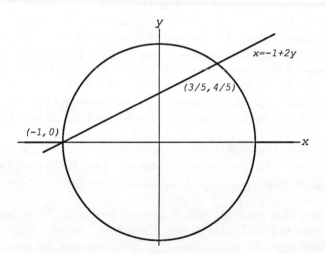

FIGURE 1.1. Method for obtaining $\mathbb{Q}$ solutions of $x^2 + y^2 = 1$

Substitution into $x^2 + y^2 = 1$ gives $(-1 + 2y)^2 + y^2 = 1$, and we are led to $1 - 4y + 4y^2 + y^2 = 1, 5y - 4 = 0, y = \frac{4}{5}, x = \frac{3}{5}$. We obtain

$(x, y) = (\frac{3}{5}, \frac{4}{5})$ as a nontrivial solution. Effectively $x = -1 + ty$ works for any rational t. We are led to

$$(x, y) = \left(\frac{t^2 - 1}{t^2 + 1}, \frac{2t}{t^2 + 1}\right). \tag{1.2}$$

Conversely any (x, y) must make a line with $(-1, 0)$ with $\frac{\Delta x}{\Delta y} =$ some t, and (x, y) is forced to be of this form.

The actual Diophantus method did not make reference to any geometry such as is Figure 1.1. The correspondence between algebra and geometry was unknown before Fermat and Descartes in the 17th century.

Diophantus used this analysis to parametrize **Pythagorean triples**: The problem is to find all relatively prime solutions of $a^2 + b^2 = c^2$ over $\mathbb{Z}$. For such a solution (a, b, c) with $a \neq 0$,

$$\frac{b}{a} = \frac{b/c}{a/c} = \frac{2t/(t^2 + 1)}{(t^2 - 1)/(t^2 + 1)} = \frac{2t}{t^2 - 1} = \frac{2m/n}{(m/n)^2 - 1} = \frac{2mn}{m^2 - n^2}.$$

So $(a, b, \pm c) = K(m^2 - n^2, 2mn, m^2 + n^2)$ with $K \in \mathbb{Q}$. Below we shall tidy matters up and come to the following conclusion.

Theorem 1.2 (Diophantus). In any Pythagorean triple (a, b, c), exactly one of a and b is odd. If a is odd and c is positive, then there exist integers m and n such that

$$(a, b, c) = (m^2 - n^2, 2mn, m^2 + n^2). \tag{1.3}$$

The integers m and n are relatively prime and not both odd. Conversely any pair (m, n) with these properties yields a Pythagorean triple via (1.3).

PROOF. The given equation taken modulo 4 shows that a and b cannot both be odd. Thus suppose a is odd and b is even, and write as above

$$(a, b, \pm c) = \frac{u}{v}(m^2 - n^2, 2mn, m^2 + n^2)$$

with u/v in lowest terms. Clearing fractions, we see that u divides $a, b,$ and c. Hence $u = \pm 1$ and we may take $u = +1$. We now have

$$v(a, b, \pm c) = (m^2 - n^2, 2mn, m^2 + n^2).$$

We can eliminate the ambiguous sign by redefining (m, n, v), leaving it alone if the sign is $+$ or replacing it by $(-n, m, -v)$ if the sign is $-$. Then we obtain

$$v(a, b, c) = (m^2 - n^2, 2mn, m^2 + n^2). \tag{1.4}$$

Since $c > 0$, $v > 0$. If m and n have any common factor, the square of the factor will appear in v, and we can cancel. Thus we may assume that m and n are relatively prime.

If p is an odd prime dividing v, then p divides both $m^2 - n^2$ and $m^2 + n^2$, hence both $2m^2$ and $2n^2$, hence both m and n. Thus no odd prime divides v.

If 2 divides v, then the evenness of vb implies that 4 divides $2mn$. Thus one of m or n must be even, and then the other must be odd. Hence $m^2 - n^2$ is odd, and va is odd, contradiction. Thus $v = 1$, and (1.4) reduces to (1.3).

In (1.3) it is clear that m and n are not both odd. Conversely, such a relatively prime pair (m, n) trivially yields a Pythagorean triple.

The method of Diophantus readily generalizes. Take $f(x,y) = 0$ to be any quadratic in (x,y). Suppose (p,q) is a solution. Then $y = q+t(x-p)$ and $x = p$ give all lines through (p,q). Substitute for y. Get a quadratic equation for x with a rational solution (for each t), namely $x = p$. Then the other solution must be rational, etc. Without even carrying out the computations, we can draw an important qualitative conclusion:

Proposition 1.3. If a quadratic equation $f(x, y) = 0$ over $\mathbb{Q}$ has one $\mathbb{Q}$ solution, it has infinitely many $\mathbb{Q}$ solutions.

EXERCISE: Parametrize the relatively prime integer solutions (a, b, c) of $2a^2 + b^2 = c^2$.

Case of degree 3.

First we consider this case projectively. There is no normal form in this generality. We shall address the same questions as in the case of degree 2:

(I) Do $\mathbb{Q}$ solutions exist?
(II) If $\mathbb{Q}$ solutions exist, how are they parametrized?

The second of these questions was considered briefly by Diophantus. The following example shows for Question I that the Hasse Principle fails.

EXAMPLE (Selmer): The projective equation $3x^3 + 4y^3 + 5z^3 = 0$ has a solution over $\mathbb{R}$ and over the p-adics for each p, but it has no $\mathbb{Q}$ solution.

Now let us discuss Question II. We begin by normalizing our cubic suitably. Taking Question I as settled, suppose that the homogeneous cubic $F(x, y, w) = 0$ has a $\mathbb{Q}$ solution. Suppose in particular that it

I: OVERVIEW

has an inflection point over $\mathbf{Q}$ in a sense that will be made precise in §II.3. If this inflection point is mapped to $(x, y, w) = (0, 1, 0)$ by a linear transformation in such a way that the tangent line becomes the line at ∞, and if the variables are scaled suitably, then the equation (in affine form) becomes

$$y^2 + a_1 xy + a_3 y = x^3 + a_2 x^2 + a_4 x + a_6.$$

(Notice that $(x, y, w) = (0, 1, 0)$ does solve this equation projectively.) Since $\mathbf{Q}$ does not have characteristic 2, we can complete the square in y and reduce matters to

$$y^2 = P(x), \qquad \text{where } \deg P = 3.$$

EXAMPLE: $u^3 + y^3 = w^3$. Put $\frac{u}{w} = \frac{3x}{y}, \frac{v}{w} = \frac{y-9}{y}$. The inverse is

$$x = \frac{3u}{w-v}, \qquad y = \frac{9w}{w-v}.$$

Then the equation becomes $y^2 - 9y = x^3 - 27$.

Anyway we have a solution (the one at ∞), and we want to know all solutions. Diophantus knew how to generate some solutions from others.

EXAMPLE: Given a number (e.g., 6), divide it into two pieces whose product is a cube diminished by its root.

SOLUTION: Let the pieces be y and $6 - y$. We want

$$y(6 - y) = x^3 - x.$$

The point $(x, y) = (-1, 0)$ is a trivial solution. It lies on a line $x = 2y - 1$. Substitution gives

$$6y - y^2 = (2y - 1)^3 - (2y - 1) = 8y^3 - 12y^2 + 4y.$$

Division by y gives a quadratic equation with irrational roots. We would surely have had rational roots if we could have made $4y$ on the right match $6y$ on the left. The 4 comes from twice the coefficient 2 of y. To get $6y$, we should use coefficient 3 for y. So $x = 3y - 1$. Then

$$6y - y^2 = 27y^3 - 27y^2 + 6y,$$
$$27y = 26.$$

So $y = \frac{26}{27}$ and $x = 3y - 1 = \frac{17}{9}$.

Newton's explanation of the method is as follows: Start with any line through $(-1, 0)$. Rotate it to be tangent. Then it intersects the given cubic twice at rational points; so the third intersection point must be rational. The picture in Figure 1.2 in the context of solutions over **R** makes matters much clearer.

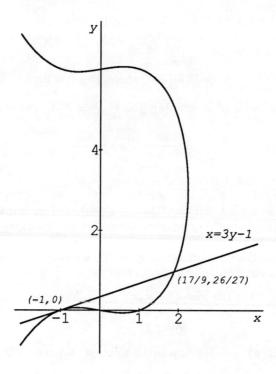

FIGURE 1.2. Newton's explanation of the Diophantus method

In summary, for this (and other) cubic equations, we have a systematic method to generate new solutions from old. The method does not give parametric families. As with $u^3 + v^3 = w^3$ we may not even get infinitely many solutions this way.

Let us analyze the method of Diophantus further, postponing detailed verifications to Chapters II and III. The above construction takes a rational solution P and produces another one, denoted PP or $P \cdot P$, by means of a tangent line. More generally, if P and Q are distinct rational solutions as in Figure 1.3, so is R. We write $R = PQ = P \cdot Q$. We have a systematic way of getting a third point from two. This operation is called the **chord-tangent composition rule**.

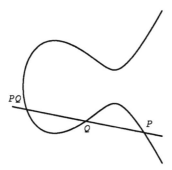

FIGURE 1.3. Chord-tangent composition rule

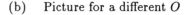

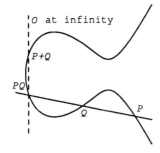 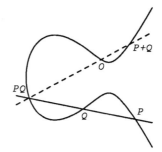

FIGURE 1.4. Group laws relative to different base points O

The chord-tangent composition rule is not applicable to certain pairs of points for a general cubic. But with an additional assumption of nonsingularity that we consider below, the rule is everywhere applicable and can be modified so as to give us an abelian group operation, as in Figure 1.4. For this purpose fix a point O on the curve. If $(x, y, w) = (0, 1, 0)$ is on the curve (i.e., if there are no y^3 terms), one frequently takes $O = (0, 1, 0)$. With O fixed, we define

$$P + Q = O \cdot PQ.$$

Let us address the group axioms. Associativity of + is complicated and is postponed to Chapter III. But commutativity is clear. Also O is an identity since
$$P + O = O \cdot PO = P.$$
To define negatives, form OO as the third point on the line tangent at O. Then $-P = OO \cdot P$. In fact,
$$P + (-P) = O \cdot P(-P) = O \cdot OO = O.$$
The geometry of negatives is especially nice if $(0, 1, 0)$ is on the curve (i.e., again, if there are no y^3 terms). Then it turns out that the choice $O = (0, 1, 0)$ leads to $OO = O$ if there are no xy^2 and x^2y terms. Comparative pictures of negatives are in Figure 1.5.

(a) Picture for $O = (0, 1, 0)$ 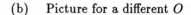 (b) Picture for a different O

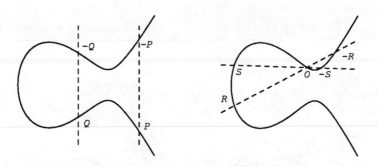

FIGURE 1.5. Negatives relative to the group law

As we indicated above, there are difficulties if the curve is "singular." Figure 1.6 illustrates two types of singular behavior. In each case, $P = (0, 0)$, and we consider $P + P$. In (1), we have to take $PP = P$, and then $P + P = OP$ is undefined since the line through O and P neither meets the curve at a third point nor is tangent to the curve at O or P. In (2), we do not have a unique definition of PP, since there is not a unique tangent line at P. The problem is that both curves look bad at $P = (0,0)$. We cannot solve for y or x in terms of the other. The condition of the Implicit Function Theorem fails at $(0,0)$ in that $\frac{\partial}{\partial x} = 0 = \frac{\partial}{\partial y}$. We say that $(0,0)$ is a singular point and that the curve is singular. A precise definition of singularities will be given in §II.2.

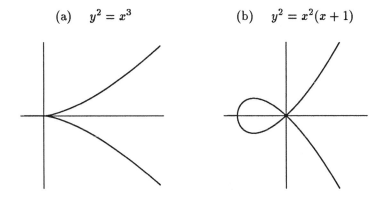

FIGURE 1.6. Singular behavior

Theorem 1.4 (Poincaré). *For a nonsingular cubic curve with a specified rational point O, the operation $+$ is well defined and makes the set of rational solutions into an abelian group with identity element O. If a different identity element O' is chosen, then the two operations are related by $P+'Q = P+Q-O'$, and the group structures are isomorphic.*

The geometry is especially simple when O is the same as OO, i.e., when O is an "inflection point." In this case we shall see that we can map O out to $(x, y, w) = (0, 1, 0)$ by a linear mapping with coefficients in $\mathbb{Q}$ and we can arrange for the tangent line to be $w = 0$. After scaling, we get

$$y^2 + a_1 xy + a_3 y = x^3 + a_2 x^2 + a_4 x + a_6,$$

which is called a **Weierstrass form** of the curve.

An **elliptic curve** over $\mathbb{Q}$ is a nonsingular cubic curve in Weierstrass form, with rational coefficients. Wider classes of curves can be reduced to elliptic curves in various ways, and elliptic curves are sometimes defined as one of these more general curves.

To visualize an elliptic curve over $\mathbb{Q}$ we can complete the square in the Weierstrass form and change variables in y to obtain $y^2 = P(x)$, where $P(x)$ is a monic cubic with distinct roots. We can graph $u = P(x)$ and obtain a picture as in Figure 1.7a or 1.7b. (Alternatively the graph might have its dip completely below the x-axis, or it might have no maximum or minimum.) Then we can scale by $y = \pm\sqrt{u}$ and obtain the pictures in Figures 1.7c and 1.7d.

(a) $u = x(x^2 - 1)$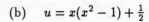

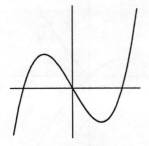

(b) $u = x(x^2 - 1) + \frac{1}{2}$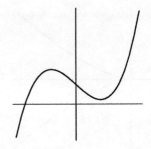

(c) $y^2 = x(x^2 - 1)$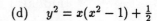

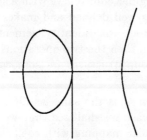

(d) $y^2 = x(x^2 - 1) + \frac{1}{2}$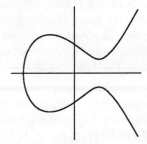

FIGURE 1.7. Graphs of **R** points of the elliptic curve $y^2 = P(x)$

These pictures are of the **R** points in the affine plane. Projectively we must include the point at infinity, $(0,1,0)$. Then topologically (and smoothly) we get one or two circles. This turns out to mean that the group of **R** points is S^1 or $S^1 \oplus \mathbf{Z}_2$. So the group of **Q** points is a subgroup of S^1 or $S^1 \oplus \mathbf{Z}_2$. The first main structure theorem about the group of **Q** points is as follows.

Theorem 1.5 (Mordell). *The group of* **Q** *points of an elliptic curve over* **Q** *is finitely generated. Hence it is* $\mathbf{Z}^r \oplus F$ *with* F *finite abelian.*

This theorem will be proved in Chapter IV. Since the finite abelian group F is contained in S^1 or $S^1 \oplus \mathbf{Z}_2$, we must have $F = \mathbf{Z}_n$ or $F = \mathbf{Z}_{2n} \oplus \mathbf{Z}_2$. For any particular curve, F can be determined. We can change variables to make the curve be $y^2 = x^3 + Ax + B$, with $A, B \in \mathbf{Z}$. It simplifies calculations to take $O = (0, 1, 0)$. Then the elements of order 2 are those with $y = 0$, at most three in number. For the rest we can use

Theorem 1.6 (Lutz-Nagell). For an elliptic curve $y^2 = x^3 + Ax + B$ over $\mathbf{Q}$ with $O = (0, 1, 0)$, the finite abelian group F can be computed as follows. If $P = (x(P), y(P))$ is in F but does not have order 1 or 2, then
  (a) $x(P)$ and $y(P)$ are in $\mathbf{Z}$
  (b) $y(P)^2$ divides $4A^3 + 27B^2$.

This theorem will be proved in Chapter V. To compute F, we simply can try all integers y whose square divides $4A^3 + 27B^2$. For each such, we seek an integer solution x of $y^2 = x^3 + Ax + B$. We can tabulate all integer pairs (x, y) obtained in this way, and we have a bound for the order $|F|$. Then we can check which elements (x, y) are of finite order since their orders (if finite) cannot exceed $|F|$.

As the elliptic curve varies, we can ask what finite abelian groups F arise in Mordell's Theorem. The following deep theorem addresses this question.

Theorem 1.7 (Mazur). The torsion subgroup of the $\mathbf{Q}$ points of an elliptic curve over $\mathbf{Q}$ is one of the 15 groups

$$\mathbf{Z}_n \text{ with } n = 1, 2, 3, \ldots, 9, 10, 12$$

$$\mathbf{Z}_{2n} \oplus \mathbf{Z}_2 \text{ with } n = 1, 2, 3, 4.$$

We turn to a discussion of the rank r of the group of $\mathbf{Q}$ points on an elliptic curve over $\mathbf{Q}$. Elliptic curves are known for which r is any of the numbers $0, 1, 2, \ldots, 12$. The proof of Mordell's Theorem gives an upper bound for r for any particular curve.

Still, no effective way is known for deciding whether $r = 0$ for a particular curve, and it is not known whether r can be arbitrarily large.

The motivation for what affects rank is based on computer calculations and on the hope expressed in the Hasse Principle. Let us consider an elliptic curve E given by
$$y^2 = F(x), \tag{1.5}$$
where $F(x)$ is a cubic polynomial with integer coefficients and with distinct roots r_1, r_2, r_3 in $\mathbf{C}$. The discriminant of F is $\prod_{i<j}(r_i - r_j)^2$ and is a nonzero integer. The group of $\mathbf{Q}$ points is denoted $E(\mathbf{Q})$.

Guided by the Hasse principle, we consider (1.5) modulo a prime p. The curve (1.5) over $\mathbf{Z}_p$ will be nonsingular (hence will be an elliptic curve) except when $p = 2$ or when p divides the discriminant of F. For nonexceptional p, we let $N(p)$ be the number of $\mathbf{Z}_p$ solutions (x, y), including the point at infinity. For example, $y^2 = x^3 + 3x$ over $\mathbf{Z}_5$ has solutions

$$(0,0) \quad (1,\pm 2), \quad (2,\pm 2), \quad (3,\pm 1), \quad (4,\pm 1), \quad \infty,$$

and thus $N(5) = 10$.

In (1.5), we always have

$$1 \leq N(p) \leq 2p + 1. \tag{1.6}$$

In fact, the lower bound is a consequence of counting the point at infinity. For the upper bound, we note that each x from 0 to $p-1$ gives at most 2 values of y, and we also must count the point at infinity. We can rewrite (1.6) as

$$|p + 1 - N(p)| \leq p. \tag{1.7}$$

So $N(p)$ is centered about $p+1$. A theorem of Hasse says that it is more closely centered about $p + 1$ than (1.7) indicates.

Theorem 1.8 (Hasse). $|p + 1 - N(p)| < 2\sqrt{p}$.

The philosophy of the Hasse Principle is that local solutions should influence global solutions. So a curve is more likely to have many global solutions if $N(p)$ is relatively large for many p. In Table 1.1, we consider the equation $y^2 = x^3 + ax$ for several choices of the integer a. For each a, we form the product

$$\prod_{p \leq R}{}' N(p)/p$$

as a function of R. The symbol $'$ means that the exceptional primes (2 and those dividing the discriminant) are not to be included. The table lists the values of this product for some R's in approximately a geometric progression.

a	r	R = 17	31	53	97	173	313	557	997
−3	0	.469	.624	.576	.698	1.030	.870	.824	.550
−4	0	1.926	1.506	1.533	1.139	1.820	1.433	1.466	1.411
+4	0	1.540	2.409	1.603	1.649	2.081	2.732	1.833	2.240
+6	0	.451	.706	.477	.384	.331	.459	.712	.846
−8	0	.602	.612	1.034	1.584	1.004	.838	.819	.852
+3	1	5.867	5.964	5.271	8.855	10.131	8.937	10.500	14.159
−6	1	1.625	1.270	2.807	3.342	5.229	4.891	4.963	5.194
+8	1	5.416	7.199	7.338	7.600	12.176	9.610	9.569	14.060
−14	1	1.095	1.456	1.230	1.532	2.221	3.077	3.689	4.204
+14	2	5.476	5.566	10.491	18.097	24.192	22.829	27.771	29.577
−82	3	7.823	12.234	22.167	41.142	53.847	106.282	124.512	134.117

TABLE 1.1 Values of $\prod'_{p \leq R} N(p)/p$ for various R for the curve $y^2 = x^3 + ax$

What we can see from Table 1.1 is that $\prod'_{p \leq R} N(p)/p$ is roughly constant in x for the cases that $E(\mathbf{Q})$ has rank 0 and that it increases for the cases that $E(\mathbf{Q})$ has rank > 0. Moreover, the rate of increase is distinctly larger when $E(\mathbf{Q})$ has larger rank. More extensive calculation with many elliptic curves leads to the following first form of a conjecture.

Conjecture 1.9 (Rough form of Birch and Swinnerton-Dyer Conjecture). For any elliptic curve E over $\mathbf{Q}$, the rank r of $E(\mathbf{Q})$ is given by the formula

$$\prod_{p \leq R}' \frac{N(p)}{p} \sim (\text{const})(\log R)^r.$$

Again $N(p)$ is the number of $\mathbf{Z}_p$ solutions (x, y), including the point at infinity, and the finite set of primes for which the reduction modulo p is singular are excluded from the product.

To get a formulation of the conjecture that fits better into an established framework, one works with the L-series of the elliptic curve, which for current purposes is given by

$$L(s) = \prod_p' \frac{1}{1 - \frac{p+1-N(p)}{p^s} + \frac{p}{p^{2s}}}, \tag{1.8}$$

the finite set of exceptional primes being excluded from the product. It follows easily from Theorem 1.8 that the infinite product (1.8) is convergent for Re $s > 3/2$. Note that if we put $s = 1$ in (1.8), then the product formally becomes the reciprocal of $\prod' N(p)/p$. The second form of the conjecture is stated in terms of the L-series.

Conjecture 1.10 (Birch and Swinnerton-Dyer). For any elliptic curve over $\mathbf{Q}$, the L-series $L(s)$ extends to be entire for $s \in \mathbf{C}$, and the order of vanishing of $L(s)$ at $s = 1$ equals the rank r.

A third, more detailed, form of the Birch and Swinnerton-Dyer Conjecture includes factors in $L(s)$ for the exceptional primes and addresses the value of the coefficient of the first nonzero term in the power series for $L(s)$ about $s = 1$. We shall not give a formal statement of this third version of the conjecture.

Conjectures about $L(1)$ do not even make sense without the analytic continuation in Conjecture 1.10. This analytic continuation is far from trivial and is itself the subject of the following two conjectures. The precise statements of these conjectures require some terms that will be defined later in the book.

Conjecture 1.11 (Hasse-Weil). $L(s)$, modified by elementary factors to a function $\Lambda(s)$, extends to be entire, and the extension satisfies the functional equation $\Lambda(2 - s) = \pm\Lambda(s)$. If N is the conductor of the elliptic curve and m is any integer prime to N, then the same kind of conclusion is true of $L(s)$ when $L(s)$ is twisted by a Dirichlet character modulo m.

Conjecture 1.12 (Taniyama-Weil). Every elliptic curve over $\mathbf{Q}$ is modular in the sense of being parametrized by modular functions in a specific way.

It turns out that Conjecture 1.12 is decidable for any particular curve. Moreover, Conjecture 1.12 implies Conjecture 1.11, and the converse implication is valid under a minor technical assumption on the curve. The depth of these conjectures is illustrated by the following theorem.

Theorem 1.13 (Frey-Serre-Ribet). The Taniyama-Weil Conjecture implies Fermat's Last Theorem.

CHAPTER II

CURVES IN PROJECTIVE SPACE

1. Projective Space

Let k be a field. The **projective plane** $P_2(k)$ over k is defined as the quotient of $\{(x,y,w) \in k^3 - \{(0,0,0)\}\}$ by a relation $\sim$, where $(x',y',w') \sim (x,y,w)$ if $(x',y',w') = \lambda(x,y,w)$ for some $\lambda \in k^\times$. When there is a need to be careful, we shall write $[(x,y,w)]$ for the member of $P_2(k)$ corresponding to (x,y,w) in k^3. But usually there will not be such a need, and we shall simply refer to (x,y,w) as a member of $P_2(k)$.

A **line** defined over k is a nonzero polynomial $L = ax + by + cw$ in $k[x,y,w]$. We regard L and a line $L' = a'x + b'y + c'w$ to be the **same line** if (a',b',c') is a multiple of (a,b,c). The locus

$$L(k) = \{(x,y,w) \,|\, ax + by + cw = 0\}$$

is well defined in $P_2(k)$ and is called the set of k **points** or k **rational points** of L. It will be handy to refer to $L(k)$ as a **line in** $P_2(k)$. If L is defined over k and K is an extension field of k, it is meaningful to speak of $L(K)$ as a subset of $P_2(K)$.

The affine plane $k^2 = \{(x,y)\}$ has a standard one-one imbedding into $P_2(k)$. Namely we map (x,y) into $[(x,y,1)]$. The set that is missed by the image is the set where $w = 0$, which is the set of k rational points of a line called the **line at infinity**.

The points with $w = 0$ are called the **points at infinity**. Except for the line at infinity, lines in $P_2(k)$ correspond under restriction exactly to lines in k^2.

In certain ways the geometry of $P_2(k)$ is simpler than the geometry of k^2:

(1) Two distinct lines in $P_2(k)$ intersect in a unique point. In fact, we set up the system of equations

$$\begin{pmatrix} a & b & c \\ a' & b' & c' \end{pmatrix} \begin{pmatrix} x \\ y \\ w \end{pmatrix} = \begin{pmatrix} 0 \\ 0 \end{pmatrix}.$$

Since the lines are distinct, the coefficient matrix has rank 2. Thus the kernel has dimension 1, and there is just one point in the intersection.

(2) Two distinct points in $P_2(k)$ lie on a unique line. In fact, we set up the system of equations

$$\begin{pmatrix} x & y & w \\ x' & y' & w' \end{pmatrix} \begin{pmatrix} a \\ b \\ c \end{pmatrix} = \begin{pmatrix} 0 \\ 0 \end{pmatrix}$$

and argue in similar fashion.

Let Φ be in $GL(3,k)$. Then Φ maps the set k^3 of all column vectors in one-one fashion onto k^3 and passes to a one-one map of $P_2(k)$ onto $P_2(k)$ called the **projective transformation** corresponding to Φ. Two Φ's give the same map of $P_2(k)$ if and only if they are multiples of one another. The group action of $GL(3,k)$ on $P_2(k)$ is transitive because $GL(3,k)$ acts transitively on $k^3 - \{(0,0,0)\}$.

If L is the line whose coefficients are given by the row vector $(a \ b \ c)$ and if Φ is a projective transformation, then the row vector $(a \ b \ c)\Phi^{-1}$ defines a new line L^Φ, and the k rational points of L^Φ are given by

$$L^\Phi(k) = \Phi(L(k)).$$

In fact, let $\begin{pmatrix} x \\ y \\ w \end{pmatrix}$ be in $L(k)$. Then $\begin{pmatrix} x' \\ y' \\ w' \end{pmatrix} = \Phi \begin{pmatrix} x \\ y \\ w \end{pmatrix}$ is in $\Phi(L(k))$ and satisfies

$$(a \ b \ c)\Phi^{-1} \begin{pmatrix} x' \\ y' \\ w' \end{pmatrix} = 0;$$

hence it is in $L^\Phi(k)$. Conversely if $\begin{pmatrix} x' \\ y' \\ w' \end{pmatrix}$ is in $L^\Phi(k)$, then $\begin{pmatrix} x \\ y \\ w \end{pmatrix} = \Phi^{-1}\begin{pmatrix} x' \\ y' \\ w' \end{pmatrix}$ satisfies

$$(a \ b \ c)\begin{pmatrix} x \\ y \\ w \end{pmatrix} = (a \ b \ c)\Phi^{-1}\begin{pmatrix} x' \\ y' \\ w' \end{pmatrix} = 0,$$

and thus $\begin{pmatrix} x' \\ y' \\ w' \end{pmatrix}$ is Φ of something in $L(k)$.

1. PROJECTIVE SPACE

About any point in $P_2(k)$ we can introduce various systems of **affine local coordinates**.

Namely fix $[(x_0, y_0, w_0)]$ in $P_2(k)$. Choose (by transitivity) some Φ in $GL(3, k)$ with $\Phi(x_0, y_0, w_0) = (0, 0, 1)$. Then we can define local coordinates on $\Phi^{-1}(k \times k \times \{1\})$ to k^2 by the one-one map

$$\varphi(\Phi^{-1}(x, y, 1)) = (x, y).$$

This definition generalizes the standard imbedding of the affine plane k^2 in $P_2(k)$ earlier; that imbedding was the case $\Phi = I$.

EXAMPLE 1. Suppose $(x_0, y_0, w_0) = (x_0, y_0, 1)$. We can choose $\Phi = \begin{pmatrix} 1 & 0 & -x_0 \\ 0 & 1 & -y_0 \\ 0 & 0 & 1 \end{pmatrix}$. Then

$$\Phi \begin{pmatrix} x \\ y \\ 1 \end{pmatrix} = \begin{pmatrix} 1 & 0 & -x_0 \\ 0 & 1 & -y_0 \\ 0 & 0 & 1 \end{pmatrix} \begin{pmatrix} x \\ y \\ 1 \end{pmatrix} = \begin{pmatrix} x - x_0 \\ y - y_0 \\ 1 \end{pmatrix}.$$

In this case, the local coordinates are defined on

$$\Phi^{-1}(k \times k \times 1) = k \times k \times 1$$

and are given by

$$\varphi(x, y, 1) = \varphi(\Phi^{-1}(\Phi(x, y, 1)))$$
$$= \varphi(\Phi^{-1}(x - x_0, y - y_0, 1)) = (x - x_0, y - y_0).$$

This Φ is handy for reducing behavior about $(x_0, y_0, 1)$ in $P_2(k)$ to behavior about $(0, 0)$ in k^2.

EXAMPLE 2. Suppose $(x_0, y_0, w_0) = (0, 1, 0)$. We can choose $\Phi = \begin{pmatrix} 0 & 0 & 1 \\ 1 & 0 & 0 \\ 0 & 1 & 0 \end{pmatrix}$. Then

$$\Phi \begin{pmatrix} x \\ 1 \\ w \end{pmatrix} = \begin{pmatrix} 0 & 0 & 1 \\ 1 & 0 & 0 \\ 0 & 1 & 0 \end{pmatrix} \begin{pmatrix} x \\ 1 \\ w \end{pmatrix} = \begin{pmatrix} w \\ x \\ 1 \end{pmatrix}$$

and

$$\varphi(x, 1, w) = \varphi(\Phi^{-1}(\Phi(x, 1, w))) = \varphi(\Phi^{-1}(w, x, 1)) = (w, x).$$

This Φ will be handy for studying behavior near the (unique) point at infinity on a cubic in Weierstrass form.

Affine local coordinates are useful in studying homogeneous polynomials in three variables. We say that a nonzero $F \in k[x,y,w]$ is **homogeneous** of degree d if every monomial in F has total degree d, and we write $k[x,y,w]_d$ for the set of such polynomials. Each such F satisfies

$$F(\lambda x, \lambda y, \lambda w) = \lambda^d F(x,y,w) \qquad \text{for } x,y,w,\lambda \in k. \tag{2.1}$$

Conversely, homogeneous polynomials over an infinite field can be detected by this property, according to Proposition 2.2 below.

Lemma 2.1. If k is an infinite field, then a nonzero polynomial $f \in k[x_1, \ldots, x_n]$ is nonzero at some point.

PROOF. We induct on n, the result for $n = 1$ being well known. Assume the result for $n - 1$, and suppose $f(c_1, \ldots, c_n) = 0$ for all $(c_1, \ldots, c_n)$. By induction,

$$f(x_1, \ldots, x_{n-1}, c) \text{ is the 0 polynomial in } n-1 \text{ variables}, \tag{2.2}$$

for each choice of c. Fix a monomial $x_1^{a_1} \cdots x_{n-1}^{a_{n-1}}$ in $n-1$ variables. The monomials in n variables containing this $(n-1)$-variable monomial contribute

$$\sum_{j \geq 0} b_j x_1^{a_1} \cdots x_{n-1}^{a_{n-1}} x_n^j$$

to f, and (2.2) says that $\sum_{j \geq 0} b_j c^j = 0$ for all c. Therefore all the b_j are 0. Repeating this argument for all monomials in $n-1$ variables, we see that f is the 0 polynomial.

Proposition 2.2. If k is an infinite field, then a nonzero polynomial $F \in k[x,y,w]$ satisfying (2.1) is homogeneous of degree d.

PROOF. Write F as the sum of homogeneous terms of different degrees: $F = \sum_{j=1}^n F_j$ with F_j of degree d_j and with $d_1 = d$. Then (2.1) gives

$$\lambda^d F(x_0, y_0, w_0) = \lambda^d F_1(x_0, y_0, w_0) + \lambda^{d_2} F_2(x_0, y_0, w_0)$$
$$+ \cdots + \lambda^{d_n} F_n(x_0, y_0, w_0)$$

for all λ. Since k is infinite, this equality for all λ implies $F(x_0, y_0, w_0) = F_1(x_0, y_0, w_0)$ and also $F_j(x_0, y_0, w_0) = 0$ for all $j \geq 2$. Letting (x_0, y_0, w_0) vary and applying Lemma 2.1 to $F - F_1$ and to F_j for $j \geq 2$, we obtain the conclusion of the proposition.

1. PROJECTIVE SPACE

A homogeneous polynomial $F \neq 0$ of degree > 0 is *not* a function on $P_2(k)$. Nevertheless we can examine the behavior of F near a point (x_0, y_0, w_0) in $k^3 - \{(0,0,0)\}$ by choosing Φ in $GL(3, k)$ with $\Phi(x_0, y_0, w_0) = (0, 0, 1)$ and defining
$$f(x, y) = F(\Phi^{-1}(x, y, 1)).$$

EXAMPLE 1. Suppose $(x_0, y_0, w_0) = (x_0, y_0, 1)$ and
$$\Phi = \begin{pmatrix} 1 & 0 & -x_0 \\ 0 & 1 & -y_0 \\ 0 & 0 & 1 \end{pmatrix}.$$

Then
$$\Phi^{-1} \begin{pmatrix} x \\ y \\ 1 \end{pmatrix} = \begin{pmatrix} x + x_0 \\ y + y_0 \\ 1 \end{pmatrix}$$
$$f(x, y) = F(x + x_0, y + y_0, 1).$$

For
$$F(x, y, w) = x^2 y + xyw + w^3,$$
the corresponding $f(x, y)$ splits into homogeneous terms as
$$f(x, y) = (x_0^2 y_0 + x_0 y_0 + 1) + (x_0^2 y + 2x_0 y_0 x + x_0 y + y_0 x)$$
$$+ (y_0 x^2 + 2x_0 xy + xy) + (x^2 y).$$

We shall use this splitting in §2 in examining singularities and tangent lines.

EXAMPLE 2. Suppose $(x_0, y_0, w_0) = (0, 1, 0)$ and $\Phi = \begin{pmatrix} 0 & 0 & 1 \\ 1 & 0 & 0 \\ 0 & 1 & 0 \end{pmatrix}$.

Then
$$\Phi^{-1} \begin{pmatrix} x \\ y \\ 1 \end{pmatrix} = \begin{pmatrix} y \\ 1 \\ x \end{pmatrix}$$
$$f(x, y) = F(y, 1, x).$$

For the same F as in Example 1, namely
$$F(x, y, w) = x^2 y + xyw + w^3,$$
we obtain
$$f(x, y) = (y^2 + xy) + (x^3)$$
with no terms of total order 0 or 1.

Conversely if d and Φ are given and if f in $k[x,y]$ has degree d, we can reconstruct F. In Example 2 above with $d=3$ and with f as above, define $G(x,y,w)$ by inserting powers of w to make all terms of degree 3:

$$G(x,y,w) = y^2 w + xyw + x^3.$$

Then put $F = G \circ \Phi$ and recover $F(x,y,w) = x^2 y + xyw + w^3$.

In the special cases that k is $\mathbf{R}$ or $\mathbf{C}$, $P_2(\mathbf{R})$ and $P_2(\mathbf{C})$ are smooth manifolds. It is helpful as motivation to recall the argument for $P_2(\mathbf{R})$. To do so, we need to give compatible charts (U, φ), where $\varphi : U \to \mathbf{R}^2$ is a homeomorphism onto an open set and where the sets U cover $P_2(\mathbf{R})$. At points (x,y,w) on the set $U_3 = \{w \neq 0\}$, we can use

$$\varphi_3(x,y,w) = \left(\frac{x}{w}, \frac{y}{w}\right).$$

This is just our affine local coordinate system about $(x_0, y_0, w_0) = (0,0,1)$ with $\Phi = 1$. On the set $U_2 = \{y \neq 0\}$, we can use

$$\varphi_2(x,y,w) = \left(\frac{x}{y}, \frac{w}{y}\right).$$

This corresponds to using $(x_0, y_0, w_0) = (0,1,0)$ and $\Phi = \begin{pmatrix} 1 & 0 & 0 \\ 0 & 0 & 1 \\ 0 & 1 & 0 \end{pmatrix}$.
Similarly we can define a chart (U_1, φ_1). To have a manifold, we are to check that functions like $\varphi_2 \circ \varphi_3^{-1}$ are smooth. In fact, these functions are rational with nonvanishing denominators. For example,

$$\varphi_2 \circ \varphi_3^{-1}(x,y) = \varphi_2(x,y,1) = \left(\frac{x}{y}, \frac{1}{y}\right),$$

with domain $\{(x,y) \mid y \neq 0\}$. More generally, all our systems of affine local coordinates (defined via (x_0, y_0, w_0) and Φ) yield compatible charts.

Once $P_2(\mathbf{R})$ and $P_2(\mathbf{C})$ are smooth manifolds, partial derivatives are available. We shall see that partial derivatives provide a useful tool even when the field k is not $\mathbf{R}$ or $\mathbf{C}$.

2. Curves and Tangents

A **plane curve** or **projective plane curve** defined over k is simply a nonzero homogeneous polynomial $F \in k[x,y,w]_d$ for some $d > 0$, except that we regard two such F's as the **same curve** if they are multiples

2. CURVES AND TANGENTS

of each other. As we remarked in §1, F is not well defined on $P_2(k)$. Nevertheless, the zero locus of F yields a set in projective space. Namely, if K is an extension field of k, then the locus

$$F(K) = \{(x,y,w) \mid F(x,y,w) = 0\}$$

is well defined in $P_2(K)$ since F is homogeneous. This locus is called the set of K **points** or K **rational points** of the curve. In the special cases that $d = 1, 2$, or 3, the curve is called a **line**, **conic**, or **cubic**, respectively.

If F is a plane curve and if we have an affine coordinate system given by Φ with $\Phi(x_0, y_0, w_0) = (0, 0, 1)$, then the corresponding **affine curve** is

$$f(x,y) = F(\Phi^{-1}(x,y,1)) \quad \text{in } k[x,y].$$

Among the k rational points of a curve, we shall distinguish between singular points and nonsingular points. Thus let $F \neq 0$ be in $k[x,y,w]_d$, fix $(x_0, y_0, w_0) \in F(k)$, and choose affine local coordinates about (x_0, y_0, w_0) given by some Φ with $\Phi(x_0, y_0, w_0) = (0, 0, 1)$. Let

$$f(x,y) = F(\Phi^{-1}(x,y,1)) \in k[x,y].$$

For example, the situation in Example 1 of §1 had $w_0 = 1$ and

$$F(x,y,w) = x^2 y + xyw + w^3,$$

and we were led to

$$\begin{aligned} f(x,y) &= (x_0^2 y_0 + x_0 y_0 + 1) + (x_0^2 y + 2x_0 y_0 x + x_0 y + y_0 x) \\ &\quad + (y_0 x^2 + 2x_0 xy + xy) + (x^2 y) \\ &= f_0(x,y) + f_1(x,y) + f_2(x,y) + f_3(x,y). \end{aligned}$$

The constant term f_0 is 0 since $(x_0, y_0, 1)$ is in $F(k)$. In this example and in general, f is the sum of homogeneous terms of degree 1 through d, say $f = f_1 + \cdots + f_d$ with $f_1, \ldots, f_d$ depending on (x_0, y_0, w_0) and Φ. We say (x_0, y_0, w_0) is a **nonsingular point** if f_1 is not the 0 polynomial in $k[x,y]$. Otherwise (x_0, y_0, w_0) is a **singular point**. In our example with $w_0 = 1$, we are to consider

$$f_1(x,y) = (2x_0 y_0 + y_0)x + (x_0^2 + x_0)y.$$

The singular points in $F(k)$ are those where the coefficients of x and y are both 0. The coefficients are 0 when $(x_0, y_0) = (0, 0)$ and $(x_0, y_0) =$

$(-1,0)$. (Back in $P_2(k)$, these are $(0,0,1)$ and $(-1,0,1)$.) But neither of these is in $F(k)$. Hence F is nonsingular at every point of $F(k)$ for which $w_0 = 1$.

We need to check that nonsingularity does not depend on the choice of Φ. Thus suppose also that $\Psi(x_0, y_0, w_0) = (0, 0, 1)$. Then
$$\Psi \circ \Phi^{-1} = \begin{pmatrix} a & b & 0 \\ c & d & 0 \\ r & s & 1 \end{pmatrix} \text{ with } \begin{pmatrix} a & b \\ c & d \end{pmatrix} \text{ invertible. Write } f(x,y) =$$
$F(\Phi^{-1}(x,y,1))$ and $g(x,y) = F(\Psi^{-1}(x,y,1))$. We write heuristically

$$\begin{aligned}
f(x,y) &= (F \circ \Psi^{-1})(\Psi \circ \Phi^{-1})(x,y,1) \\
&= (F \circ \Psi^{-1})(ax + by, cx + dy, rx + sy + 1) \\
&= (F \circ \Psi^{-1})\left((rx + sy + 1)\left(\frac{ax + by}{rx + sy + 1}, \frac{cx + dy}{rx + sy + 1}, 1\right)\right) \\
&= (rx + sy + 1)^d g\left(\frac{ax + by}{rx + sy + 1}, \frac{cx + dy}{rx + sy + 1}\right) \\
&= (rx + sy + 1)^d (g_1 + \cdots + g_d)\left(\frac{ax + by}{rx + sy + 1}, \frac{cx + dy}{rx + sy + 1}\right) \\
&= (rx + sy + 1)^{d-1} g_1(ax + by, cx + dy) \\
&\quad + \cdots + g_d(ax + by, cx + dy).
\end{aligned} \qquad (2.3)$$

This computation is valid in the quotient field $k(x,y)$. By expanding the various powers of $(rx + sy + 1)$ and regrouping by homogeneity in (x,y), we can read off
$$f_1(x,y) = g_1(ax + by, cx + dy). \qquad (2.4)$$
Similarly
$$g_1(x,y) = f_1(\alpha x + \beta y, \gamma x + \delta y) \text{ with } \begin{pmatrix} \alpha & \beta \\ \gamma & \delta \end{pmatrix} = \begin{pmatrix} a & b \\ c & d \end{pmatrix}^{-1}.$$
So f_1 and g_1 are both zero or both nonzero.

Proposition 2.3. Suppose $F \in k[x,y,w]_m$ and $G \in k[x,y,w]_n$ are plane curves. If (x_0, y_0, w_0) is in $F(k) \cap G(k)$, then (x_0, y_0, w_0) is a singular point of FG.

PROOF. Choose affine local coordinates with $\Phi(x_0, y_0, w_0) = (0, 0, 1)$, and define $f(x,y) = F(\Phi^{-1}(x,y,1))$ and $g(x,y) = G(\Phi^{-1}(x,y,1))$. Then we can write $f = f_1 + \cdots + f_m$ and $g = g_1 + \cdots + g_n$. Since
$$f(x,y)g(x,y) = (FG)(\Phi^{-1}(x,y,1)),$$
and since fg has no first-degree terms, it follows that FG has a singular point at (x_0, y_0, w_0).

2. CURVES AND TANGENTS

We say that the plane curve F over k is **nonsingular** (or **smooth**) if F is nonsingular at every point of $F(\bar{k})$, where $\bar{k}$ is the algebraic closure of k. Otherwise we say F is **singular**. Nonsingularity at all points of $F(k)$ does not imply F is nonsingular. For example, the curve

$$F(x, y, w) = x^3 - 6xw^2 + 6yw^2 - y^3$$

is defined over $k = \mathbb{Q}$, is nonsingular at every point of $F(\mathbb{Q})$, and has a singular point at $(x, y, w) = (\sqrt{2}, \sqrt{2}, 1)$. So the curve is singular.

Nonsingularity of the curve F implies irreducibility over the algebraic closure $\bar{k}$, according to Corollary 2.5 below. The corollary depends on Bezout's Theorem, whose proof is deferred to §5.

Theorem 2.4 (Bezout's Theorem). Suppose $F \in k[x, y, w]_m$ and $G \in k[x, y, w]_n$ are plane curves. Then $F(\bar{k}) \cap G(\bar{k})$ is nonempty. If it has more than mn points, then F and G have as a common factor some homogeneous polynomial of degree > 0.

Corollary 2.5. Suppose F is a plane curve and is reducible over $\bar{k}$. Then the factors are homogeneous polynomials, and F is singular.

PROOF. Write $F = F_1 F_2$ nontrivially. Let d_1 and e_1 be the highest and lowest degrees of terms in F_1, and let d_2 and e_2 be the highest and lowest degrees of terms in F_2. The product of the terms of degree d_1 in F_1 and the terms of degree d_2 in F_2 is nonzero and is the $d_1 d_2$ degree part of F. The product of the terms of degree e_1 in F_1 and the terms of degree e_2 in F_2 is nonzero and is the $e_1 e_2$ degree part of F. Since F is homogeneous, $d_1 d_2 = e_1 e_2$. It follows that $d_1 = e_1$ and $d_2 = e_2$; thus F_1 and F_2 are homogeneous. Bezout's Theorem says that $F_1(\bar{k}) \cap F_2(\bar{k})$ is nonempty, and Proposition 2.3 says that any point in this intersection is a singular point for F. Hence F is singular.

Suppose that (x_0, y_0, w_0) is a nonsingular point of $F(k)$ and that $\Phi(x_0, y_0, w_0) = (0, 0, 1)$. We have written $f(x, y) = F(\Phi^{-1}(x, y, 1))$ with $f = f_1 + \cdots + f_d$. Then $f_1(x, y) = px + qy$ with p and q in k, not both 0. Consider f_1 as a polynomial in three variables $\tilde{f}_1(x, y, w)$ independent of w. We lift back to a nonzero member $L = \tilde{f}_1 \circ \Phi$ of $k[x, y, w]_1$, and the result is called the **tangent line** to F at (x_0, y_0, w_0).

Let us see that the tangent line is independent of the choice of affine local coordinates. Thus suppose also that $\Psi(x_0, y_0, w_0) = (0, 0, 1)$. Form $g(x, y) = F(\Psi^{-1}(x, y, 1))$ and $\Psi \circ \Phi^{-1} = \begin{pmatrix} a & b & 0 \\ c & d & 0 \\ r & s & 1 \end{pmatrix}$, and let the

respective tangent lines be L_Φ and L_Ψ. Then we have
$$L_\Phi(x,y,w) = \tilde{f}_1(\Phi(x,y,w)), \qquad (2.5)$$
$$\begin{aligned}\tilde{f}_1(x,y,w) &= f_1(x,y) = g_1(ax+by, cx+dy)\\ &= \tilde{g}_1(ax+by, cx+dy, rx+sy+w) = \tilde{g}_1(\Psi \circ \Phi^{-1}(x,y,w)),\end{aligned}$$
$$\begin{aligned}L_\Psi(x,y,w) &= \tilde{g}_1(\Psi(x,y,w))\\ &= \tilde{g}_1(\Psi \circ \Phi^{-1}(\Phi(x,y,w))) = \tilde{f}_1(\Phi(x,y,w)).\end{aligned} \qquad (2.6)$$

Comparing (2.5) and (2.6), we see that $L_\Phi(x,y,w) = L_\Psi(x,y,w)$.

If F is a plane curve and Φ is a projective transformation, then $F^\Phi = F \circ \Phi^{-1}$ is another plane curve, and the k rational points of F^Φ are given by
$$F^\Phi(k) = \Phi(F(k)),$$
by the same kind of reasoning as in §1 for the case that F is a line. This reasoning shows also that if (x_0, y_0, w_0) is a nonsingular point for F, then $\Phi(x_0, y_0, w_0)$ is a nonsingular point for F^Φ.

From calculus we expect that nonsingularity can be decided in terms of partial derivatives when k is $\mathbf{R}$ or $\mathbf{C}$. In fact, the argument that works for $\mathbf{R}$ or $\mathbf{C}$ works also for general k. All that is needed is a definition of partial derivatives, along with a few elementary properties. The definition of a partial derivative of a polynomial is clear, and it is routine to check that the product rule and the several-variable Chain Rule are valid in the context of polynomials. We find, at a nonsingular point, that the tangent line is given by a familiar formula from calculus, as in the following proposition.

Proposition 2.6. Let F be a plane curve over k. If (x_0, y_0, w_0) is on the curve, then (x_0, y_0, w_0) is a nonsingular point if and only if at least one of $\frac{\partial F}{\partial x}, \frac{\partial F}{\partial y}, \frac{\partial F}{\partial w}$ is nonzero at (x_0, y_0, w_0). In this case the tangent line L to F at (x_0, y_0, w_0) is given by
$$L = \left[\frac{\partial F}{\partial x}\right]_{(x_0,y_0,w_0)} x + \left[\frac{\partial F}{\partial y}\right]_{(x_0,y_0,w_0)} y + \left[\frac{\partial F}{\partial w}\right]_{(x_0,y_0,w_0)} w.$$

PROOF. Choose affine local coordinates with $\Phi(x_0, y_0, w_0) = (0,0,1)$. Since (x_0, y_0, w_0) is on the curve, $F \circ \Phi^{-1}(x,y,w)$ vanishes upon substitution of $(0,0,1)$. Hence every monomial in $F \circ \Phi^{-1}$ contains a factor of x or y, and it follows that
$$\frac{\partial^n}{\partial w^n}(F \circ \Phi^{-1})(0,0,1) = 0 \qquad \text{for every } n. \qquad (2.7)$$

2. CURVES AND TANGENTS

We write $f(x,y) = F(\Phi^{-1}(x,y,1))$ with $f = f_1 + \cdots + f_d$. The coefficients of the linear term f_1 are $\frac{\partial f}{\partial x}(0,0)$ and $\frac{\partial f}{\partial y}(0,0)$. Since

$$f = F \circ \Phi^{-1} \circ ((x,y) \to (x,y,1)),$$

the Chain Rule gives

$$\left(\frac{\partial f}{\partial x} \quad \frac{\partial f}{\partial y}\right)_{(0,0)} = \left(\frac{\partial F}{\partial x} \quad \frac{\partial F}{\partial y} \quad \frac{\partial F}{\partial w}\right)_{(x_0,y_0,w_0)} \Phi^{-1} \begin{pmatrix} 1 & 0 \\ 0 & 1 \\ 0 & 0 \end{pmatrix}. \quad (2.8)$$

Let (x', y', w') be given. Multiplying on the right by $\begin{pmatrix} x' \\ y' \end{pmatrix}$, we obtain

$$\tilde{f}_1(x', y', w') = \left(\frac{\partial f}{\partial x} \quad \frac{\partial f}{\partial y}\right)_{(0,0)} \begin{pmatrix} x' \\ y' \end{pmatrix}$$

$$= \left(\frac{\partial F}{\partial x} \quad \frac{\partial F}{\partial y} \quad \frac{\partial F}{\partial w}\right)_{(x_0,y_0,w_0)} \Phi^{-1} \begin{pmatrix} x' \\ y' \\ 0 \end{pmatrix}. \quad (2.9)$$

By (2.7) for $n = 1$, the third entry of

$$\left(\frac{\partial F}{\partial x} \quad \frac{\partial F}{\partial y} \quad \frac{\partial F}{\partial w}\right)_{(x_0,y_0,w_0)} \Phi^{-1}$$

is 0. Thus the right side of (2.9) is

$$= \left(\frac{\partial F}{\partial x} \quad \frac{\partial F}{\partial y} \quad \frac{\partial F}{\partial w}\right)_{(x_0,y_0,w_0)} \Phi^{-1} \begin{pmatrix} x' \\ y' \\ w' \end{pmatrix}.$$

Putting $(x', y', w') = \Phi(x, y, w)$, we obtain

$$L(x, y, w) = \tilde{f}_1 \circ \Phi(x, y, w)$$

$$= \left(\frac{\partial F}{\partial x} \quad \frac{\partial F}{\partial y} \quad \frac{\partial F}{\partial w}\right)_{(x_0,y_0,w_0)} \begin{pmatrix} x \\ y \\ w \end{pmatrix}. \quad (2.10)$$

By (2.8), we have a singular point if all first partials of F are 0. Otherwise (2.10) shows that $\tilde{f}_1$ is not 0; hence f_1 is not 0 and the point is nonsingular. In this case, (2.10) gives the tangent line.

The quadratic term $\tilde{f}_2 \circ \Phi(x,y,w)$ can be identified similarly, provided the characteristic of k is not 2. But the expression is more complicated and depends on the choice of Φ. It involves also the **Hessian matrix** of F, defined as

$$H = H(x_0, y_0, w_0) = \begin{pmatrix} \dfrac{\partial^2 F}{\partial x^2} & \dfrac{\partial^2 F}{\partial x \partial y} & \dfrac{\partial^2 F}{\partial x \partial w} \\ \dfrac{\partial^2 F}{\partial x \partial y} & \dfrac{\partial^2 F}{\partial y^2} & \dfrac{\partial^2 F}{\partial y \partial w} \\ \dfrac{\partial^2 F}{\partial x \partial w} & \dfrac{\partial^2 F}{\partial y \partial w} & \dfrac{\partial^2 F}{\partial w^2} \end{pmatrix}_{(x_0, y_0, w_0)}. \quad (2.11)$$

Proposition 2.7. Suppose the characteristic of k is not 2. Let F be a plane curve over k of degree d, let (x_0, y_0, w_0) be a nonsingular point on the curve, and choose affine local coordinates with $\Phi(x_0, y_0, w_0) = (0,0,1)$. Put $f(x,y) = F \circ \Phi^{-1}(x,y,1)$, let $f_2(x,y)$ be the quadratic term in the expansion of $f(x,y)$ about 0, and let $\tilde{f}_2(x,y,w) = f_2(x,y)$. Then $Q_\Phi(x,y,w) = \tilde{f}_2 \circ \Phi(x,y,w)$ is given by

$$2Q_\Phi(x,y,w) = \begin{pmatrix} x & y & w \end{pmatrix} H \begin{pmatrix} x \\ y \\ w \end{pmatrix} - 2(d-1) L'_\Phi(x,y,w) L(x,y,w), \quad (2.12)$$

where H is the Hessian matrix of F at (x_0, y_0, w_0), $L(x,y,w)$ is the tangent line at (x_0, y_0, w_0), and $L'_\Phi(x,y,w)$ is a line with $L'_\Phi(x_0, y_0, w_0) = 1$.

PROOF. Let H^Φ be the Hessian matrix of $F^\Phi = F \circ \Phi^{-1}$ at $(0,0,1)$. We first show that

$$H^\Phi = \Phi^{-1^{\mathrm{tr}}} H \Phi^{-1}. \quad (2.13)$$

In fact, the Chain Rule gives

$$\begin{pmatrix} \dfrac{\partial F^\Phi}{\partial x} & \dfrac{\partial F^\Phi}{\partial y} & \dfrac{\partial F^\Phi}{\partial w} \end{pmatrix}_{(x,y,w)} = \begin{pmatrix} \dfrac{\partial F}{\partial x} & \dfrac{\partial F}{\partial y} & \dfrac{\partial F}{\partial w} \end{pmatrix}_{\Phi^{-1}(x,y,w)} \Phi^{-1}.$$

Transposing gives

$$\begin{pmatrix} \dfrac{\partial F^\Phi}{\partial x} \\ \dfrac{\partial F^\Phi}{\partial y} \\ \dfrac{\partial F^\Phi}{\partial w} \end{pmatrix}_{(x,y,w)} = \Phi^{-1^{\mathrm{tr}}} \begin{pmatrix} \dfrac{\partial F}{\partial x} \\ \dfrac{\partial F}{\partial y} \\ \dfrac{\partial F}{\partial w} \end{pmatrix}_{\Phi^{-1}(x,y,w)}.$$

2. CURVES AND TANGENTS

Now we can apply the Chain Rule to each row of this equation and evaluate at $(x, y, w) = (0, 0, 1)$. When the results are assembled, we obtain (2.13).

The matrix of second partials of $f(x,y) = F \circ \Phi^{-1}(x, y, 1)$ at $(0, 0)$ is obtained by taking the upper left 2-by-2 block of H^Φ and putting $w = 1$. Thus

$$2f_2(x', y') = (x' \ \ y') \begin{pmatrix} \frac{\partial^2 f}{\partial x^2} & \frac{\partial^2 f}{\partial x \partial y} \\ \frac{\partial^2 f}{\partial x \partial y} & \frac{\partial^2 f}{\partial y^2} \end{pmatrix}_{(0,0)} \begin{pmatrix} x' \\ y' \end{pmatrix}$$

$$= (x' \ \ y' \ \ 0) H^\Phi \begin{pmatrix} x' \\ y' \\ 0 \end{pmatrix}$$

$$= (x' \ \ y' \ \ w') H^\Phi \begin{pmatrix} x' \\ y' \\ w' \end{pmatrix}$$

$$- 2w'(0 \ \ 0 \ \ 1) H^\Phi \begin{pmatrix} x' \\ y' \\ w' \end{pmatrix} - w'^2 \left[\frac{\partial^2 F^\Phi}{\partial w^2}\right]_{(0,0,1)}.$$

The last term on the right is 0, by (2.7) for $n = 2$. Putting $\begin{pmatrix} x' \\ y' \\ w' \end{pmatrix} = \Phi \begin{pmatrix} x \\ y \\ w \end{pmatrix}$, we obtain

$$2Q_\Phi(x, y, w) = (x \ \ y \ \ w) H \begin{pmatrix} x \\ y \\ w \end{pmatrix} - 2w'(0 \ \ 0 \ \ 1) H^\Phi \begin{pmatrix} x' \\ y' \\ w' \end{pmatrix}$$

$$= (x \ \ y \ \ w) H \begin{pmatrix} x \\ y \\ w \end{pmatrix}$$

$$- 2w' \left(\left[\frac{\partial^2 F^\Phi}{\partial x \partial w}\right]_{(0,0,1)} x' + \left[\frac{\partial^2 F^\Phi}{\partial y \partial w}\right]_{(0,0,1)} y' \right). \quad (2.14)$$

With d as the degree of F, the only monomial of F^Φ that makes a contribution to $\left[\frac{\partial^2 F^\Phi}{\partial x \partial w}\right]_{(0,0,1)}$ is xw^{d-1}, and the contribution is $d - 1$ times what $\left[\frac{\partial F^\Phi}{\partial x}\right]_{(0,0,1)}$ gives. A similar remark applies to $\left[\frac{\partial^2 F^\Phi}{\partial y \partial w}\right]_{(0,0,1)}$.

Thus the second term on the right of (2.14) is

$$= -2(d-1)w'\left(\left[\frac{\partial F^\Phi}{\partial x}\right]_{(0,0,1)} x' + \left[\frac{\partial F^\Phi}{\partial y}\right]_{(0,0,1)} y'\right)$$

$$= -2(d-1)w'f_1(x',y') = -2(d-1)w'\tilde{f}_1(x',y',w')$$

$$= -2(d-1)w'\tilde{f}_1(\Phi(x,y,w)) = -2(d-1)w'L(x,y,w).$$

The value of w' is the third entry of $\Phi\begin{pmatrix}x\\y\\w\end{pmatrix}$, and this is a nonzero linear expression in x,y,w. We define it to be $L'_\Phi(x,y,w)$.

3. Flexes

In this section we shall define the **intersection multiplicity** of a line and a plane curve at a point. The curve may be nonsingular or singular at the point. The notation for intersection multiplicity will be $i(P,L,F)$, where $P = (x_0, y_0, w_0)$ is in $F(k) \cap L(k)$, F is in $k[x,y,w]_d$, and L is in $k[x,y,w]_1$.

We introduce affine local coordinates. Choose Φ with $\Phi(x_0, y_0, w_0) = (0,0,1)$ and put

$$f(x,y) = F(\Phi^{-1}(x,y,1)) = f_1(x,y) + \cdots + f_d(x,y)$$

$$l(x,y) = L(\Phi^{-1}(x,y,1)).$$

Since $l(0,0) = 0$, $l(x,y) = bx - ay$ for some constants a and b not both 0. Then $\varphi(t) = \begin{pmatrix}at\\bt\end{pmatrix}$ parametrizes $l(x,y) = 0$. The expression $f(\varphi(t))$ is a polynomial in t with $f(\varphi(0)) = 0$. In fact,

$$f(\varphi(t)) = f_1(at,bt) + f_2(at,bt) + \cdots + f_d(at,bt)$$
$$= tf_1(a,b) + t^2 f_2(a,b) + \cdots + t^d f_d(a,b).$$

There are two possibilities. If $f \circ \varphi$ is not the 0 polynomial, then $f(\varphi(t))$ has a zero of some finite order at $t = 0$, and this order is defined to be the intersection multiplicity $i(P,L,F)$. If $f \circ \varphi$ is the 0 polynomial, we say that $i(P,L,F) = +\infty$. It will be convenient to define $i(P,L,F) = 0$ if P is not in $L(k) \cap F(k)$.

We need to check that $i(P, L, F)$ does not depend on the choice of Φ. Thus suppose that $\Psi(x_0, y_0, w_0) = (0, 0, 1)$. Write

$$\Psi \circ \Phi^{-1} = \begin{pmatrix} \alpha & \beta & 0 \\ \gamma & \delta & 0 \\ r & s & 1 \end{pmatrix}$$

$$f'(x,y) = F(\Psi^{-1}(x,y,1)) = f'_1(x,y) + \cdots + f'_d(x,y)$$
$$l'(x,y) = L(\Psi^{-1}(x,y,1)) = b'x - a'y \tag{2.15}$$
$$\varphi'(t) = \begin{pmatrix} a't \\ b't \end{pmatrix}.$$

From (2.4), we have

$$l(x, y) = l'(\alpha x + \beta y, \gamma x + \delta y).$$

Since $l(x, y) = bx - ay$, (2.15) gives

$$b = b'\alpha - a'\gamma \quad \text{and} \quad -a = b'\beta - a'\delta.$$

Putting $\Delta = \alpha\delta - \beta\gamma$, we thus have

$$\begin{pmatrix} a \\ b \end{pmatrix} = \begin{pmatrix} \delta & -\beta \\ -\gamma & \alpha \end{pmatrix} \begin{pmatrix} a' \\ b' \end{pmatrix} \quad \text{and} \quad \begin{pmatrix} a' \\ b' \end{pmatrix} = \Delta^{-1} \begin{pmatrix} \alpha & \beta \\ \gamma & \delta \end{pmatrix} \begin{pmatrix} a \\ b \end{pmatrix}.$$

Now (2.3) gives

$$f(x,y) = (rx+sy+1)^{d-1} f'_1(\alpha x+\beta y, \gamma x+\delta y) + \cdots + f'_d(\alpha x+\beta y, \gamma x+\delta y).$$

So

$$f(\varphi(t)) = (art + bst + 1)^{d-1} \Delta f'_1(a't, b't)$$
$$+ (art + bst + 1)^{d-2} \Delta^2 f'_2(a't, b't) + \cdots + \Delta^d f'_d(a't, b't).$$

From this equation we see that $i(P, L, F)$ is the same when computed from f as when computed from f'.

Proposition 2.8. *If a line L and a plane curve F meet at a point P, then $i(P, L, F) = +\infty$ if and only if L divides F.*

PROOF. If L divides F, then $l(x,y)$ divides $f(x,y)$. Since $l(\varphi(t))$ is the 0 polynomial, so is $f(\varphi(t))$.

Conversely suppose $f(\varphi(t))$ is the 0 polynomial, so that $f_r(a,b) = 0$ for all r with $1 \leq r \leq d$. Without loss of generality, suppose $b \neq 0$. The equality

$$0 = f_r(a,b) = c_0 a^r + c_1 a^{r-1} b + \cdots + c_r b^r$$
$$= b^r \left(c_0 \left(\frac{a}{b}\right)^r + c_1 \left(\frac{a}{b}\right)^{r-1} + \cdots + c_r \right)$$

says that $X - \dfrac{a}{b}$ is a factor of $b^r(c_0 X^r + c_1 X^{r-1} + \cdots + c_r)$. If we write

$$b^r(c_0 X^r + \cdots + c_r) = \left(X - \frac{a}{b} \right) u(X),$$

then

$$b^r f_r(x,y) = y^r b^r \left(c_0 \left(\frac{x}{y}\right)^r + \cdots + c_r \right)$$
$$= y^r \left(\frac{x}{y} - \frac{a}{b} \right) u\left(\frac{x}{y}\right) = b^{-1} l(x,y) \left(y^{r-1} u\left(\frac{x}{y}\right) \right).$$

Hence $l(x,y)$ divides $f_r(x,y)$. It follows that $l(x,y)$ divides $f(x,y)$ and then that L divides F.

As was the case with the tangent line, there is an expression for intersection multiplicity that avoids affine local coordinates. For current purposes this expression, which we give as Proposition 2.9, is not too helpful, because it actually represents a disguised use of a particular system of affine local coordinates that may not be the most convenient one for applications. We shall not use this proposition until Chapter V.

Proposition 2.9. Suppose $F \in k[x,y,w]_m$ is a plane curve, $L \in k[x,y,w]_1$ is a line, and $P = (x_0, y_0, w_0)$ is a point on L. If (x', y', w') is any point on L with $[(x', y', w')] \neq [(x_0, y_0, w_0)]$, then $i(P, L, F)$ equals the order of vanishing at $t = 0$ of

$$\psi(t) = F(x_0 + tx', y_0 + ty', w_0 + tw').$$

PROOF. Let (x'', y'', w'') be a point with $L(x'', y'', w'') = 1$, and choose affine local coordinates Φ so that

$$\Phi^{-1}\begin{pmatrix} 0 \\ 0 \\ 1 \end{pmatrix} = \begin{pmatrix} x_0 \\ y_0 \\ w_0 \end{pmatrix}, \quad \Phi^{-1}\begin{pmatrix} 0 \\ 1 \\ 0 \end{pmatrix} = \begin{pmatrix} x' \\ y' \\ w' \end{pmatrix}, \quad \Phi^{-1}\begin{pmatrix} 1 \\ 0 \\ 0 \end{pmatrix} = \begin{pmatrix} x'' \\ y'' \\ w'' \end{pmatrix}.$$

This is possible since the above three image vectors for Φ^{-1} are linearly independent. The corresponding local expression for L is then simply

$$l(x,y) = L \circ \Phi^{-1} \begin{pmatrix} x \\ y \\ 1 \end{pmatrix} = x.$$

The usual parametrization of $l(x,y) = 0$ comes from $\varphi(t) = \begin{pmatrix} 0 \\ t \end{pmatrix}$. Thus we can compute the local expression for F as

$$f(\varphi(t)) = F \circ \Phi^{-1} \begin{pmatrix} 0 \\ t \\ 1 \end{pmatrix} = F(x_0 + tx', y_0 + ty', w_0 + tw').$$

Thus $\psi(t) = f(\varphi(t))$, and the proposition follows.

Suppose that P is in $L(k) \cap F(k)$, that P is a nonsingular point of F, and that L_T is the tangent line to F at P. In the above notation, the local expression for L_T is

$$l_T(x,y) = f_1(x,y).$$

Thus

$$\begin{aligned} i(P,L,F) = 1 &\iff f_1(a,b) \neq 0 \\ &\iff (a,b) \notin L_T(k) \\ &\iff \text{image } \varphi \not\subseteq L_T(k) \\ &\iff L(k) \not\subseteq L_T(k) \\ &\iff L \text{ is not the same as } L_T. \end{aligned} \quad (2.16)$$

Normally we should expect the tangent line at P to have intersection multiplicity 2 with F. We say that a nonsingular point P of F is a **flex** or **inflection point** of $F(k)$ if the intersection multiplicity of the tangent line to F at P is ≥ 3.

Let us see how we would identify a flex from the definition. Suppose F is given and we want to check $(x_0, y_0, 1)$. As usual, we can take $\Phi = \begin{pmatrix} 1 & 0 & -x_0 \\ 0 & 1 & -y_0 \\ 0 & 0 & 1 \end{pmatrix}$, and then $f(x,y) = F(\Phi^{-1}(x,y,1))$ is simply given by $f(x,y) = F(x+x_0, y+y_0, 1)$. We rewrite f as a sum of homogeneous terms in x and y. Then $(x_0, y_0, 1)$ is on the curve exactly if the constant term is 0, the point is nonsingular if also the first order term $f_1(x,y)$ is not 0, and the point is a flex if also the second order term $f_2(x,y)$ satisfies $f_2(a,b) = 0$, where a and b are defined by $f_1(x,y) = bx - ay$. Proposition 2.8 may suggest that the condition on f_2 is equivalent with the condition that $f_1(x,y)$ divide $f_2(x,y)$. In fact, we can see this equivalence algebraically. [If $f_2(x,y) = cx^2 + dxy + ey^2$, suppose

$$cx^2 + dxy + ey^2 = (bx - ay)(rx + sy).$$

Then $c = br$, $d = -ar + bs$, $e = -as$; hence $ca^2 + dab + eb^2 = 0$. Conversely if $ca^2 + dab + eb^2 = 0$, we define $r = c/b$ and $s = -e/a$, checking separately the cases that $b = 0$ or $a = 0$.]

EXAMPLE. Let $F(x,y,w) = x^2y + xyw + w^3$. About $(x_0, y_0, 1)$ we have seen that

$$f(x,y) = (x_0^2 y_0 + x_0 y_0 + 1) + ((2x_0 y_0 + y_0)x + (x_0^2 + x_0)y) \\ + (y_0 x^2 + (2x_0 + 1)xy) + x^2 y.$$

Assume that the point is on the curve, i.e., that

$$x_0^2 y_0 + x_0 y_0 + 1 = 0. \tag{2.17}$$

Then

$$f_1(x,y) = (2x_0 y_0 + y_0)x + (x_0^2 + x_0)y$$
$$f_2(x,y) = (y_0)x^2 + (2x_0 + 1)xy.$$

If we put $b = 2x_0 y_0 + y_0$ and $a = -(x_0^2 + x_0)$, the condition for a flex is

$$0 = f_2(a,b) = y_0 a^2 + (2x_0 + 1)ab,$$

i.e.,

$$y_0(x_0^2 + x_0)^2 - (2x_0 + 1)(x_0^2 + x_0)(2x_0 y_0) = 0. \tag{2.18}$$

We can check that (2.18) is equivalent with the condition that $f_1(x,y)$ divide $f_2(x,y)$.

In the above example, one way to find all flexes $(x_0, y_0, 1)$ would be to solve (2.17) and (2.18) simultaneously, discarding any singular points at the end. However, there is a more elegant way that does not involve passing to affine local coordinates; the only additional condition is that the field is not to have characteristic 2. We require two lemmas.

3. FLEXES

Lemma 2.10 (Principal Axis Theorem). *If char $k \neq 2$ and if A is a symmetric square matrix over k, then there exists a nonsingular square matrix Ψ over k such that $\Psi^{tr} A \Psi$ is diagonal.*

PROOF. We induct on the size, the case of size 1-by-1 being trivial. Suppose the result is known for size $(n-1)$-by-$(n-1)$. If A has size n-by-n, write $A = \begin{pmatrix} a & b \\ b^{tr} & d \end{pmatrix}$ with a symmetric of size $(n-1)$-by-$(n-1)$ and with d of size 1-by-1. If $d \neq 0$, let $x = -d^{-1}b$. Then

$$\begin{pmatrix} I & x \\ 0 & 1 \end{pmatrix} \begin{pmatrix} a & b \\ b^{tr} & d \end{pmatrix} \begin{pmatrix} I & 0 \\ x^{tr} & 1 \end{pmatrix} = \begin{pmatrix} * & 0 \\ 0 & d \end{pmatrix},$$

and the result is reduced to the $(n-1)$-by-$(n-1)$ case and follows by induction. If $d = 0$, we may assume inductively that $b \neq 0$, say $b_i \neq 0$. Let y be an $(n-1)$-dimensional row vector with i^{th} entry δ and with other entries 0. Then

$$\begin{pmatrix} I & 0 \\ y & 1 \end{pmatrix} \begin{pmatrix} a & b \\ b^{tr} & 0 \end{pmatrix} \begin{pmatrix} I & y^{tr} \\ 0 & 1 \end{pmatrix} = \begin{pmatrix} * & * \\ * & yay^{tr} + b^{tr}y^{tr} + yb \end{pmatrix}$$
$$= \begin{pmatrix} * & * \\ * & \delta^2 a_{ii} + 2\delta b_i \end{pmatrix}.$$

Since $2b_i \neq 0$, there is some value of δ for which $\delta^2 a_{ii} + 2\delta b_i \neq 0$. Thus we are reduced to the case $d \neq 0$, which we have already handled.

Lemma 2.11. *Let $A = (A_{ij})$ be a nonzero 3-by-3 symmetric matrix over a field k with char $k \neq 2$. Then the conic*

$$C(x,y,w) = (x \quad y \quad w) A \begin{pmatrix} x \\ y \\ w \end{pmatrix}$$

associated to A is reducible over the algebraic closure $\bar{k}$ if and only if $\det A = 0$.

PROOF. If $C(x,y,w)$ is reducible, Corollary 2.5 shows that the curve has a singular point (x_0, y_0, w_0). By Proposition 2.6

$$\frac{\partial C}{\partial x} = \frac{\partial C}{\partial y} = \frac{\partial C}{\partial w} = 0 \quad \text{at} \quad (x_0, y_0, w_0).$$

Since A is symmetric, these conditions say that $A \begin{pmatrix} x_0 \\ y_0 \\ w_0 \end{pmatrix} = \begin{pmatrix} 0 \\ 0 \\ 0 \end{pmatrix}$.

Hence $\det A = 0$.

Conversely let $\det A = 0$. By Lemma 2.10, choose a nonsingular Φ such that $A' = \Phi^{-1^{tr}} A \Phi^{-1}$ is diagonal. Without loss of generality we may assume that $A'_{33} = 0$. Put $\begin{pmatrix} x' \\ y' \\ w' \end{pmatrix} = \Phi \begin{pmatrix} x \\ y \\ w \end{pmatrix}$. Then

$$C(x,y,w) = \begin{pmatrix} x' & y' & w' \end{pmatrix} A' \begin{pmatrix} x' \\ y' \\ w' \end{pmatrix} = A'_{11} x'^2 + A'_{22} y'^2.$$

The right side is reducible over $\bar{k}$. Substituting back in terms of (x, y, w), we see that $C(x, y, w)$ is reducible over $\bar{k}$.

Proposition 2.12. Suppose the characteristic of k is not 2. Let F be a plane curve over k of degree $d \geq 2$, and let (x_0, y_0, w_0) be a nonsingular point on the curve. Then (x_0, y_0, w_0) is a flex if and only if the Hessian matrix of F satisfies $\det H(x_0, y_0, w_0) = 0$.

PROOF. Without loss of generality, we may suppose k is algebraically closed. Let L be the tangent line at (x_0, y_0, w_0). Choose affine local coordinates with $\Phi(x_0, y_0, w_0) = (0, 0, 1)$, put $f(x, y) = F(\Phi^{-1}(x, y, 1))$, and let $f_1(x, y)$ and $f_2(x, y)$ be the first-order and second-order parts. The condition that (x_0, y_0, w_0) be a flex is that $f_1(x, y)$ divide $f_2(x, y)$, hence that L divide the conic Q_Φ of Proposition 2.7.

Suppose that (x_0, y_0, w_0) is a flex. Then L divides Q_Φ, and (2.12) shows that L divides the conic defined by the matrix $H(x_0, y_0, w_0)$. By Lemma 2.11, $\det H(x_0, y_0, w_0) = 0$.

Conversely suppose $\det H(x_0, y_0, w_0) = 0$. By Lemma 2.11 the conic $C(x, y, w)$ defined by the matrix $H(x_0, y_0, w_0)$ is reducible. Let us say that $C = L_1 L_2$, where L_1 and L_2 are lines. We now divide matters into cases.

Suppose $d > 2$. In affine local coordinates defined by Φ, Q_Φ becomes homogeneous quadratic in (x, y), and the term

$$-2(d-1)L'_\Phi(x, y, w)L(x, y, w)$$

becomes

$$-2(d-1)f_1(x, y) + \text{higher order}.$$

Therefore C has $2(d-1)L$ as tangent at (x_0, y_0, w_0). But $C = L_1 L_2$ implies C has tangent L_1 or L_2 at (x_0, y_0, w_0), up to a constant factor. Hence L is a multiple of L_1 or L_2, and L divides C. By (2.12), L divides Q_Φ. Thus (x_0, y_0, w_0) is a flex.

3. FLEXES

Now suppose $d = 2$. We still have $C = L_1 L_2$. The second partials of C are twice the entries of $H(x_0, y_0, w_0)$. Since C and F are both conics, $C = 2F$. Then $F = \frac{1}{2} L_1 L_2$, and easy computation shows that the nonsingular points are the points of $F(k)$ not in $L_1(k) \cap L_2(k)$ and that each such is a flex.

EXAMPLE. Let $F(x, y, w) = x^2 y + xyw + w^3$. The Hessian matrix at $(x_0, y_0, 1)$ is

$$H(x_0, y_0, 1) = \begin{pmatrix} 2y_0 & 2x_0 + 1 & y_0 \\ 2x_0 + 1 & 0 & x_0 \\ y_0 & x_0 & 6 \end{pmatrix}.$$

The condition in Proposition 2.11 that the determinant vanish says that

$$2y_0(x_0^2 + x_0) - 6(2x_0 + 1)^2 = 0. \tag{2.19}$$

This is ostensibly different from (2.18). But use of (2.17) reduces both equations consistently to

$$3(2x_0 + 1)^2 = -1.$$

Corollary 2.13. Over an algebraically closed field of characteristic not equal to 2, a nonsingular conic has no flex, while a nonsingular plane curve of degree > 2 has at least one flex.

PROOF. Let the curve be F. Since F is nonsingular, Proposition 2.12 says that the flexes are all the common solutions of

$$F(x, y, w) = 0$$

and

$$\det H(x, y, w) = 0.$$

If $\deg F > 2$, Bezout's Theorem says there is a solution. If $\deg F = 2$, then $H(x, y, w)$ is a constant matrix H, and we have to show that $\det H$ is not 0. Let $C(x, y, w)$ be the conic defined by H. At the end of the proof of Proposition 2.12, we saw that $C = 2F$. If $\det H = 0$, Lemma 2.11 says that C is reducible, and thus F is reducible, in contradiction to Corollary 2.5.

If F has degree $d \geq 2$, then $\det H$ has degree $3(d-2)$. The converse half of Bezout's Theorem says that an irreducible F of degree d (over a field of characteristic not 2) has $\leq 3d(d-2)$ flexes unless F and $\det H$ have a common factor. One can show that a plane curve F and its $\det H$ have no common factor unless F is a product of lines; in particular, F and $\det H$ have no common factor if F is irreducible.

4. Application to Cubics

The most general cubic is

$$F(x,y,w) = c_{yyy}y^3 + c_{xyy}xy^2 + c_{xxy}x^2y + c_{yyw}y^2w + c_{xyw}xyw$$
$$+ c_{yww}yw^2 + c_{xxx}x^3 + c_{xxw}x^2w + c_{xww}xw^2 + c_{www}w^3. \quad (2.20)$$

We examine the algebraic effect of some geometric conditions that we might impose on such a curve. We shall make calculations in local coordinates, despite the availability of Proposition 2.6, because we do not want to exclude k of characteristic 2 when we deal with flexes.

CONDITION 1. The curve is to pass through $(x,y,w) = (0,1,0)$. Equivalently

$$c_{yyy} = 0.$$

CONDITION 2. In addition, the curve is to have $(x,y,w) = (0,1,0)$ as a nonsingular point. To examine this condition, we choose $\Phi = \begin{pmatrix} 0 & 0 & 1 \\ 1 & 0 & 0 \\ 0 & 1 & 0 \end{pmatrix}$ as the map that carries $(x_0, y_0, w_0) = (0,1,0)$ to $(0,0,1)$. Then

$$f(x,y) = F(\Phi^{-1}(x,y,1)) = F(y,1,x)$$
$$= c_{xyy}y + c_{xxy}y^2 + c_{yyw}x + c_{xyw}xy + c_{yww}x^2$$
$$+ c_{xxx}y^3 + c_{xxw}y^2 + c_{xww}yx^2 + c_{www}x^3. \quad (2.21)$$

Hence

$$f_1(x,y) = c_{yyw}x + c_{xyy}y.$$

The condition is that

$$c_{yyw} \neq 0 \quad \text{or} \quad c_{xyy} \neq 0.$$

CONDITION 3. In addition, the curve is to have $L(x,y,w) = w$ as tangent line at $(0,1,0)$. (Recall that this line is the line at infinity.) Referring to (2.21), we have

$$\tilde{f}_1(x,y,w) = c_{yyw}x + c_{xyy}y.$$

Hence
$$L(x,y,w) = \tilde{f}_1(\Phi(x,y,w)) = \tilde{f}_1(w,x,y) = c_{yyw}w + c_{xyy}x.$$

The condition is that
$$c_{xyy} = 0.$$

In view of Condition 2, we automatically have
$$c_{yyw} \neq 0,$$

and then the tangent line is the same line as w.

CONDITION 4. In addition, the curve is to have a flex at $(0,1,0)$. Referring to (2.21), we have
$$f_2(x,y) = c_{yww}x^2 + c_{xyw}xy + c_{xxy}y^2$$

and
$$f_1(x,y) = c_{yyw}x.$$

The condition that f_1 divide f_2 is that
$$c_{xxy} = 0.$$

The condition that $i((0,1,0), w, F)$ be defined is that w not divide F, hence that
$$c_{xxx} \neq 0.$$

When all four conditions are satisfied, the cubic becomes
$$\begin{aligned} F(x,y,w) &= c_{yyw}y^2w + c_{xyw}xyw + c_{yww}yw^2 \\ &+ c_{xxx}x^3 + c_{xxw}x^2w + c_{xww}xw^2 + c_{www}w^3 \end{aligned} \quad (2.22)$$

with $c_{yyw} \neq 0$ and $c_{xxx} \neq 0$.

Proposition 2.14. If F is a cubic over k such that $F(k)$ has a flex (x_0, y_0, w_0), then there exists a projective transformation Φ such that F^Φ is the same curve as
$$y^2w + a_1xyw + a_3yw - x^3 - a_2x^2w - a_4xw^2 - a_6w^3. \quad (2.23)$$

PROOF. Choose Φ_1 so that $\Phi_1(x_0, y_0, w_0) = (0, 1, 0)$. Then F^{Φ_1} has a flex at $(0,1,0)$. Let $L(x,y,w) = \alpha x + \beta y$ be the tangent to F^{Φ_1} at

$(0,1,0)$, choose a nonsingular matrix $\begin{pmatrix} a & b \\ c & d \end{pmatrix}$ with $\alpha a + \beta c = 0$, and define
$$\Phi_2^{-1} = \begin{pmatrix} a & 0 & b \\ 0 & 1 & 0 \\ c & 0 & d \end{pmatrix}.$$

Then Φ_2 has the property that L^{Φ_2} is the same line as w. Hence $(F^{\Phi_1})^{\Phi_2} = F^{\Phi_2 \Phi_1}$ has a flex at $(0,1,0)$ and has w as tangent. The discussion above shows that $F^{\Phi_2 \Phi_1}$ has the form (2.22). Put

$$\Phi_3^{-1} = \begin{pmatrix} t & 0 & 0 \\ 0 & t & 0 \\ 0 & 0 & 1 \end{pmatrix}$$

with t to be specified. Then $(F^{\Phi_2 \Phi_1})^{\Phi_3} = F^{\Phi_3 \Phi_2 \Phi_1}$ has

$$F^{\Phi_3 \Phi_2 \Phi_1}(x,y,w) = F^{\Phi_2 \Phi_1}(tx, ty, w)$$
$$= c_{yyw}(t^2 y^2 w) + \cdots + c_{xxx}(t^3 x^3) + \cdots.$$

Taking $t = c_{yyw}/c_{xxx}$, we see that the coefficients of $y^2 w$ and x^3 in $F^{\Phi_3 \Phi_2 \Phi_1}$ are the same. Thus $\Phi = \Phi_3 \Phi_2 \Phi_1$ is the required transformation.

A cubic of the form (2.23) is said to be in **Weierstrass form**. The corresponding **affine Weierstrass form**, achieved by putting $w = 1$, is written as an equation

$$y^2 + a_1 xy + a_3 y = x^3 + a_2 x^2 + a_4 x + a_6. \tag{2.24}$$

The notation in (2.24) is standard; the subscripts will be seen in Chapter III to indicate the degree of homogeneity of the corresponding term under a certain change of variables.

By Corollary 2.13, a nonsingular cubic over an algebraically closed field of characteristic not equal to 2 necessarily has a flex. Hence it can be put in Weierstrass form by a projective transformation.

An **elliptic curve** over k is a nonsingular cubic over k that is in Weierstrass form. In testing a curve (2.23) for nonsingularity, note that $(0,1,0)$ is the only point at infinity on the curve, and it is nonsingular by Condition 2. Hence only points $(x_0, y_0, 1)$ need to be checked.

4. APPLICATION TO CUBICS

Proposition 2.15. Let F be a nonsingular cubic over k, and let L be a line. Then $\sum_P i(P, L, F)$ is 0, 1, or 3.

REMARKS. (1) If $P \neq Q$ are in $F(k)$, let L be the line containing P and Q. Then the proposition says there is a third point on $L(k) \cap F(k)$ if we count multiplicities. (This third point may coincide with P or Q.) We define PQ to be the third point.

(2) To define PP, let L be the tangent to F at P. Then $i(P, L, F) \geq 2$. The proposition says that there is a third point on $L(k) \cap F(k)$ if we count multiplicities. This third point will be P if P is a flex of F, and it will be some different point if P is not a flex. In either case, we define PP to be this third point. The rule for determining PQ in this paragraph or the previous one is called the **chord-tangent composition law**.

PROOF. Let F be as in (2.20). Without loss of generality the sum is not 0. Thus take $P = (x_0, y_0, w_0)$ to be in $L(k) \cap F(k)$. Since everything is invariant under projective transformations, the same argument as in the proof of Proposition 2.14 shows we may assume that $(x_0, y_0, w_0) = (0, 1, 0)$ and that the tangent line there is w. Conditions 1 and 3 show that $c_{yyy} = c_{xyy} = 0$. We consider some possibilities for L.

(1) $L = w$. Putting $w = 0$ in F gives

$$F(x, y, 0) = c_{xxy} x^2 y + c_{xxx} x^3. \tag{2.25}$$

(1a) Suppose $c_{xxy} = c_{xxx} = 0$. Then w divides F, and Corollary 2.5 shows that F is singular, contradiction.

(1b) Suppose $c_{xxx} \neq 0$ and $c_{xxy} = 0$. Then Condition 4 shows that $(0, 1, 0)$ is a flex with $i(P, L, F)$ defined and ≥ 3. Since $i(P, L, F)$ cannot be > 3 for a cubic, $i(P, L, F) = 3$. The expression (2.25) cannot be 0 without $x = 0$, and so there are no other points in $L(k) \cap F(k)$. So $L(k) \cap F(k)$ contains P with multiplicity 3, and it contains no other point.

(1c) Suppose $c_{xxy} \neq 0$. Then Condition 4 shows that $(0, 1, 0)$ is not a flex, hence that $i(P, L, F) = 2$. Meanwhile (2.25) produces a unique additional point Q on L for which (2.25) is 0; we can write $Q = (1, y_1, 0)$. To complete the study of this case, we need to show that $i(Q, L, F) \neq 0$.

Define $\Phi = \begin{pmatrix} 0 & 0 & 1 \\ -y_1 & 1 & 0 \\ 1 & 0 & 0 \end{pmatrix}$, so that $\Phi(1, y_1, 0) = (0, 0, 1)$. Then

$$f(x, y) = F(\Phi^{-1}(x, y, 1)) = F(1, y + y_1, x)$$
$$l(x, y) = L(\Phi^{-1}(x, y, 1)) = L(1, y + y_1, x) = x.$$

Calculating $f(x,y)$, we find that
$$f_1(x,y) = (c_{yyw}y_1^2 + c_{xyw}y_1 + c_{xxw})x + (c_{xxy})y.$$
Since $c_{xxy} \neq 0$, $f_1(x,y)$ is not a multiple of $l(x,y)$. Hence L is not tangent to F and Q, and $i(Q,L,F) = 1$ as required.

(2) L is not the same as w. Since L goes through $P = (0,1,0)$ and is not the tangent, we have $i(P,L,F) = 1$. Assume that $i(Q,L,F) \geq 1$ for some $Q \neq P$. We may write $Q = (x_0, y_0, 1)$. Transforming by
$$\Phi = \begin{pmatrix} 1 & 0 & -x_0 \\ 0 & 1 & -y_0 \\ 0 & 0 & 1 \end{pmatrix},$$
which has $\Phi(P) = P$ and $\Phi(Q) = (0,0,1)$, we may assume from the outset that Q is $(0,0,1)$. Then L, which passes through both P and Q, is the line $L = x$. Using $\Phi = I$, we form
$$f(x,y) = F(x,y,1) = c_{xxy}x^2 y + c_{yyw}y^2 + c_{xyw}xy + c_{yww}y$$
$$+ c_{xxx}x^3 + c_{xxw}x^2 + c_{xww}x$$
and
$$l(x,y) = L(x,y,1) = x.$$
Then $l(x,y) = bx - ay$ with $a = 0$ and $b = 1$. Putting $\varphi(t) = \begin{pmatrix} at \\ bt \end{pmatrix}$, we have
$$f(\varphi(t)) = t(c_{xww}a + c_{yww}b) + t^2(c_{xxw}a^2 + c_{xyw}ab + c_{yyw}b^2)$$
$$+ t^3(c_{xxx}a^3 + c_{xxy}a^2 b)$$
$$= tc_{yww} + t^2 c_{yyw}. \qquad (2.26)$$

Condition 3 shows that $c_{yyw} \neq 0$. If $c_{yww} = 0$, then (2.26) shows that $i(Q,L,F) = 2$ and that $f(x,y)$ vanishes on the locus $l(x,y) = 0$ only at $(x,y) = (0,0)$; hence P and Q are the only points contributing to the sum in question, and the sum is 3. If $c_{yww} \neq 0$, then (2.26) shows that $i(Q,L,F) = 1$ and that there is just one more point R where $i(R,L,F) > 0$; interchanging the roles of Q and R shows that $i(R,L,F) = 1$ and hence that the sum is 3.

5. Bezout's Theorem and Resultants

Before returning to Bezout's Theorem, we introduce the **resultant** $R(f,g)$ of two polynomials. Let A be a unique factorization domain, e.g., $k[x_1, \ldots, x_r]$. For f and g in $A[X]$ with
$$f(X) = a_0 + a_1 X + \cdots + a_m X^m$$
$$g(X) = b_0 + b_1 X + \cdots + b_n X^n, \qquad (2.27)$$

let $[R(f,g)]$ be the $(m+n)$-by-$(m+n)$ matrix

$$= \begin{pmatrix} a_0 & a_1 & \cdots & a_{m-1} & a_m & 0 & 0 & \cdots & 0 \\ 0 & a_0 & \cdots & a_{m-2} & a_{m-1} & a_m & 0 & \cdots & 0 \\ \vdots & & & & & & & & \vdots \\ 0 & \cdots & & & a_0 & & & \cdots & a_m \\ b_0 & b_1 & & & b_{n-1} & b_n & 0 & \cdots & 0 \\ 0 & b_0 & & & b_{n-2} & b_{n-1} & b_n & \cdots & 0 \\ \vdots & & & & & & & & \vdots \\ 0 & \cdots & & b_0 & b_1 & & \cdots & & b_n \end{pmatrix},$$

in which there are n rows above the b_0 in the first column and there are m remaining rows. Let $R(f,g)$ be the determinant

$$R(f,g) = \det[R(f,g)].$$

Proposition 2.16. With f and g in $A[X]$ of the form (2.27), the following are equivalent:

(1) f and g have a common factor of degree > 0.
(2) $af + bg = 0$ for some nonzero a and b in $A[X]$ with $\deg a < n$ and $\deg b < m$.
(3) $R(f,g) = 0$.

When $R(f,g) \neq 0$, there exist nonzero a and b in $A[X]$ with $\deg a < n$, $\deg b < m$, and $a(X)f(X) + b(X)g(X) = R(f,g)$. (Here $R(f,g)$ is to be regarded as a polynomial of degree 0.)

PROOF. (1) $\Rightarrow$ (2). If $u \mid f$ and $u \mid g$, write $f = bu$ and $g = -au$. Then (2) holds.

(2) $\Rightarrow$ (1). Factor f and bg. If (1) fails, then the prime factors of f of degree > 0 must occur in the factorization of b, with multiplicities. So $\deg f \leq \deg b$, contradiction.

(2) $\Leftrightarrow$ (3). For any a and b of the form

$$\begin{aligned} a(X) &= \alpha_0 + \alpha_1 X + \cdots + \alpha_{n-1} X^{n-1} \\ b(X) &= \beta_0 + \beta_1 X + \cdots + \beta_{m-1} X^{m-1} \end{aligned} \quad (2.28)$$

with coefficients in the quotient field $\tilde{A}$ of A, we have

$$\begin{pmatrix} \alpha_0 & \alpha_1 & \cdots & \alpha_{n-1} & \beta_0 & \cdots & \beta_{m-1} \end{pmatrix} [R(f,g)] \\ = \begin{pmatrix} c_0 & c_1 & \cdots & c_{m+n-1} \end{pmatrix}, \quad (2.29a)$$

where
$$a(X)f(X) + b(X)g(X) = c(X) = c_0 + c_1 X + \cdots + c_{m+n-1} X^{m+n-1}. \tag{2.29b}$$
If nontrivial $a(X)$ and $b(X)$ exist with coefficients in A such that $c(X) = 0$, then $\ker[R(f,g)]^{\mathrm{tr}} \neq 0$ and $R(f,g) = 0$. Conversely if $R(f,g) = 0$, then $\ker[R(f,g)]^{\mathrm{tr}} \neq 0$ and there exist nontrivial $a(X)$ and $b(X)$ with coefficients in $\tilde{A}$ such that $c(X) = 0$. Clearing denominators, we obtain $a(X)$ and $b(X)$ with coefficients in A such that $c(X) = 0$.

Last statement: If $R(f,g) \neq 0$, then the cofactor formula for the inverse of $[R(f,g)]$ with entries in $\tilde{A}$ says that
$$[R(f,g)]^{-1} = R(f,g)^{-1}[S(f,g)],$$
where $[S(f,g)]$ has entries in A. Thus we can define elements
$$\alpha_0, \ldots, \alpha_{n-1}, \beta_0, \ldots, \beta_{m-1}$$
in A by
$$\begin{pmatrix} \alpha_0 & \alpha_1 & \cdots & \alpha_{n-1} & \beta_0 & \cdots & \beta_{m-1} \end{pmatrix}$$
$$= \begin{pmatrix} R(f,g) & 0 & \cdots & 0 \end{pmatrix} [R(f,g)]^{-1}.$$
If we define $a(X)$ and $b(X)$ in $A[X]$ by (2.28), then (2.29) shows that $a(X)f(X) + b(X)g(X) = R(f,g)$.

Proposition 2.17. If $A = k[x_1, \ldots, x_r]$ and if f and g are as in (2.27) with a_j homogeneous of degree $m - j$ and b_j homogeneous of degree $n - j$, then $R(f,g)$ is homogeneous of degree mn.

PROOF. For an exponent u obtained by factoring powers
$$(t^n, t^{n-1}, \ldots, t^m, t^{m-1}, \ldots)$$
of t from the rows of the matrix displayed below and for v obtained by factoring powers $(t^{m+n}, t^{m+n-1}, \ldots)$ from the columns, we have
$$t^u R(f,g)(tx_1, \ldots, tx_r) = \det \begin{pmatrix} t^n t^m a_0 & t^n t^{m-1} a_1 & \cdots & t^n a_m & \cdots \\ 0 & t^{n-1} t^m a_0 & \cdots & & \\ \cdots & & & & \\ t^m t^n b_0 & t^m t^{n-1} b_1 & \cdots & t^m b_n & \cdots \\ 0 & t^{m-1} t^n b_0 & \cdots & & \end{pmatrix}$$
$$= t^v R(f,g)(x_1, \ldots, x_r).$$
So $R(f,g)(tx) = t^{v-u} R(f,g)$. Computing u and v, we find that $u = \frac{1}{2}m(m+1) + \frac{1}{2}n(n+1)$ and $v = \frac{1}{2}(m+n)(m+n+1)$, so that $v - u = mn$. By Proposition 2.2 applied to $\bar{k}$, $R(f,g)$ is homogeneous of degree mn.

5. BEZOUT'S THEOREM AND RESULTANTS

Let us restate Bezout's Theorem, originally given as Theorem 2.4.

Theorem 2.18 (Bezout's Theorem). Suppose $F \in k[x,y,w]_m$ and $G \in k[x,y,w]_n$ are plane curves. Then $F(\bar{k}) \cap G(\bar{k})$ is nonempty. If it has more than mn points, then F and G have as a common factor some homogeneous polynomial of degree > 0.

PROOF OF FIRST CONCLUSION. Write F and G in the form

$$F(x,y,w) = a_0 + a_1 w + \cdots + a_m w_m \quad \text{with } a_j \in \bar{k}[x,y]_{m-j}$$
$$G(x,y,w) = b_0 + b_1 w + \cdots + b_n w_n \quad \text{with } b_j \in \bar{k}[x,y]_{n-j}.$$
(2.30)

Regarding F and G as polynomials in w, with coefficients in $A = \bar{k}[x,y]$, we form $R(F,G)$, which Proposition 2.17 identifies as a member of $\bar{k}[x,y]_{mn}$.

Since $R(F,G)(x,y)$ is homogeneous and $\bar{k}$ is algebraically closed, we can choose a point $(x_0, y_0) \neq (0,0)$ where $R(F,G)(x_0, y_0) = 0$. Then the resultant of $F(x_0, y_0, w)$ and $G(x_0, y_0, w)$ is 0, and Proposition 2.16 says that these two polynomials in w have a common factor. Since $\bar{k}$ is algebraically closed, this common factor vanishes at some $w = w_0$, and then we must have $F(x_0, y_0, w_0) = G(x_0, y_0, w_0) = 0$.

PROOF OF SECOND CONCLUSION. Suppose $F(\bar{k}) \cap G(\bar{k})$ contains $mn + 1$ points. Join these points by lines, and pick a point defined over $\bar{k}$ that is not on any of the lines. Applying a projective transformation, we may assume the point is $(0,0,1)$. Write F and G in the form (2.30). Regarding F and G as polynomials in w, with coefficients in $A = \bar{k}[x,y]$, we again form $R(F,G)$, which Proposition 2.17 identifies as a member of $\bar{k}[x,y]_{mn}$. For fixed (x_0, y_0), Proposition 2.16 says that $R(F,G)(x_0, y_0) = 0$ if and only if $F(x_0, y_0, w)$ and $G(x_0, y_0, w)$ have a common factor (hence necessarily some $w - w_0$ factor since $\bar{k}$ is algebraically closed), if and only if $F(x_0, y_0, w_0) = G(x_0, y_0, w_0) = 0$ for some w_0. So at each of our $mn + 1$ points, say (x_i, y_i, w_i), we have $R(F,G)(cx_i, cy_i) = 0$ for all c. Since $(x_i, y_i) \neq (0,0)$, $R(F,G)$ vanishes on the line $y_i x - x_i y = 0$. Consequently $y_i x - x_i y$ divides $R(F,G)$.

Suppose $(x_i, y_i) = (x_j, y_j)$. Then (x_i, y_i, w_i) and (x_j, y_j, w_j) both satisfy $y_i x - x_i y = 0$. Since $(0,0,1)$ satisfies this also and since $(0,0,1)$ is not to be on any of the connecting lines, we obtain a contradiction.

Thus the $mn+1$ factors $y_i x - x_i y$ are distinct primes in $\bar{k}[x,y]$ dividing $R(F,G)$. By unique factorization, their product divides $R(F,G)$. Since $\deg R(F,G) = mn$, we conclude $R(F,G) = 0$. Then Proposition 2.16 shows that F and G have a common factor in $\bar{k}[x,y][w] = \bar{k}[x,y,w]$. The common factor is homogeneous by the first conclusion of Corollary 2.5 (which does not depend on Bezout's Theorem).

Corollary 2.19. Suppose $F \in k[x,y,w]_d$ is a plane curve and L is a line. If $F(k) \cap L(k)$ has more than d elements, then L divides F. This condition is satisfied if F vanishes on $L(k)$ and k has $\geq d$ elements.

PROOF. The number of points on $L(k)$ is one more than the number of elements of k, hence is $\geq d+1$. Then $F(\bar{k}) \cap L(\bar{k})$ has $\geq d+1$ points, and Bezout's Theorem says L and F have a common factor. Since L is prime and $k[x,y,w]$ has unique factorization, L divides F.

The full-strength version of Bezout's Theorem says that two plane curves F and G of degrees m and n meet in $\leq mn$ points even when multiplicities are counted, and that the number is $= mn$ if k is algebraically closed and multiplicities are counted. We do not need the full-strength version of Bezout's Theorem and have consequently not defined intersection multiplicity in full generality. The result that we shall prove instead is the corresponding sharpening of the special case of Corollary 2.19 in which one of the curves is a line.

Corollary 2.20. If $F \in k[x,y,w]_d$ is a plane curve and L is a line such that $\sum_p i(p, L, F) > d$, then L divides F.

PROOF. Without loss of generality, we may assume k is algebraically closed and is in particular infinite. Suppose L does not divide F. Then Corollary 2.19 shows that $F(k) \cap L(k)$ is finite, and it follows from Proposition 2.8 that $\sum_p i(p, L, F)$ is finite.

Possibly by applying a projective transformation, we may assume that the line L is $w = 0$. Then the points P_j with $i(P_j, L, F) > 0$ are of the form $[(x_j, y_j, 0)]$. Possibly by applying a second projective transformation, one that translates the y variable, we may assume that no y_j is 0. Then we can write $P_j = [(r_j, 1, 0)]$ with r_j in k. Now $F(x, 1, 0)$ is a polynomial in x of degree $\leq d$. We shall prove that

$$i((r,1,0), w, F) = \text{multiplicity of } r \text{ as a root of } F(x,1,0). \quad (2.31)$$

Since k is infinite, there is some r not in $F(k) \cap L(k)$, and this r in (2.31) shows that $F(x, 1, 0)$ is not the 0 polynomial. Hence $F(x, 1, 0)$ has $\leq d$ roots, counting multiplicities, and it follows from (2.31) that $\sum_p i(p, L, F) \leq d$, as required.

To prove (2.31), we introduce affine local coordinates about $(r, 1, 0)$, using $\Phi^{-1} = \begin{pmatrix} 1 & 0 & r \\ 0 & 0 & 1 \\ 0 & 1 & 0 \end{pmatrix}$, so that $\Phi(r, 1, 0) = (0, 0, 1)$. The local versions f of F and l of $L = w$ are

$$f(x,y) = F(\Phi^{-1}(x,y,1)) = F(x+r, 1, y)$$
$$l(x,y) = L(\Phi^{-1}(x,y,1)) = y.$$

5. BEZOUT'S THEOREM AND RESULTANTS

Thus $l(x,y)$ is of the form $bx - ay$ with $a = -1$ and $b = 0$. We can parametrize l by $\varphi(t) = \begin{pmatrix} at \\ bt \end{pmatrix} = \begin{pmatrix} -t \\ 0 \end{pmatrix}$, and then

$$f(\varphi(t)) = f(-t, 0) = F(-t+r, 1, 0).$$

The order of vanishing of $f(\varphi(t))$ at $t = 0$, which is $i((r,1,0), L, F)$, thus equals the order of the 0 of $F(-t+r, 1, 0)$ at $t = 0$, which equals the multiplicity of r as a root of $F(x, 1, 0)$. This proves (2.31), and the corollary follows.

CHAPTER III

CUBIC CURVES IN WEIERSTRASS FORM

1. Examples

We begin with some examples of plane curves, showing how to bring them into Weierstrass form and discussing some of their solutions.

A. Diophantus cubic example.

The equation from Chapter 1 in affine form is

$$y(6-y) = x^3 - x. \tag{3.1}$$

Following the procedure in the proof of Proposition 2.13, we replace y by $-y$ and x by $-x$ to obtain

$$y^2 + 6y = x^3 - x.$$

If we complete the square and replace $y+3$ by y, we are led to

$$y^2 = x^3 - x + 9. \tag{3.2}$$

The trivial solutions of (3.1) were $x=0$ or ± 1 with $y=0$ or 6. These correspond in (3.2) to $x=0$ or ± 1 with $y=\pm 3$. Diophantus's nontrivial solution of (3.1) began with $P=(-1,0)$ and amounted to using $PP = (\frac{17}{9}, \frac{26}{27})$; here PP is the result of applying the chord-tangent composition law (§II.4) to P and P. In terms of (3.2) we would take the corresponding trivial solution $P=(1,3)$ and use $PP = (-\frac{17}{9}, \frac{55}{27})$.

B. Cubic case of Fermat's Last Theorem.

The equation in projective form is

$$x^3 + y^3 = w^3. \tag{3.3}$$

Since we can always clear denominators, integer solutions and rational solutions amount to the same thing. With $F(x,y,w) = x^3 + y^3 - w^3$, we have

$$\frac{\partial F}{\partial x} = 3x^2, \quad \frac{\partial F}{\partial y} = 3y^2, \quad \frac{\partial F}{\partial w} = -3w^2.$$

1. EXAMPLES

The only place where all first partials are 0 is $(0,0,0)$, which does not correspond to a point in projective space. By Proposition 2.6, F is nonsingular. To look for flexes, we compute the Hessian determinant

$$\det H(x,y,w) = \det \begin{pmatrix} 6x & 0 & 0 \\ 0 & 6y & 0 \\ 0 & 0 & -6w \end{pmatrix} = -6^3 xyw.$$

We seek simultaneous solutions of (3.3) and

$$xyw = 0. \tag{3.4}$$

For example, we can use $(x_0, y_0, w_0) = (0, 1, 1)$. To map this to $(0, 1, 0)$, which is the point at infinity of a curve in Weierstrass form, we apply successively $\begin{pmatrix} 1 & 0 & 0 \\ 0 & 1 & -1 \\ 0 & 0 & 1 \end{pmatrix}$ to carry $(0, 1, 1)$ to $(0, 0, 1)$ and then $\begin{pmatrix} 1 & 0 & 0 \\ 0 & 0 & 1 \\ 0 & 1 & 0 \end{pmatrix}$ to carry $(0, 0, 1)$ to $(0, 1, 0)$. The tangent $y - w$ maps first to y and then to w. Hence

$$\Phi = \begin{pmatrix} 1 & 0 & 0 \\ 0 & 0 & 1 \\ 0 & 1 & 0 \end{pmatrix} \begin{pmatrix} 1 & 0 & 0 \\ 0 & 1 & -1 \\ 0 & 0 & 1 \end{pmatrix} = \begin{pmatrix} 1 & 0 & 0 \\ 0 & 0 & 1 \\ 0 & 1 & -1 \end{pmatrix}$$

maps F into Weierstrass form except for scaling. To carry out the computations, we put $(a\ b\ c) = \Phi(x\ y\ w)$. Then

$$\begin{pmatrix} x \\ y \\ w \end{pmatrix} = \Phi^{-1} \begin{pmatrix} a \\ b \\ c \end{pmatrix}$$

$$= \begin{pmatrix} 1 & 0 & 0 \\ 0 & 1 & 1 \\ 0 & 1 & 0 \end{pmatrix} \begin{pmatrix} a \\ b \\ c \end{pmatrix} = \begin{pmatrix} a \\ b+c \\ b \end{pmatrix}.$$

Substituting in (3.3), we obtain

$$a^3 + (b+c)^3 = b^3.$$

Passing to affine form by replacing (a, b, c) by $(x, y, 1)$, we obtain

$$3y^2 + 3y = -x^3 - 1.$$

We scale by changing y to $-3y$ and x to $-3x$, and the result is
$$y^2 - \tfrac{1}{3}y = x^3 - \tfrac{1}{27}.$$
We make a further adjustment to clear denominators: We change y to y/u^3 and x to x/u^2 with $u = 3$. Then the equation becomes
$$y^2 - 9y = x^3 - 27. \tag{3.5}$$

Fermat showed that the only **Q**-rational (projective) solutions of (3.3) are the trivial ones: (x, y, w) must be one of $(1, 0, 1)$, $(0, 1, 1)$, and $(1, -1, 0)$. The corresponding solutions of (3.5) are $(3, 9)$, ∞, and $(3, 0)$.

For future reference let us record the composite transformations from (3.3) to (3.5) and back. If (X, Y) satisfies $X^3 + Y^3 = 1$ and (x, y) satisfies $y^2 - 9y = x^3 - 27$, the relationships are

$$\left. \begin{array}{l} X = \dfrac{3x}{y} \\[4pt] Y = \dfrac{y-9}{y} \end{array} \right\} \quad \text{and} \quad \left\{ \begin{array}{l} x = \dfrac{3X}{1-Y} \\[4pt] y = \dfrac{9}{1-Y}. \end{array} \right. \tag{3.6}$$

These are the transformations that we used in discussing $u^3 + v^3 = w^3$ in Chapter I.

Equation (3.5) can be transformed some more. Completing the square eliminates the y term and leads to
$$y^2 = x^3 - \tfrac{27}{4}. \tag{3.7}$$
Scaling by $(x, y) \to (x/u^2, y/u^3)$ with $u = 2$ leads to
$$y^2 = x^3 - 2^4 \cdot 3^3. \tag{3.8}$$

C. Congruent numbers.

A positive integer n is **congruent** if n is the area of a rational right triangle. For example, the Pythagorean triple $(3, 4, 5)$ has area 6. So $n = 6$ is congruent.

Congruent numbers were studied as early as 984 A.D., were taken up by Fibonacci in 1225, and were later studied by Fermat. The problem is: Given n, decide whether n is congruent. Results for small square-free n are as follows:

1	not congruent	Fermat
2	not congruent	Fermat
3	not congruent	Fermat
5	congruent for $(\tfrac{3}{2}, \tfrac{20}{3}, \tfrac{41}{6})$	Fibonacci
6	congruent for $(3, 4, 5)$.	

Part of the connection between the congruent-number problem and elliptic curves is given in the following proposition. We shall study the relationship further in Chapter IV, and we shall prove there that $n = 1, 2, 3$ are not congruent.

1. EXAMPLES

Proposition 3.1. If n is a square-free positive integer, then the following are equivalent:

(1) n is congruent: $n = \frac{1}{2}ab$, where (a, b, c) is a rational Pythagorean triple.
(2) There exist three rational squares in arithmetic progression with common difference n.

Moreover, these conditions imply

(3) There exists a $\mathbb{Q}$ rational point (x, y) on

$$y^2 = x^3 - n^2 x \tag{3.9}$$

other than $(-n, 0)$, $(0, 0)$, $(n, 0)$, and ∞.

EXAMPLES. Arithmetic progressions meeting the second condition for the congruent numbers 5 and 6 are

$$n = 6 \quad \left(\frac{1}{4}, \frac{25}{4}, \frac{49}{4}\right)$$

$$n = 5 \quad \left(\left(\frac{29}{12}\right)^2, \left(\frac{41}{12}\right)^2, \left(\frac{49}{12}\right)^2\right).$$

PROOF. (1) $\Rightarrow$ (2). Given (a, b, c), put $x = c^2/4$. Then $(a-b)^2/4 = x - n$ and $(a+b)^2/4 = x + n$, so that $x - n$ and x and $x + n$ are all squares of rational numbers.

(2) $\Rightarrow$ (1). Given x such that $x - n$ and x and $x + n$ are all squares, put

$$a = (x + n)^{1/2} + (x - n)^{1/2}$$
$$b = (x + n)^{1/2} - (x - n)^{1/2}$$
$$c = 2x^{1/2}.$$

Then a, b, c are rational, and $a^2 + b^2 = c^2$.

(2) $\Rightarrow$ (3). If x is the middle number in the progression, then the product of the three is $x^3 - n^2 x$ and is a square. Hence (3.9) holds with x as the middle number of the progression. The progression cannot be $-n, 0, n$ since n is square-free. Thus the $\mathbb{Q}$ rational point in question satisfies (3).

REMARK. In Corollary 4.3 we shall prove conversely that (3) $\Rightarrow$ (2).

D. Quartic case of Fermat's Last Theorem.

Fermat showed that $x^4 + y^4 = z^4$ has no nontrivial integer solutions by showing that

$$u^4 + v^4 = w^2 \tag{3.10}$$

has no nontrivial integer solutions. We give his proof as Proposition 4.1. Equation (3.10) is not homogeneous, but we can still study it as $\left(\dfrac{u}{v}\right)^4 + 1 = \left(\dfrac{w}{v^2}\right)^2$. Thus the equation to study for $\mathbb{Q}$-rational points is

$$y^2 = x^4 + 1. \tag{3.11}$$

If we make the nonprojective change of variables $x \to x$ and $y \to y + x^2$, we are led to the cubic equation

$$\begin{aligned} y^2 + 2x^2 y &= 1 \quad \text{affine} \\ y^2 w + 2x^2 y &= w^3 \quad \text{projective.} \end{aligned} \tag{3.12}$$

We can verify that (3.12) is nonsingular and can look for $\mathbb{Q}$-rational flexes in the same way as for (3.3). The only $\mathbb{Q}$-rational flex is at $(x, y, w) = (1, 0, 0)$, where the tangent line is $y = 0$. We apply $\Phi_1 = \begin{pmatrix} 0 & 1 & 0 \\ 1 & 0 & 0 \\ 0 & 0 & 1 \end{pmatrix}$ in order to send $(1, 0, 0)$ to $(0, 1, 0)$ with the tangent becoming $x = 0$, and then we apply $\Phi_2 = \begin{pmatrix} 0 & 0 & 1 \\ 0 & 1 & 0 \\ 1 & 0 & 0 \end{pmatrix}$ in order to fix $(0, 1, 0)$ and to make the tangent become $w = 0$. The resulting equation is

$$2y^2 = x^3 - x.$$

Scaling by sending y to $2y$ and x to $2x$ changes the equation to

$$y^2 = x^3 - \tfrac{1}{4}x.$$

We can make the coefficients be integers by sending y to $\tfrac{1}{8}y$ and x to $\tfrac{1}{4}x$; the result is the elliptic curve

$$y^2 = x^3 - 4x. \tag{3.13}$$

Since (3.10) has only trivial solutions, the only $\mathbb{Q}$-rational solutions of (3.13) are ∞, $(0, 0)$, and $(\pm 2, 0)$. We shall prove this result directly in §IV.7.

1. EXAMPLES

For future reference let us record the composite transformations from (3.11) to (3.13) and back. If (X, Y) satisfies $Y^2 = X^4 + 1$ and (x, y) satisfies $y^2 = x^3 - 4x$, the relationships are

$$\left. \begin{array}{l} X = \dfrac{y}{2x} \\ Y = \dfrac{y^2 + 8x}{4x^2} \end{array} \right\} \text{ and } \left\{ \begin{array}{l} x = \dfrac{2}{Y - X^2} \\ y = \dfrac{4X}{Y - X^2}. \end{array} \right. \tag{3.14}$$

The quartic Fermat equation $x^4 + y^4 = z^4$ can be treated also as $u^4 + v^2 = w^4$. This equation becomes $y^2 = x^4 - 1$, and an analysis like the one above reduces it to the elliptic curve

$$y^2 = x^3 + 4x. \tag{3.15}$$

The **Q**-rational solutions of (3.15) are ∞, $(0,0)$, and $(2, \pm 4)$.

E. Fermat's problem for Mersenne.

In 1643 Fermat posed to Mersenne the problem of finding a relatively prime integer-valued Pythagorean triple (X, Y, Z) such that the hypotenuse Z is a square and the sum of the legs is a square. The answer that Fermat had in mind is

$$\begin{aligned} X &= 1061652293520 \\ Y &= 4565486027761 \\ Z &= 4687298610289, \end{aligned} \tag{3.16}$$

and this is the smallest nontrivial solution.

In algebraic terms, we are given $X^2 + Y^2 = Z^2$, $Z = b^2$, $X + Y = a^2$. If we write $e = X - Y$, then X and Y are $\frac{1}{2}(a^2 \pm e)$. Thus

$$b^4 = Z^2 = X^2 + Y^2 = \frac{1}{2}(a^4 + e^2)$$

and

$$e^2 = 2b^4 - a^4. \tag{3.17}$$

Dividing by a^4, we are led to the affine curve

$$v^2 = 2u^4 - 1. \tag{3.18}$$

If (u, v) satisfies (3.18) and if (x, y) is defined by

$$\begin{aligned} x &= \frac{2(v + 2u^2 - 1)}{(u-1)^2} \\ y &= \frac{4[(2u-1)v + 2u^3 - 1]}{(u-1)^3}, \end{aligned} \tag{3.19}$$

then (x, y) satisfies
$$y^2 = x^3 + 8x. \tag{3.20}$$
Conversely if (x, y) satisfies (3.20), then (u, v) may be recovered from
$$\begin{aligned} u &= \frac{y - 2x - 8}{y - 4x + 8} \\ v &= \frac{y^2 - 24x^2 + 48y - 16x - 64}{(y - 4x + 8)^2} \end{aligned} \tag{3.21}$$

so as to satisfy (3.18). For example, $(x, y) = (0, 0)$, $(1, 3)$, and $(1, -3)$ correspond to $(u, v) = (-1, -1)$, $(-1, 1)$, and $(-13, -239)$. But there are some singularities: $(u, v) = (13, 239)$ maps to $(x, y) = (8, 24)$, but $(x, y) = (8, 24)$ maps to a removable singularity. Also $(x, y) = (8, -24)$ maps to $(u, v) = (1, -1)$, but $(u, v) = (1, -1)$ maps to a removable singularity. Large points (x, y) correspond to points near $(u, v) = (1, 1)$, and large points (u, v) correspond to points near $x = 4 \pm \sqrt{8}$.

We shall not be concerned at this time with how to obtain the transformations (3.19) and (3.21). What we shall see, apart from the singularities, is that a $\mathbf{Q}$ rational point on the elliptic curve (3.20) leads to a $\mathbf{Q}$-rational solution (u, v) of (3.18), hence to an integral solution (a, b, e) of (3.17) with a and b relatively prime. Then a and b must both be odd, and so must e. We can thus define an integer tuple (X, Y, Z) by

$$X = \frac{1}{2}(a^2 + e), \quad Y = \frac{1}{2}(a^2 - e), \quad Z = b^2, \tag{3.22}$$

and the only question will be whether X and Y are positive. Unfortunately small solutions like $(a, b, e) = (1, 13, 239)$ lead to Y negative.

We shall return to Fermat's solution to the problem—and to the interpretation in terms of elliptic curves—in Chapter IV.

2. Weierstrass Form, Discriminant, j-invariant

A cubic over k in Weierstrass form is given projectively by

$$y^2 w + a_1 xyw + a_3 yw^2 = x^3 + a_2 x^2 w + a_4 xw^2 + a_6 w^3, \tag{3.23a}$$

with coefficients in k. We know that the only $\bar{k}$ rational point on the line at infinity $w = 0$ is $(x, y, w) = (0, 1, 0)$. Moreover, we saw in §II.4 that $(0, 1, 0)$ is a nonsingular point and is an inflection point, the tangent line being the line at infinity. Since the behavior at $(0, 1, 0)$ is so well

2. WEIERSTRASS FORM, DISCRIMINANT, j-INVARIANT

understood, we can study much of the behavior of the curve by working with the affine form

$$y^2 + a_1 xy + a_3 y = x^3 + a_2 x^2 + a_4 x + a_6, \qquad (3.23b)$$

This form has the advantage that the notation is simpler. The significance of the subscripts will be explained later in this section.

In the examples in §1, specific plane curves of the form (3.23b) simplified under changes of variables. Actually these changes of variables are valid more generally, under only an assumption on the characteristic of k. The expressions for how the coefficients of (3.23b) are affected by these changes of variables do not depend on the characteristic and are as follows. The notation is absolutely standard:

$$\begin{aligned} b_2 &= a_1^2 + 4a_2 \\ b_4 &= 2a_4 + a_1 a_3 \\ b_6 &= a_3^2 + 4a_6 \\ b_8 &= a_1^2 a_6 + 4a_2 a_6 - a_1 a_3 a_4 + a_2 a_3^2 - a_4^2 \end{aligned} \qquad (3.24)$$

and

$$\begin{aligned} c_4 &= b_2^2 - 24 b_4 \\ c_6 &= -b_2^3 + 36 b_2 b_4 - 216 b_6. \end{aligned} \qquad (3.25)$$

The first simplification of (3.23b) assumes that char$(k) \neq 2$. We complete the square in (3.23b), replacing $y + \frac{1}{2}(a_1 x + a_3)$ by $\frac{1}{2}y$, and the result is

$$y^2 = 4x^3 + b_2 x^2 + 2b_4 x + b_6 \qquad (3.26)$$

with b_2, b_4, b_6 as in (3.24). (The coefficient b_8 will play a role later in this section and also in §4.)

The second simplification assumes in addition that char$(k) \neq 3$. We replace (x, y) in (3.26) by $\left(\dfrac{x - 3b_2}{36}, \dfrac{y}{108} \right)$, and the result is

$$y^2 = x^3 - 27 c_4 x - 54 c_6 \qquad (3.27)$$

with c_4 and c_6 as in (3.25).

Although our chief interest is in elliptic curves over $\mathbb{Q}$, it is not enough to consider only equations (3.27). The reason lies in what happens when we reduce curves over $\mathbb{Q}$ modulo a prime p. The following example illustrates.

EXAMPLE. Cubic case of Fermat's Last Theorem. The Fermat equation (3.3) led to the Weierstrass form in (3.5):

$$y^2 - 9y = x^3 - 27. \tag{3.28}$$

Further changes of variables then led to the equation

$$y^2 = x^3 - 2^4 3^3, \tag{3.29}$$

which is of the form (3.27). Following Theorem 3.2, we shall see that (3.28) is nonsingular modulo all primes but 3, while (3.29) is nonsingular modulo all primes but 2 and 3. Thus the passage from (3.28) to (3.29) loses information that might be obtained by reduction modulo 2.

For any field k, we introduce the **discriminant** Δ of the curve (3.23b) or (3.26) by the formula

$$\Delta = -b_2^2 b_8 - 8b_4^3 - 27b_6^2 + 9b_2 b_4 b_6 \tag{3.30}$$

with b_2, b_4, b_6, b_8 as in (3.24). When $\mathrm{char}(k)$ is not 2 or 3, we can solve for Δ from the relation

$$1728\Delta = c_4^3 - c_6^2. \tag{3.31}$$

WARNING. If (3.27) is used as our *original* equation, its Δ is $2^6 3^9 (c_4^3 - c_6^2)$, and this is off by a factor of $2^{12} 3^{12}$ from Δ in (3.31). The factor is introduced by the scaling in passing from (3.23) to (3.26) and then (3.27).

Theorem 3.2. The cubic (3.23) is singular if and only if $\Delta = 0$.

This theorem will be proved later in this section. By way of examples, note that $\Delta = -3^9$ for (3.28), and $\Delta = -2^{12} 3^9$ for (3.29), so that (3.29) is singular modulo 2 but (3.28) is not. To get at the proof of the theorem, we introduce the discriminant of a cubic polynomial in one variable.

Let

$$f(x) = x^3 - \alpha x^2 + \beta x - \gamma = (x - r_1)(x - r_2)(x - r_3) \tag{3.32}$$

be a monic cubic polynomial over k with roots in $\bar{k}$. Here α, β, and γ are given by the elementary symmetric polynomials

$$\alpha = r_1 + r_2 + r_3, \quad \beta = r_1 r_2 + r_1 r_3 + r_2 r_3, \quad \gamma = r_1 r_2 r_3.$$

2. WEIERSTRASS FORM, DISCRIMINANT, j-INVARIANT

We can check that

$$\det \begin{pmatrix} 1 & 1 & 1 \\ r_1 & r_2 & r_3 \\ r_1^2 & r_2^2 & r_3^2 \end{pmatrix} = (r_3 - r_2)(r_3 - r_1)(r_2 - r_1) \tag{3.33a}$$

and that

$$\begin{pmatrix} 1 & 1 & 1 \\ r_1 & r_2 & r_3 \\ r_1^2 & r_2^2 & r_3^2 \end{pmatrix} \begin{pmatrix} 1 & r_1 & r_1^2 \\ 1 & r_2 & r_2^2 \\ 1 & r_3 & r_3^2 \end{pmatrix} = \begin{pmatrix} 3 & \sigma_1 & \sigma_2 \\ \sigma_1 & \sigma_2 & \sigma_3 \\ \sigma_2 & \sigma_3 & \sigma_4 \end{pmatrix}, \tag{3.33b}$$

where $\sigma_i = r_1^i + r_2^i + r_3^i$ for $1 \le i \le 4$. The **discriminant** d of $f(x)$ is given by

$$d = (r_1 - r_2)^2(r_1 - r_3)^2(r_2 - r_3)^2 \tag{3.34}$$

Proposition 3.3. The discriminant d of the polynomial $f(x)$ in (3.32) is given by

$$d = \det \begin{pmatrix} 3 & \sigma_1 & \sigma_2 \\ \sigma_1 & \sigma_2 & \sigma_3 \\ \sigma_2 & \sigma_3 & \sigma_4 \end{pmatrix},$$

where

$$\sigma_1 = \alpha$$
$$\sigma_2 = \alpha^2 - 2\beta$$
$$\sigma_3 = \alpha^3 - 3\alpha\beta + 3\gamma$$
$$\sigma_4 = \alpha^4 - 4\alpha^2\beta + 2\beta^2 + 4\alpha\gamma.$$

PROOF. The determinant formula follows from (3.33) and (3.34). Then $\sigma_1, \sigma_2, \sigma_3, \sigma_4$ are symmetric polynomials in r_1, r_2, r_3 and hence are polynomials in the elementary symmetric polynomials α, β, γ. There is an algorithm for finding the polynomials in α, β, γ, and application of it yields the above expressions for $\sigma_1, \sigma_2, \sigma_3, \sigma_4$. These expressions can be verified by direct computation.

Corollary 3.4. For the cubic polynomial $f(x) = x^3 + px + q$, the discriminant is $d = -4p^3 - 27q^2$.

PROOF. This is the special case of Proposition 3.3 in which $\alpha = 0$, $\beta = p$, and $\gamma = -q$.

EXAMPLE. Cubic polynomial $f(x)$ defined over **R**. The discriminant d is 0 if and only if f has a repeated root. If the three roots are real, then $d \geq 0$. If f has one real root r_1 and one pair of complex conjugate roots r_2 and $r_3 = \bar{r}_2$, then $(r_1 - r_2)(r_1 - \bar{r}_2)$ is real and $(r_2 - \bar{r}_2)$ is imaginary; since d is the square of the product, d is ≤ 0.

In the general case, $d = 0$ for the cubic polynomial in (3.32) if and only if at least two of the roots are equal. For a cubic polynomial that is not monic, we define d to be the same as for the multiple that is monic. If we then replace x by $\frac{x}{C}$ in a cubic, the discriminant gets multiplied by C^6 (since each root gets multiplied by C).

The relevance of the discriminant d to detecting singularities of cubics in Weierstrass form is as follows.

Proposition 3.5. If C is a nonzero element of k and if $\text{char}(k) \neq 2$, then the plane curve

$$y^2 = C(x^3 - \alpha x^2 + \beta x - \gamma) \tag{3.35}$$

is nonsingular if and only if $f(x) = C(x^3 - \alpha x^2 + \beta x - \gamma)$ has distinct roots in $\bar{k}$.

PROOF. Since §II.4 showed that there can be no singularity on the line at infinity, Proposition 2.6 says that the curve is singular if and only if there exists a $\bar{k}$ rational point $(x_0, y_0, 1)$ on the curve where the following three equations are all satisfied:

$$\frac{\partial}{\partial x}: \quad 0 = 3x_0^2 - 2\alpha x_0 + \beta \tag{3.36a}$$

$$\frac{\partial}{\partial y}: \quad 2y_0 = 0 \tag{3.36b}$$

$$\frac{\partial}{\partial w}: \quad y_0^2 = C(-\alpha x_0^2 + 2\beta x_0 - 3\gamma) \tag{3.36c}$$

Equations (3.35), (3.36a), and (3.36b) are equivalent with

$$0 = y_0 = f(x_0) = f'(x_0), \tag{3.37}$$

and (3.36c) is redundant, giving the extra condition $3f(x_0) - x_0 f'(x_0) = 0$. Thus the only candidates for singular points over $\bar{k}$ are $(x_0, 0, 1)$, where x_0 is a root of f, and such a candidate $(x_0, 0, 1)$ is singular if and only if x_0 is a multiple root of f.

2. WEIERSTRASS FORM, DISCRIMINANT, j-INVARIANT

Proposition 3.6. If $\operatorname{char}(k) \neq 2$, let d_b and d_c be the discriminants of the cubic polynomials on the right sides of (3.26) and (3.27), respectively. Then
$$d_c = 2^{12} 3^{12} d_b \tag{3.38a}$$
and
$$\Delta = 2^4 d_b. \tag{3.38b}$$

PROOF IF $\operatorname{char}(k) \neq 3$. Apart from translations, (3.27) is obtained from (3.26) by replacing x by x/C with $C = 6^2$, and we have seen that the effect on discriminants is to multiply them by C^6. Thus (3.38a) follows. By Corollary 3.34 and (3.31), we have
$$d_c = -4(-27c_4)^3 - 27(-54c_6)^2 = 2^2 \cdot 3^9 \cdot 12^3 \Delta.$$
Then (3.38b) follows from (3.38a).

PROOF IF $\operatorname{char}(k) = 3$. It is immediate from Corollary 3.34 that $d_c = 0$, and hence (3.38a) is valid. The discriminant d_b of (3.26) is the same as that of (3.35) with
$$\alpha = -\frac{b_2}{4}, \qquad \beta = \frac{b_4}{2}, \qquad \gamma = -\frac{b_6}{4}.$$
Since $3 = 0$, Proposition 3.3 says that
$$d_b = 2\sigma_1 \sigma_2 \sigma_3 - \sigma_2^3 - \sigma_1^2 \sigma_4 = -\sigma_1 \sigma_2 \sigma_3 - \sigma_2^3 - \sigma_1^2 \sigma_4,$$
where
$$\sigma_1 = \alpha$$
$$\sigma_2 = \alpha^2 + \beta$$
$$\sigma_3 = \alpha^3$$
$$\sigma_4 = \alpha^4 - \alpha^2 \beta - \beta^2 + \alpha\gamma$$
with
$$\alpha = -b_2, \qquad \beta = -b_4, \qquad \gamma = -b_6.$$
Substituting gives
$$d_b = -\beta^3 + \alpha^2 \beta^2 - \alpha^3 \gamma = b_4^3 + b_2^2 b_4^2 - b_2^3 b_6. \tag{3.39}$$
Meanwhile, in any characteristic we can check that
$$4b_8 = b_2 b_6 - b_4^2, \tag{3.40}$$
an equality that was already used to show that (3.30) implies (3.31). Therefore in characteristic 3,
$$\Delta = -b_2^2 b_8 + b_4^3 = -b_2^3 b_6 + b_2^2 b_4^2 + b_4^3.$$
Comparing this equation with (3.39), we see that $\Delta = d_b$, and (3.38b) follows.

PROOF OF THEOREM 3.2.

Case 1. Suppose $\text{char}(k) \neq 2$. Then (3.23) is singular if and only if (3.26) is singular, if and only if the right side of (3.26) has a repeated root (by Proposition 3.5), if and only if $d_b = 0$, if and only if $\Delta = 0$ (by (3.39)).

Case 2. Suppose $\text{char}(k) = 2$. Then Δ reduces to

$$\Delta = b_2^2 b_8 + b_6^2 + b_2 b_4 b_6$$
$$= a_1^6 a_6 + a_1^5 a_3 a_4 + a_1^4 a_2 a_3^2 + a_1^4 a_4^2 + a_3^4 + a_1^3 a_3^3. \tag{3.41}$$

Meanwhile, just as in the first paragraph of the proof of Proposition 3.5, (3.23) can have singularities only at $\bar{k}$ rational points $(x_0, y_0, 1)$ on the curve, and it has a singularity at such a point if and only if

$$\frac{\partial}{\partial x}: \quad 0 = a_1 y_0 + x_0^2 + a_4 \tag{3.42a}$$

$$\frac{\partial}{\partial y}: \quad 0 = a_1 x_0 + a_3 \tag{3.42b}$$

$$\frac{\partial}{\partial w}: \quad 0 = y_0^2 + a_1 x_0 y_0 + a_2 x_0^2 + a_6. \tag{3.42c}$$

Equation (3.42c) is redundant, being the sum of the curve, x_0 times (3.42a), and y_0 times (3.42b).

Suppose $a_1 = 0$. Then $\Delta = 0$ if and only if $a_3 = 0$, if and only if (3.42b) holds. To complete this case, it is enough to show that the system

$$y_0^2 = x_0^3 + a_2 x_0^2 + a_4 x_0 + a_6$$
$$0 = x_0^2 + a_4$$

has a solution in $\bar{k}$. But we have only to choose $x_0 \in \bar{k}$ so that the second equation holds, substitute into the first equation, and choose $y_0 \in \bar{k}$ so that the first equation holds.

Suppose $a_1 \neq 0$. Then (3.42b) and (3.42a) successively give

$$x_0 = a_1^{-1} a_3 \quad \text{and} \quad y_0 = a_1^{-3} a_3^2 + a_1^{-1} a_4.$$

Substitution of these values for x and y in the difference of the two sides of (3.23b) gives

$$(a_1^{-6} a_3^4 + a_1^{-2} a_4^2) + (a_1^{-3} a_3^3 + a_1^{-1} a_3 a_4) + (a_1^{-3} a_3^3 + a_1^{-1} a_3 a_4)$$
$$+ a_1^{-3} a_3^3 + a_1^{-2} a_2 a_3^2 + a_1^{-1} a_3 a_4 + a_6,$$

and (3.41) says this is just $a_1^{-6} \Delta$. Thus (x_0, y_0) satisfies (3.23b), yielding $(x_0, y_0, 1)$ as a singular point, if and only if $\Delta = 0$. This completes the proof of the theorem.

2. WEIERSTRASS FORM, DISCRIMINANT, j-INVARIANT

An **admissible change of variables** in a Weierstrass equation (3.23b) is one of the form

$$x = u^2 x' + r \quad \text{and} \quad y = u^3 y' + su^2 x' + t \tag{3.43a}$$

with u, r, s, t in k and $u \neq 0$. In matrix form it is given by

$$\begin{pmatrix} x \\ y \\ w \end{pmatrix} = \Phi^{-1} \begin{pmatrix} x' \\ y' \\ w' \end{pmatrix} \quad \text{with} \quad \Phi^{-1} = \begin{pmatrix} u^2 & 0 & r \\ su^2 & u^3 & t \\ 0 & 0 & 1 \end{pmatrix} \tag{3.43b}$$

It fixes $[(0,1,0)]$ and carries the tangent $w = 0$ to the same line. The cubic F in Weierstrass form gets carried to the same curve as a cubic F^Φ still in Weierstrass form. Up to a constant, admissible changes of variables are the most general linear transformations with these properties. We have already used such changes of variables on several occasions to normalize Weierstrass equations in various ways. Two elliptic curves that are related by an admissible change of variables are said to be **isomorphic**.

Under the change of variables (3.43) and the normalization that makes the coefficients of wy^2 and x^3 be 1, let F get carried to F^Φ. In the special case that $r = s = t = 0$, this passage $F \to F^\Phi$ multiplies the coefficients $a_1, a_3, a_2, \ldots, b_2, \ldots, c_4, c_6$ by powers of u. The significance of a subscript i is that that coefficient has been multiplied by u^{-i}. We say that the coefficient has **weight** i.

More generally when r, s, t are not all 0, the above passage $F \to F^\Phi$ maps the coefficients $a_1, a_3, a_2, \ldots, b_2, \ldots, c_4, c_6$ into new coefficients $a_1', a_3', a_2', \ldots, b_2', \ldots, c_4', c_6'$. Then a primed coefficient is u^{-i} times an expression in $r, s, t, a_1, \ldots, c_6$ that does not involve u. In the case of c_4 and c_6, we have $c_4' = u^{-4} c_4$ and $c_6' = u^{-6} c_6$ with r, s, t not involved.

When coefficients are multiplied, their weights add. Linear combinations of terms with the same weight again have that weight. In this sense the discriminant Δ has weight 12. Since Δ can be expressed in terms of c_4 and c_6 (at least when k has characteristic not equal to 2 or 3), Δ is unaffected by r, s, and t.

Elliptic Curve	Δ	c_4	c_6
$y^2 + y = x^3 - x^2$	-11	16	-152
$y^2 + xy = x^3 - 2x^2 + x$	-15	1	-161
$y^2 + y = x^3 + x^2 + x$	-19	-32	8
$y^2 + xy + y = x^3$	-26	-23	-181
$y^2 + y = x^3$	-27	0	-216
$y^2 + xy - y = x^3$	-28	25	-253
$y^2 + y = x^3 + x^2 - x$	-35	64	-568
$y^2 + xy = x^3 + x^2 + x$	-39	-23	235
$y^2 + y = x^3 + x^2$	-43	16	-280
$y^2 = x^3 - x^2 + x$	-48	-32	-224
$y^2 = x^3 + x^2 + x$	-48	-32	224
$y^2 + xy + y = x^3 - x^2$	-53	-15	-297
$y^2 + 3xy - y = x^3$	-54	153	-1917
$y^2 + xy = x^3 - x^2 + x$	-55	-39	-189
$y^2 + xy = x^3 - 2x + 1$	-61	97	-1009
$y^2 + xy = x^3 + x$	-63	-47	71
$y^2 = x^3 + x$	-64	-48	0
$y^2 + y = x^3 - 5x^2 - 4x - 1$	-67	592	14408
$y^2 + 3xy + y = x^3 - x^2$	-83	-47	199
$y^2 + xy - y = x^3 + x^2$	-89	49	-521
$y^2 + y = x^3 + x$	-91	-48	-216
$y^2 + 4xy - y = x^3$	-91	352	-6616

TABLE 3.1. Some elliptic curves with small negative discriminant

To get some feeling for elliptic curves (i.e., nonsingular cubics (3.23)) without relying on classical examples or random selection, one can look for elliptic curves of small discriminant. These, it turns out, are an infrequent occurrence. They remain nonsingular, when reduced modulo p, for all but a small number of primes p, and thus they supply a good source of examples for many purposes. Tables 3.1 and 3.2 list, up to admissible changes of variables over $\mathbb{Q}$, all elliptic curves with $|\Delta| < 100$ whose coefficients are all integers ≤ 9 in absolute value. The triple (Δ, c_4, c_6) can be used to detect equivalence under admissible changes of variables over $\mathbb{Q}$: It is always true that two elliptic curves over $\mathbb{Q}$ with the same (Δ, c_4, c_6) are related by a transformation (3.43), since the changes of variables that pass from (3.23) to (3.27) can both be inverted. Conversely if two elliptic curves are related by a transformation

2. WEIERSTRASS FORM, DISCRIMINANT, j-INVARIANT

(3.43), their discriminants stand in the ratio of u^{12}, and this is possible with integer coefficients and with $|\Delta| < 100$ only when $u = \pm 1$. Hence (Δ, c_4, c_6) is a complete invariant of the equivalence class for curves limited as in the tables.

Elliptic Curve	Δ	c_4	c_6
$y^2 + 7xy + 2y = x^3 + 4x^2 + x$	15	3841	-238049
$y^2 + 3xy = x^3 + x$	17	33	-81
$y^2 + y = x^3 - x$	37	48	-216
$y^2 + 2xy - 3y = x^3 - 1$	37	160	-2008
$y^2 + xy = x^3 - 3x^2 + x$	57	73	539
$y^2 + 9xy - 9y = x^3 + 9x^2 - 5x - 3$	62	15873	-1999809
$y^2 = x^3 - x$	64	48	0
$y^2 + xy = x^3 - x$	65	49	-73
$y^2 + xy = x^3 - x^2 - x$	73	57	243
$y^2 + 3xy - y = x^3 - x^2$	79	97	-881
$y^2 = x^3 - 2x + 1$	80	96	-864
$y^2 = x^3 - 2x - 1$	80	96	864
$y^2 = x^3 + x^2 - x$	80	64	-352
$y^2 = x^3 - x^2 - x$	80	64	352
$y^2 + xy = x^3 + x^2 - x$	89	73	-485
$y^2 + 5xy + y = x^3$	98	505	-11341

TABLE 3.2. Some elliptic curves with small positive discriminant

For an elliptic curve the j-**invariant** is defined by

$$j = c_4^3/\Delta. \tag{3.44}$$

This is well defined by Theorem 3.2, and it has weight 0. It is unaffected by r, s, and t and thus is invariant under all admissible changes of variables.

EXAMPLES. Consider $y^2 = x^3 + px + q$ over a field with characteristic not equal to 2 or 3. We have $c_4 = -48p$ and $c_6 = -864q$. (See the Warning before Theorem 3.2.) By (3.31),

$$\Delta = \frac{(-48p)^3 - (-864q)^2}{1728} = -2^4(4p^3 + 27q^2).$$

Thus (3.44) gives

$$j = 1728 \frac{4p^3}{4p^3 + 27q^2}. \qquad (3.45)$$

Thus, for example, every elliptic curve $y^2 = x^3 + ax$ has $j = 1728$; this includes those in Subsections C, D, and E of §1. Similarly every elliptic curve $y^2 = x^3 + a$ has $j = 0$; this includes the one in Subsection B of §1. Finally the Diophantus example in Subsection A of §1 was $y^2 = x^3 - x + 9$. This has $p = -1$ and $q = 9$. Thus (3.45) gives $j = -\dfrac{2^8 \cdot 3^3}{2183}$.

Proposition 3.7. Suppose that k has characteristic not equal to 2 or 3.

(a) If two elliptic curves are related by an admissible change of variables, then they have the same j-invariant.

(b) If $j_0 \in k$ is given, then there exists an elliptic curve over k with j-invariant j_0.

(c) If k is algebraically closed and two elliptic curves have the same j-invariant, then they are related by an admissible change of variables.

PROOF. (a) This was observed above.

(b) The cases $j_0 = 0$ and $j_0 = 1728$ were handled in the examples above. For other values we can specialize the above example to

$$y^2 = x^3 - \frac{27}{4} \frac{j_0}{j_0 - 1728} x - \frac{27}{4} \frac{j_0}{j_0 - 1728},$$

in which $p = q = -\dfrac{27}{4} \dfrac{j_0}{j_0 - 1728}$, and then we can apply (3.45).

(c) We can normalize the two curves to be $y^2 = x^3 + Ax + B$ and $y^2 = x^3 + A'x + B'$, by (a). Here A has weight 4 and B has weight 6. Thus if we put $y = y'/u^3$ and $x = x'/u^2$ in the first equation, we get

$$y^2 = x^3 + Au^4 x + Bu^6.$$

What we want to arrange is that

$$A' = Au^4 \quad \text{and} \quad B' = Bu^6. \qquad (3.46)$$

Equality of j-invariants means that

$$\frac{4A^3}{4A^3 + 27B^2} = \frac{4A'^3}{4A'^3 + 27B'^2},$$

from which we see that
$$A^3 B'^2 = A'^3 B^2. \qquad (3.47)$$

Now we distinguish cases.

(i) $A = 0$. Then $B \neq 0$ by nonsingularity, and $A' = 0$ by (3.47). If we take $u = (B'/B)^{1/6}$, then (3.46) holds.

(ii) $B = 0$. Then $A \neq 0$ by nonsingularity, and $B' = 0$ by (3.47). If we take $u = (A'/A)^{1/4}$, then (3.46) holds.

(iii) $AB \neq 0$. First we take $u = (A'/A)^{1/4}$ to get $A' = Au^4$. Then (3.47) says $B'^2 = (A'/A)^3 B^2 = u^{12} B^2$. Hence $B' = Bu^6$ or $B' = -Bu^6$. In the first case, (3.46) has been verified. In the second case we replace u by $\sqrt{-1}\, u$ and recompute to find that (3.46) holds.

3. Group Law

For a nonsingular cubic curve with a specified k rational point O, we noted in Chapter I that the operation $P + Q = O \cdot PQ$, given in terms of the chord-tangent composition law of §II.4, makes the points on the curve into an abelian group with O as identity. In particular, the conclusion applies to elliptic curves, which are nonsingular cubics in Weierstrass form. In this section, we shall state this result as a theorem and give a geometric proof.

Theorem 3.8 (Poincaré). Let F be a nonsingular cubic, and let O be in $F(k)$. Then the operation $P + Q = O(PQ)$, given in terms of the chord-tangent composition law, makes $F(k)$ into an abelian group with O as identity and with negatives given by $-P = (OO)P$. If K is a field extension of k, then the inclusion $F(k) \subseteq F(K)$ is a group homomorphism. If a different base point O' is chosen, then the two operations are related by $P +' Q = P + Q - O'$, and the group structures are isomorphic.

The heart of the matter is to prove associativity, whose essence is captured by the following lemma. We shall prove the lemma after showing how the lemma implies the theorem.

Lemma 3.9. For any points P, Q, P', Q' in $F(k)$,
$$(PP')(QQ') = (PQ)(P'Q'). \qquad (3.48)$$

PROOF THAT LEMMA 3.9 IMPLIES THEOREM 3.8. We saw in Chapter I that + is commutative, that O is an identity, and that $-P$ is an additive inverse. For associativity the lemma gives the second equality of

$$P(O(QR)) = (PQ \cdot Q)(O \cdot QR) = (PQ \cdot O)(Q \cdot QR) = (O(PQ))R.$$

Thus
$$P(Q + R) = (P + Q)R.$$

If we apply O to both sides, we get

$$P + (Q + R) = (P + Q) + R,$$

as required. Thus $F(k)$ is an abelian group under $+$. Clearly the inclusion $F(k) \subseteq F(K)$ is a group homomorphism.

Let O' be given. Then the lemma gives the second equality of

$$\begin{aligned}(P+Q) - O' &= O[(O \cdot PQ)(OO \cdot O')] \\ &= O[(O \cdot OO)(PQ \cdot O')] \\ &= O[O \cdot (P +' Q)] = P +' Q.\end{aligned} \quad (3.49)$$

Define a map $\varphi : (F(k), +) \to (F(k), +')$ by $\varphi(P) = P - O'$. Then

$$\varphi(P + Q) = (P + Q) - O' = P +' Q$$

by (3.49). Thus φ is a homomorphism. It is clearly one-one and onto.

OVERVIEW OF PROOF OF LEMMA 3.9. To prove (3.48) and thus Lemma 3.9, we shall divide matters into a nondegenerate case and a degenerate case, and we shall use quite different methods for the two cases. The distinction between "nondegenerate" and "degenerate" will depend on which, if any, of the 8 points

$$P, \; P', \; Q, \; Q', \; PQ, \; P'Q', \; PP', \; QQ'$$

are equal to one another. Specifically we arrange these points as 8 of the entries of a 3-by-3 matrix

$$\begin{pmatrix} P & P' & PP' \\ Q & Q' & QQ' \\ PQ & P'Q' & \end{pmatrix} \quad (3.50)$$

We say that we are in the **nondegenerate case** if none of these points is equal to a point in a different row and column (e.g., P should not equal Q', QQ', or $P'Q'$). Otherwise we are in the **degenerate case**. The proof in the nondegenerate case simplifies a little if the 8 points are actually distinct, but it does not simplify much.

3. GROUP LAW

PROOF OF (3.48) IN THE NONDEGENERATE CASE. Let L'_1, L'_2, L'_3 be the lines that meet $F(k)$ at the points of the respective rows of (3.50), with the usual convention that if a point appears more than once in a row, then the line is supposed to be tangent to the curve at the point. Let L''_1, L''_2, L''_3 be the similar lines determined by the columns of (3.50). Figure 3.1 illustrates matters, showing L'_1, L'_2, L'_3 as solid lines and L''_1, L''_2, L''_3 as dashed lines.

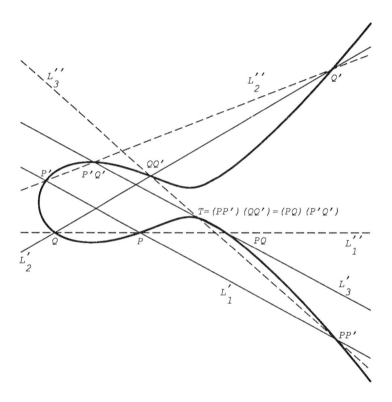

FIGURE 3.1. Configuration for $(PP')(QQ') = (PQ)(P'Q')$

If we are thinking of L'_1, L'_2, L'_3, it is natural to include $(PQ)(P'Q')$ as the lower right entry of (3.50), while if we are thinking of L''_1, L''_2, L''_3, it is natural to include $(PP')(QQ')$ as the lower right entry. Our objective, of course, is to prove that these two candidates for the lower right entry are equal. Thus, in referring to (3.50), we shall regard the lower right entry as blank.

We claim as a consequence of our nondegeneracy assumption that

$$\text{each } L'_i \text{ is different from each } L''_j. \tag{3.51}$$

Thus $L'_i(k)$ and $L''_j(k)$ meet only in the $(ij)^{\text{th}}$ entry of (3.50) unless $i = j = 3$. To see (3.51), first suppose i and j are not both 3, say $i \neq 3$. If $L'_i = L''_j$, then the other point(s) of the jth column must occur as other point(s) of the ith row, since the ith row gives all points of $F(k) \cap L'_i$ with their multiplicities, and this occurrence of a point in two such entries of the matrix would contradict nondegeneracy. If $i = j = 3$ and $L'_3 = L''_3$, then PQ in L'_3 must occur as one of the three points of $F(k) \cap L''_3(k)$, counting multiplicities. It might be the missing lower right entry of (3.50), but then $P'Q'$ would have to coincide with PP' or QQ', and we would again have a contradiction to nondegeneracy.

Let T be the point where L'_3 and L''_3 meet; this is unique by (3.51). We shall prove (3.48) by showing that both sides of (3.48) are equal to T. In the proof we may assume that k is algebraically closed. (Actually we shall use only that k is infinite.)

Let L' and L'' be the cubics given by

$$L' = L'_1 L'_2 L'_3 \quad \text{and} \quad L'' = L''_1 L''_2 L''_3. \tag{3.52}$$

The main step of the proof will be to show that

$$aF + a'L' + a''L'' = 0 \tag{3.53}$$

for suitable a, a', a'' in k with $a \neq 0$. To define these constants, let R' be a point of $L'_1(k)$ other than P, P', and PP' (possible since k is infinite), and let R'' be a point not in $L'(k)$ (existence by Lemma 2.1). Choose a, a', a'' in k not all 0 such that

$$\begin{pmatrix} F(R') & L'(R') & L''(R') \\ F(R'') & L'(R'') & L''(R'') \end{pmatrix} \begin{pmatrix} a \\ a' \\ a'' \end{pmatrix} = \begin{pmatrix} 0 \\ 0 \end{pmatrix}, \tag{3.54}$$

and put

$$F_0 = aF + a'L' + a''L''. \tag{3.55}$$

We shall prove that $F_0 = 0$.

In doing so, we shall make repeated use of two facts about intersection multiplicities of lines and plane curves. If L is a line, p is a point, and G_1 and G_2 are plane curves, then

$$i(p, L, G_1 G_2) = i(p, L, G_1) + i(p, L, G_2). \tag{3.56}$$

3. GROUP LAW

If, in addition, G_1 and G_2 have the same degree and if $G_1 + G_2$ is not 0, then
$$i(p, L, G_1 + G_2) \geq \min\{i(p, L, G_1), i(p, L, G_2)\}. \qquad (3.57)$$
These are obvious from the definitions.

Thus suppose $F_0 \neq 0$, so that F_0 is a cubic. Let L be one of the lines $L_1', L_2', L_3', L_1'', L_2'', L_3''$ and suppose the point p is repeated N times in the corresponding row or column of (3.50), with the lower right entry not included. We shall show that
$$i(p, L, F_0) \geq N. \qquad (3.58)$$

In fact, without loss of generality, suppose L is a factor of the cubic L'. We have
$$i(p, L, F) \geq N \qquad (3.59)$$
by construction and by definition of the chord-tangent composition law. In addition, Proposition 2.8 gives
$$i(p, L, L') = +\infty. \qquad (3.60)$$

The point p appears in N columns of (3.50) and thus lies on $\geq N$ of the lines L_1'', L_2'', L_3''. By (3.56)
$$i(p, L, L'') \geq N. \qquad (3.61)$$

Putting (3.59), (3.60), and (3.61) together with the aid of (3.57), we obtain (3.58).

Equation (3.54) implies that $F_0(R') = 0$, hence that $i(R', L_1', F_0) \geq 1$. Adding this inequality to (3.58) for $L = L_1'$ and for p equal to the distinct points in the first row of (3.50), we obtain $\sum_p i(p, L_1', F_0) \geq 4$. From Corollary 2.20 we conclude that L_1' divides F_0. Let us write $F_0 = L_1' C$ for a conic C.

Suppose p appears $N \geq 1$ times in the second row of (3.50) but not at all in the first row. Then $i(p, L_2', L_1') = 0$ since the first row lists all points of $F(k) \cap L_1'(k)$. And $i(p, L_2', F_0) \geq N$ by (3.58). Hence (3.56) gives
$$i(p, L_2', C) \geq N. \qquad (3.62)$$

We shall show that (3.62) remains valid even if p does appear in the first row. In this case, $N = 1$ by the nondegeneracy assumption. Suppose that p appears ≥ 2 times in the jth column, possibly including once in the second row. By (3.58), we have $i(p, L_j'', F_0) \geq 2$. Since $i(p, L_j'', L_1') = 1$ by (3.51), (3.56) gives $i(p, L_j'', C) \geq 1$. That is, p is in

$C(k)$. Hence $i(p, L_2', C) \geq 1$, and (3.62) is proved even when p appears in the first row.

Summing (3.62) over the distinct points p of the second row of (3.50), we obtain $\sum_p i(p, L_2', C) \geq 3$. From Corollary 2.20 we conclude that L_2' divides C. Thus we can write $F_0 = L_1' L_2' M$ for a line M.

Now consider the two points of the third row of (3.50). First suppose they coincide; call their common value p. By nondegeneracy, p cannot occur in the first or second row, so that $i(p, L_3', L_1') = 0$ and $i(p, L_3', L_2') = 0$. Since (3.58) gives $i(p, L_3', F_0) \geq 2$, we conclude from (3.56) that $i(p, L_3', M) \geq 2$. By Corollary 2.20, L_3' divides M.

Next suppose that the two points of the third row of (3.50) are distinct. Let p be one of them, say the one in the jth column. If p appears N times in the jth column, (3.58) gives

$$i(p, L_j'', F_0) \geq N.$$

On the other hand,

$$i(p, L_j'', L_1' L_2') = N - 1$$

by (3.56). Another application of (3.56) then shows that $i(p, L_j'', M) \geq 1$. Thus p is in $M(k)$. Since the same argument applies equally to the other point on the third row of (3.50), we see that L_3' and M have two distinct points in common. Therefore L_3' divides M.

Thus L_3' divides M in either case, and we conclude that

$$F_0 = c L_1' L_2' L_3' = c L'$$

for a nonzero constant c. But $F_0(R'') = 0$ by (3.54), and $L'(R'') \neq 0$ by definition of R''. We have arrived at a contradiction, and we conclude that $F_0 = 0$. This proves (3.53).

We still have to show that $a \neq 0$ in (3.53). If $a = 0$, then we may assume by symmetry that $a' \neq 0$, hence that L' divides L''. Thus L_1' divides $L_1'' L_2'' L_3''$, and unique factorization shows that L_1' is the same line as some L_j'', in contradiction to (3.51). We conclude that $a \neq 0$.

Therefore

$$F = c'L' + c''L'' \qquad (3.63)$$

for suitable c' and c'' in k. Recall that T is the common intersection of L_3' and L_3''. It follows from (3.63) that $F(T) = 0$. Since T is in $F(k) \cap L_3'(k)$,

$$T \quad \text{is one of} \quad PQ, \ P'Q', \ (PQ)(P'Q'). \qquad (3.64\text{a})$$

Since T is in $F(k) \cap L_3''(k)$,

3. GROUP LAW

T is one of PP', QQ', $(PP')(QQ')$.
(3.64b)

Since, by nondegeneracy, neither of the first two points of (3.64a) can equal either of the first two points of (3.64b), we conclude either that

$$(PQ)(P'Q') = T = (PP')(QQ') \tag{3.65}$$

and we are done, or that one of the following four possibilities occurs:

$$PQ = T = (PP')(QQ') \tag{3.66}$$
$$P'Q' = T = (PP')(QQ')$$
$$PP' = T = (PQ)(P'Q')$$
$$QQ' = T = (PQ)(P'Q').$$

These four formulas are symmetric under permutation of notation, and it is enough to handle the first one.

Thus suppose (3.66) holds. Consider $(PQ)(P'Q')$. This is on L'_3, hence on L'. Also it is on F. Since $c'' \neq 0$ in (3.63) (F being nonsingular), $(PQ)(P'Q')$ must be on L''. Hence $(PQ)(P'Q')$ is on at least one of L''_1, L''_2, and L''_3.

Suppose $(PQ)(P'Q')$ is on L''_1. Since it is on L'_3, we must have

$$(PQ)(P'Q') = PQ.$$

Substituting into (3.66), we obtain (3.65), and we are done.

Suppose $(PQ)(P'Q')$ is on L''_2. Since it is on L'_3, we must have

$$(PQ)(P'Q') = P'Q'. \tag{3.67}$$

If $P'Q' = PQ$, we can substitute into (3.66) and obtain (3.65), and then we are done. Thus suppose $P'Q' \neq PQ$. From (3.67) it follows that

$$i(P'Q', L'_3, F) = 2. \tag{3.68}$$

Now PQ is on L''_3 by (3.66), and it is on L''_1 automatically. Thus

$$i(PQ, L'_3, L'') \geq 2$$

by (3.56). Since

$$i(PQ, L'_3, L') = +\infty,$$

(3.63) and (3.57) give

$$i(PQ, L'_3, F) \geq 2. \tag{3.69}$$

Combining (3.68) and (3.69), we obtain $\sum_p i(p, L'_3, F) \geq 4$. By Corollary 2.20, we find that L'_3 divides F, in contradiction to the nonsingularity of F.

We conclude that $(PQ)(P'Q')$ is on L''_3. Since it is on both L'_3 and L''_3, (3.51) shows it is T. In combination with (3.66), this conclusion yields (3.65). This completes the proof of (3.48) in the nondegenerate case.

PROOF OF (3.48) IN THE DEGENERATE CASE. The assumption is that two of the 8 points in (3.50) in different rows and columns are equal. Up to symmetry the cases we need to consider are at the most

$$P = Q'$$
$$P = QQ' \quad \text{(and hence } PQ = Q'\text{)}$$
$$PP' = PQ \quad \text{(and hence } P' = Q\text{)}.$$

The identity (3.48) reduces in these cases to

$$(PP')(QP) \stackrel{?}{=} (PQ)(P'P)$$
$$(PP')P \stackrel{?}{=} Q'(P'Q')$$
$$(PP')(P'Q') \stackrel{?}{=} (PP')(P'Q').$$

The first and third of these are trivially valid, and the second is valid since both sides reduce to P'. This completes the proof of (3.48) in the degenerate case.

4. Computations with the Group Law

Fix an elliptic curve E over k, i.e., a nonsingular cubic in Weierstrass form (3.23). It is customary to choose $O = (0, 1, 0)$, the point at infinity, and then Theorem 3.8 says that $E(k)$ becomes an abelian group with O as identity element. This particular choice of O has some immediate implications:

(1) $OO = O$ since O is an inflection point.

(2) The additive inverse, usually given by $-P = OO \cdot P$, is now given by

$$-P = OP$$

as a consequence of (1). Then

$$P = (x_0, y_0) \quad \text{implies} \quad -P = (x_0, -y_0 - a_1 x_0 - a_3), \quad (3.70)$$

because the line from O to P is $x = x_0$ in affine form (or $x - x_0 w = 0$ in projective form).

(3) If $a_1 = a_3 = 0$, so that E is given in affine form by $y^2 = f(x)$ with f a monic cubic polynomial, then (3.70) specializes to the statement that

$$P = (x_0, y_0) \quad \text{implies} \quad -P = (x_0, -y_0). \quad (3.71)$$

In this case the elements of order 2, which have $P = -P$, are the points on the x-axis. There are at most three such points.

4. COMPUTATIONS WITH THE GROUP LAW

Let us derive a formula for the group law for $E(k)$. Let P_1 and P_2 be given with $P_2 \neq -P_1$. We shall calculate $P_3 = P_1 + P_2$. The notation will be

$$P_1 = (x_1, y_1), \qquad P_2 = (x_2, y_2), \qquad P_3 = (x_3, y_3)$$

$$y = mx + b = \begin{cases} \text{line through } P_1 \text{ and } P_2 & \text{if } P_1 \neq P_2 \text{ (\textbf{chord case})} \\ \text{tangent line at } P_1 & \text{if } P_1 = P_2 \text{ (\textbf{tangent case})}. \end{cases}$$

We begin by calculating m and b. The result is

$$m = \begin{cases} \dfrac{y_2 - y_1}{x_2 - x_1} & \text{in chord case} \\[1em] \dfrac{3x_1^2 + 2a_2 x_1 + a_4 - a_1 y_1}{2y_1 + a_1 x_1 + a_3} & \text{in tangent case} \end{cases} \quad (3.72\text{a})$$

$$b = \begin{cases} \dfrac{y_1 x_2 - y_2 x_1}{x_2 - x_1} & \text{in chord case} \\[1em] \dfrac{-x_1^3 + a_4 x_1 + 2a_6 - a_3 y_1}{2y_1 + a_1 x_1 + a_3} & \text{in tangent case.} \end{cases} \quad (3.72\text{b})$$

The formulas for m and b in the chord case are simply formulas for the line through P_1 and P_2 and do not involve cubics. To prove the formulas in the tangent case, we begin by computing $m = \frac{dy}{dx}$ by implicit differentiation of (3.23b):

$$2yy' + a_1 xy' + a_1 y + a_3 y' = 3x^2 + 2a_2 x + a_4.$$

Putting $y' = m$ and $(x, y) = (x_1, y_1)$, we obtain (3.72a) in the tangent case. Along the line $y = mx + b$, we have $y - y_1 = m(x - x_1)$. Thus $y = mx + (y_1 - mx_1)$, and we obtain $b = y_1 - mx_1$. In other words

$$b = \frac{2y_1^2 + a_1 x_1 y_1 + a_3 y_1 - 3x_1^3 - 2a_2 x_1^2 - a_4 x_1 + a_1 x_1 y_1}{2y_1 + a_1 x_1 + a_3}.$$

Substitution for y_1^2 from (3.23b) then gives (3.72b) in the tangent case.

Now we can calculate $P_3 = (x_3, y_3) = P_1 + P_2$. Let $F(x, y)$ be the difference of the right side and the left side of the Weierstrass equation (3.23b). Along the line $y = mx + b$, this expression becomes $F(x, mx+b)$, and it is clearly a monic cubic polynomial in x. Since P_1, P_2, and $P_1 P_2$ are on this line and are in $E(k)$ and since (3.70) shows that x_3 equals the x coordinate of $P_1 P_2$, all of x_1, x_2, x_3 are roots of

$$F(x, mx + b) = 0,$$

with multiplicities counted. Thus
$$F(x, mx+b) = (x-x_1)(x-x_2)(x-x_3).$$
Substitution of the expression for F gives
$$x^3 + a_2x^2 + a_4x + a_6 - (mx+b)^2 - a_1x(mx+b) - a_3(mx+b)$$
$$= x^3 - (x_1+x_2+x_3)x^2 + \text{lower order}.$$
We can equate the coefficients of x^2 on the left and right, obtaining
$$a_2 - m^2 - a_1m = -(x_1+x_2+x_3),$$
and the result is
$$\left.\begin{array}{l} x_3 = m^2 + a_1m - a_2 - x_1 - x_2 \\ y_3 = -(m+a_1)x_3 - b - a_3 \end{array}\right\} \text{ in chord and tangent cases.} \quad (3.73)$$

A special case of (3.73) is that $P_1 = P_2$, so that we are in the tangent case. If $P = (x,y)$, then substitution of (3.72a) into (3.73) yields a formula for the x coordinate $x(2P)$ of $2P$:
$$x(2P) = \frac{x^4 - b_4x^2 - 2b_6x - b_8}{4x^3 + b_2x^2 + 2b_4x + b_6}, \quad (3.74)$$
where b_2, b_4, b_6, b_8 are as in (3.24).

EXAMPLE 1. Cubic case of Fermat's Last Theorem. This was discussed in Subsection B of §1. Since Fermat's Last Theorem is valid in degree 3, (3.3) has only trivial solutions. Transforming to (3.7), we see that
$$y^2 = x^3 - \tfrac{27}{4}$$
has only $(3, \tfrac{9}{2})$, $(3, -\tfrac{9}{2})$, and ∞ as its $\mathbb{Q}$ solutions. Thus the group of solutions must be isomorphic to $\mathbb{Z}_3$. Let us verify that $P = (3, \tfrac{9}{2})$ has order 3. We have
$$a_1 = a_3 = a_2 = a_4 = 0 \quad \text{and} \quad a_6 = -\tfrac{27}{4}.$$
Thus
$$b_2 = b_4 = b_8 = 0 \quad \text{and} \quad b_6 = -27.$$
Hence (3.74) gives
$$x(2P) = \left.\frac{x^4 + 54x}{4x^3 - 27}\right|_{x=3} = \frac{81 + 162}{108 - 27} = 3.$$
So $2P = P$ or $2P = -P$. The conclusion $2P = P$ is false because it forces $P = O = \infty$. Thus $2P = -P$ and $3P = O$. In other words, P indeed does have order 3.

EXAMPLE 2. $y^2 + y = x^3 - x$ over $\mathbb{Q}$. This is one of the elliptic curves in Table 3.2, and it has $\Delta = 37$. The point $P = (0,0)$ is an obvious solution, and we can calculate from (3.73) that

$$\begin{array}{ll} P = (0,0) & 5P = (\frac{1}{4}, -\frac{5}{8}) \\ 2P = (1,0) & 6P = (6,14) \\ 3P = (-1,-1) & 7P = (-\frac{5}{9}, \frac{8}{27}) \\ 4P = (2,-3) & 8P = (\frac{21}{25}, -\frac{69}{125}). \end{array} \qquad (3.75)$$

It will follow from Theorem 5.1a that P generates an infinite cyclic subgroup. (In fact, P generates all solutions.)

5. Singular Points

Singular Weierstrass curves arise for certain primes p when a Weierstrass curve with integral coefficients is considered modulo p. Such curves are easy to analyze, and we shall note some of their features in this section.

Let E be a singular Weierstrass curve over a field k. The point ∞ on the curve is nonsingular, and we have only to analyze points (x, y) in the affine plane. Proposition 3.10 will show that there is only one singularity and that, under a mild restriction on k, it occurs at a k rational point (x_0, y_0). Translating (x_0, y_0) to the origin (an admissible change of variables), we are led to the projective curve (3.23a) with $a_6 = 0$. The condition that $\dfrac{\partial}{\partial x}$ give 0 at $(0,0,1)$ means that $a_4 = 0$, and the condition that $\dfrac{\partial}{\partial y}$ give 0 at $(0,0,1)$ means that $a_3 = 0$. Thus E is given in affine form by the equation

$$y^2 + a_1 xy = x^3 + a_2 x^2. \qquad (3.76)$$

We can factor
$$y^2 + a_1 xy - a_2 x^2 \qquad (3.77)$$

over $\bar{k}$, obtaining

$$(y - \alpha x)(y - \beta x) = x^3 \qquad \text{with } \alpha, \beta \in \bar{k}. \qquad (3.78)$$

We say that the singular point $(0,0)$ is a **cusp** if $\alpha = \beta$, or a **node** if $\alpha \neq \beta$. Pictures of the two kinds of behavior (with $a_1 = 0$) appear in Figure 1.6.

Proposition 3.10. For a singular Weierstrass curve E over k, there is only one singular point (x_0, y_0). It is k rational if either
  (i) char$(k) \neq 2$ or
  (ii) char$(k) = 2$ and k is closed under the operation of taking square roots (as is the case when k is a finite field of characteristic 2).
The point (x_0, y_0) is a cusp if $c_4 = 0$, or a node if $c_4 \neq 0$.

PROOF. If char$(k) \neq 2$, we can apply, without loss of generality, a projective transformation to eliminate the a_1 and a_3 terms. By Proposition 3.5 the curve will be singular if and only if the cubic polynomial f in x has a repeated root. In this case f and f' have a greatest common divisor g over k with degree ≥ 1, and the singular points are $(x_0, 0)$, where x_0 ranges over the roots of g.

If g has degree 1, its unique root x_0 is in k, and $(x_0, 0)$ is the unique singular point. If g has degree 2, its two roots x_0 and x_0' must be equal, since otherwise x_0 and x_0' would both be roots of multiplicity ≥ 2 for the cubic polynomial f. Since we can conclude $x_0 = x_0'$, x_0 is in k, and $(x_0, 0)$ is the unique singular point.

If char$(k) = 2$, the singular points are the points (x_0, y_0) on the affine curve over $\bar{k}$ satisfying (3.42a) and (3.42b). If $a_1 \neq 0$, (3.42b) uniquely determines x_0, and (3.42a) then uniquely determines y_0; the resulting (x_0, y_0) is k rational. If $a_1 = 0$, (3.42a) forces $x_0^3 + 4 = 0$. In characteristic 2, square roots are unique, and the assumption (ii) says that x_0 is in k. Since (3.42b) shows $a_3 = 0$, y_0 is given by

$$y_0^2 = x_0^3 + a_2 x_0^2 + a_4 x_0 + a_6$$

and again exists in k and is unique in $\bar{k}$ under our assumption (ii).

Return to general k. Under the translation over $\bar{k}$ that moves (x_0, y_0) to the origin, c_4 is unaffected. Thus it is enough to decide cusp vs. node in (3.76). From (3.24) and (3.25), the value of c_4 is

$$c_4 = b_2^2 - 24 b_4 = (a_1^2 + 4a_2)^2 - 24(2a_4 + a_1 a_3) = (a_1^2 + 4a_2)^2. \quad (3.79)$$

If char$(k) \neq 2$, the discriminant of (3.77) is $a_1^2 + 4a_2$, which is 0 if and only if $c_4 = 0$; hence $\alpha = \beta$ (and there is a cusp) if and only if $c_4 = 0$. If char$(k) = 2$, then

$$(y - \alpha x)^2 = y^2 + a_1 x y - a_2 x^2$$

says $a_1 = 0$ and $\alpha^2 = a_2$; hence $\alpha = \beta$ (and there is a cusp) if and only if $a_1 = 0$, which happens if and only if $c_4 = 0$.

5. SINGULAR POINTS

We can parametrize our singular Weierstrass curve (3.76) by the method that Diophantus used for conics (see Chapter I): We parametrize lines through the singular point $(0,0)$ by their slopes, and we find that most lines intersect the curve at one and only one other point. To handle exceptional lines, it is necessary to distinguish two subcases of nodes. Let us say that a node is a **split case** if the members α and β of (3.78) lie in k. In the contrary case, α and β lie in a nontrivial quadratic extension of k, and we say that the node is a **nonsplit case**.

Proposition 3.11. For the singular Weierstrass equation E in (3.76), the map
$$t \longrightarrow \left(t^2 + a_1 t - a_2,\, t(t^2 + a_1 t - a_2)\right)$$
carries $k - \{\alpha, \beta\}$ one-one onto $E(k) - \{\infty\} - \{(0,0)\}$. If k is a finite field with $|k|$ elements, the nonsingular set $E(k) - \{(0,0)\}$ therefore has

$\|k\| - 1$	elements if $(0,0)$ is a split-case node
$\|k\| + 1$	elements if $(0,0)$ is a nonsplit-case node
$\|k\|$	elements if $(0,0)$ is a cusp.

PROOF. In (3.76), $x = 0$ gives only $y = 0$, and $(0,0)$ is singular. Thus any nonsingular point of $E(k) - \{\infty\}$ has $y = tx$ for a unique member t of k. Substituting tx for y in (3.76) and using $x \neq 0$, we are led to
$$t^2 + a_1 t - a_2 = x.$$
Then also y is t times this expression for x, and k maps onto the affine solutions of (3.76). To exclude $x = 0$ from the image, we must exclude the roots of $t^2 + a_1 t - a_2$; these are α and β. Once $x = 0$ is not in the image, the map is one-one since t is recovered as y/x. The numerology if k is a finite field is then clear.

CHAPTER IV

MORDELL'S THEOREM

1. Descent

Mordell's Theorem (Theorem 4.11 below) is the statement that the group $E(\mathbf{Q})$ is finitely generated if E is an elliptic curve over $\mathbf{Q}$. The proof is motivated by a close examination of Fermat's **method of descent**. To prove that $x^4 + y^4 = z^4$ has no nontrivial integer solutions, Fermat showed that a nontrivial integer solution of $u^4 + v^4 = w^2$ leads to a strictly smaller nontrivial integer solution of the same equation, and he thereby arrived at a contradiction.

In this section we shall look at Fermat's proof in detail, and then we shall translate it into an argument about the elliptic curve $y^2 = x^3 - 4x$ via the transformations of Subsection D of §III.1. It will turn out that the descent procedure passes from the point P to a point $\frac{1}{2}P$ (or possibly $-\frac{1}{2}P$) on the elliptic curve.

With Mordell's Theorem in hand, we know that the passage $P \to \frac{1}{2}P$ cannot go on forever. But, in fact, the argument can be turned around to give a proof of Mordell's Theorem. The passage $P \to \frac{1}{2}P$ has certain obstructions, which are measured by $E(\mathbf{Q})/2E(\mathbf{Q})$. (In Fermat's proof these obstructions correspond to an occasional need to adjust the parameters in apparently trivial ways.) The hard step in the proof of Mordell's Theorem will be to prove that $E(\mathbf{Q})/2E(\mathbf{Q})$ is finite, so that the descent passage $P \to \frac{1}{2}P$ can proceed for all but a finite number of situations. The other step in the proof is to codify the notion of a "smaller" solution of $u^4 + v^4 = w^2$, or of a "simpler" solution of $y^2 = x^3 - 4x$. This is done by introducing a suitable notion of **height** such that only finitely many points on the elliptic curve have height less than any given constant C. For the notion of height $h(P)$ that we introduce, we shall have $h(\frac{1}{2}P) = \frac{1}{4}h(P)$, so that the descent process $P \to \frac{1}{2}P$ eventually carries us to height less than C. If C is taken larger than the maximum of $h(P)$ for P in a set of coset representatives for the finite group $E(\mathbf{Q})/2E(\mathbf{Q})$, then the finite set of points of height less than C will turn out to be a finite set of generators for $E(\mathbf{Q})$.

Let us begin with Fermat's argument. We use the symbol $\square$ to denote a square.

1. DESCENT

Proposition 4.1 (Fermat). $u^4 + v^4 = w^2$ has no nontrivial integer solutions.

PROOF. Arguing by contradiction, suppose we have a nontrivial integer solution. Without loss of generality, we may assume $GCD(u,v) = 1$, u is odd, and v is even. Writing

$$(u^2)^2 + (v^2)^2 = \Box$$

and applying Theorem 1.2, we obtain integers m and n with

$$u^2 = m^2 - n^2, \qquad v^2 = 2mn, \qquad GCD(m,n) = 1.$$

Applying Theorem 1.2 to $m^2 = u^2 + n^2$, in which u is odd, we obtain integers p and q with

$$m = p^2 + q^2, \qquad u = p^2 - q^2, \qquad n = 2pq, \qquad GCD(p,q) = 1.$$

Since v is even, we can write

$$\left(\frac{v}{2}\right)^2 = \Box = \frac{1}{2}mn = pq(p^2 + q^2).$$

Here p, q, and $p^2 + q^2$ are relatively prime since $GCD(p,q) = 1$. Thus

$$p = \Box, \qquad q = \Box, \qquad p^2 + q^2 = \Box,$$

and we can write

$$p = r^2, \qquad q = s^2, \qquad r^4 + s^4 = \Box.$$

Thus we have been led from the old solution $u^4 + v^4 = \Box$ to a new solution $r^4 + s^4 = \Box$. These are related by

$$\left(\frac{v}{2}\right)^2 = pq(p^2 + q^2) = r^2 s^2 (r^4 + s^4),$$

hence by

$$v = 2rs\sqrt{r^4 + s^4}. \tag{4.1}$$

If the new solution were trivial, then rs would be 0, so that v would be 0; this would mean that the old solution was trivial. We conclude that $rs \neq 0$, and then it is clear from (4.1) that $r < v$ and $s < v$. Hence

$$\max(|r|, |s|) < \max(|u|, |v|).$$

The passage from an old solution to a new solution therefore cannot go on forever, and we arrive at a contradiction.

IV: MORDELL'S THEOREM

ANALYSIS OF PROOF. Let us go backwards. We have

$$\begin{aligned}(r,s) &\longrightarrow (p,q) = (r^2, s^2) \\ &\longrightarrow (m,n) = (p^2+q^2, 2pq) = (r^4+s^4, 2r^2s^2) \\ &\longrightarrow (u,v) = (p^2-q^2, v) = (r^4-s^4, 2rs\sqrt{r^4+s^4}).\end{aligned} \quad (4.2)$$

Now let us match parameters with what is happening for the corresponding elliptic curve. Schematically we have

$$\begin{array}{ccc} (x,y) \text{ on cubic} & \longrightarrow & (u,v) \text{ on quartic} \\ & & \downarrow \\ (c,d) \text{ on cubic} & \longleftarrow & (r,s) \text{ on quartic} \end{array} \quad (4.3)$$

The equations that are satisfied by these variables are

$$\begin{array}{cc} y^2 = x^3 - 4x & u^4 + v^4 = w^2 \\ d^2 = c^3 - 4c & r^4 + s^4 = \square. \end{array}$$

We seek an expression for x in terms of c. Let $X = u/v$ and $Y = w/v^2$, so that $Y^2 = X^4 + 1$. Then (3.14) says that the top line of (4.3) is implemented by

$$\left.\begin{aligned} X &= \frac{y}{2x} \\ Y &= \frac{y^2+8x}{4x^2} \end{aligned}\right\} \quad \text{and} \quad \left\{\begin{aligned} x &= \frac{2}{Y-X^2} \\ y &= \frac{4X}{Y-X^2}. \end{aligned}\right.$$

Let $R = r/s$, so that

$$R^2 = \left(\frac{d}{2c}\right)^2 = \frac{c^3 - 4c}{4c^2} = \frac{c^2 - 4}{4c}. \quad (4.4)$$

Then

$$\begin{aligned} Y - X^2 &= \frac{w-u^2}{v^2} = \frac{(m^2+n^2) - (m^2-n^2)}{2mn} \\ &= \frac{n}{m} = \frac{2r^2s^2}{r^4+s^4} = \frac{2R^2}{R^4+1}, \end{aligned}$$

1. DESCENT

and hence

$$x = \frac{2}{Y-X^2} = \frac{R^4+1}{R^2} = \frac{\left(\frac{c^2-4}{4c}\right)^2+1}{\left(\frac{c^2-4}{4c}\right)} = \frac{(c^2+4)^2}{4c(c^2-4)}. \tag{4.5}$$

INTERPRETATION OF ANALYSIS. In the notation of (3.23) and (3.24), our elliptic curve has

$$a_1 = a_3 = a_2 = a_6 = 0 \quad \text{and} \quad a_4 = -4.$$

Thus

$$b_2 = b_6 = 0, \quad b_4 = -8, \quad b_8 = -16.$$

The point (c, d) is on the elliptic curve, and thus $c = x(P)$ for a point P. Then (3.74) says that

$$x(2P) = \frac{c^4 + 8c^2 + 16}{4c^3 - 16c},$$

and this expression coincides with the expression (4.5) for x. Computation shows also that $y(2P)$ agrees with y except for a factor of sgn c. Thus Fermat's method of descent amounts to a construction that starts with P' in $E(\mathbf{Q})$ and constructs $\pm\frac{1}{2}P'$ in $E(\mathbf{Q})$.

Closer inspection shows that we made certain adjustments to our variables as we went along, in order to be able to carry out the descent. The first one was to assume that u is odd and v is even, rather than the other way around. This is the first adjustment. The next assumption was that u and v are positive; if u and v are not positive, we adjusted them to make them positive. We chose r and s to be the positive square roots of p and q, but we could not force r to be odd and s to be even. Normalization of r and s to be of the same form as u and v was a second instance of the first adjustment. In the context of the elliptic curve, it turns out that both adjustments amount to changing the point we are studying by a coset representative of $E(\mathbf{Q})/2E(\mathbf{Q})$. The fact that at most two adjustments are needed reflects the fact that $E(\mathbf{Q})/2E(\mathbf{Q})$ is a sum of only two copies of $\mathbf{Z}_2$ in this example.

Finally we can ask why the descent has to stop. For the quartic equation this was clear, since the integer entries of the solution were going down. Let us translate this condition into the context of the elliptic curve. For a rational number $\frac{i}{j}$ in lowest terms, we define

$$\left|\frac{i}{j}\right|_\infty = \log\max(|i|, |j|).$$

Since r and s in (4.2) are relatively prime, we have

$$\left|\frac{r}{s}\right|_\infty = \log\max(|r|,|s|).$$

The construction of r and s had the property that

$$|R|_\infty = \left|\frac{r}{s}\right|_\infty < \left|\frac{u}{v}\right|_\infty = |X|_\infty.$$

With the quartic equation, the descent has to stop because only finitely many rationals have $|\cdot|_\infty$ less than a given constant. To work directly with the elliptic curve, we first re-examine (4.1) and see that in fact

$$|R|_\infty = \left|\frac{r}{s}\right|_\infty < \frac{1}{2}\left|\frac{u}{v}\right|_\infty = \frac{1}{2}|X|_\infty. \tag{4.6}$$

Now R is given in terms of c by (4.4), and a little checking shows that

$$\bigl||R|_\infty - |c|_\infty\bigr| \leq M \tag{4.7a}$$

for some computable constant M. Similarly

$$\bigl||X|_\infty - |x|_\infty\bigr| \leq M. \tag{4.7b}$$

Putting together (4.6) and (4.7), we obtain

$$|c|_\infty \leq \frac{1}{2}|x|_\infty + \frac{3}{2}M. \tag{4.8}$$

Thus

$$|c|_\infty < |x|_\infty \quad \text{as long as} \quad 3M < |x|_\infty. \tag{4.9}$$

The values of $|c|_\infty$ and $|x|_\infty$ are a naive version of the notion of height mentioned at the beginning of the chapter. Condition (4.9) does not say that the descent must stop, but it does say that it has to lead to a (nontrivial) solution of the elliptic curve with small numerator and denominator. Although this conclusion does not rule out all nontrivial solutions for this example immediately, we shall see that it does give enough information for the conclusion that $E(\mathbb{Q})$ is finitely generated.

To get a cleaner argument, we ultimately will use a more refined definition of height, and inequality (4.8) will not have the additive constant on the right side.

2. Condition for Divisibility by 2

The proof of Mordell's Theorem will involve showing that $E(\mathbf{Q})/2E(\mathbf{Q})$ is finite and will involve using a suitable notion of height to prove that the descent must stop. In this section we prove a theorem that allows us to recognize elements of $2E(\mathbf{Q})$. We continue to use the symbol $\square$ to denote a square.

Theorem 4.2. Let E be an elliptic curve over a field of characteristic not equal to 2 or 3. Suppose E is given by

$$y^2 = (x-\alpha)(x-\beta)(x-\gamma) = x^3 + rx^2 + sx + t$$

with α, β, γ in k. For (x_2, y_2) in $E(k)$, there exists (x_1, y_1) in $E(k)$ with $2(x_1, y_1) = (x_2, y_2)$ if and only if $x_2 - \alpha$, $x_2 - \beta$, and $x_2 - \gamma$ are squares in k.

PROOF OF NECESSITY. If (x_1, y_1) exists, let $y = mx + b$ be the tangent line at x_1. Since (x_1, y_1) and $(x_2, -y_2)$ both satisfy

$$(x-\alpha)(x-\beta)(x-\gamma) = y^2 = (mx+b)^2$$

and since $y = mx + b$ is actually tangent at x_1, the three roots of

$$(x-\alpha)(x-\beta)(x-\gamma) - (mx+b)^2$$

are $x_2, x_1,$ and x_1. Thus

$$(x-\alpha)(x-\beta)(x-\gamma) - (mx+b)^2 = (x-x_2)(x-x_1)^2.$$

Putting $x = \alpha$, we have $-\square = (\alpha - x_2)\square$. Since x_1 cannot be α, the square on the right is not 0. We conclude $x_2 - \alpha$ is a square. Similarly $x_2 - \beta$ and $x_2 - \gamma$ are squares.

PREPARATION FOR SUFFICIENCY. Reduction of quartics to cubics.

If we have a monic quartic polynomial equation in one variable to solve, we can translate x to eliminate the x^3 term. Then we have

$$x^4 + ax^2 + bx + c = 0,$$
$$(x^2)^2 = -ax^2 - bx - c. \tag{4.10}$$

86   IV: MORDELL'S THEOREM

With u as an unknown to be specified, (4.10) is equivalent with

$$(x^2 + u)^2 = (2u - a)x^2 - bx + (u^2 - c). \qquad (4.11)$$

We choose u so that the right side of (4.11) has a double root in x. Then

$$\begin{aligned} 0 = B^2 - 4AC &= b^2 - 4(2u-a)(u^2 - c) \\ &= b^2 - 8u^3 + 4au^2 + 8cu - 4ac, \end{aligned}$$

and hence
$$8u^3 - 4au^2 - 8cu + (4ac - b^2) = 0.$$

If we solve for u and substitute into (4.11), the result is $\square = \square$. We take square roots and are led to a quadratic equation for x.

PROOF OF SUFFICIENCY. Changing variables in x, we may assume $x_2 = 0$. We are given

$$y^2 = (x - \alpha)(x - \beta)(x - \gamma) = x^3 + rx^2 + sx + t$$

with
$$y_2^2 = -\alpha\beta\gamma = t \qquad (4.12a)$$

and with $-\alpha = \alpha_1^2$, $-\beta = \beta_1^2$, $-\gamma = \gamma_1^2$, and we are to produce (x_1, y_1). Without loss of generality, we may adjust the signs of $\alpha_1, \beta_1, \gamma_1$ so that

$$\alpha_1 \beta_1 \gamma_1 = y_2. \qquad (4.12b)$$

Take $y = mx + y_2$ as the line through $(0, y_2)$ tangent at the unknown point (x_1, y_1). Then the three roots of

$$(x - \alpha)(x - \beta)(x - \gamma) - (mx + y_2)^2 \qquad (4.13)$$

are 0, x_1, and x_1. So

$$\frac{1}{x}(x^3 + rx^2 + sx - m^2 x^2 - 2my_2 x) \qquad \text{is to be} \qquad (x - x_1)^2.$$

Thus the discriminant of

$$x^2 + (r - m^2)x + (s - 2my_2) \qquad (4.14)$$

is to be 0:
$$(r - m^2)^2 = 4(s - 2my_2). \qquad (4.15)$$

2. CONDITION FOR DIVISIBILITY BY 2

This is a quartic equation in m. If it has a root m_0 in k, then $x_1 = \frac{1}{2}(m_0^2 - r)$ is a double root of (4.14) and hence also of (4.13). Consequently $2(x_1, m_0 x_1 + y_2) = (0, -y_2)$, and $2(x_1, -m_0 x_1 - y_2) = (0, y_2)$. In other words (x_1, y_1) exists with $2(x_1, y_1) = (x_2, y_2)$.

Thus it is enough to show that (4.15) has a root in k. As in the preparation for this part of the proof, let u be an unknown to be specified. From (4.15) we have

$$(m^2 - r + u)^2 = 2um^2 - 8y_2 m + (u^2 - 2ru + 4s). \tag{4.16}$$

We try to find u such that the right side is a square as a function of m. Thus

$$0 = B^2 - 4AC = 64y_2^2 - 8u(u^2 - 2ru + 4s)$$

$$u^3 - 2ru^2 + 4su - 8y_2^2 = 0.$$

The roots of this equation are $u = -2\alpha, -2\beta, -2\gamma$. [In fact, just change variables with $u = -2v$, use the identity (4.12a), and reduce the polynomial to $-8(v^3 + rv^2 + sv + t)$.] Let us take $u = -2\alpha$. Then (4.16) gives

$$(m^2 - r - 2\alpha)^2 = -4\alpha m^2 - 8y_2 m + (4\alpha^2 + 4r\alpha + 4s).$$

Substituting for r and s in terms of α, β, γ and using (4.12b), we obtain

$$(m^2 - \alpha + \beta + \gamma)^2 = 4(\alpha_1 m - \beta_1 \gamma_1)^2,$$

$$m^2 - \alpha + \beta + \gamma = \pm 2(\alpha_1 m - \beta_1 \gamma_1),$$

$$m^2 \mp 2\alpha_1 m - \alpha = -\beta - \gamma \mp 2\beta_1 \gamma_1,$$

$$(m \mp \alpha_1)^2 = \beta_1^2 \mp 2\beta_1 \gamma_1 + \gamma_1^2 = (\beta_1 \mp \gamma_1)^2.$$

Taking square roots and denoting an independent sign by $\pm'$, we find

$$m = \pm \alpha_1 \pm' (\beta_1 \mp \gamma_1).$$

We can reverse the steps, and we see that we have found four roots m in k to the equation (4.15). This proves the theorem.

Let us digress for a moment and return to the subject of congruent numbers, first discussed in Subsection C of §III.1. Recall that n is congruent if n is the area of a rational right triangle. Using the easy half of Theorem 4.2, we obtain the following corollary, which says that (3) $\Rightarrow$ (2) in Proposition 3.1.

Corollary 4.3. Let n be a square-free positive integer, and suppose there exists a $\mathbf{Q}$ rational point (x, y) on

$$y^2 = x^3 - n^2 x$$

other than $(-n, 0)$, $(0, 0)$, $(n, 0)$, and ∞. Then there exist three rational squares in arithmetic progression with common difference n. Hence n is congruent.

PROOF. Let $P = (x_1, y_1)$ be a $\mathbf{Q}$ solution other than the trivial ones. Then $y_1 \neq 0$ and P does not have order 1 or 2. Hence $2P$ is not ∞, say $2P = (x_2, y_2)$. Then the theorem says that $x_2 + n$, x_2, and $x_2 - n$ are all squares, as required. By $(2) \Rightarrow (1)$ in Proposition 3.1, n is congruent.

Corollary 4.4 (Fermat). $n = 2$ is not a congruent number.

PROOF. If 2 were congruent, $(1) \Rightarrow (3)$ in Proposition 3.1 would say that $y^2 = x^3 - 4x$ has a nontrivial solution. Transforming via (3.14), we would obtain a nontrivial solution of the equation $u^4 + v^4 = w^2$. But this conclusion contradicts Proposition 4.1.

3. $E(\mathbf{Q})/2E(\mathbf{Q})$, Special Case

The main step in Mordell's Theorem is the proof of the following result.

Theorem 4.5. If E is an elliptic curve over $\mathbf{Q}$, then the abelian group $E(\mathbf{Q})/2E(\mathbf{Q})$ is finite.

Using an admissible change of variables, we may assume that E is of the form

$$y^2 = (x - \alpha)(x - \beta)(x - \gamma) \tag{4.17}$$

with integer coefficients. In this section we make the *additional assumption* that α, β, γ are in $\mathbf{Q}$, hence in $\mathbf{Z}$.

Since α, β, γ are in $\mathbf{Q}$, the subgroup of $E(\mathbf{Q})$ of elements of order ≤ 2 is of the form $\mathbf{Z}_2 \oplus \mathbf{Z}_2$. The picture of $E(\mathbf{R})$ is qualitatively the same as in Figure 1.7c. We consider the group $\mathbf{Q}^\times/\mathbf{Q}^{\times 2}$, writing it as

$$\mathbf{Q}^\times/\mathbf{Q}^{\times 2} = \{\pm 2^a 3^b 5^c 7^d \cdots \mid a, b, c, d, \cdots \in \{0, 1\}\} = \sum_{\pm, 2, 3, 5, 7, \ldots} \oplus \mathbf{Z}_2. \tag{4.18}$$

We continue to use the symbol $\square$ to denote a square.

3. $E(\mathbf{Q})/2E(\mathbf{Q})$, SPECIAL CASE

Proposition 4.6. Let E be the elliptic curve (4.17) over $\mathbf{Q}$ with integer roots, and define $\varphi : E(\mathbf{Q}) \to \mathbf{Q}^\times/\mathbf{Q}^{\times 2}$ by

$$\varphi(P) = \begin{cases} (x-\alpha)\mathbf{Q}^{\times 2} & \text{if } P = (x,y) \text{ with } P \neq \infty \text{ and } x \neq \alpha \\ (\alpha-\beta)(\alpha-\gamma)\mathbf{Q}^{\times 2} & \text{if } P = (\alpha,0) \\ \mathbf{Q}^{\times 2} & \text{if } P = \infty. \end{cases} \quad (4.19)$$

Then φ is a group homomorphism.

REMARK. Then $\varphi(2P) = \varphi(P)^2 \in 1 \cdot \mathbf{Q}^{\times 2}$, so that φ descends to a homomorphism

$$\varphi_\alpha : E(\mathbf{Q})/2E(\mathbf{Q}) \longrightarrow \mathbf{Q}^\times/\mathbf{Q}^{\times 2}. \quad (4.20)$$

PROOF. Let $P_1 + P_2 = P_3$. We are to show $\varphi(P_1)\varphi(P_2)\varphi(P_3)^{-1}$ is in $\mathbf{Q}^{\times 2}$. Since $\varphi(P_3) = \varphi(-P_3)$ and $\varphi(P_3) = \varphi(P_3)^{-1}$, it is enough to show that $P_1 + P_2 + P_3 = 0$ implies $\varphi(P_1)\varphi(P_2)\varphi(P_3)$ is in $\mathbf{Q}^{\times 2}$. If one of the P_i is ∞, this conclusion is trivial. Thus we may assume P_i is of the form (x_i, y_i) for $i = 1, 2, 3$.

Case 1: No (x_i, y_i) is $(\alpha, 0)$. Let $y = mx + b$ be the line on which P_1, P_2, P_3 lie. Each $P_i = (x_i, y_i)$ satisfies

$$(x-\alpha)(x-\beta)(x-\gamma) = y^2 = (mx+b)^2.$$

Hence $(x-\alpha)(x-\beta)(x-\gamma) - (mx+b)^2$ is 0 for $x = x_1, x_2, x_3$ with multiplicities. Thus

$$(x-\alpha)(x-\beta)(x-\gamma) - (mx+b)^2 = (x-x_1)(x-x_2)(x-x_3).$$

Putting $x = \alpha$ gives $(x_1-\alpha)(x_2-\alpha)(x_3-\alpha) = (m\alpha+b)^2$, and thus $\varphi(P_1)\varphi(P_2)\varphi(P_3)$ is a square in $\mathbf{Q}^\times$.

Case 2: $(x_1, y_1) = (\alpha, 0)$. Then neither (x_2, y_2) nor (x_3, y_3) is $(\alpha, 0)$, since otherwise the other point would be ∞. Let $y = mx + b$ be the line on which P_1, P_2, P_3 lie. As before,

$$(x-\alpha)(x-\beta)(x-\gamma) - (mx+b)^2$$

is 0 for $x = \alpha, x_2, x_3$ with multiplicities, and thus

$$(x-\alpha)(x-\beta)(x-\gamma) - (mx+b)^2 = (x-\alpha)(x-x_2)(x-x_3). \quad (4.21)$$

Now $x - \alpha$ has to divide $(mx+b)^2$, and hence $mx + b = m(x-\alpha)$. So (4.21) becomes

$$(x-\alpha)(x-\beta)(x-\gamma) - m^2(x-\alpha)^2 = (x-\alpha)(x-x_2)(x-x_3)$$

or

$$(x-\beta)(x-\gamma) - m^2(x-\alpha) = (x-x_2)(x-x_3).$$

Putting $x = \alpha$ gives

$$(\alpha-\beta)(\alpha-\gamma) = (\alpha-x_2)(\alpha-x_3),$$

so that $\varphi(P_1) = \varphi(P_2)\varphi(P_3)$. Thus $\varphi(P_1)\varphi(P_2)\varphi(P_3)$ is in $\mathbf{Q}^{\times 2}$.

Corollary 4.7. Let E be the elliptic curve (4.17) over $\mathbf{Q}$ with integer roots, and define

$$\varphi_\alpha : E(\mathbf{Q})/2E(\mathbf{Q}) \longrightarrow \mathbf{Q}^\times/\mathbf{Q}^{\times 2}$$

as in (4.19). Define φ_β similarly. Then the homomorphism

$$\varphi_\alpha \times \varphi_\beta : E(\mathbf{Q})/2E(\mathbf{Q}) \longrightarrow \mathbf{Q}^\times/\mathbf{Q}^{\times 2} \times \mathbf{Q}^\times/\mathbf{Q}^{\times 2} \qquad (4.22)$$

is one-one.

PROOF. Suppose $(x,y) \in E(\mathbf{Q})$ maps to $\mathbf{Q}^{\times 2}$ under φ_α and φ_β.

Case 1: Suppose (x,y) is not ∞ or $(\alpha, 0)$ or $(\beta, 0)$. Then the condition is that $x - \alpha$ and $x - \beta$ are $\square$. Since $(x-\alpha)(x-\beta)(x-\gamma) = \square$, $x - \gamma$ is $\square$. By Theorem 4.2, $(x,y) = 2(x', y')$ for some (x', y') in $E(\mathbf{Q})$. Thus (x,y) is in $2E(\mathbf{Q})$.

Case 2: Suppose $(x,y) = (\alpha, 0)$. Then

$$\square = \varphi_\alpha(\alpha, 0) = (\alpha - \beta)(\alpha - \gamma)\mathbf{Q}^{\times 2}$$
$$\square = \varphi_\beta(\alpha, 0) = (\alpha - \beta)\mathbf{Q}^{\times 2}.$$

Then $\alpha - \beta$ and $\alpha - \gamma$ are $\square$. Also 0 is a square. By Theorem 4.2, $(\alpha, 0) = 2(x', y')$ for some (x', y'), and thus $(\alpha, 0)$ is in $2E(\mathbf{Q})$.

Case 3: Suppose $(x,y) = (\beta, 0)$. This case follows from Case 2 by symmetry.

Let us write

$$\mathbf{Q}^\times/\mathbf{Q}^{\times 2} \times \mathbf{Q}^\times/\mathbf{Q}^{\times 2} = \sum_{\pm,2,3,\ldots} \oplus (\mathbf{Z}_2 \oplus \mathbf{Z}_2). \qquad (4.23)$$

It will turn out that the image of $\varphi_\alpha \times \varphi_\beta$ in $\mathbf{Q}^\times/\mathbf{Q}^{\times 2} \times \mathbf{Q}^\times/\mathbf{Q}^{\times 2}$ is 0 in almost all coordinates of (4.23). Let d be the discriminant of the cubic polynomial $(x - \alpha)(x - \beta)(x - \gamma)$, namely

$$d = (\alpha - \beta)^2(\alpha - \gamma)^2(\beta - \gamma)^2. \qquad (4.24)$$

(See Proposition 3.3 and the neighboring discussion.) If p is a prime and r is rational, we write $p^a \| r$ if $r = p^a q$ and $q \in \mathbf{Q}$ has no factor of p in its numerator or denominator.

3. $E(\mathbf{Q})/2E(\mathbf{Q})$, SPECIAL CASE

Proposition 4.8. Let E be the elliptic curve (4.17) over $\mathbf{Q}$ with integer roots, let $\varphi_\alpha \times \varphi_\beta$ be the homomorphism (4.22), and let d be the discriminant (4.24). Then the image of $\varphi_\alpha \times \varphi_\beta$ in (4.23) is confined to the coordinates $\pm$ and primes p dividing d. Thus the homomorphism

$$E(\mathbf{Q})/2E(\mathbf{Q}) \longrightarrow \sum_{\pm, p|d} \oplus (\mathbf{Z}_2 \oplus \mathbf{Z}_2) \qquad (4.25)$$

is one-one.

REMARK. The group on the right of (4.25) is finite, and thus Proposition 4.28 completes the proof of Theorem 4.5 under the special assumption that the roots α, β, γ in (4.17) are in $\mathbf{Q}$.

PROOF. Let (x,y) be a point other than ∞ in $E(\mathbf{Q})$. Assume for the moment that $x \notin \{\alpha, \beta, \gamma\}$. Fix a prime $p = 2, 3, 5, \ldots$, and define integers a, b, c by

$$p^a \| (x - \alpha), \qquad p^b \| (x - \beta), \qquad p^c \| (x - \gamma).$$

Since (4.17) says that $(x - \alpha)(x - \beta)(x - \gamma)$ is a square, we have

$$a + b + c \equiv 0 \mod 2. \qquad (4.26)$$

Suppose at least one of a, b, c is < 0. Say $a < 0$. Since α is an integer, $p^{|a|} \|$ (denominator of x). Therefore

$$p^a \| (x - \alpha), \qquad p^a \| (x - \beta), \qquad p^a \| (x - \gamma).$$

In other words, $a = b = c$. From (4.26), it then follows that

$$a \equiv b \equiv c \equiv 0 \mod 2.$$

Consequently the image of (x, y) in the pth coordinate of (4.23) is 0.

Suppose at least one of a, b, c is > 0. Say $a > 0$. If $p \nmid d$, then $p \nmid (\alpha - \beta)$ and hence p cannot occur in the numerator of

$$x - \beta = (x - \alpha) + (\alpha - \beta).$$

In other words, $b = 0$. Similarly, if $p \nmid d$, then $p \nmid (\alpha - \gamma)$, and we conclude $c = 0$. By (4.26), a is even. Thus again we have

$$a \equiv b \equiv c \equiv 0 \mod 2,$$

and again the image of (x, y) in the pth coordinate of (4.23) is 0.

We have been assuming that $x \notin \{\alpha, \beta, \gamma\}$. When x is in this set, then $\varphi_\alpha(x)$ and $\varphi_\beta(x)$ are products of $\alpha - \beta$, $\alpha - \gamma$, and $\beta - \gamma$, up to sign, and these are all prime to p if $p \nmid d$. Hence the pth coordinate of $\varphi_\alpha(x) \times \varphi_\beta(x)$ is $(0, 0)$.

Combining these conclusions with the one-oneness from Corollary 4.7, we obtain the proposition.

4. $E(\mathbf{Q})/2E(\mathbf{Q})$, General Case

So far, we have proved Theorem 4.5 in the special case that the roots α, β, γ in (4.17) are in $\mathbf{Q}$. In the general case, let k be the splitting field of the cubic polynomial $(x - \alpha)(x - \beta)(x - \gamma)$ over $\mathbf{Q}$. The rough idea is to imitate the proof in §3, replacing $\mathbf{Q}$ everywhere by this extension field k. But we encounter three complications:

(1) The modified approach will lead to information about $E(k)/2E(k)$, and we need to deduce information about $E(\mathbf{Q})/2E(\mathbf{Q})$.
(2) The decomposition (4.18) of $\mathbf{Q}^\times/\mathbf{Q}^{\times 2}$ uses unique factorization in $\mathbf{Z}$, and unique factorization may fail in the ring of algebraic integers in k.
(3) The $\pm$ coordinate of $\mathbf{Q}^\times/\mathbf{Q}^{\times 2}$ in (4.18) represents units in $\mathbf{Z}$, and the units need to be controlled in whatever ring replaces $\mathbf{Z}$ in the general case.

We can dispose of (1) rather quickly.

Proposition 4.9. Let E be the elliptic curve (4.17) over $\mathbf{Q}$, and let k be the splitting field of $(x-\alpha)(x-\beta)(x-\gamma)$ over $\mathbf{Q}$. Then the canonical homomorphism

$$E(\mathbf{Q})/2E(\mathbf{Q}) \longrightarrow E(k)/2E(k) \qquad (4.27)$$

has $\leq 2^{2[k:\mathbf{Q}]}$ elements in its kernel. Consequently if $E(k)/2E(k)$ is finite, then so is $E(\mathbf{Q})/2E(\mathbf{Q})$.

REMARK. We shall use the Galois group $\mathrm{Gal}(k/\mathbf{Q})$. Note that if $Q = (x, y)$ is in $E(k)$ and σ is in $\mathrm{Gal}(k/\mathbf{Q})$, then $Q^\sigma = (\sigma(x), \sigma(y))$ is in $E(k)$ since E is defined over $\mathbf{Q}$.

PROOF. Let
$$E[2] = \{Q' \in E(k) \mid 2Q' = 0\}.$$

The points P in $E(\mathbf{Q})$ that map to 0 under the canonical homomorphism are those in $E(\mathbf{Q}) \cap 2E(k)$. For each such P, select Q_P in $E(k)$ such that $2Q_P = P$. Then we obtain a function

$$\lambda_P : \mathrm{Gal}(k/\mathbf{Q}) \longrightarrow E[2] \qquad (4.28\mathrm{a})$$

by the definition

$$\lambda_P(\sigma) = Q_P^\sigma - Q_P. \qquad (4.28\mathrm{b})$$

4. $E(\mathbf{Q})/2E(\mathbf{Q})$, GENERAL CASE

[In fact, to see that λ_P takes its values in $E[2]$, we observe that $2(Q_P^\sigma - Q_P) = (2Q_P)^\sigma - 2Q_P = P^\sigma - P = 0$.]

Now let us show that $\lambda_P = \lambda_{P'}$ implies P' is in $P + 2E(\mathbf{Q})$. In fact, if

$$Q_P^\sigma - Q_P = \lambda_P = \lambda_{P'} = Q_{P'}^\sigma - Q_{P'}$$

for all σ, then

$$(Q_{P'} - Q_P)^\sigma = Q_{P'} - Q_P.$$

Since k is a normal extension of $\mathbf{Q}$, $k^{\mathrm{Gal}(k/\mathbf{Q})} = \mathbf{Q}$. Thus $Q_{P'} - Q_P$ is in $E(\mathbf{Q})$. Hence

$$P' - P = 2(Q_{P'} - Q_P) \in 2E(\mathbf{Q}),$$

as required.

Thus for each member of the kernel of (4.27), we select P in $E(\mathbf{Q})$ mapping into $2E(k)$, and then we can associate λ_P as in (4.28). What we showed above is that if we start with a different member of the kernel, then the resulting λ_P is a different function from $\mathrm{Gal}(k/\mathbf{Q})$ to $E[2]$. Hence the order of the kernel is $\leq$ the number of functions from $\mathrm{Gal}(k, \mathbf{Q})$ to $E[2]$, which is $4^{|\mathrm{Gal}(k,\mathbf{Q})|} = 2^{2[k:\mathbf{Q}]}$.

Complications (2) and (3) are more subtle. Let us list some examples.

EXAMPLE 1. $y^2 = x^3 + x$. The splitting field is $k = \mathbf{Q}(\sqrt{-1})$. The ring of integers $\mathbf{Z}[\sqrt{-1}]$ is a unique factorization domain, and its group of units is $\{(\sqrt{-1})^k\}_{k=0}^{3} \cong \mathbf{Z}_4$.

EXAMPLE 2. $y^2 = x^3 - 2x$. The splitting field is $k = \mathbf{Q}(\sqrt{2})$. The ring of integers $\mathbf{Z}[\sqrt{2}]$ is a unique factorization domain, and its group of units is the infinite group $\{\pm(1 \pm \sqrt{2})^k\}_{k=-\infty}^{\infty} \cong \mathbf{Z} \oplus \mathbf{Z}_2$.

EXAMPLE 3. $y^2 = x^3 + 5x$. The splitting field is $k = \mathbf{Q}(\sqrt{-5})$. The ring of integers $\mathbf{Z}[\sqrt{-5}]$ is not a unique factorization domain. In fact, the numbers $2, 3, 1 \pm \sqrt{-5}$ are all prime, and $2 \cdot 3 = (1+\sqrt{-5})(1-\sqrt{-5})$. The group of units is $\{\pm 1\} \cong \mathbf{Z}_2$.

In the first two examples, $\mathbf{Z}[\sqrt{-1}]$ and $\mathbf{Z}[\sqrt{2}]$ are unique factorization domains. So it is meaningful to write

$$k^\times / k^{\times 2} = \{\text{units}/(\text{squares of units})\} \oplus \{p_1^a p_2^b \cdots \mid a, b, \cdots \in \{0, 1\}\}$$
$$= \{\text{units}/(\text{squares of units})\} \oplus \sum_{p \text{ prime}} \mathbf{Z}_2; \quad (4.29)$$

in the sum we list only one representative from each class $\{p\epsilon \,|\, \epsilon = \text{unit}\}$ of associate primes. As in (4.19), we get a homomorphism $\varphi : E(k) \to k^\times/k^{\times 2}$ generically given by $x \to (x - \alpha)k^{\times 2}$. And as in Corollary 4.7, we obtain a one-one homomorphism

$$\varphi_\alpha \times \varphi_\beta : E(k)/2E(k) \longrightarrow \{\text{units/(squares of units)}\} \oplus \sum_{p \text{ prime}} (\mathbf{Z}_2 \oplus \mathbf{Z}_2).$$
(4.30)

As in Proposition 4.8, we can discard all primes p except those dividing the integer discriminant d given in (4.24). There are only finitely many such primes because of unique factorization. Then $E(k)/2E(k)$ has been mapped one-one into the finite group

$$\{\text{units/(squares of units)}\} \oplus \sum_{p|d}(\mathbf{Z}_2 \oplus \mathbf{Z}_2), \qquad (4.31)$$

and $E(k)/2E(k)$ is therefore finite.

For the third example, unique factorization fails for $\mathbf{Z}[\sqrt{-5}]$, and (4.29) is no longer valid. However, if we let $S = \{1, 2, 2^2, 2^3, \ldots\}$ and define $R = S^{-1}\mathbf{Z}[\sqrt{-5}]$, then R is a ring with $\mathbf{Z}[\sqrt{-5}] \subseteq R \subseteq k$, and it turns out that R has the following two properties:

(1) R is a principal idea domain, hence a unique factorization domain.
(2) The group of units in R is finitely generated (with -1 and $\frac{1}{2}$ as generators).

Since R contains $\mathbf{Z}[\sqrt{-5}]$, the quotient field of R is still k. By unique factorization, (4.29) is valid if we interpret units and primes as units and primes in R. The argument yielding a one-one homomorphism (4.30) is then valid, and again we can collapse the image to (4.31). Since the group of units in R is finitely generated, it is a direct sum of cyclic groups, and thus

$$\{\text{units/(squares of units)}\}$$

is a finite group. Therefore (4.31) is a finite group, and $E(k)/2E(k)$ has to be finite.

The same argument as above reduces the finiteness of $E(k)/2E(k)$ in the general case to the following result.

Theorem 4.10. Let k be a finite extension of $\mathbf{Q}$, and let $\mathcal{O}_k$ be the ring of algebraic integers in k. Then there exists a ring R with $\mathcal{O}_k \subseteq R \subseteq k$ such that

(1) R is a principal ideal domain, hence a unique factorization domain.
(2) The group of units in R is finitely generated.

The proof of Theorem 4.10 relies on three famous results in algebraic number theory. In §9 we shall sketch the relevant background from algebraic number theory and then give a proof of this theorem.

As we mentioned, the finiteness of $E(k)/2E(k)$ follows from Theorem 4.10. Then Proposition 4.9 implies that $E(\mathbf{Q})/2E(\mathbf{Q})$ is finite. This completes the proof of Theorem 4.5.

5. Height and Mordell's Theorem

With the hard step finished, we can now address Mordell's Theorem. In this section we state the theorem, introduce the notions of "naive height" and "canonical height," and show how Mordell's Theorem follows from Theorem 4.5.

Theorem 4.11 (Mordell). If E is an elliptic curve over $\mathbf{Q}$, then the abelian group $E(\mathbf{Q})$ is finitely generated.

Using an admissible change of variables, we may assume that E is of the form
$$y^2 = x^3 + Ax + B \tag{4.32}$$
with A and B in $\mathbf{Z}$. For $P = (x,y)$ in $E(\mathbf{Q})$, write $x = \dfrac{p}{q}$ with $\mathrm{GCD}(p,q) = 1$, and define the **naive height** of P by
$$h_0(P) = \log \max(|p|, |q|) \geq 0. \tag{4.33}$$

By convention we define
$$h_0(\infty) = 0.$$

For any given C,
$$\{P \in E(\mathbf{Q}) \mid h_0(P) \leq C\} \qquad \text{is a finite set.} \tag{4.34}$$

Proposition 4.12. The naive height satisfies
$$h_0(2P) = 4h_0(P) + O(1), \tag{4.35}$$
where $O(1)$ is bounded independently of $P \in E(\mathbf{Q})$.

PROOF. If $P^* = 2P = (x^*, y^*) \neq \infty$, we know from (3.74) that

$$x^* = \frac{x^4 - b_4 x^2 - 2b_6 x - b_8}{4x^3 + b_2 x^2 + 2b_4 x + b_6},$$

where $b_2 = 0$, $b_4 = 2A$, $b_6 = 4B$, $b_8 = -A^2$. Thus

$$x^* = \frac{x^4 - 2Ax^2 - 8Bx + A^2}{4(x^3 + Ax + B)}.$$

If $x = \frac{p}{q}$ with $\mathrm{GCD}(p, q) = 1$, then $x^* = \frac{p^*}{q^*}$ with

$$p^* = p^4 - 2Ap^2 q^2 - 8Bpq^3 + A^2 q^4$$
$$q^* = 4q(p^3 + Apq^2 + Bq^3).$$

Let $\delta = \mathrm{GCD}(p^*, q^*)$, $p^{**} = p^*/\delta$, and $q^{**} = q^*/\delta$, so that $x^* = \frac{p^{**}}{q^{**}}$ in lowest terms. Then

$$\max(|p^{**}|, |q^{**}|)$$
$$\leq \max(|p^*|, |q^*|)$$
$$\leq \max(|p|, |q|)^4 \max(1 + 2|A| + 8|B| + A^2, 4(1 + |A| + |B|))$$
$$= C_{A,B} \max(|p|, |q|)^4,$$

and

$$h_0(2P) \leq 4 h_0(P) + \log C_{A,B}. \tag{4.36}$$

To get an inequality in the reverse direction, we shall bound $\max(|p|, |q|)$ in terms of $\max(|p^*|, |q^*|)$, and we shall bound $\max(|p^*|, |q^*|)$ in terms of $\max(|p^{**}|, |q^{**}|)$. Let $d = -4A^3 - 27B^2 \neq 0$ be the discriminant of the cubic polynomial $x^3 + AX + B$. Then one can check the following identities:

$$4dq^7 = (3p^3 - 5Apq^2 - 27Bq^3)q^* - 4(3p^2 q + 4Aq^3)p^*$$
$$4dp^7 = -(A^2 Bp^3 + (5A^4 + 32AB^2)p^2 q$$
$$+ (26A^3 B + 192B^3)pq^2 - 3(A^5 + 8A^2 B^2)q^3)q^*$$
$$- 4((4A^3 + 27B^2)p^3 - A^2 Bp^2 q$$
$$+ (3A^4 + 22AB^2)pq^2 + 3(A^3 B + 8B^3)q^3)p^*.$$

$$\tag{4.37}$$

[To derive these formulas, one can use two applications of Proposition 2.16 and its proof, with $f = p^*$ and $g = q^*$. In one application, $X = p$ and the domain is $k[q]$; in the other application, $X = q$ and the domain is $k[p]$. A symbolic manipulation program is a helpful tool.]

From (4.37), we obtain

$$\max(|p|,|q|)^7 \leq C'_{A,B} \max(|p|,|q|)^3 \max(|p^*|,|q^*|)$$

and thus

$$\max(|p|,|q|)^4 \leq C'_{A,B} \max(|p^*|,|q^*|). \tag{4.38}$$

This inequality is one of the desired bounds. For the other one, we see from (4.37) that $\delta | 4dq^7$ and $\delta | 4dp^7$. Since $\gcd(p,q) = 1$, $\delta | 4d$. Hence δ is bounded, and

$$\max(|p^*|,|q^*|) \leq C'' \max(|p^{**}|,|q^{**}|).$$

Combining this inequality with (4.38), we have

$$\max(|p|,|q|)^4 \leq C'_{A,B} C'' \max(|p^{**}|,|q^{**}|). \tag{4.39}$$

Inequalities (4.36) and (4.39) together prove the proposition.

Proposition 4.13. There exists a unique function $h : E(\mathbf{Q}) \to \mathbf{R}$ satisfying

(i) $h(P) - h_0(P)$ is bounded
(ii) $h(2P) = 4h(P)$.

The function is given by

$$h(P) = \lim_{n \to \infty} \frac{h_0(2^n P)}{4^n}. \tag{4.40}$$

It has $h(P) \geq 0$ with equality if and only if P has finite order. Also $\{P \mid h(P) \leq C\}$ is a finite set for each C.

REMARK. The function h is called the **canonical height**.

PROOF. We begin with uniqueness. Let h satisfy (i) and (ii), with a bound C' in (i). Then

$$|4^n h(P) - h_0(2^n P)| = |h(2^n P) - h_0(2^n P)| \leq C'$$

by (ii) and (i). Thus

$$\left| h(P) - \frac{h_0(2^n P)}{4^n} \right| \leq \frac{C'}{4^n},$$

and h must be given by (4.40).

For existence, we first prove that $\left\{\dfrac{h_0(2^n P)}{4^n}\right\}$ is Cauchy. By Proposition 4.12, we can write

$$|h_0(2Q) - 4h_0(Q)| \leq C''$$

with C'' independent of Q. Then $N \geq M \geq 0$ implies

$$\begin{aligned}
|4^{-N} h_0(2^N P) &- 4^{-M} h_0(2^M P)| \\
&= \left|\sum_{n=M}^{N-1} \left(4^{-n-1} h_0(2^{n+1} P) - 4^{-n} h_0(2^n P)\right)\right| \\
&\leq \sum_{n=M}^{N-1} 4^{-n-1} |h_0(2^{n+1} P) - 4 h_0(2^n P)| \\
&\leq \sum_{n=M}^{N-1} 4^{-n-1} C'' \leq \frac{4^{-M} C''}{3},
\end{aligned} \qquad (4.41)$$

and the right side tends to 0 as M and N tend to ∞. Thus $h(P)$ is defined. Letting $N \to \infty$ in (4.41) and taking $M = 0$ gives

$$|h(P) - h_0(P)| \leq \frac{C''}{3},$$

which proves (i). Result (ii) is clear from (4.40). Also $h(P) \geq 0$.

If P is a torsion point, then $2^n P$ lies in a finite set. So $h(P) = 0$. In addition, (i) implies $\{P \mid h(P) \leq C\}$ is finite.

Now suppose P has infinite order. Since $\{P \mid h(P) \leq 1\}$ is finite, we must have $h(2^n P) > 1$ for some n. By (ii), $h(P) > 4^{-n} > 0$.

Proposition 4.14. The canonical height on $E(\mathbb{Q})$ satisfies

$$h(P + Q) + h(P - Q) = 2h(P) + 2h(Q). \qquad (4.42)$$

PROOF. We shall prove that

$$h(P + Q) + h(P - Q) \leq 2h(P) + 2h(Q). \qquad (4.43)$$

Once we have done so, we can apply (4.43) to $P' = P+Q$ and $Q' = P-Q$ to get

$$h(2P) + h(2Q) \leq 2h(P + Q) + 2h(P - Q).$$

5. HEIGHT AND MORDELL'S THEOREM

Dividing by 2 and using property (ii) of h, we obtain

$$2h(P) + 2h(Q) \leq h(P+Q) + h(P-Q).$$

In combination with (4.43), this inequality proves (4.42).

To prove (4.43), it is enough, by (4.40), to prove that

$$h_0(P+Q) + h_0(P-Q) \leq 2h_0(P) + 2h_0(Q) + O(1).$$

If P or Q is ∞, this inequality is trivial; if $P+Q$ or $P-Q$ is ∞, the inequality reduces to Proposition 4.12. Thus let us write

$$P = (x, y) \quad \text{with } x = \frac{p}{q} \text{ in lowest terms,}$$

$$Q = (x', y') \quad \text{with } x = \frac{p'}{q'} \text{ in lowest terms,}$$

$$|x|_\infty = \max(|p|, |q|), \qquad |x'|_\infty = \max(|p'|, |q'|), \qquad x_\pm = x(P \pm Q).$$

Since $P \neq \pm Q$, (3.72) and (3.73) in the chord case give

$$x_\pm = x(P \pm Q) = \left(\frac{y' \mp y}{x' - x}\right)^2 - x - x'.$$

Thus

$$\begin{aligned}
x_+ + x_- &= 2 \frac{y'^2 + y^2 - (x' + x)(x' - x)^2}{(x' - x)^2} \\
&= 2 \frac{(x'^3 + x^3) + A(x' + x) + 2B - (x' + x)(x'^2 - 2xx' + x^2)}{(x' - x)^2} \\
&= 2 \frac{xx'(x' + x) + A(x' + x) + 2B}{(x' - x)^2} \\
&= 2 \frac{pp'(p'q + pq') + Aqq'(p'q + pq') + 2Bq^2 q'^2}{(p'q - pq')^2} \qquad (4.44)
\end{aligned}$$

and
$$\begin{aligned}
x_+ x_- &= \frac{(y'^2 - y^2)^2}{(x'-x)^4} - 2(x+x')\frac{y'^2+y^2}{(x'-x)^2} + (x+x')^2 \\
&= \frac{(x'-x)^2(x'^2+xx'+x^2+A)^2}{(x'-x)^4} \\
&\quad - \frac{2(x+x')[(x'+x)(x'^2-xx'+x^2+A)+2B]}{(x'-x)^2} + (x+x')^2 \\
&= \frac{(x'^2+xx'+x^2+A)^2 - 2(x'^2+2xx'+x^2)(x'^2-xx'+x^2+A)}{(x'-x)^2} \\
&\quad + \frac{-4B(x+x') + (x'^2-x^2)^2}{(x'-x)^2} \\
&= \frac{(xx'-A)^2 - 4B(x+x')}{(x'-x)^2} \quad \text{after simplification} \\
&= \frac{(pp'-qq'A)^2 - 4qq'B(pq'+p'q)}{(p'q-pq')^2}.
\end{aligned} \quad (4.45)$$

Examining (4.44) and (4.45), we see that
$$x_+ + x_- = \frac{r}{t} \quad \text{and} \quad x_+ x_- = \frac{s}{t}$$
with
$$\max(|r|, |s|, |t|) \le C|x|_\infty^2 |x'|_\infty^2. \quad (4.46)$$
Then x_+ and x_- are the two roots of $X^2 - \left(\frac{r}{t}\right)X + \left(\frac{s}{t}\right) = 0$, namely $X = \frac{1}{2t}(r \pm \sqrt{r^2 - 4st})$, and we see that
$$x_+ \in \frac{\mathbb{Z}}{2t} \quad \text{and} \quad x_- \in \frac{\mathbb{Z}}{2t}. \quad (4.47)$$

Let $x_+ = \frac{p_+}{q_+}$ and $x_- = \frac{p_-}{q_-}$ in lowest terms. By (4.47), $2t = \delta_+ q_+ = \delta_- q_-$ for suitable integers δ_+ and δ_-. Then
$$(\delta_+ \delta_-)(q_+ q_-) = 4t^2. \quad (4.48)$$

From
$$\frac{r}{t} = x_+ + x_- = \frac{p_+}{q_+} + \frac{p_-}{q_-} = \frac{p_+ q_- + p_- q_+}{q_+ q_-},$$
we obtain
$$(p_+ q_- + p_- q_+)t = rq_+ q_- = \frac{4rt^2}{\delta_+ \delta_-}$$

5. HEIGHT AND MORDELL'S THEOREM

and hence
$$\delta_+\delta_-(p_+q_- + p_-q_+) = 4rt. \tag{4.49}$$

From
$$\frac{s}{t} = x_+x_- = \frac{p_+p_-}{q_+q_-},$$

we obtain
$$(p_+p_-)t = sq_+q_- = \frac{4st^2}{\delta_+\delta_-}$$

and hence
$$(\delta_+p_+)(\delta_-p_-) = 4st. \tag{4.50}$$

We shall show that $t|\delta_+\delta_-$, and then

$$\begin{array}{ll} |p_+q_- + p_-q_+| \leq 4|r| & \text{from (4.49)} \\ |p_+p_-| \leq 4|s| & \text{from (4.50)} \\ |q_+q_-| \leq 4|t| & \text{from (4.48)} \end{array} \tag{4.51}$$

To see that $t|\delta_+\delta_-$, first fix a prime p. We construct by (4.50) integers a and $b \geq 0$ such that $p^a|\delta_+p_+$, $p^b|\delta_-p_-$, and p^{a+b} is the exact power of p dividing t. Since $2t = \delta_+q_+ = \delta_-q_-$, $p^a|\delta_+q_+$ and $p^b|\delta_-q_-$. Then p^a divides $\text{GCD}(\delta_+p_+, \delta_+q_+) = \delta_+$, and p^b divides $\text{GCD}(\delta_-p_-, \delta_-q_-) = \delta_-$. So $p^{a+b}|\delta_+\delta_-$. Since p is arbitrary, $t|\delta_+\delta_-$. This proves (4.51).

We readily check the numerical inequality

$$\max(|p_+|, |q_+|)\max(|p_-|, |q_-|) \leq 2\max(|p_+q_- + p_-q_+|, |p_+p_-|, |q_+q_-|).$$

Substituting in the right side from (4.51) and then using (4.46), we obtain

$$\max(|p_+|, |q_+|)\max(|p_-|, |q_-|) \leq 8\max(|r|, |s|, |t|) \leq 8C|x|_\infty^2|x'|_\infty^2.$$

Consequently

$$h_0(P+Q) + h_0(P-Q) \leq 2h_0(P) + 2h_0(Q) + \log 8C,$$

and the proposition follows.

PROOF OF THEOREM 4.11. Since Theorem 4.5 has shown $E(\mathbf{Q})/2E(\mathbf{Q})$ to be finite, there exists some sufficiently large C such that the set
$$S = \{P \in E(\mathbf{Q}) \mid h(P) \leq C\}$$
contains a representative for each coset of $E(\mathbf{Q})/2E(\mathbf{Q})$. The set S is finite. We claim it generates $E(\mathbf{Q})$.

Assume the contrary, and let $P \in E(\mathbf{Q})$ be outside the group generated by S. Since $\{P' \in E(\mathbf{Q}) \mid h(P') \leq C'\}$ is finite, we may assume P is chosen with $h(P)$ as small as possible among all points outside the group generated by S. Choose $Q \in S$ with
$$P \equiv Q \mod 2E(\mathbf{Q}),$$
say $P = Q + 2R$. By Proposition 4.14, we have either
$$h(P+Q) \leq h(P) + h(Q)$$
or
$$h(P-Q) \leq h(P) + h(Q);$$
fix the sign of $h(P \pm Q)$ so that this happens. From $P + Q = 2(Q+R)$ and $P - Q = 2R$, we have $P \pm Q = 2P'$ for some $P' \in E(\mathbf{Q})$. Then
$$4h(P') = h(2P') = h(P \pm Q) \leq h(P) + h(Q)$$
$$\leq h(P) + C < 2h(P) \leq 4h(P).$$
So $h(P') < h(P)$. By minimality, P' is in the group generated by S. Since Q has this property and $P \pm Q = 2P'$, P has this property, contradiction.

6. Geometric Formula for Rank

We have now proved that the $\mathbf{Q}$ rational points of an elliptic curve E over $\mathbf{Q}$ form a finitely generated abelian group. Thus
$$E(\mathbf{Q}) \cong \mathbf{Z}^r \oplus F, \tag{4.52}$$
where F is a finite abelian group. The group F is uniquely defined as the torsion subgroup, and Chapter V will show how to determine it completely for any particular E. The integer r is the **rank** of $E(\mathbf{Q})$.

In this section we shall give a geometric limit formula for the rank, under the special assumption that E is of the form
$$y^2 = x^3 + Ax + B, \quad A, B \in \mathbf{Z}. \tag{4.53}$$
The tool will be the canonical height h defined in §5. We give a complete proof modulo one result from Euclidean Fourier analysis.

6. GEOMETRIC FORMULA FOR RANK

Proposition 4.15. With E as in (4.53), there exists a unique $\mathbf{Z}$ bilinear form $\langle P, Q \rangle$ on $E(\mathbf{Q})$ such that $\langle P, P \rangle = h(P)$. This form descends to $E(\mathbf{Q})/F \cong \mathbf{Z}^r$ and is positive definite there.

REMARKS. Let $\{P_i\}$ be a $\mathbf{Z}$ basis of $\mathbf{Z}^r$, and let $c_{ij} = \langle P_i, P_j \rangle$, so that the form is given by

$$\left\langle \sum_i m_i P_i, \sum_j n_j P_j \right\rangle = \sum_{i,j} m_i c_{ij} n_j. \tag{4.54}$$

Then the proof will show that the symmetric matrix (c_{ij}) is positive semidefinite and that $\langle \sum_i m_i P_i, \sum_i m_i P_i \rangle = 0$ only when all m_i are 0. Subsequently we shall see that (c_{ij}) is actually a positive definite matrix.

PROOF. If the bilinear form exists, it has to be given by

$$\langle P, Q \rangle = \tfrac{1}{2}\big(h(P+Q) - h(P) - h(Q)\big). \tag{4.55}$$

This proves uniqueness. For existence, define $\langle P, Q \rangle$ by (4.55). Then $\langle P, Q \rangle$ is certainly symmetric. For additivity in the first variable, we use Proposition 4.14 several times. First we have

$$\begin{aligned}
\langle P, -Q \rangle &= \tfrac{1}{2}\big(h(P-Q) - h(P) - h(Q)\big) \\
&= -\tfrac{1}{2}\big(h(P+Q) - h(P) - h(Q)\big) = -\langle P, Q \rangle.
\end{aligned} \tag{4.56}$$

Then we can write

$$\begin{aligned}
\langle P + P', Q \rangle &+ \langle P - P', Q \rangle \\
&= \tfrac{1}{2}\big(h(P+P'+Q) + h(P-P'+Q) \\
&\quad - h(P+P') - h(P-P') - h(Q) - h(Q)\big) \\
&= \tfrac{1}{2}\big(2h(P+Q) + 2h(P') - 2h(P) - 2h(P') - 2h(Q)\big) \\
&= 2\langle P, Q \rangle.
\end{aligned} \tag{4.57}$$

Interchanging P and P' and using (4.56) gives

$$\langle P + P', Q \rangle - \langle P - P', Q \rangle = 2\langle P', Q \rangle.$$

Adding this to (4.57), we obtain

$$\langle P + P', Q \rangle = \langle P, Q \rangle + \langle P', Q \rangle.$$

Thus the form is bilinear.

Next let us observe that

$$|\langle P,Q\rangle|^2 \leq h(P)h(Q). \tag{4.58}$$

In fact, the bilinearity proves that

$$0 \leq h(nQ - mP) = n^2 h(Q) - 2mn\langle P,Q\rangle + m^2 h(P).$$

Hence

$$h(Q)\lambda^2 - 2\langle P,Q\rangle\lambda + h(P) \geq 0$$

for all $\lambda \in \mathbf{Q}$, and the inequality extends to all $\lambda \in \mathbf{R}$ by a passage to the limit. The discriminant must then be ≤ 0, and the result is (4.58).

If Q is a torsion point, then $h(Q) = 0$, and (4.58) shows that $\langle P,Q\rangle = 0$. Hence

$$0 = \langle P,Q\rangle = \tfrac{1}{2}\bigl(h(P+Q) - h(P) - h(Q)\bigr) = \tfrac{1}{2}\bigl(h(P+Q) - h(P)\bigr),$$

and $h(P+Q) = h(P)$. It follows that the form descends to $E(\mathbf{Q})/F$.

On $\mathbf{Z}^r$ the form is given by (4.54). Since $h \geq 0$, we see that

$$\sum_{i,j} \frac{m_i}{N} c_{ij} \frac{m_j}{N} \geq 0$$

for all systems of rationals $\frac{m_1}{N}, \ldots, \frac{m_r}{N}$. Passing to the limit, we see that $\sum_{i,j} \lambda_i c_{ij} \lambda_j \geq 0$ for all reals $\lambda_1, \ldots, \lambda_r$. Thus (c_{ij}) is positive semidefinite.

Let us now sharpen the conclusion of Proposition 4.15 by showing that the matrix (c_{ij}) is positive definite. We shall use the following celebrated result of Minkowski.

Lemma 4.16 (Minkowski). If E is a compact convex set in $\mathbf{R}^r$ containing 0 and closed under negatives and having volume $> 4^r$, then E contains a nonzero member of $\mathbf{Z}^r$.

REMARK. Fine tuning of the proof would show that the same conclusion holds when the volume is merely $\geq 2^r$.

PROOF. Let N be an integer large enough so that the standard cube C with center 0 and side $4N$ contains E. Suppose E contains no nonzero member of $\mathbf{Z}^r$. We claim that the sets $\{l + \tfrac{1}{2}E\}$ for $l \in \mathbf{Z}^r$ are disjoint.

6. GEOMETRIC FORMULA FOR RANK

In fact, if $l_1 + \frac{1}{2}e_1 = l_2 + \frac{1}{2}e_2$ with $l_1 \neq l_2$, then $l_1 - l_2 = \frac{1}{2}(e_2 - e_1)$, and this is in E since e_2 and $-e_1$ are in E and E is convex.

If every coordinate of l is $\leq N$ in absolute value, then $l + \frac{1}{2}E$ is in C. Hence

$$(4N)^r = \text{vol}(C) \geq \sum_{|\text{coords } l| \leq N} \text{vol}(l + \tfrac{1}{2}E)$$
$$\geq (2N)^r \text{vol}(\tfrac{1}{2}E) = 2^{-r}(2N)^r \text{vol}(E).$$

Since $\text{vol}(E) > 4^r$ by assumption, this is a contradiction.

Proposition 4.17. With E as in (4.53), the symmetric positive semidefinite matrix (c_{ij}) in (4.54) is actually positive definite.

PROOF. Since (c_{ij}) is symmetric and positive semidefinite, we can choose column vectors $v^{(1)}, \ldots, v^{(r)}$ that form an orthonormal basis of eigenvectors of (c_{ij}) with respective eigenvalues

$$\lambda^{(1)} \geq \cdots \geq \lambda^{(r)} \geq 0.$$

We are to prove that $\lambda^{(r)} > 0$. Arguing by contradiction, suppose $\lambda^{(r)} = 0$.

Let $\{e^{(i)}\}$ be the standard orthonormal basis of column vectors, and write $v^{(k)} = \sum_i v_i^{(k)} e^{(i)}$. We shall identify $P = \sum_i m_i P_i$ with the column vector $\sum_i m_i e^{(i)}$, so that the $\mathbb{Z}^r$ quotient of $E(\mathbb{Q})$ gets identified with the standard lattice of column vectors. The form $\langle \cdot, \cdot \rangle$ then transfers to column vectors with the definition

$$\left\langle \sum_i a_i e^{(i)}, \sum_j a'_j e^{(j)} \right\rangle = \sum_{i,j} a_i c_{ij} a'_j.$$

Then

$$\left\langle \sum_k b_k v^{(k)}, \sum_l b_l v^{(l)} \right\rangle = \left\langle \sum_{k,i} b_k v_i^{(k)} e^{(i)}, \sum_{l,j} b_l v_j^{(l)} e^{(j)} \right\rangle$$
$$= \sum_{k,l} b_k b_l \sum_{i,j} v_i^{(k)} c_{ij} v_j^{(l)}$$
$$= \sum_{k,l} b_k b_l \sum_i \lambda^{(l)} v_i^{(k)} v_i^{(l)}$$
$$= \sum_k \lambda^{(k)} b_k^2 \qquad (4.59)$$

since the $v^{(k)}$ are orthonormal.

Let ϵ be the minimum value of $h(P) = \langle P, P \rangle$ for $P \notin F$. This exists and is > 0 since $\{P \,|\, h(P) \leq C\}$ is always finite. Let E be the compact convex set of column vectors

$$E = \left\{ \sum_{k=1}^{r} b_k v^{(k)} \,\bigg|\, \max_{1 \leq k \leq r-1} |b_k| \leq \left(\frac{\epsilon}{2r\lambda^{(1)}}\right)^{1/2} \text{ and } |b_r| \leq M \right\},$$

with M chosen so large that $\text{vol}(E) > 4^r$. By Lemma 4.16, E contains a nonzero lattice point $\sum b_k v^{(k)} \leftrightarrow P$. But then (4.59) gives

$$h(P) = \left\langle \sum_k b_k v^{(k)}, \sum_l b_l v^{(l)} \right\rangle = \sum_{k=1}^{r} \lambda^{(k)} b_k^2$$

$$= \sum_{k=1}^{r-1} \lambda^{(k)} b_k^2 \qquad \text{since } \lambda^{(r)} = 0$$

$$\leq \sum_{k=1}^{r-1} \lambda^{(1)} \left(\frac{\epsilon}{2r\lambda^{(1)}}\right) \qquad \text{since } \sum b_k v^{(k)} \text{ is in } E$$

$$= \frac{\epsilon}{2},$$

and we have a contradiction. Thus $\lambda^{(r)} > 0$, and (c_{ij}) is definite.

The matrix $(c_{ij}) = (\langle P_i, P_j \rangle)$ depends on the choice of the basis $\{P_i\}$ of $\mathbf{Z}^r$, but its determinant does not. The **elliptic regulator** of E over $\mathbf{Q}$ is the number

$$R_{E/\mathbf{Q}} = \det(\langle P_i, P_j \rangle).$$

Proposition 4.17 shows that $R_{E/\mathbf{Q}} > 0$. This number enters into the statement of the third form of the Birch and Swinnerton-Dyer Conjecture discussed in Chapter I.

We can now give the geometric interpretation of rank. We use the symbol $\sim$ to denote "asymptotic to," in the sense that the ratio tends to 1.

Proposition 4.18. With E as in (4.53), the following formula is valid as $T \to \infty$:

$$\#\{(x,y) \in E(\mathbf{Q}) \,|\, |x|_\infty \leq T\} \begin{cases} = \#(F) & \text{if } r = 0 \\ \sim \dfrac{\#(F)\Omega_r}{R_{E/\mathbf{Q}}^{1/2}} (\log T)^{r/2} & \text{if } r > 0. \end{cases}$$

Here $\#(\cdot)$ refers to the number of elements in a set, and Ω_r is the volume of the unit ball in $\mathbf{R}^r$.

PROOF. We may assume $r > 0$. Let (c_{ij}) be the positive definite matrix in (4.54). With $e^{(i)}$ as the standard basis of $\mathbf{R}^r$, we borrow the following fact from Euclidean Fourier analysis: The number of lattice points $\sum n_i e^{(i)}$ with $\sum n_i c_{ij} n_j \le t$ is

$$\sim \frac{\Omega_r t^{r/2}}{\det(c_{ij})^{1/2}}$$

as $t \to \infty$. Consequently the number of points $\sum n_i P_i + f$ in $E(\mathbf{Q}) = \mathbf{Z}^r \oplus F$ with $h(\sum n_i P_i + f) \le t$ is

$$\sim \frac{\#(F)\Omega_r t^{r/2}}{R_{E/\mathbf{Q}}^{1/2}}$$

as $t \to \infty$. Since $h - h_0$ is bounded and $r > 0$, the same asymptotic estimate is valid for h_0. Taking $t = \log T$ and using the definition of h_0, we obtain the proposition.

7. Upper Bound on the Rank

Although Proposition 4.18 does give a limit formula for the rank of $E(\mathbf{Q})$, the formula is not very practical as a computational tool, even for getting an idea of what the rank is. In this section we shall take a different approach to estimating the rank, namely by sharpening the bound on the order of $E(\mathbf{Q})/2E(\mathbf{Q})$.

For simplicity we return to the situation of §3, where the elliptic curve E is given by (4.17) and the roots α, β, γ of the cubic polynomial are integers. We shall obtain an upper bound for the rank that is sharp in a number of cases. The same principle applies in the more general setting of §4 (where α, β, γ are not necessarily integers), but the upper bound is inclined to be too big.

In the situation of §3, let us notice an easy upper bound for the rank r. Proposition 4.8 shows that the order of $E(\mathbf{Q})/2E(\mathbf{Q})$ is bounded above by 2 to the power

$$s = 2 + 2\#\{\text{primes } p \mid p \mid d\},$$

d being the discriminant. On the other hand, the torsion subgroup F of $E(\mathbf{Q})$ contains $\mathbf{Z}_2 \oplus \mathbf{Z}_2$ as a subgroup and, because of the structure of $E(\mathbf{R})$, contains no more than two cyclic summands of even order. It is thus the direct sum of a group of odd order and two cyclic groups of even order. Consequently F contributes 2^2 to the order of $E(\mathbf{Q})/2E(\mathbf{Q})$. Meanwhile the free abelian part $\mathbf{Z}^r$ of $E(\mathbf{Q})$ contributes 2^r to the order of $E(\mathbf{Q})/2E(\mathbf{Q})$. Thus

$$r + 2 \leq s = 2 + 2\#\{\text{primes } p \mid p \mid d\},$$

and our easy upper bound is

$$r \leq 2\#\{\text{primes } p \mid p \mid d\}. \tag{4.60}$$

Our objective is to sharpen (4.60). Let us call

p	**good**	if $p \nmid d$
p	**fairly bad**	if p divides exactly one of $\alpha - \beta, \beta - \gamma, \alpha - \gamma$
p	**very bad**	if p divides all three of $\alpha - \beta, \beta - \gamma, \alpha - \gamma$.

These notions are related to the discussion in §III.5, but we omit the details. Let

$$n_1 = \text{number of fairly bad primes}$$
$$n_2 = \text{number of very bad primes}$$

Our first improvement of (4.60) is contained in the following proposition.

Proposition 4.19. Let E be the elliptic curve (4.17) over $\mathbf{Q}$ with integer roots α, β, γ, and let r be the rank of $E(\mathbf{Q})$. Then

$$r \leq n_1 + 2n_2 - 1. \tag{4.61}$$

PROOF. We go over the proof of Proposition 4.8 in order to reduce the image of

$$E(\mathbf{Q})/2E(\mathbf{Q}) \longrightarrow \sum_{\pm, p \mid d} \oplus (\mathbf{Z}_2 \oplus \mathbf{Z}_2) \tag{4.62}$$

while keeping the mapping one-one. We shall show that the image in the $\pm$ coordinate lies in a subgroup $\mathbf{Z}_2$ of $\mathbf{Z}_2 \oplus \mathbf{Z}_2$ and the same is true of the pth coordinate if p is fairly bad. Then the same analysis that gave (4.60) will now give (4.61).

First consider the $\pm$ coordinate. The roots α, β, γ are integers and can be ordered. Let us say $\alpha < \beta < \gamma$, so that

$$x - \alpha > x - \beta > x - \gamma. \tag{4.63}$$

7. UPPER BOUND ON THE RANK

If $x \notin \{\alpha, \beta, \gamma\}$, the possible signs in (4.63) are apparently $+++$, $++-$, $+--$, and $---$. For a point (x,y) in $E(\mathbf{Q})$, the product $(x-\alpha)(x-\beta)(x-\gamma)$ equals a square (namely y^2), and thus only $+++$ and $+--$ can occur. Therefore the $\pm$ coordinate of the image of φ_α in (4.20) and (4.25) is 0, and the first factor of $\mathbf{Z}_2$ can be dropped. [Note that if (4.25) used $\varphi_\alpha \times \varphi_\gamma$ or $\varphi_\beta \times \varphi_\gamma$, then the image in the $\pm$ coordinate would still be in a $\mathbf{Z}_2$ subgroup of $\mathbf{Z}_2 \oplus \mathbf{Z}_2$, namely $0 \oplus \mathbf{Z}_2$ and $\text{diag}(\mathbf{Z}_2)$ in the respective cases.]

Actually the above argument applies only when $x \notin \{\alpha, \beta, \gamma\}$. When $x = \alpha$, (4.19) shows that the $\pm$ component of $\varphi_\alpha(x)$ is obtained from $\text{sgn}[(\alpha - \beta)(\alpha - \gamma)] = +1$ and hence is trivial. When x is β or γ, then the $\pm$ component of $\varphi_\alpha(x)$ is obtained from $\text{sgn}(x - \alpha) = +1$ and hence is trivial.

Now consider the p coordinate when p is fairly bad. First let us assume that $p \mid (\alpha - \beta)$. Let (x,y) be a point other than ∞ on $E(\mathbf{Q})$, and suppose temporarily that $x \notin \{\alpha, \beta, \gamma\}$. Define integers a, b, c by

$$p^a \| (x - \alpha), \qquad p^b \| (x - \beta), \qquad p^c \| (x - \gamma).$$

In (4.26) we saw that $a + b + c \equiv 0 \mod 2$. Also we saw that if any of a, b, c is < 0, then $a \equiv b \equiv c \equiv 0 \mod 2$, so that the image of $\varphi_\alpha \times \varphi_\beta$ in the pth coordinate is $(a \mod 2, b \mod 2) = (0, 0)$.

Suppose $a > 0$, so that $p \mid (x - \alpha)$ in the sense that the numerator of $x - \alpha$ has p as a factor. Since $p \nmid (\alpha - \gamma)$, we have $p \nmid (x - \gamma)$. Thus $c = 0$ and $a + b \equiv 0 \mod 2$, and the image of $\varphi_\alpha \times \varphi_\beta$ in the pth coordinate is contained in $\text{diag}(\mathbf{Z}_2)$. Suppose instead that $b > 0$, so that $p \mid (x - \beta)$. Since $p \nmid (\beta - \gamma)$, we have $p \nmid (x - \gamma)$. Thus $c = 0$, $a + b \equiv 0 \mod 2$, and the image is in $\text{diag}(\mathbf{Z}_2)$. Finally suppose instead that $c > 0$, so that $p \mid (x - \gamma)$. Since $p \nmid (\gamma - \alpha)$ and $p \nmid (\gamma - \beta)$, we have $p \nmid (x - \alpha)$ and $p \nmid (x - \beta)$. Thus $a = b = 0$, and the image 0 is in $\text{diag}(\mathbf{Z}_2)$.

Exceptional cases occur when $x \in \{\alpha, \beta, \gamma\}$. If $x = \alpha$, then $\varphi_\alpha(x) = (\alpha - \beta)(\alpha - \gamma)$ and $\varphi_\beta(x) = \beta - \alpha$. Since $\varphi_\alpha(x)/\varphi_\beta(x)$ is not divisible by p, the image of $\varphi_\alpha(x) \times \varphi_\beta(x)$ in the pth coordinate is again contained in $\text{diag}(\mathbf{Z}_2)$. If $x = \beta$ instead, then $\varphi_\alpha(x) = \beta - \alpha$ and $\varphi_\beta(x) = (\beta - \alpha)(\beta - \gamma)$. Since $\varphi_\beta(x)/\varphi_\alpha(x)$ is not divisible by p, the image of $\varphi_\alpha(x) \times \varphi_\beta(x)$ in the pth coordinate is contained in $\text{diag}(\mathbf{Z}_2)$. If $x = \gamma$ instead, then $\varphi_\alpha(x) = (\gamma - \alpha)$ and $\varphi_\beta(x) = (\gamma - \beta)$ are not divisible by p. So $a = b = 0$, and the image 0 is in $\text{diag}(\mathbf{Z})$.

We have been assuming that $p \mid (\alpha - \beta)$. If $p \mid (\beta - \gamma)$ instead, the pth coordinate of the image of $\varphi_\alpha \times \varphi_\beta$ is in $0 \oplus \mathbf{Z}_2$, while if $p \mid (\alpha - \gamma)$, the pth coordinate of the image of $\varphi_\alpha \times \varphi_\beta$ is in $\mathbf{Z}_2 \oplus 0$. In each case the image is confined to a $\mathbf{Z}_2$ subgroup of $\mathbf{Z}_2 \oplus \mathbf{Z}_2$. This completes the proof.

Once again let us return to congruent numbers. The integer n is **congruent** if n is the area of a rational right triangle. Let us borrow the following lemma from Theorem 5.2.

Lemma 4.20. If n is a square-free integer and E is the elliptic curve
$$y^2 = x^3 - n^2 x,$$
then the torsion subgroup of $E(\mathbf{Q})$ is $\mathbf{Z}_2 \oplus \mathbf{Z}_2$.

Proposition 4.21. A square-free integer n fails to be congruent if and only if the elliptic curve
$$E : y^2 = x^3 - n^2 x,$$
has the property that $E(\mathbf{Q})$ has rank 0.

PROOF. If $E(\mathbf{Q})$ has rank 0, it is a torsion group, and Lemma 4.20 shows it to be $\mathbf{Z}_2 \oplus \mathbf{Z}_2$. By (1) $\Rightarrow$ (3) in Proposition 3.1, n is not congruent. Conversely if n is not congruent, Corollary 4.3 shows that E has no $\mathbf{Q}$ rational point other than $(-n, 0)$, $(0, 0)$, $(n, 0)$ and ∞. Hence $E(\mathbf{Q})$ has rank 0.

Corollary 4.22 (Fermat). $n = 1$ is not a congruent number.

PROOF. The elliptic curve E with $y^2 = x^3 - x$ is of the form considered in Proposition 4.19, with $\{\alpha, \beta, \gamma\} = \{-1, 0, 1\}$. All primes are good except $p = 2$, which is fairly bad. Thus $n_1 = 1$ and $n_2 = 0$. By Proposition 4.19, $E(\mathbf{Q})$ has rank 0. By Proposition 4.21, $n = 1$ is not congruent.

We already saw in Corollary 4.4 that $n = 2$ is not a congruent number. If we attempt to reprove this result from Propositions 4.19 and 4.21, we run into a problem. The relevant elliptic curve is $y^2 = x^3 - 4x$, with $\{\alpha, \beta, \gamma\} = \{-2, 0, 2\}$. All primes are good except $p = 2$, which is very bad. Thus $n_1 = 0$ and $n_2 = 1$, and Proposition 4.19 gives us the estimate $r \leq 1$ for the rank. We do not immediately get $r = 0$.

However, an even more careful analysis of the proof of Proposition 4.8 will give the sharper estimate. We need to study the relationship between the $\pm$ coordinate of $\varphi_\alpha \times \varphi_\beta$ and the $p = 2$ coordinate. In Table 4.1 below, we list the effect of $\varphi = \varphi_\alpha, \varphi_\beta, \varphi_\gamma$ on an $x(P)$ other than α, β, γ. Each column for $\pm$ gives a conceivable configuration of signs, and each column for $p = 2$ gives a conceivable set of residues modulo 2 for the power of 2 in $\varphi(x)$. We keep in mind that $(x + 2)x(x - 2)$ has to be a square.

7. UPPER BOUND ON THE RANK

Map	Image mod $\mathbb{Q}^{\times 2}$	$\pm$		2			
φ_α	$x+2$	$+$	$+$	0	0	1	1
φ_β	x	$+$	$-$	0	1	0	1
φ_γ	$x-2$	$+$	$-$	0	1	1	0

TABLE 4.1. Image of φ for $y^2 = x^3 - 4x$ with $x \notin \{-2, 0, 2\}$

At all other primes the image of each φ is a square. Conceivably all 8 combinations of a column for $\pm$ and a column for $p = 2$ could occur; actually we shall see that there is a restriction.

The idea is to adjust P by an element of order 2 to make the $p = 2$ column trivial. For $P = (\alpha, 0), (\beta, 0), (\gamma, 0)$ the corresponding information is assembled in Table 4.2.

Map	Image of $(-2,0)$	$\pm$	2	Image of $(0,0)$	$\pm$	2	Image of $(2,0)$	$\pm$	2
φ_α	8	$+$	1	2	$+$	1	4	$+$	0
φ_β	-2	$-$	1	-4	$-$	0	2	$+$	1
φ_γ	-4	$-$	0	-2	$-$	1	8	$+$	1

TABLE 4.2. Image of φ for $y^2 = x^3 - 4x$ with $x \in \{-2, 0, 2\}$

Since all nontrivial columns appear for $p = 2$ and since each φ is a homomorphism, we can adjust any point P under study by adding an element of $\mathbb{Z}_2 \oplus \mathbb{Z}_2$ so that φ_α, φ_β, and φ_γ all map the adjusted point into $(0, 0)$ in the $p = 2$ coordinate. For the adjusted point, which we still call $P = (x(P), y(P))$, the image of φ is as in Table 4.3.

Map	Image mod $\mathbb{Q}^{\times 2}$	$\pm$		2	Total	
φ_α	$x+2$	$+$	$+$	0	□	□
φ_β	x	$+$	$-$	0	□	$-$ □
φ_γ	$x-2$	$+$	$-$	0	□	$-$ □

TABLE 4.3. Adjusted image of φ for $y^2 = x^3 - 4x$

There are only two possibilities, one of two columns from $\pm$ and the column for $p = 2$.

We shall show that the second possibility, namely

$$x + 2 = \square, \qquad x = -\square, \qquad x - 2 = -\square, \qquad (4.64)$$

does not occur. Then it will follow that $r = 0$. Arguing by contradiction, suppose (4.64) occurs. Subtracting the first two equations gives

$$2 = \square + \square.$$

If these squares have even denominators, then the power of two is the same for each and is even; also the numerators must be odd. Clearing denominators, we obtain

$$0 \equiv \square + \square \mod 8$$

with both squares odd integers, and this is a contradiction. Thus the first two squares in (4.64) have odd denominators. Subtracting the last two equations in (4.64) gives

$$2 = -\square + \square,$$

and the first of these squares has an odd denominator. Therefore so does the second. Clearing denominators, we obtain

$$2m^2 \equiv -\square + \square \mod 8 \qquad (4.65)$$

with m an odd integer and both squares equal to integers. Then $m^2 \equiv 1 \mod 8$, and (4.65) is impossible. Thus (4.64) cannot occur, and $y^2 = x^3 - 4x$ has rank 0. This argument has given a different proof that $n = 2$ is not congruent.

A similar technique gives refined information about the rank of $E(\mathbb{Q})$ for $y^2 = x^3 - p^2 x$, where p is an odd prime. The result is as follows.

Proposition 4.23. Let p be an odd prime, and let E be the elliptic curve

$$y^2 = x^3 - p^2 x.$$

Then the rank r of $E(\mathbb{Q})$ satisfies

$$\begin{aligned} r &\leq 2 && \text{if } p \equiv 1 \mod 8 \\ r &= 0 && \text{if } p \equiv 3 \mod 8 \\ r &\leq 1 && \text{if } p \equiv 5 \text{ or } 7 \mod 8. \end{aligned}$$

Consequently any $p \equiv 3 \mod 8$ is not a congruent number.

7. UPPER BOUND ON THE RANK

REMARK. The proof will use some facts about quadratic residues from elementary number theory.

PROOF. We argue as above, taking $\{\alpha, \beta, \gamma\} = \{-p, 0, p\}$. All primes are good except 2 and p; 2 is fairly bad, and p is very bad. The information analogous to what is in Table 4.1 is assembled as Table 4.4.

Map	Image mod $\mathbb{Q}^{\times 2}$	$\pm$		2		p			
φ_α	$x + p$	+	+	0	1	0	0	1	1
φ_β	x	+	−	0	0	0	1	0	1
φ_γ	$x - p$	+	−	0	1	0	1	1	0

TABLE 4.4. Image of φ for $y^2 = x^3 - p^2 x$ with $x \notin \{-p, 0, p\}$

If we make a table like Table 4.2 that lists what happens for the torsion points $(-p, 0)$, $(0, 0)$, and $(p, 0)$, we find that the torsion points account for all nontrivial columns for the prime p in Table 4.4. Since each φ is a homomorphism, we can adjust any point P under study so that φ_α, φ_β, and φ_γ all map the adjusted point into $(0, 0)$ in the p coordinate. For the adjusted point, which we still call $P = (x(P), y(P))$, Table 4.5 shows the four possibilities for the combined contribution from a column for $\pm$ and a column for 2. The point is that certain columns listed under "Total," depending on p, cannot occur. If only two columns can occur, then $r \leq 1$. If only one column can occur, then $r = 0$.

Map	Image mod $\mathbb{Q}^{\times 2}$	Total			
φ_α	$x + p$	□	□	2□	2□
φ_β	x	□	−□	□	−□
φ_γ	$x - p$	□	−□	2□	−2□

TABLE 4.5. Adjusted image of φ for $y^2 = x^3 - p^2 x$

Suppose $\begin{pmatrix} \Box \\ -\Box \\ -\Box \end{pmatrix}$ occurs. Subtracting the expressions for $x+p$ and $x-p$, we obtain $2p = \Box + \Box$. If p occurs in the numerator of one square, then p^2 occurs in the numerators of both squares and $p^2|2p$ gives a contradiction. Thus the numerators of the squares are prime to p. Clearing fractions and reducing modulo p, we obtain $0 \equiv \Box + \Box$ mod p, with both squares prime to p. Then -1 is a square modulo p, and it follows that $p \equiv 1$ mod 4. Thus

$$\begin{pmatrix} \Box \\ -\Box \\ -\Box \end{pmatrix} \text{ occurring} \implies p \equiv 1 \text{ or } 5 \mod 8. \tag{4.66}$$

Suppose $\begin{pmatrix} 2\Box \\ \Box \\ 2\Box \end{pmatrix}$ occurs. The first two entries give $p = 2\Box - \Box$. Arguing as in the previous paragraph, we are led to $0 \equiv 2\Box - \Box$ mod p, with both squares prime to p. Then 2 is a square modulo p, and it follows that $p \equiv \pm 1$ mod 8. Thus

$$\begin{pmatrix} 2\Box \\ \Box \\ 2\Box \end{pmatrix} \text{ occurring} \implies p \equiv 1 \text{ or } 7 \mod 8. \tag{4.67}$$

Suppose $\begin{pmatrix} 2\Box \\ -\Box \\ -2\Box \end{pmatrix}$ occurs. The second and third entries give $p = 2\Box - \Box$, and the previous paragraph again shows that $p \equiv 1$ or 7 mod 8. The first and second entries give $p = 2\Box + \Box$. Arguing as above, we conclude that -2 is a square modulo p, and it follows that $p \equiv 1$ or 3 mod 8. Thus

$$\begin{pmatrix} 2\Box \\ -\Box \\ -2\Box \end{pmatrix} \text{ occurring} \implies p \equiv 1 \mod 8. \tag{4.68}$$

In view of how the number of occurring columns affects the rank, the implications (4.66), (4.67), and (4.68) prove the proposition.

8. Construction of Points in $E(\mathbf{Q})$

The refinements to the method of descent in §7 are not good enough to produce members of $E(\mathbf{Q})$ with certainty, but they are excellent at cutting down the possibilities enough to make a computer search manageable. We shall consider two examples in this section. The first example will go in the converse direction to Proposition 4.23 by showing that all small primes $p \equiv 5 \bmod 8$ are congruent numbers (i.e., $r = 1$ in Proposition 4.23). The second example will solve Fermat's problem for Mersenne that was introduced in Subsection E of §III.1.

A. Congruent numbers

For the first example, fix a prime $p \equiv 5 \bmod 8$, According to (4.66), if p is congruent, then we can write

$$\begin{pmatrix} x+p \\ x \\ x-p \end{pmatrix} = \begin{pmatrix} \Box \\ -\Box \\ -\Box \end{pmatrix}$$

for suitable squares in $\mathbf{Q}$ and for some $x = x(P)$. To quantify matters write

$$\begin{pmatrix} x+p \\ x \\ x-p \end{pmatrix} = \begin{pmatrix} a^2 \\ -b^2 \\ -c^2 \end{pmatrix}. \tag{4.69}$$

Then we must have $a^2 - c^2 = -2b^2$, which we write as

$$a^2 = c^2 - 2b^2. \tag{4.70}$$

We parametrize the solutions by the method of Diophantus in Chapter I: We change the equation from projective to affine form by writing it as

$$1 = u^2 - 2v^2. \tag{4.71}$$

Then $(u, v) = (1, 0)$ gives one solution. A line through $(1, 0)$ is $u = tv+1$ for any choice of t in $\mathbf{Q}$. Substituting, we obtain

$$1 = (tv + 1)^2 - 2v^2,$$

which is solved by $v = 0$ and $v = \dfrac{2t}{2 - t^2}$. Thus our parametrization of solutions of (4.71) is

$$(u, v) = \left(\frac{2 + t^2}{2 - t^2}, \frac{2t}{2 - t^2} \right), \quad t \in \mathbf{Q}.$$

Write $t = r/s$ in lowest terms. Then

$$\frac{1}{a}(a,b,c) = (1,v,u) = \frac{1}{2s^2 - r^2}(2s^2 - r^2, 2rs, 2s^2 + r^2),$$

and

$$\begin{aligned} a &= (2s^2 - r^2)\lambda \\ b &= (2rs)\lambda \\ c &= (2s^2 + r^2)\lambda \end{aligned} \qquad (4.72)$$

for some $\lambda \in \mathbb{Q}$.

We shall show that λ is of the form $\frac{1}{n}$. In fact, say $\lambda = \frac{m}{n}$ in lowest terms. Then (4.72) shows that m divides the integers na and nb. Hence m divides a and b. Referring to (4.69), we see that m^2 divides $a^2 + b^2 = p$. Thus $m = \pm 1$ and we may take $m = 1$.

Since $\lambda = \frac{1}{n}$, the equation

$$p = a^2 + b^2 = (r^4 + 4s^4)\lambda^2$$

shows that

$$r^4 + 4s^4 = pn^2. \qquad (4.73)$$

At this stage we have thinned out the possibilities considerably. We look for a solution (r, s, n) of (4.73), and it determines (a, b, c) in (4.72), since $\lambda = \frac{1}{n}$. Then (a, b, c) determines an x of the correct form so that (4.69) holds, and x is of the form $x(P)$ for a point P in $E(\mathbb{Q})$. By letting r and s range from 1 to 50 and seeing whether the square root of $p^{-1}(r^4 + 4s^4)$ is very close to an integer, we are led to all the entries in Table 4.6 except the one for $p = 53$. Since r and s (and ultimately $x(P)$) exist for each p in the table, all these numbers p are congruent.

8. CONSTRUCTION OF POINTS IN $E(\mathbb{Q})$

p	r	s	n	$x(P)$
5	1	1	1	-4
13	1	3	5	$-\dfrac{36}{25}$
29	7	5	13	$-\dfrac{4900}{169}$
37	1	21	145	$-\dfrac{1764}{21025}$
53	286	119	11890	$-\dfrac{1158313156}{35343025}$
61	41	39	445	$-\dfrac{10227204}{198025}$
109	6	7	10	$-\dfrac{1764}{25}$
149	14	17	50	$-\dfrac{56644}{625}$

TABLE 4.6. Some primes $p \equiv 5 \mod 8$ that are congruent numbers

To discover (r, s, n) for $p = 53$ in the table, we refine the argument still further. We are trying to solve

$$(r^2)^2 + 4(s^2)^2 = 53n^2.$$

Consider the affine curve

$$j^2 + 4k^2 = 53$$

over $\mathbb{Q}$. One solution is $(j, k) = (7, 1)$, and the method of Diophantus in Chapter I will allow us to parametrize all solutions. A line through $(7, 1)$ is $(k - 1) = t(j - 7)$ for any choice of t. Substituting for k, we obtain a formula for j in terms of t. Then we can recover k. The result is

$$(j, k) = \left(\frac{28t^2 - 8t - 7}{4t^2 + 1}, \frac{-4t^2 - 14t + 1}{4t^2 + 1} \right),$$

which we can reparametrize as

$$(j, k) = \left(\frac{7t^2 - 4t - 7}{t^2 + 1}, \frac{-t^2 - 7t + 1}{t^2 + 1} \right).$$

Putting $t = d/e$ in lowest terms gives

$$(r^2, s^2, n) = \mu(7d^2 - 4de - 7e^2, -d^2 - 7de + e^2, d^2 + e^2). \qquad (4.74)$$

From the third entry, we see that $\mu > 0$. The numerator of μ can be taken to be 1 since $\text{GCD}(r, s) = 1$. If the denominator is h, then h divides

$$7d^2 - 4de - 7e^2 + 7(-d^2 - 7de + e^2) = -53de \qquad \text{and} \qquad d^2 + e^2.$$

Any prime factor of h must divide $53de$. If it divides d or e, then it divides $d^2 + e^2$ and hence both d and e, contradiction. Thus $\mu = \frac{1}{53}$ or $\mu = 1$.

Handling $\mu = \frac{1}{53}$ does not look very easy. So let us look for a solution with $\mu = 1$. Our equality is then

$$(r^2, s^2, n) = (7d^2 - 4de - 7e^2, -d^2 - 7de + e^2, d^2 + e^2). \qquad (4.75)$$

We analyze

$$-d^2 - 7de + e^2 = \square \quad (= s^2)$$

by the method of Diophantus. The affine equation is

$$-d'^2 - 7d'e' + e'^2 = 1,$$

and $(d', e') = (0, 1)$ is a solution. The line $e' = d't + 1$ leads to

$$(d', e') = \left(\frac{2t - 7}{-t^2 + 7t + 1}, \frac{t^2 + 1}{-t^2 + 7t + 1} \right).$$

If $t = \xi/\eta$ in lowest terms, then

$$(d, e, s) = \nu(2\xi\eta - 7\eta^2, \xi^2 + \eta^2, -\xi^2 + 7\xi\eta + n^2). \qquad (4.76)$$

In the same way as for μ in (4.74), we find $\nu = \frac{1}{53}$ or $\nu = 1$.

We look for a solution with $\nu = 1$. Using a short computer search with small values of ξ and η and with d and e determined by (4.76), we check whether the first entry $7d^2 - 4de - 7e^2$ of (4.75) is a square, so that we can take it as r^2. For $\xi = 10$ and $\eta = 3$, we do get a square, and the result is what is listed in Table 4.6.

The reader may wish to try such an analysis for $p = 101$, which is the only $p \equiv 5 \mod 8$ that is < 150 and is missing from Table 4.6.

8. CONSTRUCTION OF POINTS IN $E(\mathbb{Q})$

B. Fermat's problem to Mersenne

We seek an integer-valued relatively prime Pythagorean triple (X, Y, Z) such that the hypotenuse Z is a square and the sum of the legs is a square. In other words we want

$$X^2 + Y^2 = Z^2, \qquad Z = b^2, \qquad X + Y = a^2.$$

With $e = X - Y$, we obtain X and Y as $\frac{1}{2}(a^2 \pm e)$, and (3.17) showed the problem comes down to solving

$$2b^4 - a^4 = e^2. \tag{4.77}$$

A solution in the style of Fermat proceeds as follows. We can work a little sloppily, apparently discarding possibilities, because all we need is one solution. From $X^2 + Y^2 = Z^2$, we write

$$X = m^2 - n^2, \qquad Y = 2mn, \qquad Z = m^2 + n^2 \tag{4.78}$$

by Theorem 1.2. (Actually this form of a solution assumes that X is odd; but if it works, it works.) From $b^2 = m^2 + n^2$ we have

$$m = r^2 - s^2, \qquad n = 2rs, \qquad b = r^2 + s^2. \tag{4.79}$$

Meanwhile, the Diophantus method of Chapter I leads from

$$a^2 = X + Y = (m+n)^2 - 2n^2$$

to

$$m + n = t^2 + 2u^2, \qquad n = 2tu, \qquad a = t^2 - 2u^2,$$

hence to

$$m = t^2 - 2tu + 2u^2, \qquad n = 2tu, \qquad a = t^2 - 2u^2. \tag{4.80}$$

Equating the two expressions for n in (4.79) and (4.80), we obtain $rs = tu$. Thus $\frac{r}{t} = \frac{u}{s}$, and we can write this as $\frac{d}{c}$ in lowest terms. Then

$$r = kd, \qquad t = kc, \qquad u = ld, \qquad s = lc. \tag{4.81}$$

Equating the two expressions for m in (4.79) and (4.80), we obtain

$$r^2 - s^2 = m = t^2 - 2tu + 2u^2.$$

Substitution from (4.81) gives
$$k^2d^2 - l^2c^2 = k^2c^2 - 2klcd + 2l^2d^2$$
and hence
$$0 = (c^2 + 2d^2)\left(\frac{l}{k}\right)^2 - 2cd\left(\frac{l}{k}\right) + (c^2 - d^2). \qquad (4.82)$$

For l/k to be in $\mathbf{Q}$, the discriminant must be a square in $\mathbf{Q}$. Thus
$$c^2d^2 - (c^2 + 2d^2)(c^2 - d^2) = \Box,$$
which we rewrite as
$$2d^4 - c^4 = f^2. \qquad (4.83)$$

This equation is of the same form as (4.77), but the integers are usually much smaller here. At first it looks as if the method of descent is leading us to a proof that there is no solution. But there is a bottom nontrivial solution with $(c,d) = (1,1)$. Substituting into (4.82), we obtain $\frac{l}{k} = 0$ or $\frac{2}{3}$. The solution $\frac{l}{k} = 0$ leads back to $(a,b) = (1,1)$, which is not interesting. From $\frac{l}{k} = \frac{2}{3}$, we have $l = 2$, $k = 3$. Going backwards, we calculate

$$\begin{array}{llll}
r = 3 & m = 5 & X = -119 \\
t = 3 & n = 12 & Y = 120 \\
u = 2 & a = 1 & Z = 169 \\
s = 2 & b = 13 & e = -239.
\end{array}$$

We must discard the result of this trial since X is negative.

But we can continue. The (a,b) from this trial becomes the (c,d) of our next trial. Putting $(c,d,f) = (1, 13, -239)$, we have
$$\frac{l}{k} = \frac{cd \pm f}{c^2 + 2d^2} = \frac{13 \mp 239}{1 + 2 \cdot 169} = -\frac{2}{3} \text{ and } \frac{84}{113}.$$

From $\frac{l}{k} = -\frac{2}{3}$, we have $l = -2$, $k = 3$. Going backwards, we calculate

$$\begin{array}{llll}
r = 39 & m = 1517 & X = 2276953 \\
t = 3 & n = -156 & Y = -473304 \\
u = -26 & a = -1343 & Z = 2325625 \\
s = -2 & b = 1525 & e = 2750257.
\end{array}$$

8. CONSTRUCTION OF POINTS IN $E(\mathbf{Q})$

This time we have a minus sign in Y and discard the result.
We try instead $\frac{84}{113}$, using $l = 84$, $k = 113$. Then

$$
\begin{aligned}
r &= 1469 & m &= 2150905 & X &= 4565486027761 \\
t &= 113 & n &= 246792 & Y &= 1061652293520 \\
u &= 1092 & a &= -2372159 & Z &= 4687298610289 \\
s &= 84 & b &= 2165017 & e &= 3503833734241.
\end{aligned}
$$

This is the solution given in (3.16).

To analyze the solution technique, we work backwards, obtaining a formula for (a, b) in terms of (c, d). Actually we had a choice of a sign for $\frac{l}{k}$, and we may regard (a, b, e) as depending on (c, d, f). The correspondence will be nicer if we choose the ambiguous sign in $\frac{l}{k}$ so that

$$\frac{l}{k} = \frac{cd - f}{c^2 + 2d^2}.$$

With this choice made, (a, b, e) becomes a function of (c, d, f). Meanwhile the tuples (a, b, e) and (c, d, f), which give $\mathbf{Z}$ solutions of (3.17), then give $\mathbf{Q}$ solutions of (3.18) and, by virtue of (3.19) and its inverse (3.21), $\mathbf{Q}$ rational points on the elliptic curve $y^2 = x^3 + 8x$.

Table 4.7 shows what the effect of $(c, d, f) \to (a, b, e)$ is, in terms of the group law for the elliptic curve. The table uses the abbreviations $P_0 = (8, -24)$ and $T = (0, 0)$. The point P_0 generates an infinite cyclic subgroup of $E(\mathbf{Q})$, and T has order two.

c	d	f	$E(\mathbf{Q})$ Point	a	b	c	$E(\mathbf{Q})$ Point
1	1	1	O	1	1	1	O
1	1	-1	P_0	1	13	-239	$2P_0$
1	-1	1	$P_0 + T$	1	13	-239	$2P_0$
1	-1	-1	T	1	1	1	O
1	13	239	$-P_0$	-1343	1525	2750257	$-2P_0$
1	13	-239	$2P_0$	-2372159	2165017	3503833734241	$4P_0$
1	-13	239	$2P_0 + T$	-2372159	2165017	3503833734241	$4P_0$
1	-13	-239	$-P_0 + T$	-1343	1525	2750257	$-2P_0$

TABLE 4.7. Effect on $E : y^2 = x^3 + 8x$ of Fermat's descent for $2b^4 - a^4 = e^2$

The table shows that Fermat's descent genuinely corresponds to the passage $P \to \frac{1}{2}P$ for the points in question. The reader may wish to carry out the symbolic manipulations appropriate for general P.

9. Appendix on Algebraic Number Theory

The object of this section is to prove Theorem 4.10, which provides a tool that was used in the proof of Mordell's Theorem. Theorem 4.10 requires a certain amount of algebraic number theory as preparation, including three basic theorems. We shall state the necessary preliminary results with most of the proofs omitted. References are given in the chapter of "Notes."

We regard $\bar{\mathbf{Q}}$ as a subfield of $\mathbf{C}$, the subfield of elements that are roots of polynomials $P(X)$ with $\mathbf{Z}$ coefficients. The members of $\bar{\mathbf{Q}}$ are called **algebraic numbers**. A **number field** K is a subfield of $\bar{\mathbf{Q}}$ that is finite-dimensional over $\mathbf{Q}$.

An **algebraic integer** is an algebraic number that is the root of a monic polynomial $P(X)$ over $\mathbf{Z}$. The set of such is denoted $\mathcal{O}$. If $\theta \in \bar{\mathbf{Q}}$ is a root of

$$a_n X^n + \cdots + a_1 X + a_0,$$

then $a_n \theta$ is a root of

$$X^n + a_{n-1} X^{n-1} + a_{n-2} a_n X^{n-2} + \cdots + a_1 a_n^{n-2} X + a_0 a_n^{n-1}.$$

Hence

$$\theta \in \bar{\mathbf{Q}} \quad \Longrightarrow \quad k\theta \in \mathcal{O} \text{ for some } k \in \mathbf{Z}. \tag{4.84}$$

One step in the proof of unique factorization for $\mathbf{Z}[X]$ shows that if a monic polynomial in $\mathbf{Z}[X]$ has a monic factor in $\mathbf{Q}[X]$, then that factor is actually in $\mathbf{Z}[X]$; consequently the minimal polynomial over $\mathbf{Q}$ of any $\theta \in \mathcal{O}$ has integer coefficients.

The set $\mathcal{O}$ of algebraic integers is actually a ring. The standard proof of this fact uses the theory of symmetric polynomials. A simpler proof uses the following lemma, which is handy also for proving (4.88) below.

Lemma 4.24. Let V be the additive group generated by complex numbers $x_1, \ldots, x_l$ that are not all 0. If α is a complex number such that $\alpha x \in V$ whenever $x \in V$, then α is an algebraic integer.

Let us write "left-by-α" for the operation of left multiplication by α. For an example of a group V as in the lemma, let $\alpha \in \mathcal{O}$ satisfy $\alpha^n + a_{n-1}\alpha^{n-1} + \cdots + a_0 = 0$. Then left-by-$\alpha$ maps the $\mathbf{Z}$ span of

9. APPENDIX ON ALGEBRAIC NUMBER THEORY

$\alpha^{n-1}, \ldots, \alpha, 1$ into itself. When there are two such elements in $\mathcal{O}$ and we use the products of their powers to generate V, the result is the following proposition.

Proposition 4.25. $\mathcal{O}$ is a ring.

Fix a number field K. It follows from the proposition that $\mathcal{O}_K = \mathcal{O} \cap K$ is a ring, the ring of **algebraic integers in** K. From (4.84) it follows that K is the field of fractions of $\mathcal{O}_K$. The fact that the only rational roots of a monic polynomial in $\mathbf{Z}[X]$ are in $\mathbf{Z}$ translates into the equality

$$\mathcal{O} \cap \mathbf{Q} = \mathbf{Z}. \tag{4.85}$$

Let $n = [K : \mathbf{Q}]$. If $\{\alpha_1, \ldots, \alpha_n\}$ is a basis of K over $\mathbf{Q}$ and if α is in K, we can write $\alpha \alpha_i = \sum_j a_{ij} \alpha_j$ with $a_{ij} \in \mathbf{Q}$. We define

$$N_{K/\mathbf{Q}}(\alpha) = \det(a_{ij}) \quad \text{(\textbf{norm} of } \alpha)$$
$$\text{Tr}_{K/\mathbf{Q}}(\alpha) = \text{Tr}(a_{ij}) \quad \text{(\textbf{trace} of } \alpha)$$

These are functions from K to $\mathbf{Q}$. It is easy to check that

(1) $N_{K/\mathbf{Q}}$ and $\text{Tr}_{K/\mathbf{Q}}$ are independent of basis.
(2) $N_{K/\mathbf{Q}}$ is a multiplicative homomorphism from $K^\times$ to $\mathbf{Q}^\times$, and $\text{Tr}_{K/\mathbf{Q}}$ is an additive homomorphism from K to $\mathbf{Q}$.
(3) $\text{Tr}_{K/\mathbf{Q}}(1) = n \neq 0$.

There are exactly n distinct isomorphisms of K into $\bar{\mathbf{Q}}$ fixing $\mathbf{Q}$, and they are given as follows. The Theorem of the Primitive Element says that $K = \mathbf{Q}(\alpha)$ for some $\alpha \in K$. Let

$$P(X) = X^n + c_{n-1} X^{n-1} + \cdots + c_0$$

be the minimal polynomial of α over $\mathbf{Q}$, and let $\alpha_1 = \alpha, \alpha_2, \ldots \alpha_n$ be the (necessarily distinct) roots of P in $\bar{\mathbf{Q}}$. Then

$$\sigma_j(c_{n-1}\alpha^{n-1} + \cdots + c_1\alpha + c_0) = c_{n-1}\alpha_j^{n-1} + \cdots + c_1\alpha_j + c_0.$$

If β is in K, the elements $\sigma_j(\beta)$ (with $\sigma_1(\beta) = \beta$) are called the **conjugates** of β in $\bar{\mathbf{Q}}$. They need not lie in K. However, each $\sigma_j(\beta)$ has the same minimal polynomial as β over $\mathbf{Q}$. Consequently

$$\sigma_j \text{ carries } \mathcal{O}_K \text{ into } \mathcal{O}. \tag{4.86}$$

To bring the conjugates of β into the theory, one uses the following proposition.

Proposition 4.26. For $\beta \in K$ let g be the minimal polynomial of β over $\mathbf{Q}$. Then $(\deg g) \mid n$. Also left-by-β as a $\mathbf{Q}$ linear map from K to K satisfies

$$\det(xI - \text{left-by-}\beta) = g(x)^r = \prod_{j=1}^{n}(x - \sigma_j(\beta)),$$

where $r = n/(\deg g)$.

Using Proposition 4.26, one can show that

$$\begin{aligned}\text{N}_{K/\mathbf{Q}}(\beta) &= \prod_{j=1}^{n}\sigma_j(\beta) \\ \text{Tr}_{K/\mathbf{Q}}(\beta) &= \sum_{j=1}^{n}\sigma_j(\beta).\end{aligned} \tag{4.87}$$

By (4.85) and (4.86), $\text{N}_{K/\mathbf{Q}}$ and $\text{Tr}_{K/\mathbf{Q}}$ carry $\mathcal{O}_K$ into $\mathbf{Z}$.

A **unit** in $\mathcal{O}_K$ is an invertible element. The set of units is denoted $\mathcal{O}_K^\times$. The first of the three basic theorems in beginning algebraic number theory is the Dirichlet Unit Theorem, which gives the structure of $\mathcal{O}_K^\times$. We shall give a weak form of the theorem as Theorem 4.29.

Lemma 4.27. The units in $\mathcal{O}_K$ are the members ε of $\mathcal{O}_K$ with $\text{N}_{K/\mathbf{Q}}(\varepsilon) = \pm 1$.

PROOF. If ε is a unit, then $\text{N}_{K/\mathbf{Q}}(\varepsilon^{-1}) = \text{N}_{K/\mathbf{Q}}(\varepsilon)^{-1} \in \mathbf{Z}$ shows that $\text{N}_{K/\mathbf{Q}}(\varepsilon)$ is an invertible integer, hence is ± 1. Conversely we use (4.86), remembering that $\sigma_1 = 1$. Then

$$\text{N}_{K/\mathbf{Q}}(\varepsilon) = \varepsilon \sigma_2(\varepsilon) \cdots \sigma_n(\varepsilon)$$

shows that

$$\varepsilon^{-1} = \sigma_2(\varepsilon) \cdots \sigma_n(\varepsilon) \text{N}_{K/\mathbf{Q}}(\varepsilon)^{-1}.$$

If $\text{N}_{K/\mathbf{Q}}(\varepsilon) = \pm 1$, then the right side is in $\mathcal{O}$.

Lemma 4.28. The torsion subgroup of $\mathcal{O}_K^\times$ consists of all N^{th} roots of unity $e^{2\pi i l/N}$ for some N that is bounded in terms of $n = [K : \mathbf{Q}]$.

9. APPENDIX ON ALGEBRAIC NUMBER THEORY

One proof of Lemma 4.28 uses properties of cyclotomic polynomials and the formula for the Euler φ function.

The isomorphisms $\sigma_j : K \to \bar{\mathbb{Q}} \subseteq \mathbb{C}$, $1 \leq j \leq n$ are of two types:
(a) those carrying K into $\mathbb{R}$. Say there are r_1 of these.
(b) those carrying K into $\mathbb{C}$ but not $\mathbb{R}$. These come in pairs σ and $\bar{\sigma}$, where the bar refers to complex conjugation. Say there are $2r_2$ of these.

Then $r_1 + 2r_2 = n$. Let $\sigma_1, \ldots, \sigma_{r_1}$ be of the first kind and
$$\sigma_{r_1+1}, \bar{\sigma}_{r_1+1}, \ldots, \sigma_{r_1+r_2}, \bar{\sigma}_{r_1+r_2}$$
be of the second kind. We use each σ_j to form an absolute value on K:
$$\|x\|_j = |\sigma_j(x)| \qquad \text{for } 1 \leq j \leq r_1 + r_2.$$
Then the mapping
$$\mathrm{Log} : x \longrightarrow (\log \|x\|_1, \ldots, \log \|x\|_{r_1+r_2})$$
is a group homomorphism of $K^\times$ into $\mathbb{R}^{r_1+r_2}$.

Theorem 4.29 (Dirichlet Unit Theorem). On $\mathcal{O}_K^\times$, the kernel of Log is the (finite) torsion subgroup, and the image is a discrete subgroup of the subspace of $\mathbb{R}^{r_1+r_2}$ where
$$x_1 + \cdots + x_{r_1} + 2x_{r_1+1} + \cdots + 2x_{r_1+r_2} = 0.$$
Consequently $\mathcal{O}_K^\times$ is a finitely generated group of rank $\leq r_1 + r_2 - 1$.

The finiteness of the torsion subgroup is by Lemma 4.28. The version of the theorem stated here is not very hard to prove. The full version of the Dirichlet Unit Theorem says that the rank of $\mathcal{O}_K^\times$ equals $r_1 + r_2 - 1$, and its proof takes considerably more effort. But we do not need this stronger version.

We turn our attention now to the structure of ideals. If $\alpha_1, \ldots, \alpha_n$ are in K, their **discriminant** is the element of $\mathbb{Q}$ given by
$$\Delta_{K/\mathbb{Q}}(\alpha_1, \ldots, \alpha_n) = \det\left(\mathrm{Tr}_{K/\mathbb{Q}}(\alpha_i \alpha_j)\right).$$
If $\alpha_1, \ldots, \alpha_n$ are in $\mathcal{O}_K$ then it is clear that their discriminant is in $\mathbb{Z}$. It is not too hard to establish the following additional properties of $\Delta_{K/\mathbb{Q}}$:

(1) $\Delta_{K/\mathbb{Q}}(\alpha_1, \ldots, \alpha_n) \neq 0$ if and only if $\{\alpha_1, \ldots, \alpha_n\}$ is a basis of K over $\mathbb{Q}$.
(2) If $\{\alpha_1, \ldots, \alpha_n\}$ and $\{\beta_1, \ldots, \beta_n\}$ are bases of K over $\mathbb{Q}$ and if $\alpha_i = \sum_j a_{ij} \beta_j$ with all $a_{ij} \in \mathbb{Z}$, then
$$\Delta_{K/\mathbb{Q}}(\alpha_1, \ldots, \alpha_n) = [\det(a_{ij})]^2 \Delta_{K/\mathbb{Q}}(\beta_1, \ldots, \beta_n).$$
(3) $\Delta_{K/\mathbb{Q}}(\alpha_1, \ldots, \alpha_n) = [\det(\sigma_j(\alpha_i))]^2.$

Proposition 4.30.

(a) Every nonzero ideal I in $\mathcal{O}_K$ (including $\mathcal{O}_K$ itself) is additively a free abelian group of rank n whose $\mathbf{Q}$ span is K.

(b) If I is a nonzero ideal in $\mathcal{O}_K$, then $\Delta_{K/\mathbf{Q}}(\alpha_1, \ldots, \alpha_n)$ is the same for every $\mathbf{Z}$ basis $\alpha_1, \ldots, \alpha_n$ of I.

(c) Every nonzero ideal in $\mathcal{O}_K$ contains a nonzero member of $\mathbf{Z}$.

(d) Every nonzero ideal in $\mathcal{O}_K$ has $|\mathcal{O}_K/I| < \infty$.

(e) $\mathcal{O}_K$ is a Noetherian ring, i.e., every ascending sequence of ideals terminates.

(f) Every nonzero proper prime ideal in $\mathcal{O}_K$ is maximal.

We define a **product** operation on nonzero ideals I and J in $\mathcal{O}_K$ by

$$IJ = \{\text{all sums of products from } I \text{ and } J\}.$$

This product is associative and commutative, and $\mathcal{O}_K$ is an identity.

If I and J are nonzero ideals in $\mathcal{O}_K$, we say that I and J are **equivalent** if $(\alpha)I = (\beta)J$ for some nonzero principal ideals (α) and (β) in $\mathcal{O}_K$. It is easy to check that this notion of "equivalent" has the following properties:

(1) it is an equivalence relation
(2) the product operation is a class property
(3) the principal ideals form a single equivalence class, which acts as an identity.

The number h_K of classes of nonzero ideals in $\mathcal{O}_K$ is called the **class number** of K. The class number equals one if and only if $\mathcal{O}_K$ is a principal ideal domain. The following is the second of the three basic theorems in beginning algebraic number theory.

Theorem 4.31. The class number of a number field is finite.

The proof requires a lemma in diophantine approximation and an elementary argument using the notions above. For the case that $[K : \mathbf{Q}] = 2$, the classes of ideals correspond to certain classes of binary quadratic forms, and the finiteness of h_K (as well as its value) is more apparent.

Corollary 4.32. Multiplication of classes of nonzero ideals in $\mathcal{O}_K$ is a group operation in which the identity is the class of principal ideals.

9. APPENDIX ON ALGEBRAIC NUMBER THEORY

The group in the statement of the corollary is called the **ideal class group**; its order is the class number of K. For the proof of the corollary, one first shows for nonzero ideals I and J in $\mathcal{O}_K$ and for $\alpha \neq 0$ in $\mathcal{O}_K$ that

$$(\alpha)I = JI \implies (\alpha) = J. \tag{4.88}$$

Lemma 4.24 is a tool in the argument. Then the finiteness of the class number forces two powers I^i and I^j of an ideal I to be equivalent, and finally (4.88) allows one to show that $I^{|j-i|-1}$ is an inverse of the class of I.

Next we give the third of the three basic theorems in beginning algebraic number theory.

Theorem 4.33 (Unique factorization of ideals). Every proper ideal I in $\mathcal{O}_K$ is uniquely a product $I = \prod_{i=1}^N P_i^{k_i}$, where the P_i are proper nonzero prime ideals and the k_i are integers > 0. The integers k_i are characterized by the property: $P_i^{k_i} \supseteq I$ but $P_i^{k_i+1} \not\supseteq I$.

The proof of Theorem 4.33 begins with two preliminary facts. The first says for three nonzero ideals A, B, C of $\mathcal{O}_K$ that

$$AB = AC \implies B = C. \tag{4.89}$$

The other preliminary fact says about ideals: *To contain is to divide.* Specifically if A and B are nonzero ideals of $\mathcal{O}_K$, then

$$B \supseteq A \implies \text{there exists an ideal } C \text{ with } A = BC. \tag{4.90}$$

The existence of factorization of ideals uses (4.90) and parts (d) and (e) of Proposition 4.30. Uniqueness uses also (4.89), Proposition 4.30f, the finite class group (Theorem 4.31 and Corollary 4.32), and the inequality $|N_{K/\mathbb{Q}}(\alpha)| \geq 2$ for $\alpha \in \mathcal{O}_K$ that is not a unit (Lemma 4.27).

PROOF OF THEOREM 4.10. Write $h = h_K$. Let $I_1, \ldots, I_h$ be representatives of the finitely many classes of ideals (Theorem 4.31), and let us say $I_1 = (1)$. Let u_j be a nonzero element of I_j, and put $u = u_1 \cdots u_h$. Then u is in I_j for $1 \leq j \leq h$. Define $S = \{1, u, u^2, \ldots\}$. Notice that

(1) $1 \in S$ and $0 \notin S$
(2) S is closed under multiplication.

From these properties it follows that $S^{-1}\mathcal{O}_K = \{s^{-1}\alpha \mid s \in S, \alpha \in \mathcal{O}_K\}$ is a ring. We shall prove that $S^{-1}\mathcal{O}_K$ is a principal ideal domain and that its group of units is finitely generated as an abelian group.

If I is an ideal in $\mathcal{O}_K$, then $\tilde{I} = S^{-1}\mathcal{O}_K$ is clearly an ideal in $S^{-1}\mathcal{O}_K$. Every ideal I_S of $S^{-1}\mathcal{O}_K$ arises in this way. In fact, $I = I_S \cap \mathcal{O}_K$ is an ideal in $\mathcal{O}_K$ and $\tilde{I} = S^{-1}(I_S \cap \mathcal{O}_K)$ coincides with I_S. [To verify this equality, let $i \in \tilde{I}$. Then $i = s^{-1}\alpha$ with $\alpha \in I_S \cap \mathcal{O}_K$. Since $S^{-1} \subseteq S^{-1}\mathcal{O}_K$ and since I_S is an ideal in $S^{-1}\mathcal{O}_K$, $i = s^{-1}\alpha$ is in I_S. Conversely if i is in I_S, write $i = s^{-1}\alpha$ with $\alpha \in \mathcal{O}_K$. Then $\alpha = si$ is also in I_S, hence is in $I_S \cap \mathcal{O}_K$, and $i = s^{-1}\alpha$ exhibits i as in $S^{-1}(I_S \cap \mathcal{O}_K) = \tilde{I}$.]

If I_S is an ideal in $S^{-1}\mathcal{O}_K$, we are to show that I_S is principal. Put $I = I_S \cap \mathcal{O}_K$. Then I is equivalent with some I_j, $1 \leq j \leq h$; let us write

$$(\alpha)I = (\beta)I_j.$$

Since u is in $I_j \cap S$, we have $S^{-1}I_j = S^{-1}\mathcal{O}_K$, and thus

$$(\alpha)_S I_S = S^{-1}(\alpha)S^{-1}I = S^{-1}(\beta)I_j = S^{-1}\mathcal{O}_K(\beta) = (\beta)_S, \quad (4.91)$$

where $(\alpha)_S$ and $(\beta)_S$ denote principal ideals in $S^{-1}\mathcal{O}_K$.

From (4.91) let us see that

$$\beta/\alpha \text{ is in } S^{-1}\mathcal{O}_K \quad \text{and} \quad I_S = (\beta/\alpha)_S. \quad (4.92)$$

In fact, (4.91) allows us to write $\beta = \alpha i_0$ for some $i_0 \in I_S$. Hence $\beta/\alpha = i_0 \in I_S \subseteq S^{-1}\mathcal{O}_K$, and $(\beta/\alpha)_S \subseteq I_S$. In the reverse direction, let $i \in I_S$ be given, and use (4.91) to write $\alpha i = \beta x$ with $x \in S^{-1}\mathcal{O}_K$. Then $i = \dfrac{\beta}{\alpha} x$ shows i is in $(\beta/\alpha)_S$, and we obtain $I_S \subseteq (\beta/\alpha)_S$. This proves (4.92).

Since I_S was an arbitrary ideal in $S^{-1}\mathcal{O}_K$, (4.92) shows that $S^{-1}\mathcal{O}_K$ is a principal ideal domain. This completes the first half of the proof.

The second half of the proof will show that $(S^{-1}\mathcal{O}_K)^\times$ is finitely generated. The argument is a little subtle, as more generators than u and $\mathcal{O}_K^\times$ are needed.

Let $u^{-s}\alpha$ be in $(S^{-1}\mathcal{O}_K)^\times$, and write $(u^{-s}\alpha)^{-1} = u^{-t}\beta$. Then $\alpha\beta = u^{s+t}$. So α is a divisor of a power ≥ 0 of u. We seek a finite set of generators for the divisors of all powers ≥ 0 of u.

Thus let $\alpha\beta = u^r$ with $\alpha, \beta \in \mathcal{O}_K$. Let

$$(u) = P_1^{k_1} \cdots P_N^{k_n} \quad (4.93)$$

be the factorization of (u) given by Theorem 4.33. Then we have

$$(\alpha)(\beta) = (u^r) = (u)^r = P_1^{k_1 r} \cdots P_N^{k_N r}.$$

By the uniquenesss in Theorem 4.33,
$$(\alpha) = P_1^{l_1} \cdots P_N^{l_N} \qquad \text{with } 0 \leq l_j \leq k_j r \text{ for } 1 \leq j \leq N.$$
By Corollary 4.32 (and Theorem 4.31), P_j^h is principal for each j. Write $P_j^h = (\gamma_j)$. For each j, write $l_j = q_j h + r_j$ with $0 \leq r_j < h$, so that
$$(\alpha) = (\gamma_1)^{q_1} \cdots (\gamma_N)^{q_N} P_1^{r_1} \cdots P_N^{r_N}. \tag{4.94}$$
Then
$$\alpha = \gamma_1^{q_1} \cdots \gamma_N^{q_N} i \qquad \text{for some } i \in P_1^{r_1} \cdots P_N^{r_N}.$$
Dividing, we see that $\dfrac{\alpha}{\gamma_1^{q_1} \cdots \gamma_N^{q_N}} (= i)$ is in $\mathcal{O}_K$. Hence we can rewrite (4.94) as
$$\gamma_1^{q_1} \cdots \gamma_N^{q_N} \left(\frac{\alpha}{\gamma_1^{q_1} \cdots \gamma_N^{q_N}} \right) = (\gamma_1^{q_1} \cdots \gamma_N^{q_N}) P_1^{r_1} \cdots P_N^{r_N}.$$
By (4.89) we conclude that
$$P_1^{r_1} \cdots P_N^{r_N} = \left(\frac{\alpha}{\gamma_1^{q_1} \cdots \gamma_N^{q_N}} \right).$$
In other words, for (4.94) to hold, the ideal $P_1^{r_1} \cdots P_N^{r_N}$ has to be principal, say
$$P_1^{r_1} \cdots P_N^{r_N} = (\delta_{r_1,\ldots,r_n}),$$
and then (4.94) is the statement that
$$\alpha = \gamma_1^{q_1} \cdots \gamma_N^{q_N} \delta_{r_1,\ldots,r_n} \varepsilon \qquad \text{with } \varepsilon \text{ in } \mathcal{O}_K^\times. \tag{4.95}$$
From (4.95), we see that $(S^{-1}\mathcal{O}_K)^\times$ is generated by $\gamma_1, \ldots, \gamma_N$, the finite number of elements $\delta_{r_1,\ldots,r_n}$ (since the r_j's are $< h$), and $\mathcal{O}_K^\times$. Theorem 4.29 shows that $\mathcal{O}_K^\times$ is finitely generated. Hence $(S^{-1}\mathcal{O}_K)^\times$ is finitely generated.

EXAMPLE. Let $K = \mathbb{Q}(\sqrt{-5})$. Then $\mathcal{O}_K = \mathbb{Z}[\sqrt{-5}]$. The theory of binary quadratic forms shows that $h_K = 2$. In the above proof we can take $I_1 = (1)$ and $I_2 = (2, 1+\sqrt{-5})$. For u we can use $u_1 = 1$, $u_2 = 2$, $u = 2$. The factorization (4.93) is
$$(2) = P_1^2 \qquad \text{with } P_1 = (2, 1+\sqrt{-5}).$$
Thus $N = 1$, and we can take $\gamma_1 = 2$. The only ideals $P_1^{r_1} \cdots P_N^{r_N}$ that need to be considered are $P_1^0 = (1)$ and P_1^1, which is not principal. Thus the only $\delta_{r_1,\ldots,r_N}$ is 1. The units of $S^{-1}\mathcal{O}_K$ are therefore generated by $\gamma_1 = 2$, $\delta_0 = 1$, and $(\mathcal{O}_K)^\times = \{\pm 1\}$.

CHAPTER V

TORSION SUBGROUP OF $E(\mathbb{Q})$

1. Overview

Let E be an elliptic curve over $\mathbb{Q}$ with notation as in the Weierstrass form (3.23). As usual, we can make an admissible change of variables so that all the coefficients are in $\mathbb{Z}$. Let $\Delta \in \mathbb{Z}$ be the discriminant of E. By Mordell's Theorem, $E(\mathbb{Q})$ is a finitely generated abelian group. Our objective in this chapter is to study the torsion subgroup $E(\mathbb{Q})_{\text{tors}}$. For any particular curve, we shall see that we can determine $E(\mathbb{Q})_{\text{tors}}$ explicitly.

The main tool in the analysis will be reduction modulo a prime p. For the curve E itself, with its $\mathbb{Z}$ coefficients, reduction modulo p will amount to considering the curve E_p over $\mathbb{Z}_p$ with the $\mathbb{Z}$ coefficients taken modulo p. It is less apparent how to get reduction modulo p to be a well defined mapping r_p defined from all of $E(\mathbb{Q})$, including points with factors of p in the denominators of the coordinates, into $E_p(\mathbb{Z}_p)$. Once we have done this in §2, we will find that $r_p : E(\mathbb{Q}) \to E_p(\mathbb{Z}_p)$ is a group homomorphism if $p \nmid \Delta$ (i.e., if E_p is nonsingular). The main theorem is the following result discovered in the 1930's independently by Lutz and Nagell.

Theorem 5.1 (Lutz-Nagell). Let E be an elliptic curve (3.23) with coefficients in $\mathbb{Z}$.

(a) If $a_1 = 0$ and if $P = (x(P), y(P), 1)$ is in $E(\mathbb{Q})_{\text{tors}}$, then $x(P)$ and $y(P)$ are integers.

(b) For any a_1, if $P = (x(P), y(P), 1)$ is in $E(\mathbb{Q})_{\text{tors}}$, then $4x(P)$ and $8y(P)$ are integers.

(c) If p is an odd prime such that $p \nmid \Delta$, then the restriction to $E(\mathbb{Q})_{\text{tors}}$ of the reduction homomorphism $r_p : E(\mathbb{Q}) \to E_p(\mathbb{Z}_p)$ is one-one. The same conclusion is valid for $p = 2$ if $2 \nmid \Delta$ and $a_1 = 0$.

(d) If $a_1 = a_3 = a_2 = 0$, so that E is given by

$$y^2 = x^3 + Ax + B, \qquad (5.1)$$

and if $P = (x(P), y(P), 1)$ is in $E(\mathbb{Q})_{\text{tors}}$, then either $y(P) = 0$ (and P has order 2) or else $y(P) \neq 0$ and $y(P)^2$ divides $d = -4A^3 - 27B^2$, the discriminant of the cubic polynomial on the right side of (5.1).

1. OVERVIEW

This theorem will be proved in §4. As was mentioned in Chapter I, a consequence is that we can determine $E(\mathbb{Q})_{\text{tors}}$ completely for any particular curve. The algorithm is to put the curve in the form (5.1), to consider each square divisor y^2 of d to give a candidate for $y(P)$ and to write down any corresponding integers x such that (x,y) is an integer solution of (5.1). There can be only finitely many such, and we obtain a bound on $|E(\mathbb{Q})_{\text{tors}}|$. For each such integer solution (x,y), we can raise it to powers up to our bound on the order, effectively checking whether it is a torsion point.

As a practical matter, this algorithm is often a little tedious. It is usually more efficient to use (c) to cut down the possibilities. Some examples will illlustrate.

EXAMPLE 1. $y^2 + y = x^3 - x^2$. There are some obvious solutions with integer coordinates, namely

$$(x,y) = (0,0), (0,-1), (1,0), (1,-1). \tag{5.2}$$

Under the doubling formula (3.74), $x = 0$ doubles to $x = 1$ and vice versa. So all four points are torsion points. For this curve, $\Delta = -11$ from Table 3.1. Rather than clear out the y and x^2 terms to be able to use Theorem 5.1d, we reduce modulo 2, since $a_1 = 0$ and $2 \nmid \Delta$. Then $E_2(\mathbb{Z}_2)$ contains the four points (5.2) and also ∞; hence $E_2(\mathbb{Z}_2) \cong \mathbb{Z}_5$. By Theorem 5.1c, $r_2 : E(\mathbb{Q})_{\text{tors}} \to E_2(\mathbb{Z}_2) \cong \mathbb{Z}_5$ is one-one. Since (5.2) has shown $E(\mathbb{Q})_{\text{tors}}$ to be nontrivial, we conclude that $E(\mathbb{Q})_{\text{tors}} \cong \mathbb{Z}_5$. The members of $E(\mathbb{Q})_{\text{tors}}$ are ∞ and the points in (5.2).

EXAMPLE 2. $y^2 + y = x^3 - x$. Again there are some apparent solutions with integer coordinates, namely

$$(x,y) = (\pm 1, 0), (\pm 1, -1), (2, 2), (2, -3), (6, -15). \tag{5.3}$$

From Table 3.2, we have $\Delta = 37$. Since $a_1 = 0$ and $2 \nmid \Delta$ and $3 \nmid \Delta$, Theorem 5.1c applies to $p = 2$ and $p = 3$. Over $\mathbb{Z}_2$ we find 5 solutions (counting ∞), while over $\mathbb{Z}_3$ we find 7 solutions. So $E(\mathbb{Q})_{\text{tors}}$ maps one-one into $\mathbb{Z}_5$ and also one-one into $\mathbb{Z}_7$. Thus $E(\mathbb{Q})_{\text{tors}} = 0$. None of the points (5.3) is a torsion point.

EXAMPLE 3. $y^2 - xy + 2y = x^3 + 2x^2$. For this curve we calculate that $\Delta = -2^7 13$. We can try to identify $E(\mathbb{Q})_{\text{tors}}$ by using parts (a) and (d) of Theorem 5.1. To find a full list of candidates for torsion points, we can rewrite E as

$$[y + (1 - \tfrac{1}{2}x)]^2 = x^3 + \tfrac{9}{4}x^2 - x + 1.$$

Making admissible changes of variables three times leads first to
$$y^2 = x^3 + \tfrac{9}{4}x^2 - x + 1,$$
then to
$$y^2 = x^3 + 9x^2 - 16x + 64,$$
and finally to
$$y^2 = x^3 - 43x + 166.$$
The discriminant of the cubic polynomial on the right side is $-2^{13}13$. Thus Theorem 5.1d says that $y(P)^2 \mid 2^{13}13$. So $y(P)^2 = 1$ or 4 or 16 or 256 or 1024 or 4096. Checking each case, we are led to the candidates
$$(3, \pm 8), \; (-5, \pm 16), \; (11, \pm 32), \; \infty.$$
Using the doubling formula, we can check that each of these is a torsion point. We can transform back to obtain the following solutions of the original equation:
$$(0,0), \; (0,-2), \; (-2,0), \; (-2,-4), \; (2,4), \; (2,-4), \; \infty. \tag{5.4}$$
The torsion subgroup of the original equation therefore consists of the points (5.4) and is isomorphic to $\mathbf{Z}_7$.

It is more efficient, however, to use Theorem 5.1c. Since $3 \nmid \Delta$, we list the solutions of the original equation modulo 3 as
$$(0,0), \; (0,1), \; (1,0), \; (1,-1), \; (-1,1), \; (-1,-1), \; \infty.$$
By Theorem 5.1c, $E(\mathbf{Q})_{\text{tors}}$ is 0 or $\mathbf{Z}_7$. In $\mathbf{Q}$ we quickly see that $(0,0)$, $(0,-2)$, and $(-2,0)$ are solutions. Using the doubling formula, we check that the doubled x coordinate of any of these makes a cycle after 3 steps. Therefore these points are torsion points, and $E(\mathbf{Q})_{\text{tors}} \cong \mathbf{Z}_7$. We can use the results of the doubling to generate the list (5.4).

EXAMPLE 4. $y^2 + xy = x^3 + 4x^2 + x$. For this curve, $\Delta = 3^2 5^2$. The point $(-\tfrac{1}{4}, \tfrac{1}{8})$ lies on the curve, and the tangent line is vertical there (since the coefficient $2y + x$ of dy is 0). Therefore $(-\tfrac{1}{4}, \tfrac{1}{8})$ has order 2 and is a torsion point with nonintegral coordinates.

To limit the size of $E(\mathbf{Q})_{\text{tors}}$, we reduce modulo 7 and find that the numbers of y's giving solutions for $x = 0, 1, \ldots, 6$ are $1, 2, 0, 1, 0, 1, 2$, respectively. Counting the point ∞, we see that $E_7(\mathbf{Z}_7)$ has order 8.

To identify $E(\mathbf{Q})_{\text{tors}}$, we replace $y + \tfrac{1}{2}x$ by y and then scale. The transformed equation is $y^2 = x^3 + 17x^2 + 16x$. In §5 we shall see that an

1. OVERVIEW

equation of the type $y^2 = x(x+r^2)(x+s^2)$ has $\mathbf{Z}_4 \oplus \mathbf{Z}_2$ as a subgroup of its torsion group. So in our case, the torsion group is exactly $\mathbf{Z}_4 \oplus \mathbf{Z}_2$. Taking the 8 solutions given in §5 and transforming back, we obtain

$$(0,0),\ (-\tfrac{1}{4},\tfrac{1}{8}),\ (-1,-1),\ (-1,2),\ (1,2),\ (1,-3),\ (-4,2),\ \infty$$

as the torsion elements for our original equation.

Since $2 \nmid \Delta$, reduction modulo 2 is a homomorphism. The group $E_2(\mathbf{Z}_2)$ consists of

$$(0,0),\ (1,0),\ (1,1),\ \infty,$$

and r_2 maps $E(\mathbf{Q})_{\text{tors}}$ onto $E_2(\mathbf{Z}_2)$ with kernel $\mathbf{Z}_2 = \{(-\tfrac{1}{4},\tfrac{1}{8}),\ \infty\}$.

Table 5.1 gives examples of elliptic curves over $\mathbf{Q}$ with 15 different torsion subgroups. Methods for generating such examples will be indicated in §5. According to Theorem 1.7 (due to Mazur), these 15 groups are the only possibilities for a torsion group over $\mathbf{Q}$. The proof of Mazur's Theorem is well beyond the scope of this book and will not be given here.

E	$E(\mathbf{Q})_{\text{tors}}$	Δ
$y^2 = x^3 + 2$	0	$-2^6 3^3$
$y^2 = x^3 + x$	$\mathbf{Z}_2$	-2^6
$y^2 = x^3 + 4$	$\mathbf{Z}_3$	$-2^8 3^3$
$y^2 = x^3 + 4x$	$\mathbf{Z}_4$	-2^{12}
$y^2 + y = x^3 - x^2$	$\mathbf{Z}_5$	-11
$y^2 = x^3 + 1$	$\mathbf{Z}_6$	$-2^4 3^3$
$y^2 - xy + 2y = x^3 + 2x^2$	$\mathbf{Z}_7$	$-2^7 13$
$y^2 + 7xy - 6y = x^3 - 6x^2$	$\mathbf{Z}_8$	$2^8 3^4 17$
$y^2 + 3xy + 6y = x^3 + 6x^2$	$\mathbf{Z}_9$	$-2^9 3^5$
$y^2 - 7xy - 36y = x^3 - 18x^2$	$\mathbf{Z}_{10}$	$-2^5 3^{10} 11^2$
$y^2 + 43xy - 210y = x^3 - 210x^2$	$\mathbf{Z}_{12}$	$2^{12} 3^6 5^3 7^4 13$
$y^2 = x^3 - x$	$\mathbf{Z}_2 \oplus \mathbf{Z}_2$	2^6
$y^2 = x^3 + 5x^2 + 4x$	$\mathbf{Z}_4 \oplus \mathbf{Z}_2$	$2^8 3^2$
$y^2 + 5xy - 6y = x^3 - 3x^2$	$\mathbf{Z}_6 \oplus \mathbf{Z}_2$	$2^2 3^6 5^2$
$y^2 = x^3 + 337x^2 + 20736x$	$\mathbf{Z}_8 \oplus \mathbf{Z}_2$	$2^{20} 3^8 5^4 7^2$

TABLE 5.1. Examples of torsion subgroups of $E(\mathbf{Q})$

There are two situations where we can decide $E(\mathbf{Q})_{\text{tors}}$ for infinitely many given curves at once (other than the situations in §5 where we construct some families just for this decidability). The two theorems that follow use Dirichlet's Theorem on primes in arithmetic progressions and will be proved in §6. Part of the first theorem was borrowed in Chapter IV and stated without proof as Lemma 4.20.

Theorem 5.2. Let E be the elliptic curve $y^2 = x^3 + Ax$ with A in $\mathbf{Z}$ and with A assumed fourth-power free. Then

$$E(\mathbf{Q})_{\text{tors}} = \begin{cases} \mathbf{Z}_2 \oplus \mathbf{Z}_2 & \text{if } -A \text{ is a square in } \mathbf{Z} \\ \mathbf{Z}_4 & \text{if } A = 4 \\ \mathbf{Z}_2 & \text{otherwise.} \end{cases}$$

Theorem 5.3. Let E be the elliptic curve $y^2 = x^3 + B$ with B in $\mathbf{Z}$ and with B assumed sixth-power free. Then

$$E(\mathbf{Q})_{\text{tors}} = \begin{cases} \mathbf{Z}_6 & \text{if } B = 1 \\ \mathbf{Z}_3 & \text{if } B = -432 = -2^4 3^3, \text{ or if } B = \square \text{ and } B \neq 1 \\ \mathbf{Z}_2 & \text{if } B = \text{cube and } B \neq 1 \\ 0 & \text{otherwise.} \end{cases}$$

2. Reduction Modulo p

We recall the definition of the p-adic norm on $\mathbf{Q}$, p being a prime. If $r \neq 0$ is in $\mathbf{Q}$, we write $r = p^n u/v$ with u and v in $\mathbf{Z}$ and with $\gcd(u,p) = \gcd(v,p) = 1$. The definition of the **p-adic norm** is $|r|_p = p^{-n}$. By convention, we define $|0|_p = 0$. This notion has the following properties:

(i) $|r + s|_p \leq \max\{|r|_p, |s|_p\}$, with equality if $|r|_p \neq |s|_p$,
(ii) $|rs|_p = |r|_p |s|_p$.

Property (ii) is clear. For (i) we write $r = p^n u/v$ and $s = p^{n'} u'/v'$, with $n \leq n'$ without loss of generality. Writing

$$r + s = p^n \left(\frac{u}{v} + p^{n'-n} \frac{u'}{v'}\right) = p^n \frac{uv' + p^{n'-n} u'v}{vv'}$$

with $\gcd(vv', p) = 1$, we obtain (i) directly.

Property (i) is called the **ultrametric inequality**. It implies $|r + s|_p \leq |r|_p + |s|_p$. If we define $d(x,y) = |x - y|_p$, then the latter inequality implies the triangle inequality for d, and d is a metric on $\mathbf{Q}$.

2. REDUCTION MODULO p

We say that $r \in \mathbb{Q}$ is p-**integral** if $|r|_p \leq 1$. By (i) and (ii), the p-integral elements form a subring of $\mathbb{Q}$ containing $\mathbb{Z}$. Those with $|r|_p < 1$ form an ideal in this subring; they of course have $|r|_p \leq p^{-1}$. The p-integral elements of $\mathbb{Q}$ can be reduced modulo p: If $r = p^n u/v$ is p-integral, i.e., if $n \geq 0$, then we define $r_p(r) \in \mathbb{Z}_p$ by

$$r_p(r) = \begin{cases} u/v \bmod p & \text{if } n = 0 \\ 0 & \text{if } n > 0. \end{cases}$$

Then $r_p : \{p\text{-integral elements}\} \to \mathbb{Z}_p$ is a ring homomorphism.

In preparation for considering plane curves, we can try to use r_p to get a map of the affine plane over $\mathbb{Q}$ to the affine plane over $\mathbb{Z}_p$, but the best we get is a map defined on

$$\{(r, s) \mid r \text{ and } s \text{ are } p\text{-integral}\}$$

as $r_p(r, s) = (r_p(r), r_p(s))$. To correct this deficiency we work with curves projectively, as follows.

To define $r_p : P_2(\mathbb{Q}) \to P_2(\mathbb{Z}_p)$, we let

$$r_p(x, y, w) = (r_p(x), r_p(y), r_p(w)), \tag{5.5}$$

where (x, y, w) are coordinates of the point in question chosen so that x, y, and w all have $|\cdot|_p \leq 1$ and at least one of them has $|\cdot|_p = 1$. Such a representative of a point in $P_2(\mathbb{Q})$ is said to be p-**reduced**. Note that if a general (x, y, w) is given, we can multiply by a suitable p^n to obtain a p-reduced representative. A p-reduced representative is unique up to a factor with $|\cdot|_p = 1$. Therefore r_p is well defined as a map of all of $P_2(\mathbb{Q})$ into $P_2(\mathbb{Z}_p)$.

Using (5.5), we can reduce projective plane curves modulo p. Let $F \in \mathbb{Q}[x, y, w]_m$ be a plane curve of degree m. Multiplying the coefficients of F by a constant, we may assume that all the coefficients have $|\cdot|_p \leq 1$ and at least one has $|\cdot|_p = 1$. Then we can reduce the coefficients modulo p, obtaining a nonzero polynomial $F_p \in \mathbb{Z}_p[x, y, w]_m$. Although F_p is not defined uniquely, it is defined uniquely up to a nonzero scalar. Therefore its zero locus $F_p(\mathbb{Z}_p)$ is well defined.

Proposition 5.4. Let $F \in \mathbb{Q}[x, y, w]_m$ be a plane curve. Under the reduction homomorphism $r_p : P_2(\mathbb{Q}) \to P_2(\mathbb{Z}_p)$ given in (5.5), the image of $F(\mathbb{Q})$ is contained in $F_p(\mathbb{Z}_p)$.

PROOF. Normalize the coefficients of F as above, and let (x, y, w) be a p-reduced representative of a point in $P_2(\mathbb{Q})$. Then

$$\begin{aligned}(x, y, w) \in F(\mathbb{Q}) &\iff F(x, y, w) = 0 \\ &\implies r_p(F(x, y, w)) = 0 \\ &\iff F_p((r_p(x), r_p(y), r_p(w))) = 0 \\ &\iff F_p(r_p(x, y, w)) = 0 \\ &\iff r_p(x, y, w) \in F_p(\mathbb{Z}_p).\end{aligned}$$

Proposition 5.5. Suppose $F \in \mathbb{Q}[x, y, w]_m$ is a plane curve, $L \in \mathbb{Q}[x, y, w]_1$ is a line, and $P = (x_0, y_0, w_0)$ is a point on L. If F_p and L_p are reductions of F and L modulo p, then the intersection multiplicities satisfy

$$i(P, L, F) \leq i(r_p(P), L_p, F_p). \tag{5.6}$$

PROOF. Without loss of generality, we may assume that (x_0, y_0, w_0) is a p-reduced representative and that the coefficients of F and L are normalized as they are supposed to be. Choose a p-reduced representative (x', y', w') of a point P' of L with $[(x', y', w')] \neq [(x_0, y_0, w_0)]$, and form

$$\begin{aligned}\psi(t) &= F(P + tP') = F(x_0 + tx', y_0 + ty', w_0 + tw') \\ &= t^r F'_r + \cdots + t^m F'_m,\end{aligned}$$

with $F'_r \neq 0$. By Proposition 2.9 the left side of (5.6) equals r. Recomputing $\psi(t)$ modulo p (i.e., in $\mathbb{Z}_p[t]$) and applying Proposition 2.9 again, we see that the right side of (5.6) is $\geq r$.

Let us apply Propositions 5.4 and 5.5 to elliptic curves E over $\mathbb{Q}$. For studying $E(\mathbb{Q})$, we may make an admissible change of variables $y \to y/u^3$, $x \to x/u^2$ to make all coefficients of E be in $\mathbb{Z}$. Then we can assume E is in Weierstrass form (3.23) with all coefficients in $\mathbb{Z}$. To apply the above theory, we consider the projective form (3.23a) of E. The coefficients of E are all in $\mathbb{Z}$ and hence are p-integral. Also wy^2 and x^3 have coefficient 1. Thus passage to E_p is given simply by writing (3.23a) with coefficients considered in $\mathbb{Z}_p$; no preliminary normalization is needed. The discriminant of E_p is clearly given by

$$\Delta_p = \Delta \mod p.$$

Thus E_p is nonsingular if and only if $p \nmid \Delta$. Our reduction map on $E(\mathbb{Q})$ is a mapping

$$r_p : E(\mathbb{Q}) \to E_p(\mathbb{Z}_p) \tag{5.7}$$

by Proposition 5.4.

3. p-ADIC FILTRATION

Proposition 5.6. *If E_p is nonsingular, then the map r_p in (5.7) is a group homomorphism.*

PROOF. Since $r_p(0,1,0) = (0,1,0)$, r_p carries O to O_p. We apply Proposition 5.5. Since the sum of intersection multiplicities over a line is ≤ 3 (by nonsingularity of E and E_p), the proposition gives $r_p(PQ) = r_p(P) \cdot r_p(Q)$. Thus

$$r_p(P+Q) = r_p(O \cdot PQ) = r_p(O) \cdot r_p(PQ) = r_p(O) \cdot (r_p(P) \cdot r_p(Q))$$
$$= O_p \cdot (r_p(P) \cdot r_p(Q)) = r_p(P) + r_p(Q),$$

and r_p is a group homomorphism.

3. p-adic Filtration

Fix an elliptic curve E with $\mathbf{Z}$ coefficients. In order to get at the integrality or almost integrality of torsion points, it is natural to work with coordinates $(x, y, 1)$, clear denominators, and see what happens. The difficulty with this approach is that it is not so easy to take advantage of the hypothesis that $(x, y, 1)$ is a torsion point.

The clever idea that works is to use coordinates $(X, 1, W)$. Proposition 5.6 shows that $r_p : E(\mathbf{Q}) \to E_p(\mathbf{Z}_p)$ is a group homomorphism if $p \nmid \Delta$. In the coordinates $(X, 1, W)$, subgroups and the homomorphism property of r_p play a more visible role, and the finite order of a torsion point becomes a usable property.

When $p \nmid \Delta$, the kernel of r_p is

$$\{(x, y, w) \in E(\mathbf{Q}) \mid r_p(x, y, w) = (0, 1, 0)\}.$$

For such a point (x, y, w), we must have $y \neq 0$. Normalizing, we may assume $y = 1$. The elements $(x, 1, w) \in E(\mathbf{Q})$ that are in $\ker(r_p)$ are those with $|x|_p < 1$ and $|w|_p < 1$. In this section we shall study the same set

$$E^{(1)}(\mathbf{Q}) = \{(x, 1, w) \in E(\mathbf{Q}) \mid |x|_p < 1 \text{ and } |w|_p < 1\},$$

but without the assumption that $p \nmid \Delta$.

Lemma 5.7. *Let $(x, 1, w)$ be in $E(\mathbf{Q})$. If $|w|_p < 1$, then $|x|_p < 1$ and $|w|_p = |x|_p^3$.*

PROOF. Putting $y = 1$ in (3.23a), we have

$$w = -a_1 xw - a_3 w^2 + x^3 + a_2 x^2 w + a_4 x w^2 + a_6 w^3. \qquad (5.8)$$

First suppose $|x|_p \geq 1$. On the right side of (5.8) the x^3 term is strictly the largest relative to $|\cdot|_p$, since

$$|-a_1 xw|_p \leq |xw|_p < |x|_p \leq |x|_p^3 = |x^3|_p$$
$$|-a_3 w^2|_p \leq |w|_p^2 < 1 \leq |x|_p^3 = |x^3|_p$$
$$|a_2 x^2 w|_p \leq |x^2 w|_p < |x|_p^2 \leq |x|_p^3 = |x^3|_p$$
$$|a_4 x w^2|_p \leq |xw^2|_p < |x|_p \leq |x|_p^3 = |x^3|_p$$
$$|a_6 w^3|_p \leq |w|_p^3 < 1 \leq |x|_p^3 = |x^3|_p.$$

By the ultrametric inequality, (5.8) gives $|w|_p = |x^3|_p$. This is a contradiction, since $|w|_p < 1$ and $|x^3|_p = |x|_p^3 \geq 1$. We conclude that $|x|_p < 1$.
Now we rewrite (5.8) as

$$w + a_1 xw + a_3 w^2 - a_2 x^2 w - a_4 x w^2 - a_6 w^3 = x^3. \qquad (5.9)$$

If $w = 0$, then $x = 0$ and $|w|_p = |x|_p^3$. If $w \neq 0$, then the w term is strictly the largest on the left side of (5.9), since

$$|-a_1 xw|_p \leq |xw|_p = |x|_p |w|_p < |w|_p$$
$$|-a_3 w^2|_p \leq |w^2|_p = |w|_p^2 < |w|_p$$
$$|a_2 x^2 w|_p \leq |x^2 w|_p = |x|_p^2 |w|_p < |w|_p$$
$$|a_4 x w^2|_p \leq |xw^2|_p = |x|_p |w|_p^2 < |w|_p$$
$$|a_6 w^3|_p \leq |w^3|_p = |w|_p^3 < |w|_p.$$

By the ultrametric inequality, (5.9) gives $|w|_p = |x^3|_p = |x|_p^3$, and the lemma follows.

For $n \geq 1$, define

$$E^{(n)}(\mathbb{Q}) = \{(x, 1, w) \in E(\mathbb{Q}) \mid |w|_p < 1 \text{ and } |x|_p \leq p^{-n}\}.$$

The *p*-adic filtration of $E^{(1)}(\mathbb{Q})$ is

$$E^{(1)}(\mathbb{Q}) \supseteq E^{(2)}(\mathbb{Q})) \supseteq E^{(3)}(\mathbb{Q}) \supseteq \ldots \quad \text{with} \bigcap_{n=1}^{\infty} E^{(n)}(\mathbb{Q}) = \{(0,1,0)\}.$$

3. p-ADIC FILTRATION

The important control on a point of $E^{(1)}(\mathbf{Q})$ other than the group identity is that it lies in $E^{(n)}(\mathbf{Q}) - E^{(n+1)}(\mathbf{Q})$ for some n. Lemma 5.7 says that
$$E^{(n)}(\mathbf{Q}) = \{(x,1,w) \in E(\mathbf{Q}) \mid |w|_p \leq p^{-3n}\}.$$

Let R be the ring of p-integral elements of $\mathbf{Q}$, i.e., members of $\mathbf{Q}$ with no factor of p in the denominator.

Proposition 5.8. The subsets $E^{(n)}(\mathbf{Q})$ of $E(\mathbf{Q})$ are subgroups. The function $P = (x,1,w) \longrightarrow x(P) = x$ gives a map $E^{(n)}(\mathbf{Q}) \to p^n R$. The composition of this map, followed by the quotient map $p^n R \to p^n R/p^{2n} R$, is a group homomorphism
$$E^{(n)}(\mathbf{Q}) \longrightarrow p^n R/p^{2n} R$$
whose kernel is contained in $E^{(2n)}(\mathbf{Q})$. Consequently the homomorphism
$$E^{(n)}(\mathbf{Q})/E^{(2n)}(\mathbf{Q}) \longrightarrow p^n R/p^{2n} R$$
is one-one.

Before coming to the proof, we mention that
$$p^n R/p^{2n} R \cong p^n \mathbf{Z}/p^{2n} \mathbf{Z} \cong \mathbf{Z}_{p^n}. \tag{5.10}$$

For the second of these isomorphisms, we observe that each group is cyclic and they both have p^n elements. For the first of the isomorphisms, we map $p^n \mathbf{Z}$ into $p^n R/p^{2n} R$ and obtain a map on the quotient
$$p^n \mathbf{Z}/p^{2n} \mathbf{Z} \longrightarrow p^n R/p^{2n} R.$$

This map is one-one since $p^n \mathbf{Z} \cap p^{2n} R \subseteq p^{2n} \mathbf{Z}$. To see that it is onto, let $p^n u/v$ be in $p^n R$. Choose a and b in $\mathbf{Z}$ with $av + bp^n = u$. Then
$$\frac{p^n u}{v} = p^n a + \frac{p^{2n} b}{v}$$
shows that $p^n a$ maps onto the coset of $p^n u/v$.

The proof of Proposition 5.8 will be preceded by the statement of a lemma. The lemma will be proved after the proposition.

Lemma 5.9. Let L be a line that meets $E(\mathbf{Q})$ in three points when multiplicities are counted, say in $P_1 = (x_1, 1, w_1)$, $P_2 = (x_2, 1, w_2)$, and $P_3 = (x_3, 1, w_3)$. If P_1 and P_2 are in $E^{(n)}(\mathbf{Q})$, then P_3 is in $E^{(n)}(\mathbf{Q})$ and

$$|x_1 + x_2 + x_3|_p \leq \begin{cases} p^{-3n} & \text{if } a_1 = 0 \\ p^{-2n} & \text{in any case.} \end{cases} \quad (5.11)$$

PROOF OF PROPOSITION 5.8. If P_1 and P_2 are in $E^{(n)}(\mathbf{Q})$, then the lemma says that $P_3 = P_1 P_2$ is in $E^{(n)}(\mathbf{Q})$. Since O is in $E^{(n)}(\mathbf{Q})$, so is $O \cdot P_1 P_2 = P_1 + P_2$. Also $O \cdot P_1 = -P_1$ is in $E^{(n)}(\mathbf{Q})$. Hence $E^{(n)}(\mathbf{Q})$ is a subgroup of $E(\mathbf{Q})$.

If P is in $E^{(n)}(\mathbf{Q})$ and $x = x(P)$, then $|x|_p \leq p^{-n}$ yields

$$|p^{-n}x|_p = |p^{-n}|_p |x|_p = p^n |x|_p \leq 1.$$

Hence $p^{-n}x$ is in R and x is in $p^n R$. Thus $P \to x(P)$ gives a map of $E^{(n)}(\mathbf{Q})$ into $p^n R$.

Let $P_1 P_2 = P_3$. Then the lemma gives

$$x(P_1) + x(P_2) + x(P_3) \in p^{2n} R. \quad (5.12)$$

If $P_3 = (x_3, 1, w_3)$, then a little computation with (3.70) shows that

$$OP_3 = -P_3 = \left(-\frac{x_3}{1 + a_1 x_3 + a_3 w_3}, 1, -\frac{w_3}{1 + a_1 x_3 + a_3 w_3}\right).$$

Hence

$$x(P_1 + P_2) = x(OP_3) = -\frac{x_3}{1 + a_1 x_3 + a_3 w_3}$$

and

$$x(P_1 + P_2) + x(P_3) = -\frac{x_3}{1 + a_1 x_3 + a_3 w_3} + x_3$$

$$= x_3 \frac{a_1 x_3 + a_3 w_3}{1 + a_1 x_3 + a_3 w_3}$$

$$\in p^{3n} R \quad (5.13)$$

since $|a_1 x_3 + a_3 w_3|_p \leq 1$ and $|1 + a_1 x_3 + a_3 w_3|_p = 1$. Combining (5.12) and (5.13), we see that

$$x(P_1 + P_2) \equiv x(P_1) + x(P_2) \mod p^{2n} R,$$

and the composition $P \to x(P) \to x(P) + p^{2n} R$ is a homomorphism.

If $P = (x, 1, w)$ is in the kernel of the resulting homomorphism on $E^{(n)}(\mathbf{Q})/E^{(2n)}(\mathbf{Q})$, then $x(P) = x$ is in $p^{2n} R$ and $|w|_p < 1$. Then $|x|_p \leq p^{-2n}$, and P is in $E^{(2n)}(\mathbf{Q})$. This completes the proof.

3. p-ADIC FILTRATION

PROOF OF LEMMA 5.9. The first part of the proof will show that the line L is of the form
$$w = mx + b \tag{5.14}$$
and will give a bound for the slope m. We begin with the equation
$$w + a_1 wx + a_3 w^2 = x^3 + a_2 x^2 w + a_4 x w^2 + a_6 w^3 \tag{5.15}$$
satisfied by the three points.

Case 1. $P_1 \neq P_2$. Here we substitute $(x_1, 1, w_1)$ and $(x_2, 1, w_2)$ into (5.15) and subtract the results. Each term of the difference is an integral multiple of
$$\begin{aligned}x_1^s w_1^t - x_2^s w_2^t &= (x_1^s - x_2^s) w_1^t + x_2^s (w_1^t - w_2^t)\\ &= (x_1 - x_2)(x_1^{s-1} + \cdots + x_2^{s-1}) w_1^t + (w_1 - w_2)(w_1^{t-1} + \cdots + w_2^{t-1}) x_2^s.\end{aligned} \tag{5.16}$$

Consider all the terms of (5.15) but w and x^3, all of which have $s+3t \geq 4$. Noting that P_1 and P_2 are in $E^{(n)}(\mathbb{Q})$, we use
$$|(x_1 - x_2)(x_1^{s-1} + \cdots + x_2^{s-1}) w_1^t|_p \\ \leq |x_1 - x_2|_p p^{-n(s-1)} p^{-3nt} \leq p^{-3n} |x_1 - x_2|_p$$
when $s > 0$, and we use
$$|(w_1 - w_2)(w_1^{t-1} + \cdots + w_2^{t-1}) x_2^s|_p \\ \leq |w_1 - w_2|_p p^{-3n(t-1)} p^{-ns} \leq p^{-n} |w_1 - w_2|_p$$
when $t > 0$. Moving the $x_1 - x_2$ terms obtained from (5.16) to the right side and the $w_1 - w_2$ terms to the left side, we can rewrite our difference of equations as
$$(w_1 - w_2)(1 + u) = (x_1 - x_2)(x_1^2 + x_1 x_2 + x_2^2 + v), \tag{5.17}$$
where $|u|_p \leq p^{-n}$ and $|v|_p \leq p^{-3n}$. Since $|u|_p < 1$, $|1 + u|_p = 1$. In particular, $x_1 = x_2$ implies $w_1 = w_2$. Since we have assumed $P_1 \neq P_2$, we conclude $x_1 \neq x_2$. Thus the line through P_1 and P_2 is of the form (5.14). In view of (5.17), the slope m is given by
$$m = \frac{w_1 - w_2}{x_1 - x_2} = \frac{x_1^2 + x_1 x_2 + x_2^2 + v}{1 + u}.$$

Hence

$$|m|_p = \left|\frac{x_1^2 + x_1 x_2 + x_2^2 + v}{1 + u}\right|_p = |x_1^2 + x_1 x_2 + x_2^2 + v|_p$$

$$\leq \max\{p^{-2n}, p^{-2n}, p^{-2n}, p^{-3n}\} = p^{-2n}. \tag{5.18}$$

Case 2. $P_1 = P_2$. The line in question is the tangent, which we know exists. Since $\mathbf{Q} \subseteq \mathbf{R}$, we can compute the tangent by implicit differentiation of (5.15). Differentiation of (5.15) and substitution of $(x, 1, w) = (x_1, 1, w_1)$ gives

$$(1 + a_1 x_1 + 2a_3 w_1 - a_2 x_1^2 - 2a_4 x_1 w_1 - 3a_6 w_1^2) dw$$
$$= (-a_1 w_1 + 3x_1^2 + 2a_2 x_1 w_1 + a_4 w_1^2) dx.$$

The coefficient of dw is of the form $1 + u'$ with $|u'|_p < 1$. Therefore $\dfrac{dw}{dx}$ is finite. The tangent line is thus of the form (5.14), and the slope m is given by $m = \dfrac{-a_1 w_1 + 3x_1^2 + 2a_2 x_1 w_1 + a_4 w_1^2}{1 + u'}$. Hence

$$|m|_p = |-a_1 w_1 + 3x_1^2 + 2a_2 x_1 w_1 + a_4 w_1^2|_p$$
$$\leq \max\{p^{-3n}, p^{-2n}, p^{-4n}, p^{-6n}\}$$
$$\leq p^{-2n}. \tag{5.19}$$

Both cases. Thus in both cases, (5.18) and (5.19) say

$$|m|_p \leq p^{-2n}. \tag{5.20}$$

Since $w_1 = mx_1 + b$, we have $b = w_1 - mx_1$, and (5.20) gives

$$|b|_p \leq p^{-3n}. \tag{5.21}$$

The three points P_1, P_2, P_3 satisfy (5.14) and (5.15) with multiplicities. Substitution of (5.14) into (5.15) gives a cubic equation

$$(mx + b) + a_1 x (mx + b) + a_3 (mx + b)^2$$
$$= x^3 + a_2 x^2 (mx + b) + a_4 x (mx + b)^2 + a_6 (mx + b)^3$$

whose roots, with multiplicities, are x_1, x_2, x_3. Then $x_1 + x_2 + x_3$ must be minus the quotient of the coefficient of x^2 by the coefficient of x^3, and we obtain

$$x_1 + x_2 + x_3 = -\frac{-a_1 m - a_3 m^2 + a_2 b + 2a_4 mb + 3a_6 m^2 b}{1 + a_2 m + a_4 m^2 + a_6 m^3}.$$

3. p-ADIC FILTRATION

By (5.20) the denominator has norm 1. Thus (5.20) and (5.21) show that

$$|x_1 + x_2 + x_3|_p \leq \max\{|a_1|_p p^{-2n}, p^{-3n}\} \leq \begin{cases} p^{-3n} & \text{if } a_1 = 0 \\ p^{-2n} & \text{in any case,} \end{cases}$$

as required. Since $|x_1| \leq p^{-n}$ and $|x_2|_p \leq p^{-n}$, we deduce that $|x_3|_p \leq p^{-n}$. From $w_3 = mx_3 + b$, we obtain $|w_3|_p \leq p^{-3n}$ by using (5.20) and (5.21). Thus P_3 is in $E^{(n)}(\mathbf{Q})$. This proves the lemma.

REMARK. The proof of Proposition 5.8 did not make use of the full strength of the lemma since it did not isolate the case $a_1 = 0$. When $a_1 = 0$, one can go over the proof of the proposition to see that p^{2n} can be replaced by p^{3n} and $E^{(2n)}$ can be replaced by $E^{(3n)}$. In particular, $a_1 = 0$ implies that there is a well defined map

$$E^{(n)}(\mathbf{Q})/E^{(3n)}(\mathbf{Q}) \longrightarrow p^n R/p^{3n} R, \tag{5.22}$$

and it is a one-one homomorphism. We shall use this observation critically in the next proposition.

Proposition 5.10. For each odd prime p, $E(\mathbf{Q})_{\text{tors}} \cap E^{(1)}(\mathbf{Q}) = 0$. This conclusion extends to $p = 2$ if $a_1 = 0$.

PROOF. Fix p. First assume $a_1 = 0$. If $E(\mathbf{Q})_{\text{tors}} \cap E^{(1)}(\mathbf{Q}) \neq 0$, the intersection contains a nonzero element P of some prime order q. Since $\bigcap_{n=1}^{\infty} E^{(n)}(\mathbf{Q}) = \{O\}$, we can find n such that P is in $E^{(n)}(\mathbf{Q})$ but not $E^{(n+1)}(\mathbf{Q})$. From $qP = 0$ and the homomorphism property of the map in Proposition 5.8 (as amended by (5.22)), we see that

$$qx(P) \quad \text{is in } p^{3n} R. \tag{5.23}$$

If $q \neq p$, then $x(P)$ must be in $p^{3n} R \subseteq p^{2n} R$, while if $q = p$, then $x(P)$ must be in $p^{3n-1} R \subseteq p^{2n} R$. In either case the one-oneness in Proposition 5.8 implies

$$P \in E^{(2n)}(\mathbf{Q}) \subseteq E^{(n+1)}(\mathbf{Q}),$$

contradiction.

Now we allow $a_1 \neq 0$ but assume p is odd. Suppose by way of contradiction that $P = (x, 1, w)$ is a nonzero element of $E(\mathbf{Q})_{\text{tors}} \cap E^{(1)}(\mathbf{Q})$ for the equation (5.15). Then P is a torsion point with $|x|_p < 1$ and $|w|_p < 1$. Also $w \neq 0$. Write

$$[(x, 1, w)] = [(\bar{x}, \bar{y}, 1)]$$

by putting $\bar{x} = \dfrac{x}{w}$ and $\bar{y} = \dfrac{1}{w}$. Then $(\bar{x}, \bar{y}, 1)$ is a torsion point with

$$\bar{y}^2 + a_1 \bar{x}\bar{y} + a_3 \bar{y} = \bar{x}^3 + a_2 \bar{x}^2 + a_4 \bar{x} + a_6. \tag{5.24}$$

We make an admissible change of variables, putting

$$\bar{x} = \frac{1}{4}\bar{x}' \qquad \text{and} \qquad \bar{y} = \frac{1}{8}(\bar{y}' - a_1 \bar{x}').$$

Substituting into (5.24), we see that $(\bar{x}', \bar{y}', 1)$ is a torsion point of

$$\bar{y}'^2 + 8a_3 \bar{y}' = \bar{x}'^3 + (4a_2 + a_1^2)\bar{x}'^2 + (16a_4 + 8a_1 a_3)\bar{x}' + 64a_6.$$

The point $(x', 1,' w')$ with $[(x', 1, w')] = [(\bar{x}', \bar{y}', 1)]$ has $\bar{x}' = \dfrac{x'}{w'}$ and $\bar{y}' = \dfrac{1}{w'}$, and it is a torsion point of

$$w' + 8a_3 w'^2 = x'^3 + (4a_2 + a_1^2)x'^2 w' + (16a_4 + 8a_1 a_3)x' w'^2 + 64a_6 w'^3. \tag{5.25}$$

This equation has integer coefficients, and its $x'w'$ term has coefficient 0. From the previous paragraph it has $E(\mathbb{Q})_{\text{tors}} \cap E^{(1)}(\mathbb{Q}) = 0$. We shall obtain a contradiction by showing that $(x', 1,' w')$ is in this intersection. In fact, we know $(x', 1, w')$ is a torsion point. Also

$$w' = \frac{w}{8 + 4a_1 x}$$

and p odd and $|x|_p < 1$ imply $|8 + 4a_1 x|_p = 1$, so that $|w'|_p = |w|_p \leq p^{-3}$. By Lemma 5.7, $|x'|_p \leq p^{-1}$. Thus $(x', 1, w')$ is in $E^{(1)}(\mathbb{Q})$. This contradiction completes the proof.

4. Lutz-Nagell Theorem

In this section we shall prove Theorem 5.1. The notation is as in §1.

PROOF OF THEOREM 5.1a. Under the assumption that $a_1 = 0$ let $(x, y, 1)$ be in $E(\mathbb{Q})_{\text{tors}}$. We are to prove that x and y are in $\mathbb{Z}$. First we handle y. Without loss of generality, we may suppose $y \neq 0$. Then $[(x, y, 1)] = [(X, 1, W)]$, where $X = x/y$ and $W = 1/y$. Fix any prime p. By Proposition 5.10, the torsion point $(X, 1, W)$ is not in $E^{(1)}(\mathbb{Q})$. By Lemma 5.7, $|W|_p \geq 1$. Thus $|y|_p = \left|\dfrac{1}{W}\right|_p \leq 1$. Since this condition holds for all p, y is an integer.

Regarding y as an integer constant in (3.23b), we see that x satisfies a monic cubic polynomial equation with integer coefficients. For such an equation, the rational roots are integers. Thus x is an integer.

PROOF OF THEOREM 5.1b. Without the assumption that $a_1 = 0$, let $(x, y, 1)$ be in $E(\mathbf{Q})_{\text{tors}}$. Then $(x', y', 1) = (4x, 8y, 1)$ satisfies

$$y'^2 + 2a_1 x'y' + 2^3 a_3 y' = x'^3 + 2^2 a_2 x'^2 + 2^4 a_4 x' + 2^6 a_6,$$

and $(x'', y'', 1) = (x', y' + a_1 x', 1)$ satisfies

$$y''^2 + 8a_3 y'' = x''^3 + (4a_2 + a_1^2)x''^2 + (16a_4 + 8a_1 a_3)x'' + 64a_6.$$

By Theorem 5.1a, x'' and y'' are integers. Tracing back, we see that x' and y' are integers. Hence $4x$ and $8y$ are integers.

PROOF OF THEOREM 5.1c. When $p \nmid \Delta$, we have $\ker(r_p) = E^{(1)}(\mathbf{Q})$. Under our hypotheses, Proposition 5.10 says that $E(\mathbf{Q})_{\text{tors}} \cap E^{(1)}(\mathbf{Q}) = 0$. Thus the restriction of r_p to $E(\mathbf{Q})_{\text{tors}}$ is one-one.

PROOF OF THEOREM 5.1d. Let E be of the form (5.1). The doubling formula (3.74) gives

$$x(2P) = \frac{x^4 - 2Ax^2 - 8Bx + A^2}{4(x^3 + Ax + B)},$$

and we write this as $x(2P) = \dfrac{\nu(x)}{4\delta(x)}$. Since $\delta(x) = y^2$, we have

$$\nu(x) = 4y^2 x(2P). \tag{5.26}$$

Now x, $x(2P)$, and y^2 are integers by Theorem 5.1a, and we see from (5.26) that $\nu(x)$ is an integer and that $y^2 \mid \nu(x)$. Thus y^2 divides both $\nu(x)$ and $\delta(x)$. Direct calculation gives

$$(3x^2 + 4A)\nu(x) - (3x^3 - 5Ax - 27B)\delta(x) = 4A^3 + 27B^2 = -d.$$

Since y^2 divides both $\nu(x)$ and $\delta(x)$, y^2 divides d.

5. Construction of Curves with Prescribed Torsion

In this section we study how to construct elliptic curves E over $\mathbf{Q}$ with prescribed features in $E(\mathbf{Q})_{\text{tors}}$. The first part of the section shows how to produce elements of a given order. Then we shall discuss how to make $\mathbf{Z}_{2n} \oplus \mathbf{Z}_2$ be a subgroup of $E(\mathbf{Q})_{\text{tors}}$. Finally we summarize matters by saying how Table 5.1 in §1 was constructed.

In studying behavior at a single point, we may translate the curve to make the special point be $P = (0,0)$ in the affine plane.

After this translation, we have $a_6 = 0$, and the curve is

$$y^2 + a_1 xy + a_3 y = x^3 + a_2 x^2 + a_4 x. \tag{5.27}$$

From §III.5 the curve is singular if and only if $a_3 = a_4 = 0$.

Assume that (5.27) is nonsingular. Then $P = (0,0)$ is of order 2 if and only if the tangent is vertical there. Vertical tangency occurs when, in the implicitly differentiated form of the curve, the coefficient of dy is 0. In (5.27), dy has coefficient $2y + a_1 x + a_3$. Thus P has order 2 if and only if $a_3 = 0$ (and therefore $a_4 \neq 0$).

Assume next that (5.27) is nonsingular and that $a_3 \neq 0$. If we make the admissible change of variables

$$(x, y) = (x', y' + a_3^{-1} a_4 x')$$

in (5.27), then P remains at $(0,0)$ and we are led to

$$y'^2 + (a_1 + 2a_3^{-1} a_4) x' y' + a_3 y' = x'^3 + (a_2 - a_1 a_3^{-1} a_4 - a_3^{-2} a_4^2) x'^2.$$

Changing notation, we can rewrite this as

$$y^2 + a_1 xy + a_3 y = x^3 + a_2 x^2 \tag{5.28}$$

with no x term. With $P = (0,0)$, the numbered formulas of §III.4 give

$$-P = (0, -a_3) \tag{5.29a}$$

and

$$2P = \left(-\frac{b_8}{b_6},\, a_1 \frac{b_8}{b_6} - a_3 \right) = (-a_2,\, a_1 a_2 - a_3). \tag{5.29b}$$

Since $3P = O$ if and only if $-P = 2P$, we see that $P = (0,0)$ in (5.28) has order 3 if and only if $a_2 = 0$.

Now assume that we have arrived at a nonsingular (5.28), i.e., $a_3 \neq 0$, and suppose $a_2 \neq 0$. Then we can eliminate one of the parameters. Namely if we make the admissible change of variables

$$(x, y) = (x'/u^2,\, y'/u^3)$$

with $u = a_3^{-1} a_2$, then P remains at $(0,0)$ and we are led to

$$y^2 + a_3^{-1} a_1 a_2 xy + a_3^{-2} a_2^3 y = x^3 + a_3^{-2} a_2^3 x^2$$

5. CURVES WITH PRESCRIBED TORSION

with equal coefficients for the y and x^2 terms. Changing notation, we can rewrite this as

$$E(b,c) \; : \; y^2 + (1-c)xy - by = x^3 - bx^2.$$

This is called the **Tate normal form**. It is nonsingular if and only if $b = 0$. Direct calculation shows that the discriminant is

$$\Delta(b,c) = (1-c)^4 b^3 - (1-c)^3 b^3 - 8(1-c)^2 b^4 + 36(1-c)b^4 - 27b^4 + 16b^5.$$

From (5.29) and (3.70), we obtain

$$\begin{aligned} P &= (0,0) & -P &= (0,b) \\ 2P &= (b, bc) & -2P &= (b, 0). \end{aligned} \qquad (5.30a)$$

Using (3.72) and (3.73), and then (3.70), we find

$$3P = (c, b-c) \qquad -3P = (c, c^2). \qquad (5.30b)$$

From (5.30) we can easily choose b and c to make $P = (0,0)$ have $4P = O$ or $5P = O$ or $6P = O$. To achieve these three conditions, we just make $2P = -2P$ or $3P = -2P$ or $3P = -3P$. From (5.30) we obtain immediately

$$\begin{aligned} 4P &= O & \iff & \quad c = 0 \\ 5P &= O & \iff & \quad b = c \\ 6P &= O & \iff & \quad b = c^2 + c. \end{aligned} \qquad (5.31)$$

Let us turn to the question how to make $\mathbf{Z}_{2n} \oplus \mathbf{Z}_2$ be a subgroup of $E(\mathbf{Q})_{\text{tors}}$. The tool will be Theorem 4.2. Since $\mathbf{Z}_2 \oplus \mathbf{Z}_2$ will have to be a subgroup, we can assume from the outset that E is of the form

$$y^2 = (x - \alpha)(x - \beta)(x - \gamma)$$

with α, β, γ in $\mathbf{Z}$. Translating $(\gamma, 0)$ to $(0, 0)$ and changing notation, we can write the curve as

$$y^2 = x(x-\alpha)(x-\beta). \qquad (5.32)$$

For $\mathbf{Z}_4 \oplus \mathbf{Z}_2$ to be contained in $E(\mathbf{Q})_{\text{tors}}$, one of the points $(x_0, 0)$ with $x_0 \in \{0, \alpha, \beta\}$ must be the double of something. Suppose $(0,0)$ is the point that is to be a double. Theorem 4.2 says that the necessary and sufficient condition is that $0-0$, $0-\alpha$, and $0-\beta$ are squares. Thus any curve

$$y^2 = x(x+r^2)(x+s^2)$$

has $\mathbf{Z}_4 \oplus \mathbf{Z}_2 \subseteq E(\mathbf{Q})_{\text{tors}}$. The subgroup $\mathbf{Z}_4 \oplus \mathbf{Z}_2$ consists of the elements

$$(0,0),\ (-r^2,0),\ (-s^2,0),\ (rs, \pm rs(r+s)),\ (-rs, \pm rs(r-s)),\ \infty.$$

This construction was used in connection with Example 4 in §1.

The application of Theorem 4.2 can be repeated, this time to be used on $(rs, rs(r+s))$, and the result is a condition for $\mathbf{Z}_8 \oplus \mathbf{Z}_2$ to be a subgroup of $E(\mathbf{Q})_{\text{tors}}$.

To summarize matters, let us say how Table 5.1 in §1 was constructed. For as many cases as possible, the examples are from Theorems 5.2 and 5.3. For $\mathbf{Z}_5$ we used (5.31), and for $\mathbf{Z}_7, \mathbf{Z}_8, \mathbf{Z}_9$ we used extensions of (5.31) that deal with $7P = O$, $8P = O$, $9P = O$. For $\mathbf{Z}_{10}$, we started with the standard form with $P = (0,0)$ of order 5 and found a case where P was the output from the doubling formula (3.74). For $\mathbf{Z}_{12}$, the same technique in principle should work with $P = (0,0)$ of order 6, but we simply quoted a known example. The examples with $\mathbf{Z}_4 \oplus \mathbf{Z}_2$ and $\mathbf{Z}_8 \oplus \mathbf{Z}_2$ were obtained by the method in the previous two paragraphs, and the example for $\mathbf{Z}_6 \oplus \mathbf{Z}_2$ was obtained by starting with the form (5.31) for $6P = 0$ and looking for a case that had three elements of order 2.

6. Torsion Groups for Special Curves

Our objective in this section is to prove Theorems 5.2 and 5.3 in §1. Each theorem requires a lemma about $E_p(\mathbf{Z}_p)$ for certain primes p. Also we shall use the form of the doubling formula (3.74) when specialized to a curve $y^2 = x^3 + Ax + B$:

$$x(2P) = \frac{x^4 - 2Ax^2 - 8Bx + A^2}{4(x^3 + Ax + B)}. \tag{5.33}$$

And we shall use Dirichlet's Theorem that there are infinitely many (positive) primes $an+b$ if $\text{GCD}(a,b) = 1$; this theorem will be proved in Chapter VII.

Lemma 5.11. Let E_p be the curve $y^2 = x^3 + Ax$ over $\mathbf{Z}_p$, and assume that $p \nmid \Delta$, $p \geq 7$, and $p \equiv 3 \mod 4$. Then $E_p(\mathbf{Z}_p)$ has exactly $p+1$ points.

6. TORSION GROUPS FOR SPECIAL CURVES

PROOF. We start from the known result that $p \equiv 3 \mod 4$, implies that -1 is not a square modulo p. For $x \neq 0$, consider the pair $\{x, -x\}$. When these elements are substituted into E, we obtain $x^3 + Ax$ and $-(x^3 + Ax)$. If the answers are 0, each one has a square root, and we get one solution from each. If they are nonzero, exactly one is a square (since -1 is not a square), and it has two square roots y. So in either case, the pair $\{x, -x\}$ gives us two solutions. Thus the nonzero x's give us $p - 1$ solutions in all. For $x = 0$, we get one more solution $(0, 0)$, and ∞ gives us one additional solution. Thus $E_p(\mathbf{Z}_p)$ has $p + 1$ points.

PROOF OF THEOREM 5.2. The main step is to show that $|E(\mathbf{Q})_{\text{tors}}|$ divides 4. By Theorem 5.1c, for all sufficiently large primes p, $|E(\mathbf{Q})_{\text{tors}}|$ divides $|E_p(\mathbf{Z}_p)|$. By Lemma 5.11, $|E(\mathbf{Q})_{\text{tors}}|$ divides $p + 1$ for all sufficiently large primes p with $p \equiv 3 \mod 4$.

Let us see that 8 does not divide $|E(\mathbf{Q})_{\text{tors}}|$. By Dirichlet's Theorem, we can choose a prime p as in the previous sentence with $p \equiv 3 \mod 8$. If 8 divides $|E(\mathbf{Q})_{\text{tors}}|$, then $8 \mid (p+1)$. But $p \equiv 3 \mod 8$ means that $p + 1 \equiv 4 \mod 8$; so $8 \nmid (p+1)$, contradiction.

Now let us see that 3 does not divide $|E(\mathbf{Q})_{\text{tors}}|$. By Dirichlet's Theorem, we can choose p large with $p \equiv 7 \mod 12$. Then $p \equiv 3 \mod 4$. Thus $3 \mid |E(\mathbf{Q})_{\text{tors}}|$ implies $3 \mid (p+1)$. But $p + 1 \equiv 8 \mod 12$ implies $p + 1 \equiv 8 \mod 3$; so $3 \nmid (p+1)$, contradiction.

Finally let us see that no odd prime $q > 3$ divides $|E(\mathbf{Q})_{\text{tors}}|$. By Dirichlet's Theorem, we can choose p large with $p \equiv 3 \mod 4q$. Then $p \equiv 3 \mod 4$. Thus $q \mid |E(\mathbf{Q})_{\text{tors}}|$ implies $q \mid (p+1)$. But $p + 1 \equiv 4 \mod 4q$ implies $p + 1 \equiv 4 \mod q$; so $q \nmid (p+1)$, contradiction.

This completes the proof that $|E(\mathbf{Q})_{\text{tors}}|$ divides 4. The torsion group will then contain $\mathbf{Z}_2 = \{(0, 0), \infty\}$ as a subgroup, and it will be $\mathbf{Z}_2 \oplus \mathbf{Z}_2$ if and only if $x^3 + Ax$ splits over $\mathbf{Q}$, i.e., if and only if $-A$ is a square. Thus the only question is when $(0, 0)$ is the double of something (so that the torsion group is $\mathbf{Z}_4$ rather than $\mathbf{Z}_2$). We can check directly for $A = 4$ that $(2, 4)$ doubles to $(0, 0)$.

Consider the equation $2(x, y) = (0, 0)$ for other A. By (5.33), we have

$$0 = x^4 - 2Ax^2 + A^2 = (x^2 - A)^2.$$

Thus $x^2 = A$. Since A is fourth-power free, x is square free. But $y^2 = x(x^2 + A) = x(x^2 + x^2) = 2x^3$ then shows that no odd prime can divide x. So $x = \pm 1$ or ± 2. Checking the possibilities, we see that $x = \pm 2$ and $A = 4$.

Lemma 5.12. Let E_p be the curve $y^2 = x^3 + B$ over $\mathbf{Z}_p$, and assume that $p \nmid \Delta$, $p \geq 5$, and $p \equiv 2 \mod 3$. Then $E_p(\mathbf{Z}_p)$ has exactly $p + 1$ points.

PROOF. Let $p = 3n+2$. The multiplicative group $\mathbf{Z}_p^\times$ has order $p-1$. Since $3 \nmid (p-1)$, no element has order 3. Therefore the homomorphism $a \to a^3$ on $\mathbf{Z}_p^\times$ is one-one, hence onto. Thus each element of $\mathbf{Z}_p$ has a unique cube root. For each y in $\mathbf{Z}_p$, the element $y^2 - B$ has a unique cube root, which we can take as x. In this way we obtain p solutions. Adjoining ∞, we see that $E_p(\mathbf{Z}_p)$ has $p+1$ points.

PROOF OF THEOREM 5.3. The main step is to show that $|E(\mathbf{Q})_{\text{tors}}|$ divides 6. By Theorem 5.1c, for all sufficiently large primes p, $|E(\mathbf{Q})_{\text{tors}}|$ divides $|E_p(\mathbf{Z}_p)|$. By Lemma 5.12, $|E(\mathbf{Q})_{\text{tors}}|$ divides $p+1$ for all sufficiently large primes p with $p \equiv 2 \mod 3$.

Let us see that 4 does not divide $|E(\mathbf{Q})_{\text{tors}}|$. By Dirichlet's Theorem, we can choose a prime p as in the previous sentence with $p \equiv 5 \mod 12$. If 4 divides $|E(\mathbf{Q})_{\text{tors}}|$, then $4 \mid (p+1)$. But $p \equiv 1 \mod 4$ means that $p + 1 \equiv 2 \mod 4$; so $4 \nmid (p+1)$, contradiction.

Now let us see that 9 does not divide $|E(\mathbf{Q})_{\text{tors}}|$. By Dirichlet's Theorem, we can choose p large with $p \equiv 2 \mod 9$. Then $p \equiv 2 \mod 3$. Thus $9 \mid |E(\mathbf{Q})_{\text{tors}}|$ implies $9 \mid (p+1)$. But $p + 1 \equiv 3 \mod 9$ implies $9 \nmid (p+1)$, contradiction.

Finally let us see that no odd prime $q > 3$ divides $|E(\mathbf{Q})_{\text{tors}}|$. By Dirichlet's Theorem, we can choose p large with $p \equiv 2 \mod 3q$. Then $p \equiv 2 \mod 3$. Thus $q \mid |E(\mathbf{Q})_{\text{tors}}|$ implies $q \mid (p+1)$. But $p + 1 \equiv 3 \mod 3q$ implies $p + 1 \equiv 3 \mod q$; so $q \nmid (p+1)$, contradiction.

This completes the proof that $|E(\mathbf{Q})_{\text{tors}}|$ divides 6. The torsion group has an element of order 2 if and only if $x^3 + B$ has a first-degree factor over $\mathbf{Q}$, i.e., if and only if B is a cube. Thus the only question is when the torsion group has elements of order 3. Such a point $P = (x, y)$ is characterized by $2P = -P$. Moreover, the x coordinate determines everything, since $2P = +P$ is impossible for $P \neq O$. By (5.33) the question is whether

$$\frac{x^4 - 8Bx}{4(x^3 + B)} = x$$

has any rational solutions x. Clearing fractions, we have

$$4x^4 + 4Bx = x^4 - 8Bx$$

$$x^4 = -4Bx.$$

One solution is $x = 0$, which gives $y^2 = B$; so $\mathbf{Z}_3$ occurs if B is a square. The only other possibility is $x^3 = -4B$. Then $y^2 = -3B$. Consequently $B < 0$. Since B is sixth-power free, the only possible prime divisors of B are 2 and 3. We readily find $B = -2^4 3^3$. So $\mathbf{Z}_3$ occurs if and only if either B is a square or $B = -2^4 3^3$.

CHAPTER VI

COMPLEX POINTS

1. Overview

We shall consider meromorphic functions on $\mathbf{C}$ periodic with two periods that are linearly independent over $\mathbf{R}$. Such functions are called **elliptic functions**. We fix the periods. Among these functions, there is a basic one $\wp(z)$, and $\wp(z)$ satisfies the differential equation

$$\wp'^2 = 4\wp^3 - g_2\wp - g_3$$

for certain constants g_2 and g_3 depending on the periods. The linear independence of the two generating periods forces the cubic $4x^3 - g_2 x - g_3$ to have nonzero discriminant.

For $t \in \mathbf{C}$ in a parallelogram Π generated by the periods, we form the parametrically defined curve

$$x = \wp(t), \quad y = \wp'(t).$$

Then we have

$$y^2 = 4x^3 - g_2 x - g_3,$$

and the discriminant is nonzero. In other words, $\wp$ provides us with a parametrization of part of a curve E over $\mathbf{C}$, and E is for all intents and purposes an elliptic curve. Taking into account poles of $\wp(t)$ and using the projective plane over $\mathbf{C}$ rather than the affine plane, we shall see that the parametrization is onto $E(\mathbf{C})$. In fact, the map is biholomorphic.

There is a natural addition on our fundamental parallelogram obtained from addition in $\mathbf{C}$, and we shall see that this operation corresponds to the group law for $E(\mathbf{C})$.

Finally we shall see that every elliptic curve over $\mathbf{C}$ can be realized by this construction. This is the inversion problem, and detailed motivation for its solution appears in §5.

2. Elliptic Functions

A function $f : \mathbf{C} \to \mathbf{C} \cup \{\infty\}$ is **doubly periodic** with periods ω_1 and ω_2 if ω_1 and ω_2 are linearly independent over $\mathbf{R}$ and if $f(z + \omega_1) = f(z) = f(z + \omega_2)$ for all $z \in \mathbf{C}$. An **elliptic function** is a meromorphic doubly periodic function.

Fix ω_1 and ω_2. The corresponding elliptic functions form a field. The parallelogram with vertices 0, ω_1, ω_2, $\omega_1 + \omega_2$ is called the **fundamental parallelogram** Π. To be more precise, we shall insist that Π contain the two sides adjacent to the origin, as well as the origin, but not the other two sides and three vertices. Any $\mathbf{C}$-translate $\alpha + \Pi$ of Π is a **period parallelogram**. With our precise definition of period parallelogram, we can conclude the following: For any period parallelogram $\alpha + \Pi$, any point in $\mathbf{C}$ is congruent modulo $\mathbf{Z}\omega_1 + \mathbf{Z}\omega_2$ to one and only one point of $\alpha + \Pi$.

Proposition 6.1 (First Liouville Theorem). There exists no nonconstant elliptic function without poles.

PROOF. Such a function would have to be bounded on the closure $\bar{\Pi}$, hence entire and bounded on $\mathbf{C}$.

Corollary 6.2. If two elliptic functions have the same poles with the same respective principal parts, then the functions differ by a constant.

Proposition 6.3 (Second Liouville Theorem). If $f(z)$ is an elliptic function with no poles on the boundary C of a period parallelogram $\alpha + \Pi$, then the sum of the residues of $f(z)$ in $\alpha + \Pi$ is zero.

PROOF. By the Residue Theorem, the sum in question equals

$$\frac{1}{2\pi i} \oint_C f(z)\,dz.$$

The integrals over opposite sides of $\alpha + \Pi$ cancel (since f is the same at congruent points and dz introduces a minus sign on one of two matching sides), and thus the integral is 0.

REMARK. If f is elliptic, it has only finitely many poles on a bounded set, and thus there is always an α for which the proposition applies.

Corollary 6.4. If $f(z)$ is a nonconstant elliptic function, then either f has more than one pole in $\alpha + \Pi$ or else f has one pole and that pole has order > 1.

PROOF. There has to be a pole, and the sum of the residues is 0.

Corollary 6.5 (Third Liouville Theorem). Suppose $f(z)$ is an elliptic function with no pole or zero on the boundary C of a period parallelogram $\alpha + \Pi$. Let $\{m_i\}$ be the orders of the various zeros in $\alpha + \Pi$, and let $\{n_j\}$ be the orders of the various poles. Then $\sum_i m_i = \sum_j n_j$.

PROOF. Since f is meromorphic, the residue of $\dfrac{f'(z)}{f(z)}$ at z_0 is

$$\begin{cases} n & \text{if } n > 0 \text{ is the order of a zero of } f \text{ at } z_0 \\ 0 & \text{if } f \text{ has no zero or pole at } z_0 \\ -n & \text{if } n > 0 \text{ is the order of a pole of } f \text{ at } z_0. \end{cases}$$

Summing over the points of the period parallelogram and applying Proposition 6.3 to $\dfrac{f'(z)}{f(z)}$, we obtain the corollary.

The **order** of an elliptic function is the number of poles in $\alpha + \Pi$, counting multiplicities. In Corollary 6.5, the order is $\sum_j n_j$. Consequently the number of zeros of $f(z)$, counting multiplicities, equals the order of $f(z)$.

Corollary 6.6. Suppose $f(z)$ is an elliptic function of order m and $\alpha + \Pi$ is a period parallelogram. For each $c \in \mathbf{C}$, f takes on the value c exactly m times, counting multiplicities.

PROOF. Apply Corollary 6.5 to $f(z) - c$.

3. Weierstrass $\wp$ Function

Fix ω_1 and ω_2 in $\mathbf{C}$ linearly independent over $\mathbf{R}$, and let Π be the fundamental parallelogram. We shall define a particular nonconstant elliptic function with periods ω_1 and ω_2, thereby proving existence of nonconstant elliptic functions. Then we shall develop some properties of this function.

Write $\Lambda = \{m\omega_1 + n\omega_2 \mid m, n \in \mathbf{Z}\}$; Λ is called the **period lattice**. The **Weierstrass $\wp$ function** relative to Λ is defined by

$$\wp(z) = \frac{1}{z^2} + \sum_{\substack{\omega \in \Lambda \\ \omega \neq 0}} \left(\frac{1}{(z-\omega)^2} - \frac{1}{\omega^2} \right). \tag{6.1}$$

We shall make repeated use of the following fact from complex variable theory: If a sequence of functions analytic in an open set converges uniformly, then the limit is analytic, and limit and derivative may be interchanged.

Lemma 6.7. If s is a real number > 2, then

$$\sum_{\substack{\omega \in \Lambda \\ \omega \neq 0}} \frac{1}{|\omega|^s}$$

converges.

PROOF. Let $\tilde{\Pi}$ be the union of the four Λ translates of Π that surround the origin:

$$\tilde{\Pi} = \Pi \cup (-\omega_1 + \Pi) \cup (-\omega_2 + \Pi) \cup (-\omega_1 - \omega_2 + \Pi).$$

The boundary of $\tilde{\Pi}$ is compact and does not contain 0, and we can choose $c > 0$ so that $|z| \geq c$ for all z in the boundary of $\tilde{\Pi}$. If $|m| \geq |n| > 0$, we then have

$$|m\omega_1 + n\omega_2| \geq |m|\left|\omega_1 + \frac{n}{m}\omega_2\right| \geq c|m|, \tag{6.2}$$

since $z = \omega_1 + \frac{n}{m}\omega_2$ lies in the boundary of $\tilde{\Pi}$.

Now consider $\sum_{\omega \neq 0} |\omega|^{-s}$. If we write $\omega = m\omega_1 + n\omega_2$, the terms with $n = 0$ contribute $2|\omega_1|^{-s} \sum_{m=1}^{\infty} m^{-s} < \infty$, and the terms with $m = 0$ contribute $2|\omega_2|^{-s} \sum_{n=1}^{\infty} n^{-s} < \infty$. By (6.2), the terms with $|m| \geq |n| > 0$ contribute

$$\sum_{|m| \geq |n| > 0} |m\omega_1 + n\omega_2|^{-s} \leq c^{-s} \sum_{|m| \geq |n| > 0} |m|^{-s} = 4c^{-s} \sum_{m=1}^{\infty} m^{-s+1} < \infty,$$

and the terms with $|n| > |m| > 0$ similarly make a finite contribution. The lemma follows.

Proposition 6.8. If F is any finite subset of Λ and if the terms corresponding to F are omitted in (6.1), then the resulting series converges absolutely uniformly on any compact subset of $\mathbb{C} - (\Lambda - F)$. Consequently $\wp(z)$ is meromorphic in $\mathbb{C}$, its only poles are double poles at the points of Λ, and $\wp'(z)$ may be computed term by term.

PROOF. We may assume F contains $\omega = 0$. The sum for $\Lambda - F$ is

$$\sum_{\omega \in \Lambda - F} \left(\frac{1}{(z-\omega)^2} - \frac{1}{\omega^2} \right) = \sum_{\omega \in \Lambda - F} \frac{2z - \frac{z^2}{\omega}}{\omega(z-\omega)^2},$$

and we have

$$\left| \frac{2z - \frac{z^2}{\omega}}{\omega(z-\omega)^2} \right| \le \frac{C}{|\omega|^3}$$

on our compact set. By Lemma 6.7, $\sum_{\omega \in \Lambda - F} \frac{C}{|\omega|^3} < \infty$. Therefore the absolute uniform convergence follows by the Weierstrass M-test. The fact from complex variable theory at the beginning of the section shows that the limit is meromorphic. The poles are as indicated because of the convergence, and the same fact from the beginning of the section shows we can compute $\wp'(z)$ term by term.

Proposition 6.9. The function $\wp(z)$ is an elliptic function with ω_1 and ω_2 as periods, and $\wp(-z) = \wp(z)$. The order of $\wp(z)$ is 2.

PROOF. We have $\wp(-z) = \wp(z)$ since the right side of (6.1) is unchanged if z is replaced by $-z$ and ω is replaced by $-\omega$. Differentiation of (6.1) gives

$$\wp'(z) = -2 \sum_{\omega \in \Lambda} \frac{1}{(z-\omega)^3},$$

and this is certainly doubly periodic. Hence $\wp'(z)$ is elliptic.

For $\omega \in \Lambda$, we differentiate $\wp(z+\omega) - \wp(z)$ and obtain $\wp'(z+\omega) - \wp'(z) = 0$. Therefore $\wp(z+\omega) - \wp(z) = C$. Evaluating at $z = -\frac{1}{2}\omega_1$ (where there is no pole, according to Proposition 6.8), we obtain

$$C = \wp(\tfrac{1}{2}\omega_1) - \wp(-\tfrac{1}{2}\omega_1) = \wp(\tfrac{1}{2}\omega_1) - \wp(\tfrac{1}{2}\omega_1) = 0.$$

Hence $\wp(z+\omega_1) = \wp(z)$. Similarly $\wp(z+\omega_2) = \wp(z)$. Since $\wp$ is meromorphic, $\wp$ is elliptic. Since $z = 0$ is the only pole of $\wp(z)$ in Π, the order of $\wp(z)$ is 2.

Theorem 6.10 Any even elliptic function (relative to periods ω_1 and ω_2) is a rational function of $\wp(z)$. Any elliptic function f is of the form $f(z) = g(\wp(z)) + \wp'(z)h(\wp(z))$ with g and h rational.

PRELIMINARY REMARKS. The second statement follows from the first since $f = f_e + f_o$ with $f_e(z) = \frac{1}{2}(f(z) + f(-z))$ even and with $f_o(z) = \frac{1}{2}(f(z) - f(-z))$ odd, and since $f_o = \wp' \cdot \frac{f_o}{\wp'}$ with $\frac{f_o}{\wp'}$ even. The proof of the first statement will be preceded by a lemma.

Lemma 6.11.

(a) For any $u \in \mathbb{C}$, the elliptic function $\wp(z) - u$ has within Π either two simple zeros or one double zero.

(b) The zeros of $\wp'(z)$ within Π are $\frac{1}{2}\omega_1$, $\frac{1}{2}\omega_2$, and $\frac{1}{2}(\omega_1 + \omega_2)$, all simple.

(c) The values $u_1 = \wp(\frac{1}{2}\omega_1)$, $u_2 = \wp(\frac{1}{2}\omega_2)$, and $u_3 = \wp(\frac{1}{2}(\omega_1+\omega_2))$ are the exact u's within Π where $\wp(z) - u$ has a double zero, and u_1, u_2, u_3 are distinct.

PROOF.

(a) Since $\wp(z)$ has order 2, we can apply Corollary 6.6.

(b) The function $\wp'(z)$ has order 3. Hence it has at most 3 zeros. Since $\wp'(z)$ is odd and periodic, we have

$$\wp'(\tfrac{1}{2}\omega_1) = -\wp'(-\tfrac{1}{2}\omega_1) = -\wp'(\omega_1 - \tfrac{1}{2}\omega_1) = -\wp'(\tfrac{1}{2}\omega_1).$$

Thus $\frac{1}{2}\omega_1$ is a zero. Similarly $\frac{1}{2}\omega_2$ and $\frac{1}{2}(\omega_1 + \omega_2)$ are zeros.

(c) If $\wp(z) - u$ has a zero at z_0, the zero is double if and only if $\wp'(z_0) = 0$. By (b), $\wp(z) - u$ can have a double zero only at $\frac{1}{2}\omega_1$, $\frac{1}{2}\omega_2$, $\frac{1}{2}(\omega_1 + \omega_2)$, and in these cases u clearly must be u_1, u_2, u_3, respectively. Conversely if $u = u_1$, then $\wp(z) - u_1$ is 0 at $z = \frac{1}{2}\omega_1$, and also $\wp'(\frac{1}{2}\omega_1) = 0$. So we have a double 0. Similar remarks apply to $\frac{1}{2}\omega_2$ and to $\frac{1}{2}(\omega_1 + \omega_2)$. Finally if at least two of u_1, u_2, u_3 were equal, say to u_0, then $\wp(z) - u_0$ would have at least two double zeros, in contradiction to (a); we conclude that u_1, u_2, u_3 are distinct.

PROOF OF THEOREM 6.10. We are to prove that any even elliptic function is a rational function of $\wp(z)$. Let $f(z)$ be even elliptic. Taking into account the evenness, we shall list "half" the zeros and poles, temporarily ignoring what happens at $z = 0$.

Let $f(a) = 0$. First suppose $a \in \Pi$ and $a \notin \{0, \frac{1}{2}\omega_1, \frac{1}{2}\omega_2, \frac{1}{2}(\omega_1 + \omega_2)\}$. Let a^* be the symmetric point, the point in Π congruent to $-a$:

$$a^* = \begin{cases} \omega_1 + \omega_2 - a & \text{if } a \text{ is interior (and not } \tfrac{1}{2}(\omega_1 + \omega_2)) \\ \omega_1 - a & \text{if } a \text{ is on the side to } \omega_1 \text{ (and not } \tfrac{1}{2}\omega_1) \\ \omega_2 - a & \text{if } a \text{ is on the side to } \omega_2 \text{ (and not } \tfrac{1}{2}\omega_2). \end{cases} \quad (6.3)$$

If a has order m, so does a^*: In fact,

$$f(a^* - z) = f(\text{period} - a - z) = f(-a - z) = f(a + z).$$

If $f(a + z) = a_m z^m + \text{higher}$, then

$$f(a^* + z) = f(a - z) = a_m(-z)^m + \text{higher}.$$

Next suppose a is one of $\frac{1}{2}\omega_1, \frac{1}{2}\omega_2, \frac{1}{2}(\omega_1+\omega_2)$. Say $a = \frac{1}{2}\omega$. Then we show a has even order. In fact,

$$f(\tfrac{1}{2}\omega - z) = f(-\tfrac{1}{2}\omega - z) = f(\tfrac{1}{2}\omega + z)$$

shows f is even about $a = \frac{1}{2}\omega$. Hence the order is even.

A similar argument applies to poles. If f has a pole at $a \in \Pi$ with $a \notin \{0, \frac{1}{2}\omega_1, \frac{1}{2}\omega_2, \frac{1}{2}(\omega_1+\omega_2)\}$, then f has a pole at a^* of the same order. Also the order of a pole at any of $\frac{1}{2}\omega_1, \frac{1}{2}\omega_2, \frac{1}{2}(\omega_1+\omega_2)$ is even.

Now we list "half" the zeros and poles of $f(z)$. Let $\{a_i\}$ be a list of the zeros of $f(z)$ in Π other than $0, \frac{1}{2}\omega_1, \frac{1}{2}\omega_2, \frac{1}{2}(\omega_1+\omega_2)$, each taken with its multiplicity, but with only one taken from each pair a, a^*. For any zero among $\frac{1}{2}\omega_1, \frac{1}{2}\omega_2, \frac{1}{2}(\omega_1+\omega_2)$, we list the point half as often as its multiplicity. Similarly let $\{b_j\}$ be a list of "half" the poles in Π other than 0.

Since all the a_i and b_j are nonzero, $\wp(a_i)$ and $\wp(b_j)$ are finite for all i and j. Thus it makes sense to define

$$g(z) = \frac{\prod_i (\wp(z) - \wp(a_i))}{\prod_j (\wp(z) - \wp(b_j))}.$$

We claim that g has the same zeros and poles as f, counting multiplicities. Since the only poles of the numerator and the denominator are at $z = 0$, the only other zeros and poles of $g(z)$ come from zeros of the numerator and the denominator.

Consider a zero z_0 of the numerator, a point where $\wp(z_0) = \wp(a_i)$. If a_i is any of $\frac{1}{2}\omega_1, \frac{1}{2}\omega_2$, or $\frac{1}{2}(\omega_1+\omega_2)$, then Lemma 6.11 says $\wp$ takes on the value $\wp(a_i)$ twice at that point and nowhere else. So $z_0 = a_i$ and $\wp(z) - \wp(a_i)$ has a zero of order two at z_0. Taking into account repeated factors for this a_i, we see that f and g have a zero of the same order at z_0.

Next suppose a_i is not $\frac{1}{2}\omega_1, \frac{1}{2}\omega_2$, or $\frac{1}{2}(\omega_1+\omega_2)$. We have $\wp(a_i^*) = \wp(a_i)$, so that $\wp(z) - \wp(a_i)$ has distinct zeros at a_i and a_i^*. The lemma says these zeros are simple. Taking into account repeated factors for this a_i, we see that f and g have a zero of the same order at z_0.

Thus f and g have the same zeros (including their orders), apart from $z = 0$. Similarly they have the same poles (including their orders), apart from $z = 0$. By Corollary 6.5, they have the same order of zero or pole at $z = 0$. Consequently f/g is entire and therefore constant, by Proposition 6.1. This completes the proof.

Theorem 6.12. The function $w = \wp(z)$ satisfies the differential equation
$$\left(\frac{dw}{dz}\right)^2 = 4w^3 - g_2 w - g_3,$$
where g_2 and g_3 are constants depending on Λ in the following way:
$$G_m = G_m(\Lambda) = \sum_{\substack{\omega \in \Lambda \\ \omega \neq 0}} \frac{1}{\omega^m} \quad \text{for } m \geq 3 \quad (= 0 \text{ for } m \text{ odd}) \tag{6.4}$$

$$\begin{aligned} g_2 &= g_2(\Lambda) = 60 G_4 \\ g_3 &= g_3(\Lambda) = 140 G_6. \end{aligned} \tag{6.5}$$

REMARKS. The series for G_m is absolutely convergent by Lemma 6.7. The theorem will be proved after Lemma 6.13.

Lemma 6.13 (Weierstrass Double Series Theorem). Suppose $f_n(z) = \sum_{k=0}^{\infty} a_k^{(n)}(z - z_0)^k$ is analytic for $|z - z_0| < r$ and $n \geq 0$, and suppose that the series
$$\begin{aligned} F(z) &= \sum_{n=0}^{\infty} f_n(z) \\ &= [a_0^{(0)} + a_1^{(0)}(z - z_0) + a_2^{(0)}(z - z_0)^2 + \ldots] \\ &\quad + [a_0^{(1)} + a_1^{(1)}(z - z_0) + a_2^{(1)}(z - z_0)^2 + \ldots] \\ &\quad + \ldots \\ &\quad + [a_0^{(n)} + a_1^{(n)}(z - z_0) + a_2^{(n)}(z - z_0)^2 + \ldots] \\ &\quad + \ldots \end{aligned}$$
is uniformly convergent for $|z - z_0| \leq \rho$ for each $\rho < r$. Then the coefficients in any column form a convergent series. Moreover, if
$$A_k = \sum_{n=0}^{\infty} a_k^{(n)}$$
is the sum of the kth column coefficients, then $\sum_{k=0}^{\infty} A_k(z - z_0)^k$ is the Taylor series for $F(z)$ about $z = z_0$, and it converges for $|z| < r$.

PROOF. By the given convergence, $F(z)$ is analytic for $|z - z_0| < r$ and hence is given by a convergent Taylor series about z_0. Its kth Taylor coefficient is
$$\frac{1}{k!} F^{(k)}(z_0) = \sum_{n=0}^{\infty} \frac{1}{k!} f_n^{(k)}(z_0) = \sum_{n=0}^{\infty} a_k^{(n)} = A_k$$
by the complex variables fact after (6.1), and the lemma follows.

3. WEIERSTRASS $\wp$ FUNCTION

PROOF OF THEOREM 6.12. We are going to apply Lemma 6.13 to

$$\wp(z) - \frac{1}{z^2} = \sum_{\substack{\omega \in \Lambda \\ \omega \neq 0}} \left(\frac{1}{(z-\omega)^2} - \frac{1}{\omega^2} \right)$$

in a small disc about $z = 0$. Here

$$\frac{1}{(z-\omega)^2} = \frac{1}{\omega^2} + 2\frac{z}{\omega^3} + 3\frac{z^2}{\omega^4} + \cdots,$$

so that

$$\wp(z) - \frac{1}{z^2} = \sum_{\substack{\omega \in \Lambda \\ \omega \neq 0}} \left(\sum_{k=1}^{\infty} \frac{(k+1)}{\omega^{k+2}} z^k \right).$$

Thus if we define G_m by (6.4) for $m \geq 3$, the lemma says that the series for G_m is convergent (which we know already) and

$$\wp(z) - \frac{1}{z^2} = \sum_{k=1}^{\infty} (k+1) G_{k+2} z^k.$$

We see directly from the definition (6.4) that $G_m = 0$ for m odd. Therefore

$$\wp(z) = \frac{1}{z^2} + 3G_4 z^2 + 5G_6 z^4 + 7G_8 z^6 + \cdots.$$

By direct calculation we find

$$\wp'(z) = -\frac{2}{z^3} + 6G_4 z + 20G_6 z^3 + 42G_8 z^5 + O(z^7)$$

$$\wp'(z)^2 = \frac{4}{z^6} - 24G_4 \frac{1}{z^2} - 80G_6 + (36G_4^2 - 168)z^2 + O(z^4)$$

$$\wp(z)^2 = \frac{1}{z^4} + 6G_4 + 10G_6 z^2 + O(z^4)$$

$$\wp(z)^3 = \frac{1}{z^6} + 9G_4 \frac{1}{z^2} + 15G_6 + (21G_8 + 27G_4^2)z^2 + O(z^4)$$

$$\wp'(z)^2 - 4\wp(z)^3 + 60G_4 \wp(z) + 140G_6 = O(z^2).$$

The left side of the last equation is therefore without poles or constant term. By Proposition 6.1, it is 0. Defining g_2 and g_3 by (6.5), we obtain the theorem.

Theorem 6.14. The map of $\mathbf{C}/\Lambda$ into $P_2(\mathbf{C})$ given by

$$z \longrightarrow \begin{cases} (\wp(z), \wp'(z), 1) & \text{for } z \notin \Lambda \\ (0, 1, 0) & \text{for } z \in \Lambda \end{cases} \tag{6.6}$$

is holomorphic and carries $\mathbf{C}/\Lambda$ one-one onto the projective plane curve $E(\mathbf{C})$, where E is given in affine form by

$$y^2 = 4x^3 - g_2 x - g_3.$$

REMARK. Recall from §II.1 that our various systems of affine local coordinates make $P_2(\mathbf{C})$ into a complex manifold.

PROOF. For $z \neq 0$, the image is $(\wp(z), \wp'(z), 1)$, which becomes $(\wp(z), \wp'(z))$ in the affine local coordinate system given by $\Phi = I$. Since each coordinate is analytic in z, our map (6.6) is holomorphic for $z \neq 0$. Near $z = 0$, we use the affine local coordinate system given by $\Phi = \begin{pmatrix} 1 & 0 & 0 \\ 0 & 0 & 1 \\ 0 & 1 & 0 \end{pmatrix}$, and the map is given by

$$z \longrightarrow \left(\frac{\wp(z)}{\wp'(z)}, 1, \frac{1}{\wp'(z)} \right) \longrightarrow \left(\frac{\wp(z)}{\wp'(z)}, \frac{1}{\wp'(z)} \right)$$

$$0 \longrightarrow (0, 1, 0) \longrightarrow (0, 0).$$

Each coordinate is analytic in a punctured disc about 0 and is continuous at 0, hence is analytic in a disc about 0. Thus (6.6) is holomorphic.

It is clear from Theorem 6.12 that the image is contained in $E(\mathbf{C})$. Suppose z_1 and z_2 in Π map to the same point. Then $\wp(z_1) = \wp(z_2)$ and $\wp'(z_1) = \wp'(z_2)$ and $z_1 \neq 0$, $z_2 \neq 0$. Since $\wp$ has order 2, $\wp(z_1) = \wp(z_2)$ with $z_1 \neq z_2$ implies $z_2 = z_1^*$, in the notation of (6.3). But then

$$\wp'(z_2) = \wp'(z_1^*) = \wp'(\text{period} - z_1) = \wp'(-z_1) = -\wp'(z_1).$$

Since also $\wp'(z_2) = \wp'(z_1)$, we see that $\wp'(z_1) = \wp'(z_2) = 0$. By Lemma 6.11, z_1 and z_2 are members of $\{\frac{1}{2}\omega_1, \frac{1}{2}\omega_2, \frac{1}{2}(\omega_1 + \omega_2)\}$. Consequently $z_1^* = z_1$, $z_2^* = z_2$. Since $z_2 = z_1^*$, we obtain $z_1 = z_2$, contradiction. We conclude that the map (6.6) is one-one.

Finally we show that the map (6.6) is onto. Let $(x, y, 1) \in E(\mathbf{C})$ be given. Since $\wp$ has order 2, we can find z (clearly $\neq 0$) with $\wp(z) = x$. Then also $\wp(z^*) = x$. The equation

$$y^2 = 4x^3 - g_2 x - g_3 = 4\wp(z)^3 - g_2 \wp(z) - g_3 = \wp'(z)^2$$

says either that $y = \wp'(z)$, in which case z maps to our point $(x, y, 1)$, or else $y = -\wp'(z)$, in which case z maps to $(x, -y, 1)$. In the latter case, $\wp'(z^*) = -\wp'(z)$ implies z^* maps to $(x, y, 1)$.

3. WEIERSTRASS $\wp$ FUNCTION

Theorem 6.15. The cubic polynomial $4w^3 - g_2 w - g_3$ factors as

$$4w^3 - g_2 w - g_3 = 4(w - u_1)(w - u_2)(w - u_3), \qquad (6.7)$$

where $u_1 = \wp(\tfrac{1}{2}\omega_1)$, $u_2 = \wp(\tfrac{1}{2}(\omega_2))$, $u_3 = \wp(\tfrac{1}{2}(\omega_1 + \omega_2))$. Consequently the plane curve

$$E \;:\; y^2 = 4x^3 - g_2 x - g_3 \qquad (6.8)$$

is nonsingular.

PROOF. For the first statement we go over the proof of Theorem 6.10 for the even elliptic function $\wp'(z)^2$. Since the only pole in Π is at $z = 0$ and since the only zeros of $\wp'(z)$ are simple zeros at u_1, u_2, u_3, the proof shows that

$$\wp'(z)^2 = C \prod_{i=1}^{3} (\wp(z) - u_i).$$

Subtracting the expression in Theorem 6.12, we find that every value $\wp(z)$ of the $\wp$ function satisfies

$$4\wp^3 - g_2 \wp - g_3 = C \prod_{i=1}^{3} (\wp - u_i).$$

This identity is impossible unless $C = 4$ (since nontrivial cubics have only 3 roots). Thus the cubics match. This proves the first statement in the theorem. The second statement follows from Lemma 6.11c and Proposition 3.5.

The nonsingularity of E in Theorem 6.15 allows us to define a complex manifold structure on $E(\mathbf{C})$ in a natural way. About $(x, y, 1)$ we define

$$E(x, y) = y^2 - (4x^3 - g_2 x - g_3).$$

We have

$$\frac{\partial E}{\partial x} = -(12x^2 - g_2), \qquad \frac{\partial E}{\partial y} = 2y.$$

When $y \neq 0$, the Implicit Function Theorem allows us to define $y = f(x)$ implicitly so that the curve is $(x, f(x), 1)$. A chart is given by $(x, f(x), 1) \to x$.

When $\dfrac{\partial E}{\partial x} \neq 0$, the Implicit Function Theorem allows us to define $x = g(y)$ implicitly so that the curve is $(g(y), y, 1)$. A chart is given by $(g(y), y, 1) \to y$.

Finally we have to consider behavior about $(0,1,0)$. About $(x,1,w)$ we define
$$e(x,w) = w - (4x^3 - g_2 w^2 x - g_3 w^3).$$
Then $\left.\dfrac{\partial e}{\partial w}\right|_{(0,1,0)} \neq 0$, so that $w = h(x)$ implicitly and the curve is $(x,1,h(x))$. A chart is given by $(x,1,w) \to x$.

We readily check that these charts make $E(\mathbf{C})$ into a complex manifold such that inclusion into $P_2(\mathbf{C})$ is holomorphic and such that any holomorphic mapping into $P_2(\mathbf{C})$ with image in $E(\mathbf{C})$ is holomorphic into $E(\mathbf{C})$. Relative to this complex structure on $E(\mathbf{C})$, we have the following corollary.

Corollary 6.16. The inverse of the one-one map $\mathbf{C}/\Lambda \to E(\mathbf{C})$ given in (6.6) is holomorphic.

PROOF. About points $(x,y,1)$ where $y \neq 0$, the coordinate function is $(x,y,1) \to x$. Thus the coordinate version of the map (6.6) is $z \to \wp(z)$ about points z_0 where $\wp'(z_0) \neq 0$. The invertibility condition for this function (by the one-dimensional Inverse Function Theorem over $\mathbf{C}$) is that $\wp'(z_0) \neq 0$. So we have invertibility.

About points $(x,y,1)$ where $y = 0$ (i.e., $\wp'(z_0) = 0$), we are considering $z = \tfrac{1}{2}\omega_1, \tfrac{1}{2}\omega_2, \tfrac{1}{2}(\omega_1+\omega_2)$. The coordinate function is $(x,y,1) \to y$. Thus the coordinate version of the map (6.6) is $z \to \wp'(z)$ about $z = \tfrac{1}{2}\omega_1, \tfrac{1}{2}\omega_2, \tfrac{1}{2}(\omega_1+\omega_2)$. The invertibility condition is that $\wp''(z_0) \neq 0$. But $\wp''(z_0) = 0$ would mean $\wp'$ has a double 0 at z_0, in contradiction to Lemma 6.11b.

About $(0,1,0)$, we are considering $z = 0$. The coordinate function is $(x,1,w) \to x$. Thus the coordinate version of the map (6.6) is $z \to \wp(z)/\wp'(z)$. This has a simple zero at $z = 0$, and thus its derivative is nonzero at $z = 0$. This completes the proof.

4. Effect on Addition

We continue to assume that ω_1 and ω_2 in $\mathbf{C}$ are linearly independent over $\mathbf{R}$, and we continue with the notation Λ, Π, and E as in §3.

We write φ for the biholomorphic map $\varphi : \mathbf{C}/\Lambda \to E(\mathbf{C})$ given by (6.6).

Theorem 6.17. The biholomorphic map $\varphi : \mathbf{C}/\Lambda \to E(\mathbf{C})$ is a group isomorphism.

4. EFFECT ON ADDITION

REMARKS. The proof will be preceded by Lemma 6.18. Had we not already proved that the operation $(P_1, P_2) \to P_1 + P_2$ in $E(\mathbf{C})$ is associative, then Theorem 6.17 would give us a proof. This was Poincaré's approach, and as a consequence he obtained associativity for the k rational points of elliptic curves defined over k, provided the field k is a subfield of $\mathbf{C}$.

Lemma 6.18. If $f : \mathbf{C}/\Lambda \times \mathbf{C}/\Lambda \to \mathbf{C}/\Lambda$ is analytic in each variable and continuous jointly, then there exist a, b, and c in $\mathbf{C}$ such that $f(z_1, z_2) \equiv az_1 + bz_2 + c \mod \Lambda$ for all $z_1, z_2 \in \mathbf{C}$.

PROOF. We lift f to $F : \mathbf{C} \times \mathbf{C} \to \mathbf{C}/\Lambda$ with F jointly continuous and separately analytic. Then we lift F to a function $\tilde{F} : \mathbf{C} \times \mathbf{C} \to \mathbf{C}$ that is jointly continuous and separately analytic. The equality

$$F(z_1 + m\omega_1 + n\omega_2, z_2) = F(z_1, z_2) \mod \Lambda$$

says that

$$\tilde{F}(z_1 + m\omega_1 + n\omega_2, z_2) = \tilde{F}(z_1, z_2) + m_1\omega_1 + n_1\omega_2$$

with m_1, n_1 in $\mathbf{Z}$ depending on z_1, z_2, m, n. For fixed m and n, m_1 and n_1 are continuous in (z_1, z_2), hence constant. Differentiation gives

$$\frac{\partial \tilde{F}}{\partial z_1}(z_1 + m\omega_1 + n\omega_2, z_2) = \frac{\partial \tilde{F}}{\partial z_1}(z_1, z_2)$$

$$\frac{\partial \tilde{F}}{\partial z_2}(z_1 + m\omega_1 + n\omega_2, z_2) = \frac{\partial \tilde{F}}{\partial z_2}(z_1, z_2).$$

Hence $\dfrac{\partial \tilde{F}}{\partial z_1}$ and $\dfrac{\partial \tilde{F}}{\partial z_2}$ are elliptic in the first variable and thus constant in z_1. Similarly they are constant in z_2 and thus globally constant. Say $\dfrac{\partial \tilde{F}}{\partial z_1} = a$ and $\dfrac{\partial \tilde{F}}{\partial z_2} = b$. Then we have $F(z_1, z_2) = az_1 + bz_2 + c$ for a suitable c, and the lemma follows.

PROOF OF THEOREM 6.17. First we note that $\varphi(0)$ is the identity element of $E(\mathbf{C})$, translation is analytic on $E(\mathbf{C})$, and addition is continuous on $E(\mathbf{C})$. (For the last of these, it is enough to prove continuity at $(0,0)$, where it is clear.)

Form $f : \mathbf{C}/\Lambda \times \mathbf{C}/\Lambda \to \mathbf{C}/\Lambda$ given by

$$f(z_1, z_2) \equiv \varphi^{-1}(\varphi(z_1) + \varphi(z_2)) \mod \Lambda.$$

This is analytic in each variable and is jointly continuous. By Lemma 6.18, $f(z_1, z_2) \equiv az_1 + bz_2 + c \mod \Lambda$. Since $f(0,0) = 0$, $c = 0$. Since $f(z, 0) = z$, $az \equiv z \mod \Lambda$ for all z and thus $a = 1$. Similarly $b = 1$. Thus $f(z_1, z_2) = z_1 + z_2$, i.e.,

$$z_1 + z_2 \equiv \varphi^{-1}(\varphi(z_1) + \varphi(z_2)) \mod \Lambda.$$

Applying φ, we obtain

$$\varphi(z_1 + z_2) = \varphi(z_1) + \varphi(z_2),$$

as required.

An analytic map $h : E(\mathbf{C}) \to E(\mathbf{C})$ fixing 0 is called an **isogeny**. (More generally an analytic map $h : E_1(\mathbf{C}) \to E_2(\mathbf{C})$ carrying identity to identity is an **isogeny**.) We can reinterpret an isogeny h by means of $\varphi : \mathbf{C}/\Lambda \to E(\mathbf{C})$, i.e., we can study $\varphi^{-1} \circ h \circ \varphi$. In Lemma 6.18, we put $f(z_1, z_2) = \varphi^{-1} \circ h \circ \varphi(z_1)$. The lemma gives $\varphi^{-1} \circ h \circ \varphi(z) = az + c$. Since $h(0) = 0$, $c = 0$. Thus our reinterpretation is

$$h(\varphi(z)) = \varphi(az).$$

The trivial maps of this sort have $a \in \mathbf{Z}$. If a is in $\mathbf{R}$ but not $\mathbf{Z}$, take $z = \omega_1$ to see that the map is not well defined. Thus the only nontrivial isogenies from E to itself have $a \in \mathbf{C}$ with a not real. In this case we say that E has **complex multiplication**.

Proposition 6.19. *E over $\mathbf{C}$ as above has complex multiplication if and only if ω_2/ω_1 lies in a quadratic extension of $\mathbf{Q}$.*

PROOF. With a as above, h cannot be well defined unless $a\Lambda \subseteq \Lambda$. Conversely if $a\Lambda \subseteq \Lambda$, let $z_1 \equiv z_2 \mod \Lambda$, say $z_1 = z_2 + \omega$. Then $az_1 = az_2 + a\omega \in az_2 + \Lambda$ shows that $z \to az$ is well defined on $\mathbf{C}/\Lambda$, i.e., we can define an isogeny h.

Now suppose $a\Lambda \subseteq \Lambda$. Then $a\omega_1 = m\omega_1 + n\omega_2$ with $m, n \in \mathbf{Z}$. If $\tau = \omega_2/\omega_1$, this says $a = m + n\tau$. Now

$$a\omega_2 = (m + n\tau)\omega_2 \quad \text{equals} \quad m'\omega_1 + n'\omega_2.$$

Dividing by ω_1 gives

$$(m + n\tau)\tau = m' + n'\tau.$$

Thus τ satisfies a quadratic equation with integer coefficients. Conversely if $r\tau^2 + s\tau + t = 0$, then $(r\tau)\tau = -t - s\tau$. Define $a = r\tau \notin \mathbf{R}$. Then $a\omega_1 = r\omega_2$ and $a\omega_2 = -t\omega_1 - s\omega_2$, so that $a\Lambda \subseteq \Lambda$.

5. Overview of Inversion Problem

In §§1-4 we have seen how to associate to a lattice $\Lambda = \mathbb{Z}\omega_1 + \mathbb{Z}\omega_2$ in $\mathbb{C}$ a nonsingular curve

$$E : y^2 = 4x^3 - g_2 x - g_3$$

with g_2 and g_3 in $\mathbb{C}$. This association is implemented by the biholomorphic mapping $\varphi : \mathbb{C}/\Lambda \to E(\mathbb{C})$ given by

$$\varphi(z) = \begin{cases} (\wp(z), \wp'(z), 1) & \text{for } z \notin \Lambda \\ (0, 1, 0) & \text{for } z \in \Lambda. \end{cases} \tag{6.9}$$

In this section we address a converse problem. We know that every elliptic curve over $\mathbb{C}$, after the linear change of variables that changes (3.23) into (3.26), is of the form

$$y^2 = 4(x-a)(x-b)(x-c) \tag{6.10}$$

with a, b, c distinct in $\mathbb{C}$. With another linear change of variables, we can even make $a+b+c = 0$, and in this case the x^2 term drops out from the right side of (6.10). Our question is: For such a curve E, does there exist a lattice Λ such that the above association carries Λ to E?

The answer is "yes." The positive answer to this inversion problem is a fairly easy consequence of the Uniformization Theorem in the theory of one complex variable. However, we prefer to take a more pedestrian approach, since it gives more information. Essentially we shall produce concrete formulas for the inverse of φ in (6.9).

In terms of parameters, we need to use E to construct a pair of generating periods ω_1 and ω_2. Then we can form the corresponding $\wp$ function, and we shall recover a, b, c as the parameters u_1, u_2, u_3 attached to this $\wp$ function.

The construction of the parameters will be a byproduct of the construction of φ^{-1}. To see the nature of φ^{-1}, let us assume that E does indeed arise from $w = \wp(z)$. Then the results of §3 give

$$\left(\frac{dw}{dz}\right)^2 = 4(w-a)(w-b)(w-c)$$

$$\frac{dw}{2\sqrt{(w-a)(w-b)(w-c)}} = dz$$

$$z(w) = \int^w \frac{dw}{2\sqrt{(w-a)(w-b)(w-c)}}. \tag{6.11}$$

This integral is called an **elliptic integral of the first kind.**

In order to assert that $w = \wp(z)$ is genuinely the inverse function of an elliptic integral of the first kind, we need to make sense out of (6.11). The first step is to make sense out of the double-valued integrand; we carry out this step in §7 after an essay on analytic continuation in §6. Then in §8 we clarify what path of integration to use in order to make sense of (6.11). Different choices of path will lead to different values of the integrand, but the different values will all lie in a translate of a lattice, namely the lattice we take as Λ.

Once we have made sense of (6.11), we can show it has a locally defined inverse, and we can show that the locally defined inverse extends to be globally meromorphic. The rest will be easy.

In §9 we shall illustrate how to make rapidly convergent numerical calculations for passing back and forth between Λ and (g_2, g_3), particularly when g_2 and g_3 are in **R**.

6. Analytic Continuation

A function element is an analytic function on a disc. If the disc is centered at $z = a$, we speak of a function element at a. We shall define a notion of "direct analytic continuation" of a function element; this will be another function element of a certain kind. Iteration of this notion will lead to the definition of "analytic continuation" of a function element along a path.

To define **direct analytic continuation,** let the first function element be

$$f(z) = a_0 + a_1(z-a) + a_2(z-a)^2 + \ldots \qquad \text{for } |z-a| < r(a). \quad (6.12)$$

Suppose s satisfies $|s - a| < r(a)$, so that the disc

$$|z - s| < r(a) - |s - a| \qquad (6.13)$$

lies completely within the disc $|z - a| < r(a)$. Then we can compute the derivatives of $f(z)$ at s to obtain

$$f(z) = b_0(s) + b_1(s)(z-s) + b_2(s)(z-s)^2 + \ldots. \qquad (6.14)$$

For fixed s satisfying $|s - a| < r(a)$, Taylor's Theorem shows that (6.14) is convergent for z as in (6.13). But more is true. If we write

$$(z-a)^n = [(z-s) + (s-a)]^n, \qquad (6.15)$$

6. ANALYTIC CONTINUATION

then we can compute the coefficients $b_0(s), b_1(s), \ldots$ from (6.15); The result is that $b_0(s), b_1(s), \ldots$ are convergent power series in $s - a$. Effectively the $b_j(s)$ are obtained by expanding (6.15), substituting into (6.12), and rearranging. We say that (6.14) arises from (6.12) as a **result of rearranging the series** at s. A second function element is a **direct analytic continuation** of the given function element if its center s satisfies $|s - a| < r(a)$ and if the series for the second function element arises from the given function element as a result of rearranging the series at s. It may be that $r(s)$ for the second function element is in fact larger than the obvious possibility $r(a) - |s - a|$ given in (6.13); in fact, this will be the case of interest for us.

Fix a and b, and let C be a path from a to b given by

$$\gamma(t), \ \alpha \le t \le \beta \quad \text{with } \gamma(\alpha) = a, \gamma(\beta) = b.$$

Let f be a function element at a. A function element g at b is said to be an **analytic continuation** of f along C if there exist

(1) a partition

$$\alpha = t_0 < t_1 < \cdots < t_n = \beta, \tag{6.16}$$

(2) discs D_j centered at $\gamma(t_j)$, $0 \le j \le n$, and
(3) function elements f_j at $\gamma(t_j)$, $0 \le j \le n$, defined on D_j

such that $f_0 = f$, $f_n = g$, $\gamma[t_{j-1}, t_j] \subseteq D_{j-1}$ for $1 \le j \le n$, and f_j is a direct analytic continuation of f_{j-1} for $1 \le j \le n$.

EXAMPLE. Let C be the unit circle $\gamma = e^{it}$, $0 \le t \le 2\pi$. Let $f(z) = \sqrt{z}$ in $|z - 1| < 1$, with the square root taken to be positive on the positive reals, and let $g(z) = \sqrt{z}$ in $|z - 1| < 1$, with the square root taken to be negative on the positive reals. It is easy to check that g is an analytic continuation of f along C. The discs D_j can be taken to be centered at the points $e^{j\pi i/4}$, $0 \le j \le 8$.

Proposition 6.20. Let C be a path from a to b, and let f be a function element at a. If g_1 and g_2 are function elements at b that are analytic continuations of f along C, then g_1 and g_2 have the same Taylor series expansions about $z = b$. (Only the radii of their discs of definition can be different.)

PROOF IF $D_0 \supseteq \gamma[\alpha, \beta]$. In terms of the partition (6.16) used to determine one of g_1 and g_2, we claim that the Taylor series for f_n is the result of rearranging the series for f_0 at $\gamma(t_n)$. If so, then certainly

g_1 and g_2 will have the same Taylor series expansions. We proceed by induction on n, the case $n = 1$ being true by definition of direct analytic continuation. Assuming the result for $n-1$, we then know that f_{n-1} is the result of rearranging the series for f_0 at $\gamma(t_{n-1})$, so that $f_0(z) = f_{n-1}(z)$ for $z \in D_0 \cap D_{n-1}$. By definition of direct analytic continuation, $f_{n-1}(z) = f_n(z)$ for $z \in D_{n-1} \cap D_n$. Thus $f_0(z) = f_n(z)$ for $z \in D_0 \cap D_{n-1} \cap D_n$. This intersection is nonempty since $\gamma(t_n)$ is in it. Thus $f_0(z) = f_n(z)$ on $D_0 \cap D_n$, and the assertion follows.

PROOF IN GENERAL CASE. The main step is to see that the partition (6.16) can be suitably refined without adjusting the series of the initial or final function element. Recall that $\gamma[t_{j-1}, t_j] \subseteq D_{j-1}$. Then

$$\delta = \min_{1 \leq j \leq n} \text{distance}(\gamma[t_{i-1}, t_i], D_{i-1}^c)$$

is > 0. Fix any positive number ρ with $\rho \leq \delta$. By uniform continuity of γ (with $\epsilon = \rho$), we can refine (6.16) to a partition

$$\alpha = s_0 < \cdots < s_m = \beta \qquad (6.17)$$

so that

$$\gamma[s_{i-1}, s_i] \subseteq \{|z - \gamma(s_{i-1})| < \rho\} \cap \{|z - \gamma(s_i)| < \rho\} \quad \text{for } 1 \leq i \leq m.$$

Define

$$\Delta_i = \{|z - \gamma(s_i)| < \rho\} \quad \text{for } 0 \leq i \leq m.$$

Let

$$t_{j-1}, t_j \qquad (6.18)$$

be two consecutive points of (6.16) and let

$$s_{k-1} = t_{j-1}, s_k, \ldots, s_l = t_j \qquad (6.19)$$

be the corresponding points of (6.17). Now the series for f_{j-1} converges in D_{j-1}, and all the discs $\Delta_{k-1}, \Delta_k, \ldots, \Delta_l$ lie in D_{j-1}, by definition of δ. Hence the special case shows that f_j results whether we use the partition (6.18) and discs D_{j-1}, D_j or the partition (6.19) and discs $\Delta_{k-1}, \Delta_k, \ldots, \Delta_l$. Putting these results together for $1 \leq j \leq n$, we see that f_n is the result of analytic continuation from f_0 using the partition (6.17) and the discs $\Delta_0, \ldots, \Delta_m$.

Now we can compare g_1 and g_2. Let us say that the above argument would work with $\delta = \delta_1$ for g_1 and with $\delta = \delta_2$ for g_2. By choosing $\rho < \min\{\delta_1, \delta_2\}$, we can choose a common refinement of both partitions and see that g_1 and g_2 are obtained from a common sequence of points and discs.

Proposition 6.21. Suppose C is a path from a to b given by $\gamma(t)$, $\alpha \leq t \leq \beta$. Suppose f is a function element at a and g is a function element at b obtained by analytic continuation of f along C. Then there exists $\epsilon > 0$ with the following property: If C' is any path from a to b given by $\psi(t)$, $\alpha \leq t \leq \beta$, that is uniformly within ϵ of $\gamma(t)$, then g is an analytic continuation of f along C'.

This proposition has the following use: We can always approximate C uniformly by polygonal paths, and thus polygonal paths give us the most general analytic continuation. We omit the proof of Proposition 6.21, as it is in much the same spirit as the previous proof.

Proposition 6.22. If g is an analytic continuation of f along a path C, then f is an analytic continuation of g along the reverse path C^{-1}.

PROOF. Without loss of generality, let g be a direct analytic continuation of f, so that the given partition is $\alpha = t_0 < t_1 = \beta$. We apply the construction in the second half of Proposition 6.20, obtaining a refinement $\alpha = s_0 < \cdots < s_m = \beta$, discs Δ_i for $0 \leq i \leq m$, and function elements (f_i, Δ_i) such that $\gamma[s_{i-1}, s_i] \subseteq \Delta_{i-1} \cap \Delta_i$, f_0 has the same series as f, f_m has the same series as g, and f_i is a direct analytic continuation of f_{i-1}. For the path C^{-1}, we use the partition

$$-\beta = r_0 < \cdots < r_m = -\alpha$$

with $r_i = -s_{m-i}$, the discs $D_i = \Delta_{m-i}$, and the function elements $g_i = f_{m-i}$. Then

$$\gamma^{-1}[r_{i-1}, r_i] = \gamma^{-1}[-s_{m-i+1}, -s_{m-i}] = \gamma[s_{m-i}, s_{m-i+1}]$$
$$\subseteq \Delta_{m-i} \cap \Delta_{m-i+1} = D_i \cap D_{i-1} \subseteq D_{i-1}.$$

Also the fact that Δ_{m-i} and Δ_{m-i+1} have the same radius shows that f_{m-i} and f_{m-i+1} are direct analytic continuations of each other, hence that g_i is a direct analytic continuation of g_{i-1}.

7. Riemann Surface of the Integrand

Let a, b, c be distinct points in $\mathbb{C}$. In this section we treat the problem of defining $\sqrt{(z-a)(z-b)(z-c)}$. Fix $z_0 \in \mathbb{C} - \{a, b, c\}$, and fix q_0 as one of the two possible values of $\sqrt{(z_0-a)(z_0-b)(z_0-c)}$. Then $\sqrt{(z-a)(z-b)(z-c)}$ at least extends to be analytic in any disc about z_0 within $\mathbb{C} - \{a, b, c\}$.

We shall define a double covering $\mathcal{R}$ of $\mathbf{C} - \{a,b,c\}$ so that $\sqrt{(z-a)(z-b)(z-c)}$ extends to be well defined on all of $\mathcal{R}$. Then we shall complete $\mathcal{R}$ to a space $\mathcal{R}^*$ over $\mathbf{C} \cup \{\infty\}$ that fills in one preimage above each of a, b, c, ∞. The space $\mathcal{R}^*$ will be a one-dimensional complex manifold (**Riemann surface**), and it will be compact. (Actually it will be a torus.)

Consider loops in $\mathbf{C} - \{a, b, c\}$ based at z_0. If such a loop C is piecewise smooth, we define its **total winding number** about a, b, and c to be

$$n(C) = n(C, a) + n(C, b) + n(C, c)$$
$$= \frac{1}{2\pi i} \left(\oint_C \frac{dz}{z-a} + \oint_C \frac{dz}{z-b} + \oint_C \frac{dz}{z-c} \right).$$

If the loop C is not piecewise smooth, all piecewise smooth loops that are uniformly sufficiently close to C can be seen to have the same total winding number, and we take this to be $n(C)$.

Homotopic loops based at z_0 have the same $n(C)$. Then $n(C)$ gives a group homomorphism from the fundamental group $\pi_1(\mathbf{C} - \{a, b, c\}, z_0)$ into $\mathbf{Z}$. This homomorphism is onto $\mathbf{Z}$ since there exists a loop that winds once around a and 0 times around b and around c. Therefore the subgroup of **even total winding number**

$$H = \{C \in \pi_1(\mathbf{C} - \{a,b,c\}, z_0) \mid n(C) \in 2\mathbf{Z}\}$$

has index 2. Let $\mathcal{R}$ be the covering space of $\mathbf{C} - \{a, b, c\}$ corresponding to H, let $e : \mathcal{R} \to \mathbf{C} - \{a, b, c\}$ be the covering map, and fix ζ_0 as a base point in $\mathcal{R}$ with $e(\zeta_0) = z_0$.

For use in §8, we record how H is generated in terms of $\pi_1(\mathbf{C} - \{a, b, c\}, z_0)$. Let $\Gamma_a, \Gamma_b, \Gamma_c$ be simple piecewise smooth loops in $\mathbf{C} - \{a, b, c\}$ based at z_0 that intersect one another only at z_0 and that have winding numbers

$$n(\Gamma_a, a) = 1, \quad n(\Gamma_a, b) = n(\Gamma_a, c) = 0$$
$$n(\Gamma_b, b) = 1, \quad n(\Gamma_b, a) = n(\Gamma_b, c) = 0 \qquad (6.20)$$
$$n(\Gamma_c, c) = 1, \quad n(\Gamma_c, a) = n(\Gamma_c, b) = 0.$$

Define

$$\Gamma_1 = \Gamma_a \Gamma_b \quad \text{and} \quad \Gamma_2 = \Gamma_b \Gamma_c. \qquad (6.21)$$

Proposition 6.23. The subgroup of $\pi_1(\mathbf{C} - \{a, b, c\}, z_0)$ of even total winding number is generated by the classes of $\Gamma_1, \Gamma_2, \Gamma_a^2, \Gamma_b^2, \Gamma_c^2$.

7. RIEMANN SURFACE OF THE INTEGRAND

PROOF. The loops $\Gamma_a, \Gamma_b, \Gamma_c$ generate $\pi_1(\mathbf{C} - \{a,b,c\}, z_0)$, and each of the purported generators is in the subgroup of even winding number. Thus let a word in $\Gamma_a, \Gamma_b, \Gamma_c$ and their inverses be given, and suppose it has even total winding number. Since $n(\cdot)$ gives $+1$ or -1 for each factor, there must be an even number of factors. Grouping them in pairs, let us show that each pair is generated in the required fashion. If the left hand member of a pair is Γ^{-1}, we replace it by $(\Gamma^2)^{-1}\Gamma$; If the right hand member of a pair is Γ^{-1}, we replace it by $\Gamma(\Gamma^2)^{-1}$. In this way we may assume that the factors of the pair come from $\Gamma_a, \Gamma_b, \Gamma_c$, not their inverses. We may assume the members of the pair are not the same, since the squares are in our list of generators. For the remaining six pairs, $\Gamma_a\Gamma_b$ and $\Gamma_b\Gamma_c$ are generators, and the other pairs are given by

$$\Gamma_b\Gamma_a = \Gamma_b^2(\Gamma_a\Gamma_b)^{-1}\Gamma_a^2$$
$$\Gamma_c\Gamma_b = \Gamma_c^2(\Gamma_b\Gamma_c)^{-1}\Gamma_b^2$$
$$\Gamma_a\Gamma_c = (\Gamma_a\Gamma_b)(\Gamma_b^2)^{-1}(\Gamma_b\Gamma_c)$$
$$\Gamma_c\Gamma_a = \Gamma_c^2(\Gamma_b\Gamma_c)^{-1}\Gamma_b^2(\Gamma_a\Gamma_b)^{-1}\Gamma_a^2.$$

This proves the proposition.

The covering space becomes a Riemann surface in a natural way. Namely let D be a disc about p in $\mathbf{C} - \{a,b,c\}$. Since D is simply connected, it is evenly covered. Thus $e^{-1}(D) = \Delta_1 \cup \Delta_2$, a disjoint union of open sets Δ_1 and Δ_2 each homeomorphic with D via e. We can take $e: \zeta \to z \in \mathbf{C}$ as a local coordinate in Δ_i, and then (e, Δ_i) becomes a chart.

We shall use the convention begun above that Greek letters refer to points and sets in $\mathcal{R}$ and the corresponding Latin letters refer to their counterparts in $\mathbf{C} - \{a,b,c\}$. We have already fixed ζ_0 with $e(\zeta_0) = z_0$. Let D_0 be a disc in $\mathbf{C}-\{a,b,c\}$ about z_0, and let Δ_0 be the component of $e^{-1}(D_0)$ to which ζ_0 belongs. Recall that $\sqrt{(z-a)(z-b)(z-c)}$ makes sense on D_0 as an analytic function with $\sqrt{(z_0-a)(z_0-b)(z_0-c)} = q_0$.

Proposition 6.24. There exists a unique analytic function $F: \mathcal{R} \to \mathbf{C}$ satisfying $F(\zeta_0) = q_0$ and having the following property: If C is any path from ζ_0 to ζ_1 in $\mathcal{R}$, if D is a disc about $z_1 = e(\zeta_1)$ in $\mathbf{C} - \{a,b,c\}$, and if $e^{-1}: D \to \Delta$ is regarded as a map from D to the component Δ of $e^{-1}(D)$ to which ζ_1 belongs, then $(F \circ e^{-1}, D)$ is an analytic continuation of

$$(\sqrt{(z-a)(z-b)(z-c)}, D_0) \qquad (6.22)$$

along $e(C)$.

REMARK. Temporarily we shall write

$$f(\zeta) = \sqrt{(\zeta - a)(\zeta - b)(\zeta - c)}$$

for the function produced by the proposition. We shall introduce better notation in (6.23).

PROOF. Uniqueness is an immediate consequence of Proposition 6.20 and the path-connectedness of $\mathcal{R}$. For existence, let $f_0(z)$ be the function element (6.22). It is clear that this f_0 continues analytically along any curve C' in $\mathbb{C} - \{a, b, c\}$ and that the result is one of the two values of the square root, because $f_0(z)^2$ extends to a global function and analytic continuation respects squaring.

Given ζ_1 in $\mathcal{R}$, let C be a path from ζ_0 to ζ_1 and let f_1 be a function element at z_1, with domain D_1, that is an analytic continuation of f_0 along $e(C)$. Let Δ_1 be the component of $e^{-1}(D_1)$ to which ζ_1 belongs, and define e^{-1} as a function from D_1 to Δ_1. Then we can define an analytic function F on Δ_1 by $F = f_1 \circ e^{-1}$.

For F to give a globally defined analytic function on $\mathcal{R}$, we need only see that F is well defined. Thus let $C^\#$ be another path from ζ_0 to ζ_1, and let $f_1^\#$ be the corresponding function element at z_1 obtained by continuation along $e(C^\#)$. We are to show that f_1 and $f_1^\#$ have the same Taylor series about $z = z_1$.

Analytic continuation of f_0 around the loop Γ_a leads to $-f_0$, and similarly for Γ_b and Γ_c. Thus analytic continuation of f_0 around a loop γ in $\mathbb{C} - \{a, b, c\}$ based at z_0 leads to $(-1)^{n(\gamma)} f_0$. Since $CC^{\#-1}$ is a loop in $\mathcal{R}$ based at ζ_0, $e(CC^{\#-1})$ leads to f_0.

Since analytic continuation of f_0 along $e(C)$ leads to f_1, analytic continuation of f_1 along $e(C^{\#-1}) = e(C^\#)^{-1}$ leads to f_0. By Proposition 6.22, analytic continuation of f_0 along $e(C^\#)$ leads to f_1. By Proposition 6.20, f_1 and $f_1^\#$ have the same Taylor series expansion. Thus F is well defined, and the proposition follows.

Now we shall complete $\mathcal{R}$ to a space $\mathcal{R}^*$. The set $\mathcal{R}^*$ is to consist of the points of $\mathcal{R}$ together with four more points $\alpha, \beta, \gamma, \infty$. We define $e : \mathcal{R}^* \to \mathbb{C} \cup \{\infty\} = P_1(\mathbb{C})$ by $e(\alpha) = a$, $e(\beta) = b$, $e(\gamma) = c$, $e(\infty) = \infty$. To topologize $\mathcal{R}^*$, we declare that $\mathcal{R}$ is to be open in $\mathcal{R}^*$ and that basic open neighborhoods of $\alpha, \beta, \gamma, \infty$ are to be defined as follows: A basic open neighborhood of α is given by $\{\alpha\} \cup e^{-1}(D^\times)$, where $D^\times$ is any punctured disc centered at a and lying in $\mathbb{C} - \{a, b, c\}$, and basic open neighborhoods of β, γ, ∞ are defined similarly. The resulting $\mathcal{R}^*$ is easily seen to be a Hausdorff regular space with a countable base,

7. RIEMANN SURFACE OF THE INTEGRAND

therefore a separable metric space. It is sequentially compact, therefore compact. With this notation the function $F(\zeta)$ of Proposition 6.24 is better written as

$$F(\zeta) = \sqrt{(\zeta - \alpha)(\zeta - \beta)(\zeta - \gamma)}. \tag{6.23}$$

This is the notation we shall normally use for F.

Let us make $\mathcal{R}^*$ into a Riemann surface. We need to define charts about $\alpha, \beta, \gamma, \infty$. In the case of α, let $D^\times$ be a punctured disc about a lying in $\mathbb{C} - \{a, b, c\}$. Then $e^{-1}(D^\times)$ has two preimages above each point of $D^\times$. The lift of a loop in $D^\times$ with winding number 1 about a is not a loop (since $\sqrt{z-a}$ is not well defined on $D^\times$) and therefore $e^{-1}(D^\times)$ is connected. It follows that $e^{-1}(D^\times)$ is a covering space of $D^\times$. Arguing as in Proposition 6.24, we can define $\sqrt{\zeta - \alpha}$ uniquely on $e^{-1}(D^\times)$ in such a way that $\sqrt{\zeta - \alpha}$ takes a prescribed value at some base point. We put $\sqrt{\zeta - \alpha} = 0$ if $\zeta = \alpha$.

Then $(\sqrt{\zeta - \alpha}, e^{-1}(D^\times) \cup \{\alpha\})$ is a continuous map into $\mathbb{C}$. Since it maps small discs to open sets, it is open. If $\sqrt{\zeta_1 - \alpha} = \sqrt{\zeta_2 - \alpha}$, then $z_1 - a = z_2 - a$ and $z_1 = z_2$. Hence ζ_1 and ζ_2 lie over the same point z. If $z = a$ then $\zeta_1 = \zeta_2 = \alpha$. Otherwise the square root changes by a factor of -1 between the two points of the same fiber. Thus $\sqrt{\zeta_1 - \alpha} = \sqrt{\zeta_2 - \alpha}$ implies $\zeta_1 = \zeta_2$, and the map is one-one. Consequently $(\sqrt{\zeta - \alpha}, e^{-1}(D^\times) \cup \{\alpha\})$ is a chart. It is clear that this chart is compatible with the charts on $\mathcal{R}$.

We treat β and γ in the same way, and we are left with handling the point ∞ of $\mathcal{R}^*$. For this we let $D^\times = \{z \mid |z| > R\}$ for $R \geq \max\{|a|, |b|, |c|\}$, and we check that $e^{-1}(D^\times)$ is connected. We define $\dfrac{1}{\sqrt{\zeta}}$ on $D^\times$, fixing the value at some base point and defining the function to be 0 at $e^{-1}(\infty)$. Then $\left(\dfrac{1}{\sqrt{\zeta}}, e^{-1}(D^\times \cup \{\infty\})\right)$ provides a compatible chart about ∞.

The function $F(\zeta)^{-1} = \dfrac{1}{\sqrt{(\zeta - \alpha)(\zeta - \beta)(\zeta - \gamma)}}$ with $F(\zeta_0)^{-1} = 1/q_0$ is analytic on $\mathcal{R}$ and is continuous from $\mathcal{R}^*$ into $\mathbb{C} \cup \{\infty\}$. By Riemann's removable singularity theorem, it is a meromorphic function on $\mathcal{R}$. Let us see its behavior near $\alpha, \beta, \gamma, \infty$.

Near α, the local coordinate is $s = \sqrt{\zeta - \alpha}$. To calculate with this relation, we cannot use the manifold variable ζ but have to use the complex variable z. Thus we have $s^2 = z - a$, $z = s^2 + a$, and

$$(z - b)(z - c) = (s^2 + a - b)(s^2 + a - c).$$

Hence
$$F(\zeta)^{-1} = s^{-1} \frac{1}{\sqrt{(s^2 + a - b)(s^2 + a - c)}} \qquad (6.24)$$

for one of the two determinations of the square root. Since the square root is analytic and nonvanishing near $s = 0$, $F(\zeta)^{-1}$ has a simple pole at $\zeta = \alpha$.

Similarly $F(\zeta)^{-1}$ has a simple pole at $\zeta = \beta$ and at $\zeta = \gamma$. Near ∞ the local coordinate is $s = \frac{1}{\sqrt{\zeta}}$. Then $z = s^{-2}$ and

$$(z - a)(z - b)(z - c) = z^3 \left(1 - \frac{a}{z}\right)\left(1 - \frac{b}{z}\right)\left(1 - \frac{c}{z}\right)$$
$$= z^3(1 - as^2)(1 - bs^2)(1 - cs^2).$$

Hence
$$F(\zeta)^{-1} = s^3 \frac{1}{\sqrt{(1 - as^2)(1 - bs^2)(1 - cs^2)}} \qquad (6.25)$$

for one of the two determinations of the square root. Since the square root is analytic and nonvanishing near $s = 0$, $f(\zeta)^{-1}$ has a triple zero at $\zeta = \infty$.

8. An Elliptic Integral

We continue with the notation of §7. Thus $F : \mathcal{R}^* \to \mathbb{C} \cup \{\infty\}$ is as in (6.23) and later, and ζ_0 is the base point of $\mathcal{R}^*$ with $e(\zeta_0) = z_0$. Our first objective is to make sense of the expression

$$w(C) = \int_C \tfrac{1}{2} F(\zeta)^{-1} d\zeta \qquad (6.26)$$

where C is any piecewise smooth path in $\mathcal{R}^*$.

Roughly the idea is that (6.26) is to be $\int_{e(C)} \tfrac{1}{2} F(z)^{-1} dz$, but the difficulty is that $F(z)^{-1}$ is not necessarily single-valued on $e(C)$. The point of the reference to $\mathcal{R}^*$ is to allow us to make a choice between $F(z)^{-1}$ or $-F(z)^{-1}$ for each point of the curve. To be precise, let C be given by $C(t)$ for $t \in I$, so that $e(C)$ is the path in $\mathbb{C} \cup \{\infty\}$ given by $e(C(t))$. Then (6.26) is to be made precise by the definition

$$w(C) = \int_{t \in I} \tfrac{1}{2} F(C(t))^{-1} e(C)'(t) \, dt. \qquad (6.27)$$

8. AN ELLIPTIC INTEGRAL

We need to check that the integral (6.27) is convergent. The only difficulty is in neighborhoods of $\alpha, \beta, \gamma, \infty$. Thus suppose C is confined to a basic neighborhood. Near α, the local parameter is $s = \sqrt{\zeta - \alpha}$, and the path $C(t)$ determines a path $s(t)$ in the coordinate patch by $s(t) = \sqrt{C(t) - \alpha}$. Then $s(t)^2 = e(C(t)) - a$, $e(C(t)) = s(t)^2 + a$, and $e(C)'(t)\, dt = 2s(t)s'(t)\, dt$. Using (6.24), we substitute into (6.27) and obtain

$$w(C) = \int_{t \in I} \frac{s'(t)\, dt}{\sqrt{(s(t)^2 + a - b)(s(t)^2 + a - c)}} \tag{6.28}$$

for a suitably small interval I. The denominator is bounded away from 0 for $t \in I$, and the integral is convergent. Similar remarks apply to β and γ.

Near ∞, the local parameter is $s = \dfrac{1}{\sqrt{\zeta}}$, and we have $s(t) = \dfrac{1}{\sqrt{C(t)}}$. Then

$$s(t)^2 = 1/e(C(t)), \qquad e(C(t)) = s(t)^{-2},$$
$$e(C)'(t)\, dt = -2s(t)^{-3} s'(t)\, dt.$$

Using (6.25), we substitute into (6.27) and obtain

$$w(C) = \int_{t \in I} \frac{-s'(t)\, dt}{\sqrt{(1 - as(t)^2)(1 - bs(t)^2)(1 - cs(t)^2)}} \tag{6.29}$$

for a suitably small interval I. Again the denominator is bounded away from 0 for $t \in I$, and the integral is convergent. Thus (6.27) is always convergent, and we can use it as a definition of (6.26).

Lemma 6.25. *If C is a piecewise smooth path in $\mathcal{R}^*$ lying completely in a basic disc with local parameter s and extending between local parameter values s_1 and s_2, then*

$$w(C) = \int_{s_1}^{s_2} \tfrac{1}{2} F(\zeta(s))^{-1} \frac{dz}{ds}\, ds. \tag{6.30}$$

In particular, the integral depends only on the endpoints, not on the path itself.

PROOF. For a disc in $\mathcal{R}$ with local parameter z, the lemma follows by applying the Cauchy Integral Theorem to a branch of

$1/\sqrt{(z-a)(z-b)(z-c)}$. For a disc centered at α, the integral is given by (6.28), which is the value of the complex line integral

$$\int \frac{ds}{\sqrt{(s^2+a-b)(s^2+a-c)}} \tag{6.31}$$

over the local version $s(t)$ of the path $C(t)$. Since the integrand of (6.31) is analytic in a disc containing the image of the path $s(t)$, the line integral depends only on the endpoints and is given as in (6.30), by the Cauchy Integral Theorem. Similar remarks apply to β and γ. Near ∞, the integral is given by (6.29), which is the value of the complex line integral

$$\int \frac{-ds}{\sqrt{(1-as^2)(1-bs^2)(1-cs^2)}}, \tag{6.32}$$

and the same considerations apply.

Lemma 6.26. If C and C' are piecewise smooth paths from ζ_1 to ζ_2 in $\mathcal{R}^*$ that are homotopic in $\mathcal{R}^*$, then $w(C) = w(C')$.

PROOF. Let the homotopy be C_t, $0 \leq t \leq 1$, where each C_t has domain interval I and each C_t extends from ζ_1 to ζ_2. Without loss of generality we may assume that each C_t is piecewise smooth. It is enough to prove, for each t_0, that $w(C_t)$ is constant for t in a neighborhood of t_0. Changing notation, we see that it is enough to prove that any piecewise smooth path C_2 that is sufficiently close to C_1 uniformly for $t \in I$ has $w(C_2) = w(C_1)$.

About each point of C_1, choose a basic disc Δ and a proper subdisc Δ'' centered at that point. The discs Δ'' cover the image of C_1, and we extract a finite subcover $\Delta_1'', \ldots, \Delta_n''$. Form the corresponding larger discs $\Delta_1, \ldots, \Delta_n$ and intermediate discs $\Delta_1', \ldots, \Delta_n'$ with

$$\Delta_i'' \subseteq \bar{\Delta}_i'' \subseteq \Delta_i' \subseteq \bar{\Delta}_i' \subseteq \Delta_i \qquad \text{for } 1 \leq i \leq n.$$

By uniform continuity of C_1, there is some $\delta > 0$ so that $C_1(t) \in \Delta_i''$ implies $C_1(t') \in \Delta_i'$ for $|t'-t| < \delta$. Let $t_0 < \cdots < t_m$ be a partition of the domain interval I of C_1 of mesh $< \delta$. For each j with $0 \leq j \leq m$, we can choose $i = i(j)$ so that $C_1(t_j)$ is in $\Delta_{i(j)}''$. Then $C_1([t_{j-1}, t_j]) \subseteq \Delta_{i(j-1)}'$ for $1 \leq j \leq m$.

Choose $\epsilon > 0$ so that $\text{dist}(\bar{\Delta}_i', \Delta_i^c) \geq \epsilon$ for $1 \leq i \leq n$. Suppose C_2 is any path from ζ_1 to ζ_2 that is uniformly within ϵ of C_1. Then we have $C_2([t_{j-1}, t_j]) \subseteq \Delta_{i(j-1)}$ for $1 \leq j \leq m$. In particular, $C_1(t_j)$ and $C_2(t_j)$ are in $\Delta_{i(j-1)} \cap \Delta_{i(j)}$, which is connected. Let S_j be a piecewise smooth

8. AN ELLIPTIC INTEGRAL

path from $C_1(t_j)$ to $C_2(t_j)$ in $\Delta_{i(j-1)} \cap \Delta_{i(j)}$, $1 \leq j \leq m-1$. Let S_0 be the constant path at $C_1(t_0)$ and let S_m be the constant path at $C_m(t_0)$. By considering cancelling integrals, we have

$$w(C_2) = \sum_{j=1}^{m} w(S_{j-1} C_2|_{[t_{j-1},t_j]} S_j). \tag{6.33}$$

But $S_{j-1} C_2|_{[t_{j-1},t_j]} S_j$ is a path in $\Delta_{i(j-1)}$ with the same endpoints as $C_1|_{[t_{j-1},t_j]}$. Thus Lemma 6.25 says that (6.33) is

$$= \sum_{j=1}^{m} w(C_1|_{[t_{j-1},t_j]}) = w(C_1).$$

This proves the lemma.

Recall from Proposition 6.23 that the classes of the loops $\Gamma_1, \Gamma_2, \Gamma_a^2, \Gamma_b^2, \Gamma_c^2$ generate the subgroup of $\pi_1(\mathbf{C}-\{a,b,c\}, z_0)$ of even total winding number. Lift these loops to loops $\tilde{\Gamma}_1, \tilde{\Gamma}_2, \Gamma_\alpha, \Gamma_\beta, \Gamma_\gamma$ in $\mathcal{R}$ based at ζ_0.

Lemma 6.27. $\int_{\Gamma_\alpha} \frac{1}{2} F(\zeta)^{-1} d\zeta = 0$, and similarly for β and γ.

PROOF. In view of Lemma 6.26 applied just within $\mathcal{R}$, the computation reduces to an integral over a loop in a basic disc of $\mathcal{R}^*$ about α. Then Lemma 6.25 applies and shows we get 0.

Define

$$\omega_1 = \int_{\tilde{\Gamma}_1} \tfrac{1}{2} F(\zeta)^{-1} d\zeta \quad \text{and} \quad \omega_2 = \int_{\tilde{\Gamma}_2} \tfrac{1}{2} F(\zeta)^{-1} d\zeta, \tag{6.34}$$

and let Λ be the subset of $\mathbf{C}$ given by

$$\Lambda = \mathbf{Z}\omega_1 + \mathbf{Z}\omega_2. \tag{6.35}$$

We give $\mathbf{C}/\Lambda$ the quotient topology, which is not yet known to be Hausdorff.

Proposition 6.28. There is a well defined map $w : \mathcal{R}^* \to \mathbf{C}/\Lambda$ with the following property. Whenever ζ is in $\mathcal{R}^*$ and C is a piecewise smooth path from ζ_0 to ζ, then

$$w(\zeta) \equiv w(C) \mod \Lambda.$$

The map w is continuous and open.

PROOF. Fix ζ in $\mathcal{R}^*$, let C be a piecewise smooth path from ζ_0 to ζ, and define
$$w(\zeta) = w(C) + \Lambda \quad \text{as a member of } \mathbf{C}/\Lambda.$$
To see that $w(\zeta)$ is well defined, let C' be another such path. We are to show that $w(C') - w(C)$ is in Λ. Without loss of generality we may assume the following: If $\zeta \notin \{\alpha, \beta, \gamma, \infty\}$, neither C nor C' meets $\{\alpha, \beta, \gamma, \infty\}$; if $\zeta \in \{\alpha, \beta, \gamma, \infty\}$, C and C' meet $\{\alpha, \beta, \gamma, \infty\}$ only at their final points. We have
$$w(C') - w(C) = w(C'C^{-1}),$$
and $C'C^{-1}$ is a loop based at ζ_0. By Lemma 6.26, we can perturb $C'C^{-1}$ near ζ, if necessary, so that $C'C^{-1}$ does not meet $\{\alpha, \beta, \gamma, \infty\}$. Afterward, $w(C'C^{-1})$ depends only on the homotopy class of $C'C^{-1}$ in $\mathcal{R}$, according to Lemma 6.26. Since Proposition 6.23 identifies the homotopy classes in $\mathcal{R}$, we see that $w(C'C^{-1})$ is a $\mathbf{Z}$ combination of $w(\tilde{\Gamma}_1), w(\tilde{\Gamma}_2), w(\Gamma_\alpha), w(\Gamma_\beta), w(\Gamma_\gamma)$. But $w(\Gamma_\alpha) = w(\Gamma_\beta) = w(\Gamma_\gamma) = 0$ by Lemma 6.27. Thus $w(C') - w(C)$ is a $\mathbf{Z}$ combination of $\omega_1 = w(\tilde{\Gamma}_1)$ and $\omega_2 = w(\tilde{\Gamma}_2)$, hence is in Λ.

Thus w is well defined. If we fix ζ_1 and a path C_1 leading to it, we can look at how w behaves in a basic disc about ζ_1 by using paths that begin with C_1 and continue completely within the basic disc. Let ζ correspond to $C_1 C$ in this way, and let $\zeta_1 \leftrightarrow s_1$ and $\zeta \leftrightarrow s$ in the local coordinate. Then Lemma 6.25 gives
$$w(\zeta) = w(C_1) + w(C) = w(C_1) + \int_{s_1}^s \tfrac{1}{2} F(\zeta(s))^{-1} \frac{dz}{ds}\, ds. \qquad (6.36)$$

This formula shows that w is continuous. Since the integral on the right is nonconstant analytic in s, w is open.

Corollary 6.29. The complex numbers ω_1 and ω_2 in (6.34) are linearly independent over $\mathbf{R}$. Therefore Λ in (6.35) is a lattice, and $\mathbf{C}/\Lambda$ is a Riemann surface.

PROOF. If ω_1 and ω_2 are dependent, then $\Lambda \subseteq \mathbf{R}\omega$ for a suitable complex number ω. The natural map $\mu : \mathbf{C}/\Lambda \to \mathbf{C}/(\mathbf{R}\omega)$ is continuous and open, with a Hausdorff image. Hence the composition $\mu \circ w : \mathcal{R}^* \to \mathbf{C}/(\mathbf{R}\omega)$, with w as in Proposition 6.28, is continuous and open. Since $\mathcal{R}^*$ is compact and $\mathbf{C}/(\mathbf{R}\omega)$ is Hausdorff, the image of $\mu \circ w$ is open, compact, and closed. Since $\mathbf{C}/(\mathbf{R}\omega)$ is connected, $\mu \circ w$ is onto. But $\mathbf{C}/(\mathbf{R}\omega)$ is noncompact, and we have a contradiction.

8. AN ELLIPTIC INTEGRAL

Theorem 6.30. The map $w : \mathcal{R}^* \to \mathbf{C}/\Lambda$ of Proposition 6.28 is one-one onto and is biholomorphic.

REMARKS. The map is well defined by Proposition 6.28, and $\mathbf{C}/\Lambda$ has a complex manifold structure as a result of Corollary 6.29. Thus it is meaningful to ask about holomorphicity of w (and also w^{-1}, once it exists).

PROOF. The map w is holomorphic as a result of (6.36). In local coordinates, (6.36) shows that w has derivative

$$\tfrac{1}{2} F(\zeta(s))^{-1} \frac{dz}{ds}. \tag{6.37}$$

At a point ζ of $\mathcal{R}$, where $s = z$ is a local parameter, (6.37) is just

$$\frac{1}{2\sqrt{(s-a)(s-b)(s-c)}},$$

which is nonvanishing. At $\zeta = \alpha$, (6.37) is the integrand of (6.31) and again is nonzero. Similar remarks apply to $\zeta = \beta$ and $\zeta = \gamma$. Finally at $\zeta = \infty$, (6.37) is the integrand of (6.32) and again is nonzero. Since the local derivative is nowhere 0, w is an immersion. An immersion between compact connected smooth manifolds of the same dimension is necessarily a covering map, and thus $w : \mathcal{R}^* \to \mathbf{C}/\Lambda$ is a covering map. If we can show w is one-one, then w^{-1} will exist, and w^{-1} will be holomorphic as a consequence of the one-dimensional complex Inverse Function Theorem.

Thus we are to show that w is one-one. Suppose $w(\zeta_1) = w(\zeta_2)$ in $\mathbf{C}/\Lambda$. Let C_1 and C_2 be piecewise smooth paths from ζ_0 to ζ_1 and ζ_2, respectively. By definition of $w(\zeta)$,

$$w(C_1) = w(C_2) + \omega,$$

for some $\omega = m\omega_1 + n\omega_2$ in Λ. Since $\Gamma = \tilde{\Gamma}_1^m \tilde{\Gamma}_2^n$ has $w(\Gamma) = \omega$, the path $C_3 = \Gamma C_2$ goes from ζ_0 to ζ_2 and has $w(C_3) = w(C_2) + \omega$. Therefore

$$w(C_1) = w(C_3). \tag{6.38}$$

Let us say that C_1 and C_3 have domain $[0, 1]$. For $0 \leq t \leq 1$, we define

$$C_1^{\#}(t) = w(C_1|_{[0,t]}) \quad \text{and} \quad C_3^{\#}(t) = w(C_3|_{[0,t]})$$

as piecewise smooth paths in $\mathbf{C}$. Define

$$C_1^{\#\#} = w \circ C_1 \quad \text{and} \quad C_3^{\#\#} = w \circ C_3 \tag{6.39}$$

as piecewise smooth paths in $\mathbb{C}/\Lambda$. Then

$$\begin{aligned}C_1{}^\#(t) \mod \Lambda &= w(C_1|_{[0,t]}) \mod \Lambda \\ &\equiv w(C_1(t)) \mod \Lambda \\ &= C_1{}^{\#\#}(t)\end{aligned}$$

and a similar computation for C_3 show that $C_1{}^\#$ and $C_3{}^\#$ are the unique lifts based at 0 of $C_1{}^{\#\#}$ and $C_3{}^{\#\#}$ from $\mathbb{C}/\Lambda$ to $\mathbb{C}$. By (6.38), $C_1{}^\#(1) = C_3{}^\#(1)$. Therefore $C_1{}^\#C_3{}^{\#-1}$ is a loop in $\mathbb{C}$, necessarily contractible. Hence $C_1{}^{\#\#}C_3{}^{\#\#-1}$ is a contractible loop in $\mathbb{C}/\Lambda$. Since w has been proved to be a covering map, (6.39) shows that $C_1 C_3^{-1}$ is a contractible loop in $\mathcal{R}^*$. Therefore C_1 and C_3 have the same endpoint in $\mathcal{R}^*$, and $\zeta_1 = \zeta_2$. We conclude that w is one-one, as required.

Theorem 6.30 completes the construction of the lattice Λ and a function $w(\zeta)$ that should have something to do with inverting a Weierstrass $\wp$ function. To make the correspondence complete, we modify $w(\zeta)$ by translation in $\mathbb{C}/\Lambda$ by the element $\int_\infty^{\zeta_0} \tfrac{1}{2} F(\zeta)^{-1} d\zeta$ of $\mathbb{C}/\Lambda$. (Recall from (6.27) and Proposition 6.28 that this element is well defined.) Thus our new definition of $w(\zeta)$ will be

$$w(\zeta) = \int_\infty^\zeta \tfrac{1}{2} F(\zeta)^{-1} d\zeta + \Lambda \quad \text{as a member of } \mathbb{C}/\Lambda. \tag{6.40}$$

The new $w : \mathcal{R}^* \to \mathbb{C}/\Lambda$ is still biholomorphic, and we let

$$w^{-1} : \mathbb{C}/\Lambda \to \mathcal{R}^*$$

be its inverse function. Let

$$\mu : \mathbb{C} \to \mathbb{C}/\Lambda$$

be the (holomorphic) quotient mapping, and recall that the extended version of e, no longer a covering map, is a function

$$e : \mathcal{R}^* \to \mathbb{C} \cup \{\infty\}$$

and is meromorphic. Define $P : \mathbb{C} \to \mathbb{C} \cup \{\infty\}$ as the meromorphic function

$$P(z) = e \circ w^{-1} \circ \mu(z). \tag{6.41}$$

8. AN ELLIPTIC INTEGRAL

Theorem 6.31. The function $P(z)$ in (6.41) is given by

$$P(z) = \wp(z) + \tfrac{1}{3}(a+b+c), \tag{6.42}$$

where $\wp$ is the Weierstrass $\wp$ function for the lattice Λ in (6.35).

PROOF. The function P is meromorphic, and it is doubly periodic since it factors through $\mathbf{C}/\Lambda$. Hence it is elliptic. Its only poles are in Λ because

$$\begin{aligned} P^{-1}(\{\infty\}) &= \mu^{-1}(w(e^{-1}(\{\infty\}))) = \mu^{-1}(w(\{\infty\})) \\ &= \mu^{-1}(0+\Lambda) \qquad \text{by (6.40)} \\ &= \Lambda. \end{aligned}$$

Now we examine the order of the pole of $P(z)$ at $z=0$. The maps μ and w^{-1} are locally invertible. Thus the order of the pole of P is the same as the order of the pole of e at ∞ in $\mathcal{R}^*$. The local parameter is $s = \frac{1}{\sqrt{\zeta}}$ near ∞ in $\mathcal{R}^*$. So $e(\zeta) = z = s^{-2}$. Thus the pole has order 2.

Next let us see that P is an even function. For z in $\mathbf{C}$, put $\zeta = w^{-1}\mu(z)$. Let C be a piecewise smooth path from ∞ to ζ in $\mathcal{R}^*$. The Riemann surface $\mathcal{R}^*$ has a holomorphic involution (namely interchange of the two sheets), and we let C' be the image curve extending from ∞ to a point ζ'. Then

$$\begin{aligned} w(\zeta') &\equiv \int_{\infty \text{ (along } C')}^{\zeta'} \tfrac{1}{2} F(\zeta)^{-1} d\zeta \mod \Lambda \\ &= -\int_{\infty \text{ (along } C)}^{\zeta} \tfrac{1}{2} F(\zeta)^{-1} d\zeta \quad \begin{array}{l} \text{since the integrands} \\ \text{in (6.27) match} \end{array} \\ &\equiv -w(\zeta) \mod \Lambda. \end{aligned}$$

Hence

$$\zeta' = w^{-1}(-w(\zeta) + \Lambda)$$

and

$$\begin{aligned} P(z) &= ew^{-1}\mu(z) = e(\zeta) = e(\zeta') \\ &= e(w^{-1}(-w(\zeta)+\Lambda)) = e(w^{-1}(-w(w^{-1}\mu(z))+\Lambda)) \\ &= e(w^{-1}(-\mu(z)+\Lambda)) = e(w^{-1}(\mu(-z)+\Lambda)) = P(-z), \end{aligned}$$

as asserted.

Our classification of poles of P shows that P has order 2. Thus it has only two zeros, counting multiplicities. The proof of Theorem 6.10 gives the structure of even elliptic functions and shows that

$$P(z) = k_1(\wp(z) - k_2). \tag{6.43}$$

Let us see that P satisfies the expected differential equation. Put $\psi = w^{-1}\mu(z) \in \mathcal{R}^*$ and restrict attention to points $\psi \in \mathcal{R}$. Then $p = e(\psi)$ is a local parameter in $\mathbb{C}$, and we have

$$P(z) = ew^{-1}\mu(z) = e(\psi) = p.$$

Since $w(\psi) = \mu(z)$, (6.40) gives

$$z \in \int_\infty^\psi \tfrac{1}{2} F(\zeta)^{-1} d\zeta + \Lambda.$$

By (6.30), we have as local expression for a derivative

$$\frac{dz}{dp} = \tfrac{1}{2} F(p)^{-1} = \frac{1}{\sqrt{4(p-a)(p-b)(p-c)}}$$

for a certain branch of the square root. Consequently

$$\frac{dp}{dz} = \sqrt{4(p-a)(p-b)(p-c)}.$$

Replacing $p = P(z)$, we see that P satisfies

$$P'^2 = 4(P-a)(P-b)(P-c). \tag{6.44}$$

Now we evaluate k_1 and k_2 in (6.43). Since P is even and has a double zero at $z = 0$, we can write

$$P(z) = \frac{c_{-2}}{z^2} + c_0 + O(z^2) \qquad (\text{as } z \to 0)$$

$$P'(z) = -\frac{2c_{-2}}{z^3} + O(z)$$

$$P'(z)^2 = \frac{4c_{-2}^2}{z^6} + O(z^{-2})$$

$$P(z) - a = \frac{c_{-2}}{z^2} + O(1)$$

$$(P(z)-a)(P(z)-b)(P(z)-c)) = \frac{c_{-2}^3}{z^6} + O(z^{-4}).$$

Thus (6.44) implies $\frac{4c_{-2}^2}{z^6} = \frac{4c_{-2}^3}{z^6}$, i.e., $c_{-2} = 1$. Thus $k_1 = 1$ in (6.43), and $P(z) = \wp(z) + k$. Substituting this formula into (6.44) gives

$$\wp'^2 = 4(\wp + k - a)(\wp + k - b)(\wp + k - c)$$
$$= 4\wp^3 + 4(3k - a - b - c)\wp^2 + \text{lower order}.$$

Since also
$$\wp'^2 = 4\wp^3 - g_2\wp - g_3,$$
we can subtract and conclude from the fact that $\wp(z)$ assumes more than two values that
$$3k - a - b - c = 0.$$
This proves the theorem.

Corollary 6.32. Every nonsingular curve
$$y^2 = 4x^3 - g_2 x - g_3$$
over $\mathbb{C}$ is of the form $\mathbb{C}/\Lambda$ for some lattice Λ and is parametrized biholomorphically by $z \to (\wp(z), \wp'(z))$, where $\wp(z)$ is the Weierstrass $\wp$ function relative to Λ.

9. Computability of the Correspondence

The theory of §§1-8 produced a correspondence $\Lambda \leftrightarrow (g_2, g_3)$ of lattices Λ in $\mathbb{C}$ with pairs (g_2, g_3) of complex numbers such that the right side of
$$E \ : \ y^2 = 4x^3 - g_2 x - g_3$$
has nonzero discriminant. The correspondence was implemented for each Λ by a biholomorphic mapping of $\mathbb{C}/\Lambda$ onto $E(\mathbb{C})$, given in terms of the Weierstrass $\wp$ function for Λ, with inverse given in terms of elliptic integrals. In this section we study the correspondence $\Lambda \leftrightarrow (g_2, g_3)$ a bit further. In particular, we shall see (at least in cases of interest) that there are rapidly convergent expressions for the correspondence that are suitable for (approximate) numerical computations.

Let us recall the exact correspondence of parameters. If Λ is given, then we obtain E by using

$$g_2 = 60 \sum_{\substack{\omega \in \Lambda \\ \omega \neq 0}} \frac{1}{\omega^4}$$

$$g_3 = 140 \sum_{\substack{\omega \in \Lambda \\ \omega \neq 0}} \frac{1}{\omega^6}$$

(6.45)

If g_2 and g_3 are given, then we let a, b, c be the roots of the right side of E, and we obtain Λ as the lattice generated by

$$\omega_1 = \int_{\tilde{\Gamma}_1} \tfrac{1}{2} F(\zeta)^{-1} d\zeta \quad \text{and} \quad \omega_2 = \int_{\tilde{\Gamma}_2} \tfrac{1}{2} F(\zeta)^{-1} d\zeta, \qquad (6.46)$$

where

α, β, γ are points in $\mathcal{R}^*$ covering a, b, c,

$F(\zeta) = \sqrt{(\zeta - \alpha)(\zeta - \beta)(\zeta - \gamma)}$,

$\tilde{\Gamma}_1$ = lift to $\mathcal{R}^*$ of loop Γ_1 around a and b,

$\tilde{\Gamma}_2$ = lift to $\mathcal{R}^*$ of loop Γ_2 around b and c.

To understand Λ, we are not concerned about minus signs on ω_1 and ω_2, and we can thus replace (6.46) by integrals in $\mathbf{C}$:

$$\begin{aligned}\omega_1 &= \int_{\Gamma_1} \frac{dz}{2\sqrt{(z-a)(z-b)(z-c)}} \\ \omega_2 &= \int_{\Gamma_2} \frac{dz}{2\sqrt{(z-a)(z-b)(z-c)}},\end{aligned} \qquad (6.47)$$

where the square root in each case is a single-valued branch in the z plane slit between the two points in question and slit also between the third point and ∞. From (6.45) and (6.47), we can read off one feature of the correspondence.

Proposition 6.33. Under the correspondence $\Lambda \leftrightarrow (g_2, g_3)$, Λ is closed under complex conjugation if and only if g_2 and g_3 are both in $\mathbf{R}$.

PROOF. If Λ is closed under conjugation, then (6.45) shows immediately that g_2 and g_3 are real. Conversely let g_2 and g_3 be real. Then either a, b, c are real, or one of them (b, say) is real and the other two are complex conjugates of each other.

Suppose a, b, c are real. Let us say that $a < b < c$. For Γ_1 we can use a curve that extends on the real axis on one side of a cut from a to b and comes back to a on the real axis on the other side of the cut with the square root having changed by a minus sign. Then

$$\omega_1 = \int_a^b \frac{dx}{\sqrt{(x-a)(x-b)(x-c)}} \qquad (6.48)$$

for one of the two determinations of the square root. The expression under the square root is positive for $a < x < b$, and therefore ω_1 is real. Similarly we can take

$$\omega_2 = \int_b^c \frac{dx}{\sqrt{(x-a)(x-b)(x-c)}} \qquad (6.49)$$

for one of the two determinations of the square root. The expression under the square root is negative for $b < x < c$, and therefore ω_2 is imaginary. Hence $\bar{\omega}_1 = \omega_1$ and $\bar{\omega}_2 = -\omega_2$. Consequently Λ is closed under conjugation.

Now suppose b is real and $a = \bar{c}$. Whatever path we use for Γ_1, we can use the complex conjugate path for Γ_2, apart from orientation. With suitable choices of the branch of the square root in each case, we can arrange that the integrands in (6.47) are pointwise complex conjugates of each other. Then it follows that ω_1 and ω_2 are complex conjugates of each other, except possibly for a sign. If indeed $\omega_1 = \bar{\omega}_2$, then Λ is closed under conjugation. If $\omega_1 = -\bar{\omega}_2$, then $\bar{\omega}_1 = -\omega_2$ and $\bar{\omega}_2 = -\omega_1$ show that Λ is closed under conjugation.

Our chief interest is in elliptic curves defined over $\mathbb{Q}$. Thus it is not unreasonable for us to restrict attention now to lattices and curves meeting the conditions of Proposition 6.33. There are rapid methods for calculating the sums in (6.45) and integrals in (6.47), and we shall use them in examining the correspondence further.

To calculate g_2 and g_3 rapidly, let us assume that ω_1/ω_2 is in the lower half plane. (If it is not, then we have only to interchange ω_1 and ω_2.) Then we shall see in from (8.10) that

$$g_2 = \frac{4\pi^4}{3\omega_2^4}\left(1 + \sum_{n=1}^{\infty} \frac{240n^3}{e^{2\pi i n \omega_1/\omega_2} - 1}\right)$$

$$g_3 = \frac{8\pi^6}{27\omega_2^6}\left(1 - \sum_{n=1}^{\infty} \frac{504n^5}{e^{2\pi i n \omega_1/\omega_2} - 1}\right). \qquad (6.50)$$

The two series in (6.50) are rapidly convergent: The numerator is a polynomial in n, but the denominator grows exponentially with n.

A general theory for handling (6.47) is available, but we shall concentrate on the case that a, b, c are real. Thus we want to investigate (6.48) and (6.49). The tool is the **arithmetic-geometric mean** of Gauss defined as follows: If a and b are positive reals with $a \geq b$, we recursively let

$$(a_0, b_0) = (a, b), \qquad (a_{i+1}, b_{i+1}) = \left(\frac{a_i + b_i}{2}, \sqrt{a_i b_i}\right).$$

Then we can see that

$$a_0 \geq a_1 \geq a_2 \geq \cdots \geq b_2 \geq b_1 \geq b_0 \tag{6.51}$$

and $\lim a_n = \lim b_n$, the convergence being quite rapid. The common value of $\lim a_n$ and $\lim b_n$ is the arithmetic-geometric mean of a and b and is denoted $M(a,b)$.

[To see the existence of $M(a,b)$, we note first that (6.51) is clear. Then

$$(a_i - b_i) - 2(a_{i+1} - b_{i+1}) = (a_i - b_i) - (a_i + b_i - 2\sqrt{a_i b_i})$$
$$= -2b_i + 2\sqrt{a_i b_i} \geq 0$$

shows that $|a_{i+1} - b_{i+1}| \leq \frac{1}{2}|a_i - b_i|$. The required convergence follows.]

Proposition 6.34. The integrals in (6.48) and (6.49) with $a < b < c$ are given in terms of the Gauss arithmetic-geometric mean by the following expressions, up to minus signs:

$$\omega_1 = \int_a^b \frac{dx}{\sqrt{(x-a)(x-b)(x-c)}} = \frac{\pi}{M(\sqrt{c-a}, \sqrt{c-b})}$$
$$\omega_2 = \int_b^c \frac{dx}{\sqrt{(x-a)(x-b)(x-c)}} = \frac{i\pi}{M(\sqrt{c-a}, \sqrt{b-a})}. \tag{6.52}$$

PROOF. For ω_1 we make the change of variables $\sqrt{x-a} = \sqrt{b-a}\sin\theta$ and obtain

$$\omega_1 = 2\int_0^{\pi/2} \frac{d\theta}{\sqrt{(c-b)\sin^2\theta + (c-a)\cos^2\theta}}.$$

For ω_2 we make the change of variables $\sqrt{x-b} = \sqrt{c-b}\cos\theta$ and obtain

$$\omega_2 = 2i\int_0^{\pi/2} \frac{d\theta}{\sqrt{(b-a)\sin^2\theta + (c-a)\cos^2\theta}}.$$

Thus the proposition will follow if we prove that $0 < r < s$ implies that $I(r,s)$, when defined by

$$I(r,s) = \int_0^{\pi/2} \frac{d\theta}{\sqrt{r^2\sin^2\theta + s^2\cos^2\theta}}, \tag{6.53}$$

9. COMPUTABILITY OF THE CORRESPONDENCE

evaluates as
$$I(r,s) = \frac{\pi}{2M(s,r)}. \tag{6.54}$$

We shall prove that
$$I(r,s) = I\left(\sqrt{rs}, \frac{r+s}{2}\right). \tag{6.55}$$

Iterating (6.55) and using the existence of $M(s,r)$ and an easy passage to the limit, we see that
$$I(r,s) = I(M(s,r), M(s,r)). \tag{6.56a}$$

But we can evaluate $I(M,M)$ trivially, and the result is
$$I(M,M) = \frac{\pi}{2M}. \tag{6.56b}$$

The equality (6.56) implies (6.54). Thus it is enough to prove (6.55).

To prove (6.55), regard $0 < r < s$ as fixed. For $0 \leq t \leq 1$, the function $\dfrac{2st}{(s+r)+(s-r)t^2}$ increases from 0 to 1. Therefore
$$\sin\theta = \frac{2s\sin\varphi}{(s+r)+(s-r)\sin^2\varphi}, \quad 0 \leq \varphi \leq \frac{\pi}{2},$$

is a legitimate change of variables, and θ extends from 0 to $\frac{\pi}{2}$. We make this change of variables in (6.53) after rewriting (6.53) as
$$I(r,s) = \int_0^{\pi/2} \frac{\cos\theta \, d\theta}{\cos^2\theta\sqrt{r^2\tan^2\theta + s^2}}. \tag{6.57}$$

We readily compute
$$\cos\theta \, d\theta = \frac{2s\cos\varphi\,[(s+r)-(s-r)\sin^2\varphi]\,d\varphi}{[(s+r)+(s-r)\sin^2\varphi]^2}$$

$$\cos^2\theta = \frac{\cos^2\varphi[(s+r)^2 - (s-r)^2\sin^2\varphi]}{[(s+r)+(s-r)\sin^2\varphi]^2}$$

$$\tan^2\theta = \frac{4s^2\sin^2\varphi}{\cos^2\varphi\,[(s+r)^2 - (s-r)^2\sin^2\varphi]}.$$

Substituting these expressions into (6.57) and simplifying, we obtain
$$I(r,s) = \int_0^{\pi/2} \frac{2\,d\varphi}{\sqrt{4rs\sin^2\varphi + (s+r)^2\cos^2\varphi}},$$

and (6.55) follows. This proves the proposition.

EXAMPLE. $y(y+1) = (x+1)x(x-1)$. This elliptic curve is listed in Table 3.2 with $\Delta = 37$. Using the transformation rule that passes from (3.23) to (3.26), we are led to

$$E \; : \; y^2 = 4x^3 - 4x + 1.$$

This curve has

$$g_2 = 4 \qquad g_3 = -1. \tag{6.58}$$

Its roots are

$$\begin{aligned} a &= -1.10716\cdots \\ b &= .269594\cdots \\ c &= .837565\cdots. \end{aligned}$$

The relevant Gauss arithmetic-geometric means are

$$M(\sqrt{c-a}, \sqrt{c-b}) = M(1.39453\cdots, .753639\cdots) = 1.04949\cdots$$
$$M(\sqrt{c-a}, \sqrt{b-a}) = M(1.39453\cdots, 1.17335\cdots) = 1.28156\cdots.$$

Thus (6.52) gives

$$\omega_1 = \frac{\pi}{M(\sqrt{c-a}, \sqrt{c-b})} = 2.99346\cdots$$
$$\omega_2 = \frac{i\pi}{M(\sqrt{c-a}, \sqrt{b-a})} = 2.45139\cdots i.$$

With the aid of (6.50), we can use ω_1 and ω_2 to estimate g_2 and g_3 as

$$\begin{aligned} g_2 &= 4.00000\cdots \\ g_3 &= -.999999\cdots \end{aligned}$$

in agreement with (6.58).

CHAPTER VII

DIRICHLET'S THEOREM

1. Motivation

Dirichlet's Theorem on primes in arithmetic progressions has the following statement.

Theorem 7.1. If m and b are relatively prime integers with $m > 0$, then there exist infinitely many primes of the form $km + b$ with k a positive integer.

This theorem was used in the proof of Theorems 5.2 and 5.3. Its proof contains the first use of L functions historically and will help as an introduction to the L functions that are the theme of the rest of this book.

It is illuminating to begin with some special cases of Theorem 7.1 as motivation. We first suppose that $m = 1$. In this case the assertion of the theorem is just that there are infinitely many primes, and there are, of course, many proofs of this result. For our purposes the interesting proof is a relatively complicated argument given by Euler in 1737. We introduce the function

$$\zeta(s) = \sum_{n=1}^{\infty} \frac{1}{n^s} \quad (7.1)$$

for real $s > 1$. (This function subsequently was named the **Riemann zeta function** and is defined and analytic for complex s with $\mathrm{Re}\, s > 1$. We postpone a more serious discussion of $\zeta(s)$ to §2.) By monotone convergence, for example, we have

$$\lim_{s \downarrow 1} \zeta(s) = +\infty. \quad (7.2)$$

We can write

$$\zeta(s) = \prod_{p \text{ prime}} \frac{1}{1 - \frac{1}{p^s}} \quad (7.3)$$

for $s > 1$. (In fact, the product for $p \leq N$ is

$$\prod_{p \leq N} \frac{1}{1-p^{-s}} = \prod_{p \leq N} \left(1 + \frac{1}{p^s} + \frac{1}{p^{2s}} + \cdots\right) = {\sum}' \frac{1}{n^s}$$

with $\sum'$ taken over all n with all prime factors $\leq N$. Letting $N \to \infty$, we obtain (7.3).)

If there were only finitely many primes, then (7.3) would yield

$$\limsup_{s \downarrow 1} \zeta(s) < \infty,$$

in contradiction to (7.2). This contradiction completes Euler's proof that there are infinitely many primes.

To consider further special cases, it is more convenient to work with

$$\log \zeta(s) = \sum_{p \text{ prime}} \log \frac{1}{1 - \frac{1}{p^s}} = \sum_{p \text{ prime}} \left(\frac{1}{p^s} + \frac{1}{2p^{2s}} + \frac{1}{3p^{3s}} + \cdots\right),$$
(7.4a)

which we shall see in the course of proving (7.30) is

$$= \sum_{p \text{ prime}} \frac{1}{p^s} + g(s) \tag{7.4b}$$

with $g(s)$ bounded as $s \downarrow 1$. Euler's proof amounts to combining (7.2) and (7.4) to see that $\sum \frac{1}{p}$ diverges, and the corollary is that there are infinitely many primes.

To handle $m = 4$ in Theorem 7.1, we want to treat primes $4k + 1$ separately from primes $4k + 3$. Dirichlet's idea for this special case amounts to working with the sum and difference of

$$\sum_{p \equiv 1 \bmod 4} \frac{1}{p^s} \quad \text{and} \quad \sum_{p \equiv 3 \bmod 4} \frac{1}{p^s}, \tag{7.5}$$

rather than the two terms separately. Tracing backwards in (7.4), we are led to consider the following expressions, which differ from the sum and difference of (7.5) by bounded terms as $s \downarrow 1$:

$$\sum_{n \text{ odd}} \frac{1}{n^s} = \prod_{\substack{p \text{ prime,} \\ p \text{ odd}}} \frac{1}{1 - p^{-s}} \tag{7.6a}$$

$$\sum_{n \text{ odd}} \frac{(-1)^{\frac{1}{2}(n-1)}}{n^s} = \left(\prod_{\substack{p \text{ prime,} \\ p=4k+1}} \frac{1}{1 - p^{-s}}\right) \left(\prod_{\substack{p \text{ prime,} \\ p=4k+3}} \frac{1}{1 + p^{-s}}\right). \tag{7.6b}$$

1. MOTIVATION

Let us write

$$\chi_0(n) = \begin{cases} 0 & \text{if } n \equiv 0 \mod 2 \\ 1 & \text{if } n \equiv 1 \mod 2 \end{cases}$$

$$\chi_1(n) = \begin{cases} 0 & \text{if } n \equiv 0 \mod 2 \\ 1 & \text{if } n \equiv 1 \mod 4 \\ -1 & \text{if } n \equiv 3 \mod 4. \end{cases} \tag{7.7}$$

With χ equal to χ_0 or χ_1, we have

$$\chi(mn) = \chi(m)\chi(n) \qquad \text{for all } m \text{ and } n.$$

Consequently the expressions (7.6) are both of the form

$$L(s,\chi) = \sum_{n=1}^{\infty} \frac{\chi(n)}{n^s} = \prod_{p \text{ prime}} \frac{1}{1 - \frac{\chi(p)}{p^s}}, \tag{7.8}$$

with χ taken as χ_0 for (7.6a) and as χ_1 for (7.6b). We can write the analog of (7.4) uniformly as

$$\log L(s,\chi) = \sum_{p \text{ prime}} \frac{\chi(p)}{p^s} + g(s,\chi)$$

with $g(s,\chi)$ bounded as $s \downarrow 1$. Therefore

$$\log(L(s,\chi_0)L(s,\chi_1)) = 2 \sum_{\substack{p \text{ prime} \\ p=4k+1}} \frac{1}{p^s} + (g(s,\chi_0) + g(s,\chi_1))$$

$$\log(L(s,\chi_0)L(s,\chi_1)^{-1}) = 2 \sum_{\substack{p \text{ prime} \\ p=4k+3}} \frac{1}{p^s} + (g(s,\chi_0) - g(s,\chi_1)). \tag{7.9}$$

For $\chi = \chi_0$, comparison of (7.3) and (7.8) shows that

$$\zeta(s) = L(s,\chi_0) \frac{1}{1 - 2^{-s}}.$$

Therefore

$$\lim_{s \downarrow 1} L(s,\chi_0) = +\infty.$$

Meanwhile the series in (7.6b) is alternating and converges for $s > 0$ by the Leibniz test. The convergence is uniform on compact sets, and the sum $L(s,\chi_1)$ is continuous for $s > 0$. Grouping the terms of this series in pairs, we see that $L(1,\chi_1) > 0$. (Actually we can even recognize the value as $\pi/4$ from the Taylor series of arctan x.) Putting these facts about $L(s,\chi_0)$ and $L(s,\chi_1)$ together, we see that both left sides of (7.9) tend to $+\infty$ as $s \downarrow 1$. It follows from (7.9) that

$$\sum_{\substack{p \text{ prime} \\ p=4k+1}} \frac{1}{p} \quad \text{and} \quad \sum_{\substack{p \text{ prime} \\ p=4k+3}} \frac{1}{p}$$

are both infinite. Hence there are infinitely many primes $4k + 1$, and there are infinitely many primes $4k + 3$.

The proof of Dirichlet's Theorem for $km + b$ will proceed in similar fashion. We return to it in §4 after a brief but systematic investigation of the kinds of series and products that we have encountered in the present section.

2. Dirichlet Series and Euler Products

A series $\sum_{n=1}^{\infty} \frac{a_n}{n^s}$ with a_n and s complex is called a **Dirichlet series**. The first result shows that the region of convergence and the region of absolute convergence are each right half planes in $\mathbb{C}$ (unless equal to the empty set or all of $\mathbb{C}$). These half planes may not be the same: $\sum \frac{(-1)^n}{n^s}$ is convergent for Re $s > 0$ and absolutely convergent for Re $s > 1$.

Proposition 7.2. Let $\sum_{n=1}^{\infty} \frac{a_n}{n^s}$ be a Dirichlet series.

(a) If the series is convergent for $s = s_0$, then it is convergent uniformly on compact sets for Re $s >$ Re s_0, and the sum of the series is analytic in this region.

(b) If the series is absolutely convergent for $s = s_0$, then it is uniformly absolutely convergent for Re $s \geq$ Re s_0.

(c) If the series is convergent for $s = s_0$, then it is absolutely convergent for Re $s >$ Re $s_0 + 1$.

(d) If the series is convergent at some s_0 and sums to 0 in a right half plane, then all the coefficients are 0.

REMARK. The proof of (a) will use the **summation by parts** formula. If $\{u_n\}$ and $\{v_n\}$ are sequences and if $U_n = \sum_{k=1}^{n} u_k$ for $n \geq 0$, then $1 \leq M \leq N$ implies

$$\sum_{n=M}^{N} u_n v_n = \sum_{n=M}^{N-1} U_n(v_n - v_{n+1}) + U_N v_N - U_{M-1} v_M. \qquad (7.10)$$

2. DIRICHLET SERIES AND EULER PRODUCTS

PROOF. For (a), we write $\dfrac{a_n}{n^s} = \dfrac{a_n}{n^{s_0}} \cdot \dfrac{1}{n^{s-s_0}} = u_n v_n$ and then apply (7.10). The given convergence means that $\{U_n\}$ is convergent, and certainly $v_n \to 0$ uniformly on any proper half plane of $\operatorname{Re} s > \operatorname{Re} s_0$. Thus the second and third terms on the right side of (7.10) tend to 0 with the required uniformity as M and N tend to ∞. For the first term, the sequence $\{U_n\}$ is bounded, and we shall show that

$$\sum_{n=1}^{\infty} |v_n - v_{n+1}| = \sum_{n=1}^{\infty} \left| \frac{1}{n^{s-s_0}} - \frac{1}{(n+1)^{s-s_0}} \right|$$

is convergent uniformly on compact sets where $\operatorname{Re} s > \operatorname{Re} s_0$. Use of (7.10) and the Cauchy criterion will complete the proof of convergence. For $n \le t \le n+1$, we have

$$|n^{-(s-s_0)} - t^{-(s-s_0)}| \le \sup_{n \le t \le n+1} \left| \frac{d}{dt}(n^{-(s-s_0)} - t^{-(s-s_0)}) \right|$$

$$= \sup_{n \le t \le n+1} \left| \frac{s - s_0}{t^{s-s_0+1}} \right| \le \frac{|s - s_0|}{n^{1+\operatorname{Re}(s-s_0)}}. \quad (7.11)$$

Thus

$$|v_n - v_{n+1}| = |n^{-(s-s_0)} - (n+1)^{-(s-s_0)}| \le \frac{|s - s_0|}{n^{1+\operatorname{Re}(s-s_0)}},$$

and the sum is uniformly convergent on compact sets with $\operatorname{Re} s > \operatorname{Re} s_0$, by the Weierstrass M-test.

Hence the given Dirichlet series is uniformly convergent on compact sets where $\operatorname{Re} s > \operatorname{Re} s_0$. Since each term is analytic in this region, the sum is analytic.

For (b), we have

$$\left| \frac{a_n}{n^s} \right| = \left| \frac{a_n}{n^{s_0}} \right| \cdot \left| \frac{1}{n^{s-s_0}} \right| \le \left| \frac{a_n}{n^{s_0}} \right|.$$

Since the sum of the right side is convergent, the desired uniform convergence follows from the Weierstrass M-test.

For (c), let $\epsilon > 0$ be given. Then

$$\left| \frac{a_n}{n^{s_0+1+\epsilon}} \right| = \left| \frac{a_n}{n^{s_0}} \right| \frac{1}{n^{1+\epsilon}}$$

with the first factor on the right bounded and the second factor contributing to a finite sum. Therefore we have absolute convergence at $s_0 + 1 + \epsilon$, and (c) follows from (b).

For (d), we may assume by (c) that there is absolute convergence at s_0. Suppose $a_1 = \cdots = a_{N-1} = 0$. By (b), $\sum_{n=N}^{\infty} \frac{a_n}{n^s} = 0$ for $\operatorname{Re} s > \operatorname{Re} s_0$. The series

$$\sum_{n=N}^{\infty} \frac{a_n}{(n/N)^s} \tag{7.12}$$

is by assumption absolutely convergent at s_0, and $\operatorname{Re} s > \operatorname{Re} s_0$ implies

$$\left|\frac{a_n}{(n/N)^s}\right| \leq \left|\frac{a_n}{(n/N)^{s_0}}\right|.$$

By dominated convergence we can take the limit of (7.12) term by term as $s \to +\infty$. The only term that survives is a_N. Since (7.12) has sum 0 for all s, we conclude $a_N = 0$. This completes the proof.

Proposition 7.3. The Riemann zeta function $\zeta(s) = \sum_{n=1}^{\infty} \frac{1}{n^s}$, initially defined and analytic for $\operatorname{Re} s > 1$, extends to be meromorphic for $\operatorname{Re} s > 0$. Its only pole is at $s = 1$, and the pole is simple.

PROOF. For $\operatorname{Re} s > 1$, we have

$$\frac{1}{s-1} = \int_1^{\infty} t^{-s}\, dt = \sum_{n=1}^{\infty} \int_n^{n+1} t^{-s}\, dt.$$

Thus $\operatorname{Re} s > 1$ implies

$$\zeta(s) = \frac{1}{s-1} + \sum_{n=1}^{\infty} \left(\frac{1}{n^s} - \int_n^{n+1} t^{-s}\, dt\right)$$

$$= \frac{1}{s-1} + \sum_{n=1}^{\infty} \int_n^{n+1} (n^{-s} - t^{-s})\, dt.$$

It is enough to show that the series on the right side converges uniformly on compact sets for $\operatorname{Re} s > 0$. Thus suppose $\operatorname{Re} s \geq \sigma > 0$ and $|s| \leq C$. By (7.11) with $s_0 = 0$, we have

$$\left|\int_n^{n+1} (n^{-s} - t^{-s})\, dt\right| \leq \int_n^{n+1} |n^{-s} - t^{-s}|\, dt \leq \frac{|s|}{n^{1+\operatorname{Re} s}} \leq \frac{C}{n^{1+\sigma}}.$$

Since $\sum n^{-(1+\sigma)} < \infty$, the desired uniform convergence follows from the Weierstrass M test.

REMARK. Actually $\zeta(s)$ extends to be meromorphic in $\mathbb{C}$ with no additional poles. We return to this point in §5.

2. DIRICHLET SERIES AND EULER PRODUCTS

We shall now examine special features of Dirichlet series that allow the series to have product expansions like the one (7.3) for $\zeta(s)$. We begin with some general facts about infinite products.

An infinite product $\prod_{n=1}^{\infty} a_n$ with $a_n \in \mathbf{C}$ and with no factor 0 is said to **converge** if the sequence of partial products converges and the limit is not 0. A necessary condition for convergence is that $a_n \to 1$.

Proposition 7.4. If $|a_n| < 1$ for all n, then the following conditions are equivalent:

(a) $\prod_{n=1}^{\infty}(1 + |a_n|)$ converges

(b) $\sum_{n=1}^{\infty} |a_n|$ converges

(c) $\prod_{n=1}^{\infty}(1 - |a_n|)$ converges.

In this case, $\prod_{n=1}^{\infty}(1 + a_n)$ converges.

PROOF. Condition (c) is equivalent with

(c') $\prod_{n=1}^{\infty}(1 - |a_n|)^{-1}$ converges.

For each of (a), (b), and (c'), convergence is equivalent with boundedness above. Since

$$1 + \sum_{n=1}^{N} |a_n| \le \prod_{n=1}^{N}(1 + |a_n|) \le \prod_{n=1}^{N} \frac{1}{1 - |a_n|},$$

we see that (c') implies (a) and that (a) implies (b). For (b) $\Rightarrow$ (c'), we may assume, without loss of generality, that $|a_n| \le \frac{1}{2}$ for all n. Since $0 \le x \le \frac{1}{2}$ implies

$$\log \frac{1}{1-x} \le |x| \sup_{0 \le t \le \frac{1}{2}} \left| \frac{d}{dt} \log \frac{1}{1-t} \right| = |x| \sup_{0 \le t \le \frac{1}{2}} \left(\frac{1}{1-t} \right) = 2|x|,$$

we have

$$\log \left(\prod_{n=1}^{N} \frac{1}{1 - |a_n|} \right) = \sum_{n=1}^{N} \log \left(\frac{1}{1 - |a_n|} \right) \le 2 \sum_{n=1}^{N} |a_n|.$$

Thus (b) implies (c').

Now suppose (a) holds. To prove that $\prod_{n=1}^{\infty}(1 + a_n)$ converges, it is enough to show that $\prod_{n=M}^{N}(1 + a_n)$ tends to 1 as M and N tend to ∞. In the expression

$$\left| \prod_{n=M}^{N} (1 + a_n) - 1 \right|,$$

we expand out the product, move the absolute values in for each term, and reassemble the product. The result is the inequality

$$\left|\prod_{n=M}^{N}(1+a_n) - 1\right| \le \prod_{n=M}^{N}(1+|a_n|) - 1.$$

By (a) the right side tends to 0 as M and N tend to ∞. Therefore so does the left side. This proves the proposition.

Consider a formal product

$$\prod_{p \text{ prime}} (1 + a_p p^{-s} + \cdots + a_{p^m} p^{-ms} + \cdots). \qquad (7.13)$$

If this product is expanded without regard to convergence, the result is the Dirichlet series $\sum_{n=1}^{\infty} \dfrac{a_n}{n^s}$, where $a_1 = 1$ and a_n is given by

$$a_n = a_{p_1^{r_1}} \cdots a_{p_k^{r_k}} \qquad \text{if } n = p_1^{r_1} \cdots p_k^{r_k}. \qquad (7.14)$$

Suppose that the Dirichlet series is in fact absolutely convergent in some right half plane. Then every rearrangement is absolutely convergent to the same sum, and the same conclusion is valid for subseries. If E is a finite set of primes and $N(E)$ is the set of positive integers requiring only members of E for their factorization, we have

$$\prod_{p \in E}(1 + a_p p^{-s} + \cdots + a_{p^m} p^{-ms} + \cdots) = \sum_{n \in N(E)} \frac{a_n}{n^s}.$$

Consequently the infinite product has a limit in the half plane of absolute convergence of the Dirichlet series, and the limiting product (7.13) equals the sum of the series. The sum of the series is 0 only if one of the factors on the left side is 0. In particular the sum of the series cannot be identically 0, by Proposition 7.2d. Thus (7.13) can equal only this one Dirichlet series.

Conversely if an absolutely convergent Dirichlet series $\sum_{n=1}^{\infty} \dfrac{a_n}{n^s}$ has the property that its coefficients are **multiplicative**, i.e.,

$$a_1 = 1 \quad \text{and} \quad a_{mn} = a_m a_n \quad \text{whenever GCD}(m,n) = 1, \qquad (7.15)$$

then we can form the product (7.13) and recover the given series by expanding (7.13) and using (7.14). In this case we say that the Dirichlet series has (7.13) as an **Euler product**. Many functions in elementary number theory give rise to multiplicative sequences; an example is $a_n = \varphi(n)$, where φ is the Euler φ function.

If the coefficients are **strictly multiplicative**, i.e.,

$$a_1 = 1 \quad \text{and} \quad a_{mn} = a_m a_n \quad \text{for all } m \text{ and } n, \tag{7.16}$$

then the pth factor of (7.13) simplifies to

$$1 + a_p p^{-s} + \cdots + (a_p p^{-s})^m + \cdots = \frac{1}{1 - \frac{a_p}{p^s}}. \tag{7.17}$$

In this case our Dirichlet series has a **first degree Euler product**:

$$\sum_{n=1}^{\infty} \frac{a_n}{n^s} = \prod_{p \text{ prime}} \frac{1}{1 - \frac{a_p}{p^s}}. \tag{7.18}$$

This is what happens with $\zeta(s)$, where all the coefficients are 1, and with $a_n = \chi_0(n)$ and $a_n = \chi_1(n)$ as in (7.7). Conversely an Euler product expansion of the form (7.18) forces the coefficients of the Dirichlet series to be strictly multiplicative.

A Dirichlet series $\sum_{n=1}^{\infty} \frac{a_n}{n^s}$ with $|a_n| \leq n^c$ for some real c is absolutely convergent for $\operatorname{Re} s > c+1$. This fact leads us to a convergence criterion for first degree Euler products.

Proposition 7.5. A first degree Euler product $\prod [1 - a_p p^{-s}]^{-1}$ with $|a_p| \leq p^c$ for some real c and all primes c defines an absolutely convergent Dirichlet series (and hence a valid identity (7.18)) for $\operatorname{Re} s > c+1$.

PROOF. The coefficients a_n are strictly multiplicative, and thus $|a_n| \leq n^c$ for all n. The absolute convergence follows.

First degree Euler products are sufficient for Dirichlet's Theorem, but other Euler products are needed in other applications, such as with elliptic curves. To isolate the notion of degree of an Euler product, let us write (7.17) as a formal identity

$$1 + a_p X + \cdots + a_p^m X^m + \cdots = \frac{1}{1 - a_p X}.$$

Here the denominator on the right is a polynomial of degree ≤ 1 with constant term 1, and it is in this sense that the Euler product (7.18) has degree 1. The expansion (7.13) is called a kth **degree Euler product** if, for each prime p, there is a polynomial $P_p(X) \in \mathbb{C}[X]$ having degree $\leq k$ and zero constant term such that

$$1 + a_p X + \cdots + a_p^m X^m + \cdots = \frac{1}{1 - P_p(X)}$$

as a formal identity. Let us factor $1 - P_p(X)$ over $\mathbb{C}$ as

$$1 - P_p(X) = (1 - r_p^{(1)} X) \cdots (1 - r_p^{(k)} X).$$

We call the complex numbers $r_p^{(j)}$ the **reciprocal roots** of $1 - P_p(X)$.

Proposition 7.6. A kth degree Euler product $\prod [1 - P_p(p^{-s})]^{-1}$ whose reciprocal roots satisfy $|r_p^{(j)}| \leq p^c$ for some real c and all primes p defines an absolutely convergent Dirichlet series for $\operatorname{Re} s > c + 1$. For such s the sum of the Dirichlet series equals the Euler product.

PROOF. We apply Proposition 7.5 to $\prod [1 - r_p^{(j)} p^{-s}]$ for each j. The product of absolutely convergent Dirichlet series can be rearranged without affecting the sum, and the result is an absolutely convergent Dirichlet series.

The final thing we need to know about Euler products is how to recognize when they are of degree k. We shall be interested only in the quadratic case, but it merely obscures the idea to make such a restriction right away.

Proposition 7.7. Let

$$1 + b_1 X + b_2 X^2 + \cdots + b_m X^m + \cdots \qquad (7.19)$$

be given. Then there is a unique polynomial

$$Q(X) = c_0 + c_1 X + \cdots + c_{k-1} X^{k-1} \qquad (7.20)$$

of degree $\leq k - 1$ such that

$$\frac{1}{1 - XQ(X)} = 1 + XQ(X) + X^2 Q(X)^2 + \cdots \qquad (7.21)$$

is congruent to (7.19) modulo X^{k+1}. For this polynomial $Q(X)$, the expressions (7.19) and (7.21) are equal if and only if

$$b_m = c_0 b_{m-1} + c_1 b_{m-2} + \cdots + c_{k-1} b_{m-k} \qquad \text{for all } m > k. \qquad (7.22)$$

PROOF. The condition that (7.19) and (7.21) be equal is the condition that $1 - XQ(X)$ be an inverse to (7.19) in the ring of formal power series over $\mathbb{C}$. Consider the equation

$$(1 - c_0 X - c_1 X^2 - \cdots - c_{k-1} X^k)(1 + b_1 X + b_2 X^2 + \cdots)$$
$$= (1 + d_1 X + d_2 X^2 + \cdots).$$

The conditions $d_1 = \cdots = d_k = 0$ uniquely determine $c_0, \ldots, c_{k-1}$ recursively. Once $c_0, \ldots, c_{k-1}$ are fixed in this fashion, d_m for $m > k$ is given as the difference of the left side minus the right side in (7.22). The result follows.

Corollary 7.8. An Euler product (7.13) with $a_1 = 1$ is of degree 2 if and only if there exists for each prime p a complex number d_p such that

$$a_{p^m} = a_p a_{p^{m-1}} + d_p a_{p^{m-2}} \tag{7.23}$$

for all $m \geq 2$. In this case the pth Euler factor is

$$\frac{1}{1 - a_p p^{-s} - d_p p^{-2s}}. \tag{7.24}$$

PROOF. Let $b_m = a_{p^m}$ in Proposition 7.7. The polynomial $Q(X)$ works out to be $c_0 + c_1 X$ with $b_1 = c_0$ and $b_2 = c_0^2 + c_1$. From (7.23) with $m = 2$, d_p has to be $a_{p^2} - a_p^2$. So we define it this way to make (7.23) valid for $m = 2$. Now

$$d_p = a_{p^2} - a_p^2 = b_2 - b_1^2 = b_2 - c_0^2 = c_1.$$

Hence (7.23) for $m > 2$ is the same equation as (7.22) for $m > 2$. Thus the corollary follows from Proposition 7.7.

3. Fourier Analysis on Finite Abelian Groups

In considering primes of the forms $4k + 1$ and $4k + 3$ in §1, we worked with the sum and difference of the expressions (7.5) and then reconstructed the individual expressions from the sum and difference. These steps correspond to doing Fourier analysis on the group $\mathbb{Z}_4^\times \cong \mathbb{Z}_2$. In this section we shall work out the Fourier analysis of $\mathbb{Z}_m^\times$ that is appropriate for handling primes of the form $km + b$.

Let G be a finite abelian group (such as $\mathbb{Z}_m^\times$). A **character** of G is a homomorphism of G into $S^1 \subseteq \mathbb{C}^\times$. The characters of G form a finite abelian group $\widehat{G}$ under pointwise multiplication:

$$(\chi\chi')(g) = \chi(g)\chi'(g).$$

Proposition 7.9. Let G be a finite abelian group. With respect to the Hermitian inner product $\langle F, F'\rangle = \sum_{g\in G} F(g)\overline{F'(g)}$ on the vector space of all complex-valued functions on G, the members of $\widehat{G}$ form an orthogonal basis, each $\chi \in \widehat{G}$ satisfying $\|\chi\|^2 = |G|$. Consequently $|\widehat{G}| = |G|$, and any function $F : G \to \mathbb{C}$ is given by the "sum of its Fourier series":

$$F(g) = \frac{1}{|G|} \sum_{\chi \in \widehat{G}} \left(\sum_{h \in G} F(h)\overline{\chi(h)} \right) \chi(g). \tag{7.25}$$

TERMINOLOGY. Equation (7.25) is called the **Fourier inversion formula**.

PROOF. First we prove that distinct characters χ and χ' are orthogonal. If χ and χ' are given with $\chi \neq \chi'$, let $\chi'' = \chi\overline{\chi'}$. Choose $g_0 \in G$ with $\chi''(g_0) \neq 1$. From

$$\chi''(g_0) \left(\sum_{g \in G} \chi''(g) \right) = \sum_{g \in G} \chi''(g_0 g) = \sum_{g \in G} \chi''(g),$$

we see that

$$[1 - \chi''(g_0)] \sum_{g \in G} \chi''(g) = 0 \quad \text{and hence} \quad \sum_{g \in G} \chi''(g) = 0.$$

Consequently

$$\langle \chi, \chi'\rangle = \sum_{g \in G} \chi(g)\overline{\chi'(g)} = \sum_{g \in G} \chi''(g) = 0,$$

and the orthogonality is proved.

It follows that the members of $\widehat{G}$ are linearly independent and hence that $|\widehat{G}| \leq |G|$. It is clear that

$$\|\chi\|^2 = \sum_{g \in G} |\chi(g)|^2 = \sum_{g \in G} 1 = |G|.$$

To see that the members of $\widehat{G}$ are a basis, we write G as a direct sum of cyclic groups. A summand $\mathbf{Z}_N$ of G has at least N distinct characters, given by $j \bmod N \to e^{2\pi i j r / N}$ for $0 \le r \le N-1$, and these characters extend to G as 1 on the other summands of G. Taking products of such characters from different summands of G, we see that $|\widehat{G}| \ge |G|$.

Therefore $\widehat{G}$ is an orthogonal basis of the space of all complex-valued functions on G. Formula (7.25) is the usual inner-product-space formula for expressing an element in terms of an orthogonal basis.

4. Proof of Dirichlet's Theorem

The proof of Dirichlet's Theorem is a direct generalization of the argument given for $m = 4$ in §1. Fix an integer $m > 1$. A **Dirichlet character modulo** m is a function $\chi : \mathbf{Z} \to S^1 \cup \{0\}$ such that

(i) $\chi(j) = 0$ if and only if $\mathrm{GCD}(j, m) > 1$
(ii) $\chi(j)$ depends only on the residue class $j \bmod m$
(iii) when regarded as a function on the residue classes modulo m, χ is a character of $\mathbf{Z}_m^\times$.

In particular, a Dirichlet character modulo m determines a character of $\mathbf{Z}_m^\times$. Conversely each character of $\mathbf{Z}_m^\times$ defines a unique Dirichlet character modulo m by lifting the character to $\{j \in \mathbf{Z} \mid \mathrm{GCD}(j, m) = 1\}$ and by defining a Dirichlet character to be 0 on the rest of $\mathbf{Z}$. It will often be notationally helpful to use the same symbol for the Dirichlet character and the character of $\mathbf{Z}_m^\times$. Because of this correspondence, the number of Dirichlet characters modulo m is $\varphi(m)$, where φ is the Euler φ function. The **principal** Dirichlet character modulo m, denoted χ_0, is the one built from the trivial character of $\mathbf{Z}_m^\times$:

$$\chi_0(j) = \begin{cases} 1 & \text{if } \mathrm{GCD}(j, m) = 1 \\ 0 & \text{if } \mathrm{GCD}(j, m) > 1. \end{cases}$$

Each Dirichlet character modulo m is strictly multiplicative, in the sense of (7.16). We assemble each as the coefficients of a Dirichlet series, the **Dirichlet L function**, by

$$L(s, \chi) = \sum_{n=1}^{\infty} \frac{\chi(n)}{n^s}. \tag{7.26}$$

Proposition 7.10.

(a) The Dirichlet series $L(s,\chi)$ is absolutely convergent for Re $s > 1$ and is given in that region by a first degree Euler product

$$L(s,\chi) = \prod_{p \text{ prime}} \frac{1}{1 - \dfrac{\chi(p)}{p^s}}. \tag{7.27}$$

(b) If χ is not principal, then the series for $L(s,\chi)$ is convergent for Re $s > 0$, and the sum is analytic in that region.

(c) For the principal Dirichlet character χ_0 modulo m, $L(s,\chi_0)$ extends to be meromorphic for Re $s > 0$. Its only pole is at $s = 1$, and the pole is simple. It is given in terms of the Riemann zeta function by

$$L(s,\chi_0) = \zeta(s) \prod_{p \mid m} \left(1 - \frac{1}{p^s}\right). \tag{7.28}$$

PROOF. For (a), the boundedness of χ implies that the series is absolutely convergent for Re $s > 1$. Since χ is strictly multiplicative, $L(s,\chi)$ has a first degree Euler product, by (7.18), and the product is convergent in the same region.

For (b), let us notice that $\chi \neq \chi_0$ implies

$$\sum_{n=1}^{m} \chi(n+b) = 0 \qquad \text{for any } b, \tag{7.29}$$

since the member of $\mathbf{Z}_m^\times$ that corresponds to χ is orthogonal to the trivial character, by Proposition 7.9. For s real and positive, let us write

$$\frac{\chi(n)}{n^s} = \chi(n) \cdot \frac{1}{n^s} = u_n v_n$$

in the notation of (7.10), putting $U_n = \sum_{k=1}^{n} u_k$. Equation (7.29) implies that $\{U_n\}$ is bounded, say $|U_n| \leq C$. By summation by parts (7.10),

$$\left| \sum_{n=M}^{N} \frac{\chi(n)}{n^s} \right| \leq \sum_{n=M}^{N-1} C \left(\frac{1}{n^s} - \frac{1}{(n+1)^s} \right) + \frac{C}{N^s} + \frac{C}{M^s} = \frac{2C}{M^s}.$$

This expression tends to 0 as M and N tend to ∞. Therefore the series (7.26) is convergent for s real and positive. By Proposition (7.2a), the series is convergent for Re $s > 0$, and the sum is analytic in this region.

For (c), let Re $s > 1$. From (7.27) with $\chi = \chi_0$, we have

$$L(s,\chi_0) = \prod_{p \nmid m} \frac{1}{1 - \dfrac{1}{p^s}}.$$

Using (7.3), we obtain (7.28). The remaining statements in (c) follow from Proposition 7.3, since the product over p with $p \mid m$ is a finite product.

4. PROOF OF DIRICHLET'S THEOREM

By Proposition 7.10b, $L(s,\chi)$ is well defined at $s = 1$ if χ is not principal. The main step in the proof of Dirichlet's Theorem is the following lemma. We defer the proof of the lemma until we have shown how the lemma implies the theorem.

Lemma 7.11. $L(1,\chi) \neq 0$ if χ is not principal.

PROOF OF THEOREM 7.1. First we show for each Dirichlet character χ modulo m that

$$\log L(s,\chi) = \sum_{p \text{ prime}} \frac{\chi(p)}{p^s} + g(s,\chi) \qquad (7.30)$$

for real $s > 1$, where $g(s,\chi)$ remains bounded as $s \downarrow 1$. In this statement we have not yet specified a branch of the logarithm, and we shall choose it presently. Fix p and define, for $s \geq 1$, a value of the logarithm of the pth factor of (7.27) by

$$\log \frac{1}{1 - \frac{\chi(p)}{p^s}} = \frac{\chi(p)}{p^s} + \frac{1}{2}\frac{\chi(p^2)}{p^{2s}} + \frac{1}{3}\frac{\chi(p^3)}{p^{3s}} + \cdots = \frac{\chi(p)}{p^s} + g(s,p,\chi). \quad (7.31)$$

Since $\left|\frac{\chi(p)}{p^s}\right| \leq \frac{1}{2}$, this logarithm satisfies

$$\left|\log \frac{1}{1-z} - z\right| \leq |z| \sup_{|w| \leq |z|} \left|\frac{d}{dw}\left(\log \frac{1}{1-w} - w\right)\right|$$

$$= |z| \sup_{|w| \leq |z|} \left|\frac{1}{1-w} - 1\right| \leq 2|z|^2$$

for $|z| \leq \frac{1}{2}$. With $z = \frac{\chi(p)}{p^s}$, we therefore have

$$|g(s,p,\chi)| \leq 2\left|\frac{\chi(p)}{p^s}\right|^2 \leq \frac{2}{p^2}.$$

Since $\sum_{p \text{ prime}} p^{-2} \leq \sum_{n=1}^{\infty} n^{-2} < \infty$, the series $\sum_p g(s,p,\chi)$ is uniformly convergent for $s \geq 1$. Let $g(s,\chi)$ be the continuous function $\sum_p g(s,p,\chi)$. Summing (7.31) over primes p, we obtain

$$\sum_p \log \frac{1}{1 - \frac{\chi(p)}{p^s}} = \sum_p \frac{\chi(p)}{p^s} + g(s,\chi).$$

By (7.27) the left side represents a branch of $\log L(s,\chi)$. This proves (7.30).

Define a function δ_b on the positive integers by

$$\delta_b(n) = \begin{cases} 1 & \text{if } n \equiv b \mod m \\ 0 & \text{otherwise.} \end{cases}$$

Proposition 7.9 gives

$$\delta_b(n) = \frac{1}{\varphi(m)} \sum_\chi \overline{\chi(b)}\chi(n).$$

Multiplying (7.30) by $\overline{\chi(b)}$ and summing on χ, we obtain

$$\varphi(m) \sum_{\substack{p \text{ prime,} \\ p=km+b}} \frac{1}{p^s} = \sum_\chi \overline{\chi(b)} \log L(s,\chi) - \sum_\chi \overline{\chi(b)} g(s,\chi). \qquad (7.32)$$

The term $\sum_\chi \overline{\chi(b)} g(s,\chi)$ is bounded as $s \downarrow 1$, by (7.30). The term $\overline{\chi_0(b)} \log L(s,\chi_0)$ is unbounded as $s \downarrow 1$, by Proposition 7.10c. For χ nonprincipal, the term $\overline{\chi(b)} \log L(s,\chi)$ is bounded as $s \downarrow 1$, by Proposition 7.10b and Lemma 7.11. Therefore the left side of (7.32) is unbounded as $s \downarrow 1$. Hence the number of primes contributing to the sum is infinite.

PROOF OF LEMMA 7.11. Let $Z(s) = \prod_\chi L(s,\chi)$. Exactly one factor of $Z(s)$ has a pole at $s = 1$, according to Proposition 7.10. If any factor has a zero at $s = 1$, then $Z(s)$ is analytic for Re $s > 0$. Assuming that $Z(s)$ is indeed analytic, we shall derive a contradiction.

Being the finite product of absolutely convergent Dirichlet series for Re $s > 1$, $Z(s)$ is given by an absolutely convergent Dirichlet series. We shall prove that the coefficients of this series are ≥ 0. More precisely we shall prove for Re $s > 1$ that

$$Z(s) = \prod_{p \nmid m} \frac{1}{\left(1 - \frac{1}{p^{f(p)s}}\right)^{g(p)}}, \qquad (7.33)$$

where $f(p)$ is the order of p in $\mathbf{Z}_m^\times$ and where $g(p) = \varphi(m)/f(p)$. The factor $(1 - p^{-f(p)s})^{-1}$ is given by a Dirichlet series with all coefficients ≥ 0. Hence so is the $g(p)^{\text{th}}$ power, and so is the product over p of the result. Thus (7.33) will prove that all coefficients of $Z(s)$ are ≥ 0.

4. PROOF OF DIRICHLET'S THEOREM

To prove (7.33), we write, for Re $s > 1$,

$$Z(s) = \prod_\chi L(s,\chi) = \prod_p \left(\prod_\chi \frac{1}{1 - \frac{\chi(p)}{p^s}} \right) = \prod_{p \nmid m} \left(\prod_\chi \frac{1}{1 - \frac{\chi(p)}{p^s}} \right).$$

Fix p not dividing m. We shall show that

$$\prod_\chi \left(1 - \frac{\chi(p)}{p^s}\right) = \left(1 - \frac{1}{p^{fs}}\right)^g, \qquad (7.34)$$

where f is the order of p in $\mathbb{Z}_m^\times$ and where $g = \varphi(m)/f$; then (7.33) will follow.

Now $\chi \to \chi(p)$ is a homomorphism of $\mathbb{Z}_m^\times$ into S^1, hence into $\{e^{2\pi i k/f}\}$ and onto some cyclic subgroup $\{e^{2\pi i k/f'}\}$ with f' dividing f. We show $f' = f$. In fact if $f' < f$, then $p^{f'} \not\equiv 1 \mod m$, while $\chi(p^{f'}) = \chi(p)^{f'} = 1$ for all χ; since $\chi(p^{f'}) = \chi(1)$ for all χ, the χ's cannot span all functions on $\mathbb{Z}_m^\times$, in contradiction to Proposition 7.9.

Thus $\chi \to \chi(p)$ is onto $\{e^{2\pi i k/f}\}$. In other words, $\chi(p)$ takes on all f^{th} roots of unity as values, and the homomorphism property ensures that each is taken on the same number of times, namely $g = \varphi(m)/f$ times. If X is an indeterminate, we then have

$$\prod_\chi (1 - \chi(p)X) = \left(\prod_{k=0}^{f-1} (1 - e^{2\pi i k/f} X) \right)^g = (1 - X^f)^g.$$

Then (7.34) follows and so does (7.33). Hence all the coefficients of the Dirichlet series of $Z(s)$ are ≥ 0.

Let us write $Z(s) = \sum_{n=1}^\infty \frac{a_n}{n^s}$ and put

$$s_0 = \inf\{s \geq 0 \mid \sum_{n=1}^\infty \frac{a_n}{n^s} \text{ converges}\}.$$

Then $s_0 \leq 1$. Suppose $s_0 > 0$. Since $\sum a_n n^{-s}$ converges uniformly on compact sets for Re $s > s_0$ (by Proposition 7.2a), we can compute its derivative term by term. Thus

$$Z^{(N)}(s_0 + 1) = \sum_{n=1}^\infty \frac{a_n(-\log n)^N}{n^{s_0+1}}. \qquad (7.35)$$

The Taylor series of $Z(s)$ about $s_0 + 1$ is

$$Z(s) = \sum_{N=0}^{\infty} \frac{1}{N!}(s - s_0 - 1)^N Z^{(N)}(s_0 + 1)$$

and is convergent at $s = \frac{1}{2}s_0$ since $Z(s)$ is analytic in the open disc centered at $s_0 + 1$ and having radius $s_0 + 1$. Thus

$$Z(\tfrac{1}{2}s_0) = \sum_{N=0}^{\infty} \frac{1}{N!}(1 + \tfrac{1}{2}s_0)^N (-1)^N Z^{(N)}(s_0 + 1),$$

with the series convergent. Substituting from (7.35), we have

$$Z(\tfrac{1}{2}s_0) = \sum_{N=0}^{\infty} \sum_{n=1}^{\infty} \frac{a_n (\log n)^N (1 + \tfrac{1}{2}s_0)^N}{N! n^{s_0+1}}.$$

This is a series with positive terms, and Fubini's Theorem allows us to interchange sums and obtain

$$Z(\tfrac{1}{2}s_0) = \sum_{n=1}^{\infty} \sum_{N=0}^{\infty} \frac{a_n (\log n)^N (1 + \tfrac{1}{2}s_0)^N}{N! n^{s_0+1}}$$

$$= \sum_{n=1}^{\infty} \frac{a_n}{n^{s_0+1}} e^{(\log n)(1+\tfrac{1}{2}s_0)}$$

$$= \sum_{n=1}^{\infty} \frac{a_n}{n^{\tfrac{1}{2}s_0}}.$$

In other words, the assumption $s_0 > 0$ led to a point between 0 and s_0 (namely $\tfrac{1}{2}s_0$) where there is convergence. This contradiction proves that $s_0 = 0$. Therefore $\sum a_n n^{-s}$ converges for Re $s > 0$.

Since the coefficients are positive, the convergence is absolute for s real and positive. By Proposition 7.2b the convergence is absolute for Re $s > 0$. Therefore the Euler product expansion (7.33) is valid for Re $s > 0$.

For $p \nmid m$ and for real $s > 0$, we have

$$\frac{1}{\left(1 - \dfrac{1}{p^{fs}}\right)^g} = (1 + p^{-fs} + p^{-2fs} + \cdots)^g$$

$$\geq 1 + p^{-fgs} + p^{-2fgs} + \cdots$$

$$= 1 + p^{-\varphi(m)s} + p^{-2\varphi(m)s} + \cdots$$

$$= \frac{1}{1 - \dfrac{1}{p^{\varphi(m)s}}}.$$

Therefore

$$\left(\prod_\chi L(s,\chi)\right)\left(\prod_{p|m}\frac{1}{1-\frac{1}{p^{\varphi(m)s}}}\right)$$

$$\geq \prod_{p\text{ prime}}(1+p^{-\varphi(m)s}+p^{-2\varphi(m)s}+\cdots) = \sum_{n=1}^\infty \frac{1}{n^{\varphi(m)s}}.$$

The sum on the right is $+\infty$ for $s = \dfrac{1}{\varphi(m)}$, while the left side is finite. This contradiction completes the proof of the lemma.

5. Analytic Properties of Dirichlet L Functions

The Riemann zeta function $\zeta(s)$ and the Dirichlet L functions $L(s,\chi)$ encode arithmetic information, and Dirichlet's Theorem comes out as a consequence of properties of these functions. The properties in question are the analytic continuation and the behavior at $s = 1$ for $\zeta(s)$ and each $L(s,\chi)$. Other theorems about primes can be deduced from other properties of these functions; for example, the Prime Number Theorem about the asymptotic number of primes $\leq N$ is a consequence of the nonvanishing of $\zeta(s)$ for Re $s = 1$.

In the case of elliptic curves over $\mathbb{Q}$, we shall introduce in a later chapter a zeta function and an L function that are defined by Euler products and encode geometric information, and deep properties of the elliptic curve come out (partly conjecturally) as a consequence of properties of these functions. This is part of a general pattern in algebraic number theory and algebraic geometry, that L functions are used to encode information prime by prime and that properties of these L functions are expected to yield deep insights into the original problem being studied.

Let us refer generally to such arithmetic or geometric L functions as **motivic**, for reasons explained in the preface. The question is how to get an analytic handle on these functions in order to exploit their properties.

For $\zeta(s)$ and $L(s,\chi)$, we obtained enough information by direct arguments to prove Theorem 7.1. But for more subtle analytic properties, one uses an indirect approach. The key idea is to relate such a function to a suitable analytic function in the upper half plane with certain transformation properties, a so-called "modular form." Each modular form has an L function, defined explicitly by a Mellin transform and given as a Dirichlet series. We shall refer generally to this kind of L function as

automorphic. Automorphic L functions have more manageable analytic properties, but they initially have little to do with algebraic number theory or algebraic geometry. The fundamental objective is to prove that motivic L functions are automorphic.

In the case of $\zeta(s)$ and $L(s,\chi)$, we shall achieve this objective in this section, using suitable theta functions as the modular forms. The analytic properties that we shall derive as a consequence are their meromorphic continuations to all of $\mathbf{C}$, the functional equations that they satisfy, and the identifications of their poles.

In the case of the L function of an elliptic curve over $\mathbf{Q}$, the assertion that this L function is always automorphic is substantially the Taniyama-Weil Conjecture. This theme will occupy us for the remaining chapters of the book. If the L function is automorphic in the expected way, it extends to an entire function and satisfies a functional equation. In particular, this $L(s)$ must extend to be defined in a neighborhood of $s = 1$. As mentioned in Chapter I, the Birch and Swinnerton-Dyer Conjecture relates the behavior of $L(s)$ near $s = 1$ to the rank of the group of $\mathbf{Q}$ points of the elliptic curve, among other things.

We begin by treating $\zeta(s)$, returning to $L(s,\chi)$ afterward. The theta function that we shall use is

$$\theta(\tau) = \sum_{n=-\infty}^{\infty} e^{in^2\pi\tau} = 1 + 2\sum_{n=1}^{\infty} e^{in^2\pi\tau}, \qquad (7.36)$$

analytic for Im $\tau > 0$. Its role in helping us to understand $\zeta(s)$ comes from the identity

$$\int_0^\infty e^{-n^2\pi\sigma} \sigma^{\frac{1}{2}s-1} \, d\sigma = \int_0^\infty e^{-x}(n^2\pi)^{-(\frac{1}{2}s-1)} x^{\frac{1}{2}s-1}(n^2\pi)^{-1} \, dx$$
$$= n^{-s}\Gamma(\tfrac{1}{2}s)\pi^{-\frac{1}{2}s}. \qquad (7.37)$$

Summing on n for $n \geq 1$ and interchanging sum and integral by Fubini's Theorem, we obtain

$$\zeta(s)\Gamma(\tfrac{1}{2}s)\pi^{-\frac{1}{2}s} = \int_0^\infty \tfrac{1}{2}[\theta(i\sigma) - 1]\sigma^{\frac{1}{2}s-1} \, d\sigma \qquad (7.38)$$

for Re $s > 1$. In the terminology that we shall use in later chapters, $\zeta(s)$ is, except for harmless factors, the "Mellin transform" of $\theta(i\sigma) - 1$, and $\theta(\tau)$ is a modular form of "level" 2 and "weight" $\tfrac{1}{2}$. Except for a growth condition, these modular form properties are the content of the following proposition.

5. ANALYTIC PROPERTIES OF DIRICHLET L FUNCTIONS

Theorem 7.12. The analytic function $\theta(\tau)$, defined for $\operatorname{Im} \tau > 0$, satisfies

(a) $\theta(\tau + 2) = \theta(\tau)$ (level 2)

(b) $\theta(-1/\tau) = (\tau/i)^{1/2}\theta(\tau)$, (weight $\frac{1}{2}$)

where the square root is the principal value cut on the negative real axis.

Before proving this proposition, let us derive some analytic consequences for $\zeta(s)$.

Corollary 7.13. The Riemann zeta function $\zeta(s)$, initially defined by (7.1) for $\operatorname{Re} s > 1$, extends to be meromorphic in $\mathbf{C}$. Its only pole is at $s = 1$, and that pole is simple. Moreover

$$\Lambda(s) = \zeta(s)\Gamma(\tfrac{1}{2}s)\pi^{-\frac{1}{2}s} \tag{7.39}$$

satisfies the functional equation

$$\Lambda(1-s) = \Lambda(s). \tag{7.40}$$

REMARK. Notice that if $\zeta(s)$ is defined only for $\operatorname{Re} s > 1$, then $\Lambda(1-s)$ and $\Lambda(s)$ have disjoint domains and (7.40) has no content. It is necessary first to extend $\zeta(s)$ to $\operatorname{Re} s > 0$. Then (7.40) is meaningful for $0 < \operatorname{Re} s < 1$ and extends to an identity on $\mathbf{C}$. The extension of $\zeta(s)$ to $\operatorname{Re} s > 0$ has already been carried out in Proposition 7.3, but a little care is required to interpret (7.38) as an identity in this whole region, not just in $\operatorname{Re} s > 1$.

PROOF. For $\sigma \geq 1$,

$$\tfrac{1}{2}[\theta(i\sigma) - 1] = \sum_{n=1}^{\infty} e^{-n^2\pi\sigma} \leq \sum_{k=1}^{\infty} e^{-k\pi\sigma} = \frac{e^{-\pi\sigma}}{1 - e^{-\pi\sigma}} \leq (1 - e^{-\pi})^{-1} e^{-\pi\sigma},$$

and thus

$$\int_1^{\infty} \tfrac{1}{2}[\theta(i\sigma) - 1]\sigma^{\frac{1}{2}s-1} \, d\sigma \tag{7.41}$$

converges for all $s \in \mathbf{C}$ and defines an entire function. For real $s > 1$, we rewrite (7.38) as

$$\Lambda(s) = \int_0^1 \tfrac{1}{2}\theta(i\sigma)\sigma^{\frac{1}{2}s-1} \, d\sigma - \tfrac{1}{2}\int_0^1 \sigma^{\frac{1}{2}s-1} \, d\sigma + \int_1^{\infty} \tfrac{1}{2}[\theta(i\sigma) - 1]\sigma^{\frac{1}{2}s-1} \, d\sigma.$$

On the right side the second term equals $-1/s$. The first term, by Theorem 7.12b, is

$$= \int_0^1 \tfrac{1}{2}\theta\left(\frac{-1}{i\sigma}\right)\left(\frac{i\sigma}{i}\right)^{-1/2} \sigma^{\frac{1}{2}s-1} d\sigma$$

$$= \int_0^1 \tfrac{1}{2}\theta\left(\frac{-1}{i\sigma}\right) \sigma^{\frac{1}{2}s-\frac{3}{2}} d\sigma$$

$$= \int_0^1 \tfrac{1}{2}\left[\theta\left(\frac{-1}{i\sigma}\right) - 1\right] \sigma^{\frac{1}{2}s-\frac{3}{2}} d\sigma + \frac{1}{s-1}.$$

The change of variables $\sigma \to 1/\sigma$ shows that this is

$$= \int_1^\infty \tfrac{1}{2}[\theta(i\sigma) - 1]\sigma^{\frac{1}{2}(1-s)-1} d\sigma - \frac{1}{1-s}.$$

Therefore

$$\Lambda(s) = \int_1^\infty \tfrac{1}{2}[\theta(i\sigma) - 1]\sigma^{\frac{1}{2}s-1} d\sigma \qquad (7.42)$$
$$+ \int_1^\infty \tfrac{1}{2}[\theta(i\sigma) - 1]\sigma^{\frac{1}{2}(1-s)-1} d\sigma - \frac{1}{1-s} - \frac{1}{s}.$$

The first two terms extend to be entire, by (7.41), and the other two terms extend to be meromorphic in $\mathbb{C}$ with poles at $s = 0$ and $s = 1$, both simple. Therefore $\Lambda(s)$ extends to be meromorphic in $\mathbb{C}$ with poles only at $s = 0$ and $s = 1$. Since (7.42) is unchanged under $s \to 1 - s$, (7.40) follows. From (7.39) and the meromorphic nature of $\Lambda(s)$, we see that $\zeta(s)$ is meromorphic in $\mathbb{C}$. Since $\Gamma(\tfrac{1}{2}s)$ is nowhere vanishing, $\zeta(s)$ can have poles only at $s = 0$ and $s = 1$. Since $\Gamma(\tfrac{1}{2}s)$ and $\Lambda(s)$ both have simple poles at $s = 0$, $\zeta(s)$ is in fact analytic at $s = 0$.

We turn to the proof of Theorem 7.12. The tool is the Poisson Summation Formula on $\mathbf{R}^1$, which is a result about the Fourier transform. The **Fourier transform** on $\mathbf{R}^1$ is the operator on Lebesgue integrable functions given by

$$\widehat{f}(u) = \int_{-\infty}^\infty e^{-2\pi i t u} f(t)\, dt.$$

If also f is continuous and $\widehat{f}$ is integrable, then the **Fourier inversion formula** says

$$f(t) = \int_{-\infty}^\infty e^{2\pi i t u} \widehat{f}(u)\, du.$$

5. ANALYTIC PROPERTIES OF DIRICHLET L FUNCTIONS

For our purposes a useful class of integrable functions is the space of **Schwartz functions**

$$\mathcal{S}(\mathbf{R}^1) = \left\{ f \in C^\infty(\mathbf{R}^1) \;\middle|\; P(t)\frac{d^k f}{dt^k}(t) \begin{array}{l} \text{is bounded for each} \\ \text{integer } k \geq 0 \text{ and} \\ \text{each polynomial } P \end{array} \right\}.$$

The Fourier transform carries $\mathcal{S}(\mathbf{R}^1)$ one-one onto $\mathcal{S}(\mathbf{R}^1)$.

Theorem 7.14 (Poisson Summation Formula). *If f is in $\mathcal{S}(\mathbf{R}^1)$, then*

$$\sum_{n=-\infty}^{\infty} f(x+n) = \sum_{n=-\infty}^{\infty} \widehat{f}(n) e^{2\pi i n x}.$$

PROOF. Define $F(x) = \sum_{n=-\infty}^{\infty} f(x+n)$. From the definition of $\mathcal{S}(\mathbf{R}^1)$, it is easy to check that this series is uniformly convergent and so is the series of kth derivatives, for each k. Consequently the function F is well defined and C^∞, and it is periodic of period one. Such a function is the sum of its Fourier series:

$$F(x) = \sum_{n=-\infty}^{\infty} \left(\int_0^1 F(t) e^{-2\pi i n t} \, dt \right) e^{2\pi i n x}. \tag{7.43}$$

The Fourier coefficient in parenthesis above is

$$\int_0^1 F(t) e^{-2\pi i n t} \, dt = \int_0^1 \sum_{k=-\infty}^{\infty} f(t+k) e^{-2\pi i n t} \, dt$$

$$= \sum_{k=-\infty}^{\infty} \int_0^1 f(t+k) e^{-2\pi i n t} \, dt$$

$$= \sum_{k=-\infty}^{\infty} \int_k^{k+1} f(t) e^{-2\pi i n t} \, dt$$

$$= \int_{-\infty}^{\infty} f(t) e^{-2\pi i n t} \, dt$$

$$= \widehat{f}(n). \tag{7.44}$$

The theorem follows by substituting (7.44) into (7.43).

For Theorem 7.12 the Poisson Summation Formula is to be applied to $f(t) = e^{-\pi t^2}$ and its dilates. In treating Dirichlet L functions, we shall apply the formula also to $f(t) = t e^{-\pi t^2}$. The relevant Fourier transforms are given in the next proposition.

Proposition 7.15.
(a) $(e^{-\pi t^2})\hat{\,} = e^{-\pi u^2}$
(b) $(te^{-\pi t^2})\hat{\,} = -iue^{-\pi u^2}$.

PROOF. For (a) we form the line integral over a rectangle in $\mathbf{C}$ of the entire function $e^{-\pi z^2}$. The rectangle extends from $-N$ to N in the x direction and from 0 to u in the y direction. The Cauchy Integral Theorem says that

$$\int_{-N}^{N} e^{-\pi x^2}\, dx - \int_{-N}^{N} e^{-\pi(x+iu)^2}\, dx + \text{(contribution from each end)} = 0.$$

As $N \to \infty$, the contribution from each end tends to 0, and we obtain

$$\int_{-\infty}^{\infty} e^{-\pi x^2}\, dx = \int_{-\infty}^{\infty} e^{-\pi(x+iu)^2}\, dx.$$

Multiplying by $e^{-\pi u^2}$, we see that

$$(e^{-\pi t^2})\hat{\,}(u) = e^{-\pi u^2} \int_{-\infty}^{\infty} e^{-\pi x^2}\, dx.$$

The integral on the right side is 1, as we see by evaluating its square $\int_{-\infty}^{\infty} \int_{-\infty}^{\infty} e^{-\pi(x^2+y^2)}\, dx\, dy$ in polar coordinates. This proves (a). Part (b) follows by differentiating both sides of the identity

$$\int_{-\infty}^{\infty} e^{-\pi t^2} e^{-2\pi i t u}\, dt = e^{-\pi u^2}$$

with respect to u.

PROOF OF THEOREM 7.12. Conclusion (a) is immediate from (7.36). For (b), let $f(t) = e^{-\pi t^2}$ and define $f_r(t) = f(r^{-1}t)$ for $r > 0$. Changing variables in the definition of Fourier transform, we immediately obtain

$$\hat{f}_r(u) = r\hat{f}(ru). \tag{7.45}$$

For our particular f, Proposition 7.15a therefore gives

$$f_r(t) = e^{-\pi r^{-2} t^2} \quad \text{and} \quad \hat{f}_r(u) = re^{-\pi r^2 u^2}.$$

Applying Theorem 7.14 to f_r with $x = 0$, we find that

$$\sum_{n=-\infty}^{\infty} e^{-\pi r^{-2} n^2} = r \sum_{n=-\infty}^{\infty} e^{-\pi r^2 n^2}.$$

With $r = \sigma^{1/2}$ and $\sigma > 0$, this identity says

$$\theta\left(-\frac{1}{i\sigma}\right) = \sigma^{1/2}\theta(i\sigma).$$

This is conclusion (b) for $\tau = i\sigma$, and (b) follows by analytic continuation.

5. ANALYTIC PROPERTIES OF DIRICHLET L FUNCTIONS

The remainder of this section will give a parallel development for Dirichlet L functions. There is one complication, namely in isolating what Dirichlet characters to consider. For example, the L function for the principal Dirichlet character χ_0 modulo m has an Euler product that is the same as the one for $\zeta(s)$ except that the factors where $p \mid m$ are dropped. So we should be content with having the functional equation for $\zeta(s)$ yield a functional equation for $L(s, \chi_0)$. Here are two subtler examples.

EXAMPLE 1. Let $m = 8$ and define $\chi(1) = \chi(5) = 1$ and $\chi(3) = \chi(7) = -1$. The resulting L function coincides with $L(s, \chi')$, where χ' is given with $m = 4$ by $\chi'(1) = 1$ and $\chi'(3) = -1$.

EXAMPLE 2. Let $m = 6$ and define $\chi(1) = 1$ and $\chi(5) = -1$. Also let $m = 3$ and define $\chi'(1) = 1$ and $\chi'(2) = -1$. Then

$$L(s, \chi) = L(s, \chi') \left(1 - \frac{1}{2^s}\right).$$

Again one should be content with having the functional equation for one of these L functions yield the functional equation for the other.

We say that two nonprincipal Dirichlet characters χ modulo m and χ' modulo m' are **associate** if $\chi(p) = \chi'(p)$ for all but finitely many primes. This is an equivalence relation. The **conductor** of an equivalence class is the least integer m'' such that the class contains a nonprincipal Dirichlet character modulo m''. In Example 1, χ and χ' are associate with conductor 4; in Example 2, χ and χ' are associate with conductor 3. A Dirichlet character modulo m is **primitive** if its conductor is m.

If χ is a Dirichlet character modulo m, we can convert χ into a Dirichlet character $\chi^\#$ modulo am by defining

$$\chi^\#(j) = \begin{cases} \chi(j) & \text{if } \operatorname{GCD}(a, j) = 1 \\ 0 & \text{if } \operatorname{GCD}(a, j) > 1 \end{cases}$$

Let us call $\chi^\#$ an **extension** of χ. If $\chi^\#$ is an extension of χ, then χ and $\chi^\#$ are associate.

Suppose χ and χ' are Dirichlet characters modulo the same m that are associate. Let us prove that $\chi = \chi'$. In fact, suppose $\operatorname{GCD}(m, b) = 1$. By Theorem 7.1 there are infinitely many primes p with $p \equiv b \bmod m$. For all but finitely many of them, hence for at least one of them, we have $\chi(p) = \chi'(p)$. Therefore $\chi(b) = \chi'(b)$. If $\operatorname{GCD}(m, b) > 1$, then $\chi(b) = \chi'(b) = 0$. Hence $\chi = \chi'$.

Consequently an equivalence class of associate Dirichlet characters contains a unique primitive Dirichlet character.

214   VII: DIRICHLET'S THEOREM

Proposition 7.16. If χ_m is a Dirichlet character modulo m and $\chi_{m'}$ is an associate Dirichlet character modulo m', then there exists a Dirichlet character χ_{GCD} modulo $\text{GCD}(m, m')$ such that χ_m and $\chi_{m'}$ are extensions of χ_{GCD}.

PROOF. Write $m = \prod p_j^{a_j}$ and $m' = \prod p_j^{a'_j}$. Put $n_j = \min(a_j, a'_j)$ and $N_j = \max(a_j, a'_j)$. By the Chinese Remainder Theorem, $\mathbf{Z}_m \cong \bigoplus \mathbf{Z}_{p_j^{a_j}}$ as rings, and thus $\mathbf{Z}_m^\times \cong \bigoplus \mathbf{Z}_{p_j^{a_j}}^\times$. The proof of Proposition 7.9 then shows that $(\mathbf{Z}_m^\times)\widehat{} \cong \bigoplus (\mathbf{Z}_{p_j^{a_j}}^\times)\widehat{}$. Let us write $\chi_m = (\ldots, \chi_{p_j^{a_j}}, \ldots)$. The extension of χ_m to $\mathbf{Z}_{\text{LCM}(m,m')}^\times$ is accomplished by extending each $\chi_{p_j^{a_j}}$ to $\mathbf{Z}_{p_j^{N_j}}^\times$. Thus we have

$$\chi_m^\# = (\ldots, \chi_{p_j^{a_j}}^\#, \ldots).$$

We argue similarly with $\chi_{m'}$, obtaining a similar formula. Since χ_m and $\chi_{m'}$ are associate, so are $\chi_m^\#$ and $\chi_{m'}^\#$. Then $\chi_m^\# = \chi_{m'}^\#$ since both Dirichlet characters occur modulo $\text{LCM}(m, m')$. This equality implies $\chi_{p_j^{a_j}}^\# = \chi_{p_j^{a'_j}}^\#$ for all j. In other words, $\chi_{p_j^{N_j}} = \chi_{p_j^{n_j}}^\#$. Putting $\chi_{\text{GCD}} = (\ldots, \chi_{p_j^{n_j}}, \ldots)$, we see that χ_m and $\chi_{m'}$ are both extensions of χ_{GCD}.

Corollary 7.17. Any Dirichlet character modulo m is an extension of the unique primitive Dirichlet character χ' with which it is associate. Consequently the conductor of χ is a divisor of m.

PROOF. We apply Proposition 7.16 to χ and χ'. Then χ and χ' are both extensions of χ_{GCD}. Since χ' is primitive, $\chi' = \chi_{\text{GCD}}$. Thus χ is an extension of χ_{GCD}.

It follows that any $L(s, \chi)$ can be obtained by deleting finitely many factors from the Euler product of some $L(s, \chi')$ with χ' primitive. For purposes of deriving functional equations, it is therefore enough to treat the primitive case. The way that "primitive" will enter the argument is through the following lemma.

Lemma 7.18. If χ is a primitive Dirichlet character modulo m and if $c(m, \chi)$ denotes the **Gauss sum**

$$c(m, \chi) = \sum_{k=0}^{m-1} e^{2\pi i k/m} \chi(k), \qquad (7.46)$$

5. ANALYTIC PROPERTIES OF DIRICHLET L FUNCTIONS 215

then
$$\sum_{k=0}^{m-1} e^{2\pi i n k/m} \chi(k) = \overline{\chi(n)} c(m, \chi) \tag{7.47}$$

for every integer n.

REMARK. It is easy to check that the nonprimitive χ in Example 2 above does not satisfy (7.47) for $n = 2$.

PROOF. If $\text{GCD}(n, m) = 1$, then

$$\sum_{k=0}^{m-1} e^{2\pi i n k/m} \chi(k)\chi(n) = \sum_{k=0}^{m-1} e^{2\pi i n k/m} \chi(kn)$$
$$= \sum_{k=0}^{m-1} e^{2\pi i k/m} \chi(k) = c(m, \chi).$$

Letting n^{-1} denote an inverse of n in $\mathbb{Z}_m^\times$, we multiply through by $\chi(n^{-1}) = \overline{\chi(n)}$ and obtain (7.47).

If $d = \text{GCD}(n, m) > 1$, we are to prove that the left side of (7.47) is 0. Let $m' = m/d$. The natural map $\mathbb{Z}_m^\times \to \mathbb{Z}_{m'}^\times$ has kernel

$$K = \{a \in \mathbb{Z}_m^\times \mid a \equiv 1 \bmod m'\}.$$

If $\chi|_K$ were to equal 1, then χ would descend to $\mathbb{Z}_{m'}^\times$ and define a Dirichlet character χ' modulo m' such that $\chi = \chi'^{\#}$. Since χ is primitive, this descent cannot happen, and we conclude that $\chi|_K$ is a nontrivial character of K.

The left side of (7.47) is

$$= \sum_{k=0}^{m-1} e^{2\pi i k/m'} \chi(k) = \sum_{r=0}^{m'-1} \sum_{\substack{0 \le k \le m-1 \\ k \equiv r \bmod m'}} e^{2\pi i k/m'} \chi(k)$$

$$= \sum_{r=0}^{m'-1} e^{2\pi i r/m'} \sum_{k \in rK} \chi(k)$$

$$= \sum_{r=0}^{m'-1} e^{2\pi i r/m'} \chi(r) \sum_{a \in K} \chi(a).$$

Since $\chi|_K$ is nontrivial, the inner sum is 0. Thus (7.47) is 0.

Theorem 7.19. Let χ be a primitive Dirichlet character modulo m, and define $\theta(\tau,\chi)$ for Im $\tau > 0$ as

$$\begin{cases} \sum_{n=-\infty}^{\infty} \chi(n)e^{in^2\pi\tau/m} = 2\sum_{n=1}^{\infty} \chi(n)e^{in^2\pi\tau/m} & \text{if } \chi(-1) = 1 \\ \sum_{n=-\infty}^{\infty} \chi(n)ne^{in^2\pi\tau/m} = 2\sum_{n=1}^{\infty} \chi(n)ne^{in^2\pi\tau/m} & \text{if } \chi(-1) = -1 \end{cases}$$

Then

(a) $\theta(\tau + 2m, \chi) = \theta(\tau, \chi)$

(b) $\theta(-1/\tau, \chi) = \dfrac{c(m,\chi)}{\sqrt{m}} (\tau/i)^{1/2} \theta(\tau, \bar\chi)$ \quad if $\chi(-1) = 1$

(b') $\theta(-1/\tau, \chi) = \dfrac{-ic(m,\chi)}{\sqrt{m}} (\tau/i)^{3/2} \theta(\tau, \bar\chi)$ \quad if $\chi(-1) = -1$

The Dirichlet L functions $L(s,\chi)$ and $L(s,\bar\chi)$ extend to be entire in s and have the following corresponding properties:

(c) If $\chi(-1) = 1$, then $\Lambda(s,\chi) = L(s,\chi)m^{\frac{1}{2}s}\Gamma(\frac{1}{2}s)\pi^{-\frac{1}{2}s}$ satisfies

$$\Lambda(s,\chi) = \frac{c(m,\chi)}{\sqrt{m}} \Lambda(1-s,\bar\chi) \tag{7.48a}$$

(c') If $\chi(-1) = -1$, then $\Lambda(s,\chi) = L(s,\chi)m^{\frac{1}{2}(s+1)}\Gamma(\frac{1}{2}(s+1))\pi^{-\frac{1}{2}(s+1)}$ satisfies

$$\Lambda(s,\chi) = -\frac{ic(m,\chi)}{\sqrt{m}} \Lambda(1-s,\bar\chi). \tag{7.48b}$$

REMARKS. Applying (c) or (c') first to χ and then to χ', we see that

$$c(m,\chi)c(m,\bar\chi) = m\chi(-1).$$

We readily check from the definition that

$$c(m,\bar\chi) = \chi(-1)\overline{c(m,\chi)}.$$

Consequently the coefficient $c(m,\chi)/\sqrt{m}$ in the functional equation (7.48) has absolute value one.

PROOF. Result (a) is clear. For (b), define

$$f(t) = \begin{cases} e^{-\pi t^2} & \text{if } \chi(-1) = 1 \\ te^{-\pi t^2} & \text{if } \chi(-1) = -1 \end{cases} \tag{7.49}$$

5. ANALYTIC PROPERTIES OF DIRICHLET L FUNCTIONS

The Poisson Summation Formula (Theorem 7.14) applied to f with $x = k/m$ gives

$$\sum_{n=-\infty}^{\infty} f\left(\frac{k}{m}+n\right) = \sum_{n=-\infty}^{\infty} \hat{f}(n)e^{2\pi ink/m}.$$

Multiplying by $\chi(k)$, summing on k from 0 to $m-1$, and using (7.47) yields

$$\sum_{n=-\infty}^{\infty}\sum_{k=0}^{m-1} \chi(k)f\left(\frac{k}{m}+n\right) = \sum_{n=-\infty}^{\infty} \hat{f}(n) \sum_{k=0}^{m-1} e^{2\pi ink/m}\chi(k)$$

$$= c(m,\chi) \sum_{n=-\infty}^{\infty} \hat{f}(n)\overline{\chi(n)}. \tag{7.50}$$

The left side of (7.50) is

$$= \sum_{n=-\infty}^{\infty}\sum_{k=0}^{m-1} \chi(k+mn)f\left(\frac{k+mn}{m}\right)$$

$$= \sum_{k=-\infty}^{\infty} \chi(k)f\left(\frac{k}{m}\right).$$

Applying (7.45) to the resulting form of (7.50), we have

$$\sum_{k=-\infty}^{\infty} \chi(k)f\left(\frac{k}{rm}\right) = c(m,\chi)\, r \sum_{n=-\infty}^{\infty} \hat{f}(rn)\overline{\chi(n)}. \tag{7.51}$$

If $\chi(-1) = 1$, we choose f as in (7.49) and put $r = \sqrt{\sigma/m}$; application of Proposition 7.15a yields

$$\sum_{k=-\infty}^{\infty} \chi(k)e^{-k^2\pi\sigma^{-1}/m} = \frac{c(m,\chi)}{\sqrt{m}}\sigma^{1/2} \sum_{n=-\infty}^{\infty} \overline{\chi(n)}e^{-n^2\pi\sigma/m}.$$

This is (b) for $\tau = i\sigma$, and (b) for general τ follows by analytic continuation.

If $\chi(-1) = -1$, we choose f as in (7.49) and again put $r = \sqrt{\sigma/m}$; application of Proposition 7.15b yields

$$\sum_{k=-\infty}^{\infty} \chi(k)\frac{k\sigma^{-1/2}}{\sqrt{m}}e^{-k^2\pi\sigma^{-1}/m}$$

$$= \frac{-ic(m,\chi)}{\sqrt{m}}\sigma^{1/2} \sum_{n=-\infty}^{\infty} \overline{\chi(n)}\frac{n\sigma^{1/2}}{\sqrt{m}}e^{-n^2\pi\sigma/m}.$$

This is (b′) for $\tau = i\sigma$, and (b′) for general τ follows by analytic continuation.

For (c), suppose $\chi(-1) = 1$. In analogy with (7.37), we have

$$\Lambda(s,\chi) = \sum_{k=1}^{\infty} \frac{\chi(k)}{k^s} m^{\frac{1}{2}s}\Gamma(\tfrac{1}{2}s)\pi^{-\frac{1}{2}s}$$

$$= \int_0^{\infty} \sum_{k=1}^{\infty} \chi(k)e^{-k^2\pi\sigma/m}\sigma^{\frac{1}{2}s-1}\,d\sigma$$

$$= \int_0^{\infty} \tfrac{1}{2}\theta(i\sigma,\chi)\sigma^{\frac{1}{2}s-1}\,d\sigma \tag{7.52}$$

for Re $s > 1$. The formal argument, disregarding convergence, is to replace σ by σ^{-1} and then apply (b). The above expression becomes

$$= \int_0^{\infty} \tfrac{1}{2}\theta(i/\sigma)\sigma^{-\frac{1}{2}s-1}\,d\sigma$$

$$= \frac{c(m,\chi)}{\sqrt{m}} \int_0^{\infty} \sigma^{\frac{1}{2}} \tfrac{1}{2}\theta(i\sigma,\bar{\chi})\sigma^{-\frac{1}{2}s-1}\,d\sigma$$

$$= \frac{c(m,\chi)}{\sqrt{m}} \Lambda(1-s,\bar{\chi})$$

by (7.52). To make this argument precise, we observe by the same argument as for (7.41) that

$$\int_1^{\infty} \tfrac{1}{2}\theta(i\sigma,\chi)\sigma^{\frac{1}{2}s-1}\,d\sigma \tag{7.53}$$

converges for all $s \in \mathbf{C}$ and defines an entire function. For Re $s > 1$, we rewrite (7.52) as

$$\Lambda(s,\chi) = \int_0^1 \tfrac{1}{2}\theta(i\sigma,\chi)\sigma^{\frac{1}{2}s-1}\,d\sigma + \int_1^{\infty} \tfrac{1}{2}\theta(i\sigma,\chi)\sigma^{\frac{1}{2}s-1}\,d\sigma$$

and transform just the first term, replacing σ by σ^{-1} and applying (b). The result is

$$\Lambda(s,\chi) = \frac{c(m,\chi)}{\sqrt{m}} \int_1^{\infty} \tfrac{1}{2}\theta(i\sigma,\bar{\chi})\sigma^{\frac{1}{2}(1-s)-1}\,d\sigma + \int_1^{\infty} \tfrac{1}{2}\theta(i\sigma,\chi)\sigma^{\frac{1}{2}s-1}\,d\sigma. \tag{7.54}$$

5. ANALYTIC PROPERTIES OF DIRICHLET L FUNCTIONS

In view of (7.53), both terms extend to entire functions of s, and therefore $\Lambda(s, \chi)$ extends to be entire in s. Using (7.54) with $\bar\chi$, we have

$$\frac{c(m,\chi)}{\sqrt{m}}\Lambda(1-s,\bar\chi) = \frac{c(m,\chi)c(m,\bar\chi)}{m}\int_1^\infty \tfrac{1}{2}\theta(i\sigma,\chi)\sigma^{\frac{1}{2}s-1}\,d\sigma$$

$$+ \frac{c(m,\chi)}{\sqrt{m}}\int_1^\infty \tfrac{1}{2}\theta(i\sigma,\bar\chi)\sigma^{\frac{1}{2}(1-s)-1}\,d\sigma. \tag{7.55}$$

To prove (c), we need to show that the right sides of (7.54) and (7.55) are equal, and this comes down to proving that

$$c(m,\chi)c(m,\bar\chi) = m\chi(-1) \tag{7.56}$$

(as was asserted in the remarks before the proof). To prove (7.56), we write

$$c(m,\chi)c(m,\bar\chi) = \sum_{k=0}^{m-1} e^{2\pi ik/m}\chi(k)c(m,\bar\chi)$$

$$= \sum_{k=0}^{m-1} e^{2\pi ik/m} \sum_{l=0}^{m-1} e^{2\pi ikl/m}\overline{\chi(l)} \qquad \text{by Lemma 7.18}$$

$$= \sum_{k=0}^{m-1} e^{2\pi ik/m} \sum_{l=0}^{m-1} e^{-2\pi ikl/m}\overline{\chi(-l)}$$

$$= \chi(-1)\sum_{l=0}^{m-1}\overline{\chi(l)}\sum_{k=0}^{m-1}e^{2\pi ik/m}e^{-2\pi ikl/m}$$

$$= \chi(-1)\sum_{l=0}^{m-1}\overline{\chi(l)}\delta_{l,1}m \qquad \text{by Proposition 7.9}$$

$$= m\chi(-1).$$

This proves (7.56). The analyticity of $L(s,\chi)$ is obtained from that of $\Lambda(s,\chi)$ by the same argument used with $\zeta(s)$ and $\Lambda(s)$ in Corollary 7.13, and (b) follows.

For (c'), suppose $\chi(-1) = -1$. In analogy with (7.37) and (7.52), we have

$$\Lambda(s,\chi) = \sum_{k=1}^{\infty} \frac{\chi(k)}{k^s} m^{\frac{1}{2}(s+1)}\Gamma(\tfrac{1}{2}(s+1))\pi^{-\frac{1}{2}(s+1)}$$

$$= \int_0^{\infty} \sum_{k=1}^{\infty} \chi(k) k e^{-k^2\pi\sigma/m} \sigma^{\frac{1}{2}(s+1)-1} \, d\sigma$$

$$= \int_0^{\infty} \tfrac{1}{2}\theta(i\sigma,\chi)\sigma^{\frac{1}{2}(s+1)-1} \, d\sigma \tag{7.57}$$

for Re $s > 1$. The formal argument, disregarding convergence, is to replace σ by σ^{-1} and then apply (b'). The above expression becomes

$$= \int_0^{\infty} \tfrac{1}{2}\theta(i/\sigma,\chi)\sigma^{-\frac{1}{2}(s+1)-1} \, d\sigma$$

$$= \frac{-ic(m,\chi)}{\sqrt{m}} \int_0^{\infty} \sigma^{\frac{3}{2}} \tfrac{1}{2}\theta(i\sigma,\bar{\chi})\sigma^{-\frac{1}{2}(s+1)-1} \, d\sigma$$

$$= \frac{-ic(m,\chi)}{\sqrt{m}} \Lambda(1-s,\bar{\chi})$$

by (7.57). This argument is made precise in the same way as for (c), by means of (7.56). The analyticity of $L(s,\chi)$ is obtained by the same argument used for $\zeta(s)$ in Corollary 7.13, and (c') follows.

CHAPTER VIII

MODULAR FORMS FOR $SL(2, \mathbf{Z})$

1. Overview

In Chapter VI we established a correspondence $\Lambda \leftrightarrow E$ of lattices in $\mathbf{C}$ with elliptic curves defined over $\mathbf{C}$. By way of introduction to the subject of modular forms, we shall now let the lattice vary. Then G_4, G_6, and Δ lead to analytic functions on the upper half plane with special transformation properties (under the group $SL(2, \mathbf{Z})$) and certain growth conditions that we list in §2. Analytic functions of this kind will be defined in §2 to be modular forms. Cusp forms will be modular forms with an additional vanishing property, and Δ will be an example.

To each cusp form, we shall associate in §3 an L function by means of a Mellin transform. This L function will be given by a Dirichlet series and will extend to be entire on $\mathbf{C}$ and to satisfy a functional equation. The proof will be much like that for the Dirichlet L functions in Theorem 7.19, and the functional equation will be of the type in (7.48).

The remainder of the chapter will introduce and use Hecke operators as a tool for expanding the L functions of selected cusp forms as Euler products. In Chapter IX we shall extend the theory of Chapter VIII to modular forms and cusp forms whose transformation properties are relative to certain special subgroups of $SL(2, \mathbf{Z})$.

After that point, we return to elliptic curves. For an elliptic curve with integer coefficients, we count the number of solutions for each finite field and assemble this information into an L function for the elliptic curve, which is given as an Euler product and is hard to use. The result is that we have two kinds of L functions, the kind from cusp forms that we understand very well and the kind from elliptic curves that contains a great deal of information.

Eichler-Shimura theory observes that cusp forms with special properties can be used to parametrize the points of elliptic curves over $\mathbf{Q}$. Moreover, the L function of the cusp form coincides with the L function of the elliptic curve. Elliptic curves that can be parametrized in this way are called **modular**.

The Taniyama-Weil Conjecture is the assertion that every elliptic curve is modular. There are many equivalent formulations, and there is an algorithm for deciding the conjecture for any particular curve.

If the conjecture is true, then restrictions on the nature of modular forms can be brought to bear on the theory of elliptic curves. A dramatic example of this process is the theorem that the Taniyama-Weil Conjecture implies Fermat's Last Theorem. A counterexample to Fermat's Last Theorem would yield an elliptic curve with remarkable properties, and one proves with some effort that such an elliptic curve cannot be modular.

2. Definitions and Examples

Let us recall the correspondence $\Lambda \leftrightarrow E$ of lattices in $\mathbf{C}$ with elliptic curves defined over $\mathbf{C}$. In (6.4) and (6.5) we defined

$$G_{2k}(\Lambda) = \sum_{\substack{\omega \in \Lambda \\ \omega \neq 0}} \frac{1}{\omega^{2k}} \quad \text{for } k \geq 2$$

$$g_2(\Lambda) = 60 G_4(\Lambda)$$

$$g_3(\Lambda) = 140 G_6(\Lambda). \tag{8.1}$$

According to (3.30), the curve

$$y^2 = 4x^3 + b_2 x^2 + 2 b_4 x + b_6$$

has

$$\Delta = -b_2^2 b_8 - 8 b_4^3 - 27 b_6^2 + 9 b_2 b_4 b_6.$$

Therefore the curve

$$y^2 = 4x^3 - g_2(\Lambda) x - g_3(\Lambda)$$

of Chapter VI has

$$\Delta(\Lambda) = g_2(\Lambda)^3 - 27 g_3(\Lambda)^2. \tag{8.2}$$

Similarly the j invariant is given by

$$j(\Lambda) = 1728 g_2(\Lambda)^3 / \Delta(\Lambda). \tag{8.3}$$

The dependence on Λ can be simplified a little. If α is in $\mathbf{C}^\times$, we can compute

$$G_{2k}(\alpha \Lambda) = \sum_{\substack{\omega \in \alpha \Lambda \\ \omega \neq 0}} \frac{1}{\omega^{2k}} = \sum_{\substack{\omega' \in \Lambda \\ \omega' \neq 0}} \frac{1}{(\alpha \omega')^{2k}} = \alpha^{-2k} G_{2k}(\Lambda)$$

2. DEFINITIONS AND EXAMPLES

and similarly

$$\Delta(\alpha\Lambda) = \alpha^{-12}\Delta(\Lambda)$$
$$j(\alpha\Lambda) = j(\Lambda).$$

Let Λ be generated by w_1 and w_2, so that $\Lambda = \mathbf{Z}w_1 \oplus \mathbf{Z}w_2$. Possibly by interchanging w_1 and w_2, we may assume that $\mathrm{Im}(w_2/w_1) > 0$. Introducing $\tau = w_2/w_1$ and $\Lambda_\tau = \mathbf{Z} \oplus \mathbf{Z}\tau$, we have

$$\Lambda = w_1\left(\mathbf{Z} \oplus \mathbf{Z}\left(\frac{w_2}{w_1}\right)\right) = w_1\Lambda_\tau.$$

Thus G_{2k}, Δ, and j are determined by their effects on $\Lambda = \Lambda_\tau$. We define

$$G_{2k}(\tau) = G_{2k}(\Lambda_\tau), \qquad \Delta(\tau) = \Delta(\Lambda_\tau), \qquad j(\tau) = j(\Lambda_\tau). \qquad (8.4)$$

We can make the same computation with a different basis $\{w_1', w_2'\}$ for Λ. We have

$$\begin{pmatrix} w_2' \\ w_1' \end{pmatrix} = \begin{pmatrix} a & b \\ c & d \end{pmatrix}\begin{pmatrix} w_2 \\ w_1 \end{pmatrix}$$

with a, b, c, d in $\mathbf{Z}$. Invertibility of this relation over $\mathbf{Z}$ implies $ad - bc = \pm 1$. Let us see the effect on the associated Λ_τ. We may as well start with $\{w_1, w_2\} = \{1, \tau\}$. Then $\{w_1', w_2'\}$ is given by

$$\begin{pmatrix} w_2' \\ w_1' \end{pmatrix} = \begin{pmatrix} a & b \\ c & d \end{pmatrix}\begin{pmatrix} \tau \\ 1 \end{pmatrix} = \begin{pmatrix} a\tau + b \\ c\tau + d \end{pmatrix},$$

and the associated τ' is

$$\tau' = \frac{a\tau + b}{c\tau + d}.$$

Since $\mathrm{Im}\,\tau > 0$ and $\mathrm{Im}\,\tau' > 0$, we must have $ad - bc = +1$, not -1. We are thus led to consider the action of $SL(2, \mathbf{R})$ on the upper half plane and the special role of the subgroup $SL(2, \mathbf{Z})$.

The group $SL(2, \mathbf{R})$ of 2-by-2 real matrices of determinant one acts on the upper half plane $\mathcal{H} = \{\mathrm{Im}\,\tau > 0\}$ in the usual way by linear fractional transformations, with $g = \begin{pmatrix} a & b \\ c & d \end{pmatrix}$ acting by $g\tau = \frac{a\tau + b}{c\tau + d}$. Let $SL(2, \mathbf{Z})$ be the subgroup with integer entries. What we saw above for an element $\gamma = \begin{pmatrix} a & b \\ c & d \end{pmatrix}$ of $SL(2, \mathbf{Z})$ was that $\Lambda_\tau \supseteq (c\tau + d)\Lambda_{\tau'}$, hence that $\Lambda_\tau = (c\tau + d)\Lambda_{\tau'}$. Therefore

$$\Lambda_{\gamma\tau} = (c\tau + d)^{-1}\Lambda_\tau.$$

The resulting transformation laws for the functions we have been considering are as follows:

$$G_{2k}(\tau) = \sum_{(m,n)\neq(0,0)} \frac{1}{(m\tau+n)^{2k}} \qquad G_{2k}(\gamma\tau) = (c\tau+d)^{2k} G_{2k}(\tau)$$

$$g_2(\tau) = 60 G_4(\tau) \qquad\qquad g_2(\gamma\tau) = (c\tau+d)^4 g_2(\tau)$$

$$g_3(\tau) = 140 G_6(\tau) \qquad\qquad g_3(\gamma\tau) = (c\tau+d)^6 g_3(\tau)$$

$$\Delta(\tau) = g_2(\tau)^3 - 27 g_3(\tau)^2 \qquad \Delta(\gamma\tau) = (c\tau+d)^{12} \Delta(\tau)$$

$$j(\tau) = 1728 g_2(\tau)^3 / \Delta(\tau) \qquad j(\gamma\tau) = j(\tau).$$

Fix an integer k. An analytic function on $\mathcal{H}$ that satisfies

$$f(\gamma\tau) = (c\tau+d)^k f(\tau) \quad \text{for all } \gamma = \begin{pmatrix} a & b \\ c & d \end{pmatrix} \in SL(2,\mathbf{Z}) \qquad (8.5)$$

is called an **unrestricted modular form** of **weight** k (for the full modular group $SL(2,\mathbf{Z})$). To have $f \not\equiv 0$, k must be even. We shall drop the word "unrestricted" if an additional condition below is satisfied.

Each unrestricted modular form f has a "q expansion" as follows. Taking $\gamma = \begin{pmatrix} 1 & 1 \\ 0 & 1 \end{pmatrix}$ in (8.5), we see that $f(\tau) = f(\tau+1)$. Putting $\tau = \rho + i\sigma$, we expand in Fourier series in the ρ variable. Since f is smooth,

$$f(\tau) = \sum_{n=-\infty}^{\infty} a_n(\sigma) e^{2\pi i \rho n} = \sum_{n=-\infty}^{\infty} a_n(\sigma) e^{2\pi n \sigma} e^{2\pi i n \tau},$$

where

$$a_n(\sigma) = \int_{-\frac{1}{2}}^{\frac{1}{2}} f(\rho + i\sigma) e^{-2\pi i n \rho} \, d\rho.$$

Then

$$a_n(\sigma) e^{2\pi n \sigma} = \int_{-\frac{1}{2}}^{\frac{1}{2}} f(\rho + i\sigma) e^{-2\pi i n(\rho+i\sigma)} \, d\rho. \qquad (8.6)$$

Imagine a rectangle in the τ plane extending in the horizontal direction from $\rho = -\frac{1}{2}$ to $+\frac{1}{2}$ and in the vertical direction from $\sigma_1 = \sigma$ to some σ_2. The integral (8.6) then represents the bottom portion of the line integral

$$\oint f(\tau) e^{-2\pi i n \tau} \, d\tau$$

2. DEFINITIONS AND EXAMPLES

over this rectangle. The total line integral is 0, by the Cauchy Integral Theorem, and the sides cancel because $f(\tau) = f(\tau+1)$. Thus the bottom portion equals the top portion when they are oriented the same way. In other words, (8.6) is a constant as a function of σ, say constantly equal to c_n. We conclude that

$$f(\tau) = \sum_{n=-\infty}^{\infty} c_n q^n \quad \text{with } q = e^{2\pi i \tau}. \tag{8.7a}$$

Here

$$c_n = \int_{-\frac{1}{2}}^{\frac{1}{2}} f(\tau) e^{-2\pi i n \tau} \, d\rho \quad \text{for any } \sigma > 0. \tag{8.7b}$$

Expression (8.7) is called the **q expansion** of f, or, for reasons discussed in §3, the expansion of f at ∞.

We say that an unrestricted modular form f is **holomorphic at ∞** and is a **modular form** if its q expansion has $c_n = 0$ for $n < 0$. If also $c_0 = 0$, we call f a **cusp form**.

In §3 we shall give a geometric interpretation for these conditions. But first we consider our standard examples G_{2k}, Δ, and j.

Proposition 8.1. For $k \geq 2$, the q expansion of $G_{2k}(\tau)$ is given by

$$G_{2k}(\tau) = 2\zeta(2k) + \frac{2(2\pi i)^{2k}}{(2k-1)!} \sum_{n=1}^{\infty} \sigma_{2k-1}(n) q^n,$$

where $\sigma_l(n) = \sum_{\substack{d|n \\ d>0}} d^l$. Consequently $G_{2k}(\tau)$ is a modular form of weight $2k$.

PROOF. We take as known the identity

$$\pi \cot \pi \tau = \frac{1}{\tau} + \sum_{m=1}^{\infty} \left(\frac{1}{\tau+m} + \frac{1}{\tau-m} \right), \tag{8.8}$$

in which the convergence is uniform on compact sets. With $q = e^{2\pi i \tau}$ and $\mathrm{Im}\, \tau > 0$ (so that $|q| < 1$), we have

$$\pi \cot \pi \tau = \pi \frac{\cos \pi \tau}{\sin \pi \tau} = i\pi \frac{q+1}{q-1} = i\pi - \frac{2\pi i}{1-q} = i\pi - 2\pi i \sum_{d=0}^{\infty} q^d.$$

Thus (8.8) gives

$$\frac{1}{\tau} + \sum_{m=1}^{\infty} \left(\frac{1}{\tau+m} + \frac{1}{\tau-m} \right) = i\pi - 2\pi i \sum_{d=0}^{\infty} q^d.$$

Differentiating $2k - 1$ times gives

$$\sum_{m=-\infty}^{\infty} \frac{1}{(\tau+m)^{2k}} = \frac{1}{(2k-1)!}(2\pi i)^{2k} \sum_{d=1}^{\infty} d^{2k-1}q^d. \qquad (8.9)$$

Hence

$$G_{2k}(\tau) = \sum_{(m,n)\neq(0,0)} \frac{1}{(n\tau+m)^{2k}}$$

$$= \sum_{m\neq 0} \frac{1}{(m)^{2k}} + \sum_{n\neq 0} \sum_{m=-\infty}^{\infty} \frac{1}{(n\tau+m)^{2k}}$$

$$= 2\zeta(2k) + 2\sum_{n=1}^{\infty} \sum_{m=-\infty}^{\infty} \frac{1}{(n\tau+m)^{2k}}.$$

Applying (8.9) with τ replaced by $n\tau$, we see that this expression is

$$= 2\zeta(2k) + 2\frac{1}{(2k-1)!}(2\pi i)^{2k} \sum_{d=1}^{\infty} \sum_{a=1}^{\infty} d^{2k-1}q^{da}$$

$$= 2\zeta(2k) + \frac{2(2\pi i)^{2k}}{(2k-1)!} \sum_{n=1}^{\infty} \sigma_{2k-1}(n)q^n,$$

as required.

REMARK. The next-to-last line of the proof shows that also

$$G_{2k}(\tau) = 2\zeta(2k) + \frac{2(2\pi i)^{2k}}{(2k-1)!} \sum_{d=1}^{\infty} \frac{d^{2k-1}q^d}{1-q^d}. \qquad (8.10)$$

The series here is rapidly convergent and allows for approximate numerical calculations. If we replace τ with ω_2/ω_1 and sort matters out, we are led to the formulas (6.50) for g_2 and g_3 that were used for the calculations in §VI.9.

Corollary 8.2. $\Delta(\tau)$ is a cusp form of weight 12 and is nonvanishing on $\mathcal{H}$. Also $j(\tau)$ is an unrestricted modular form of weight 0 with q expansion

$$j(\tau) = \frac{1}{q} + 744 + \sum_{n=1}^{\infty} c_n q^n.$$

PROOF. The nonvanishing of $\Delta(\tau)$ follows from Theorem 6.15. From (8.2) and (8.1), we have

$$\Delta(\tau) = g_2(\tau)^3 - 27g_3(\tau)^2 = 60^3 G_4(\tau)^3 - 27 \cdot 140^2 G_6(\tau)^2.$$

Taking as known the identities

$$\zeta(4) = \frac{\pi^4}{90} \quad \text{and} \quad \zeta(6) = \frac{\pi^6}{3^3 \cdot 5 \cdot 7},$$

we obtain from Proposition 8.1

$$G_4(\tau) = \frac{\pi^4}{45} + \frac{(2\pi)^4}{3}(q + 9q^2 + 28q^3 + 73q^4 + \dots)$$

$$G_6(\tau) = \frac{2\pi^6}{3^3 \cdot 5 \cdot 7} - \frac{(2\pi)^6}{60}(q + 33q^2 + 244q^3 + 1057q^4 + \dots).$$

Therefore

$$\Delta(\tau) = (2\pi)^{12}(q - 24q^2 + 252q^3 - 1472q^4 + \dots), \quad (8.11)$$

and $\Delta(\tau)$ is a cusp form. For the q expansion of $j(\tau)$, we have

$$j(\tau) = \frac{1728 \cdot 60^3 G_4(\tau)^3}{\Delta(\tau)}$$

$$= \frac{1 + 720q + 179280q^2 + 16954560q^3 + \dots}{q - 24q^2 + 252q^3 - 1472q^4 + \dots}$$

$$= \frac{1}{q} + 744 + 196884q + 21493760q^2 + \dots \quad (8.12)$$

as required.

3. Geometry of the q Expansion

Armed with examples, let us discuss the geometry connected with the q expansion. Let R be the closed subset of $\mathcal{H}$ described in Figure 8.1. Proposition 8.5 below addresses the fact that R is a fundamental domain for the action of $SL(2, \mathbb{Z})$ in $\mathcal{H}$. This fact needs a little care in its formulation, since $\begin{pmatrix} -1 & 0 \\ 0 & -1 \end{pmatrix}$ acts as the identity on $\mathcal{H}$. The group $SL(2, \mathbb{R}) / \left\{ \pm \begin{pmatrix} 1 & 0 \\ 0 & 1 \end{pmatrix} \right\}$ acts effectively, and R is really a fundamental domain for the group

$$\Gamma = SL(2, \mathbb{Z}) / \left\{ \pm \begin{pmatrix} 1 & 0 \\ 0 & 1 \end{pmatrix} \right\}.$$

Let T and S be the images in Γ of the members $\begin{pmatrix} 1 & 1 \\ 0 & 1 \end{pmatrix}$ and $\begin{pmatrix} 0 & 1 \\ -1 & 0 \end{pmatrix}$ of $SL(2,\mathbf{Z})$. These elements are given by

$$T(\tau) = \tau + 1 \quad \text{and} \quad S(\tau) = -1/\tau, \tag{8.13}$$

and S has order 2 (not 4). Before proving that R is a fundamental domain, we examine these elements more closely.

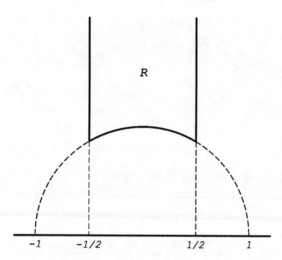

FIGURE 8.1. Fundamental domain for $SL(2,\mathbf{Z})$

Proposition 8.3. The elements $\begin{pmatrix} 1 & 1 \\ 0 & 1 \end{pmatrix}$ and $\begin{pmatrix} 0 & 1 \\ -1 & 0 \end{pmatrix}$ generate $SL(2,\mathbf{Z})$.

PROOF. Assume the contrary. Let $\tilde{\Gamma}$ be the subgroup of $SL(2,\mathbf{Z})$ generated by $\begin{pmatrix} 1 & 1 \\ 0 & 1 \end{pmatrix}$ and $\begin{pmatrix} 0 & 1 \\ -1 & 0 \end{pmatrix}$. Among all members $\begin{pmatrix} a & b \\ c & d \end{pmatrix}$ of $SL(2,\mathbf{Z})$ not in $\tilde{\Gamma}$, choose one with $\max\{|a|,|c|\}$ as small as possible. Then with $\max\{|a|,|c|\}$ fixed, we may assume $\min\{|a|,|c|\}$ is as small as possible. Multiplying if necessary by $\begin{pmatrix} 0 & 1 \\ -1 & 0 \end{pmatrix}$ on the left, we may assume that $|a| \leq |c|$. Now

$$\begin{pmatrix} 0 & 1 \\ -1 & 0 \end{pmatrix}\begin{pmatrix} 1 & 1 \\ 0 & 1 \end{pmatrix}\begin{pmatrix} 0 & 1 \\ -1 & 0 \end{pmatrix}^{-1} = \begin{pmatrix} 1 & 0 \\ -1 & 1 \end{pmatrix}$$

3. GEOMETRY OF THE q EXPANSION

shows that $\begin{pmatrix} 1 & 0 \\ -1 & 1 \end{pmatrix}$ is in $\tilde{\Gamma}$. Thus

$$\begin{pmatrix} 1 & 0 \\ \pm 1 & 1 \end{pmatrix}\begin{pmatrix} a & b \\ c & d \end{pmatrix} = \begin{pmatrix} a & b \\ c \pm a & d \pm b \end{pmatrix}$$

is not in $\tilde{\Gamma}$. Unless $a = 0$, one of $c \pm a$ has $|c \pm a| < |c|$ and contradicts our construction of $\begin{pmatrix} a & b \\ c & d \end{pmatrix}$. Thus we may assume $a = 0$. Multiplying on the left by $\begin{pmatrix} 0 & 1 \\ -1 & 0 \end{pmatrix}$ or its inverse $\begin{pmatrix} 0 & -1 \\ 1 & 0 \end{pmatrix}$, we see that $SL(2, \mathbf{Z})$ contains an element $\begin{pmatrix} 1 & * \\ 0 & * \end{pmatrix}$ not in $\tilde{\Gamma}$. This element is a power of $\begin{pmatrix} 1 & 1 \\ 0 & 1 \end{pmatrix}$, and we have a contradiction.

Since $\begin{pmatrix} 1 & 1 \\ 0 & 1 \end{pmatrix}$ and $\begin{pmatrix} 0 & 1 \\ -1 & 0 \end{pmatrix}$ are in $SL(2, \mathbf{Z})$, an unrestricted modular form f of weight k satisfies

$$\begin{aligned} f(\tau + 1) &= f(\tau) \\ f(-1/\tau) &= (-\tau)^k f(\tau). \end{aligned} \tag{8.14}$$

On the other hand, Proposition 8.3 implies the following converse to the validity of (8.14).

Corollary 8.4. An analytic function f on $\mathcal{H}$ that satisfies (8.14) is an unrestricted modular form.

PROOF. For $g = \begin{pmatrix} a & b \\ c & d \end{pmatrix}$ in $SL(2, \mathbf{R})$, let

$$\delta(g, \tau) = c\tau + d. \tag{8.15}$$

Direct computation gives

$$\delta(g_1 g_2, \tau) = \delta(g_1, g_2\tau)\delta(g_2, \tau) \tag{8.16a}$$

and

$$\delta(g^{-1}, \tau) = \delta(g, g^{-1}\tau)^{-1}. \tag{8.16b}$$

If g_1 and g_2 are in $SL(2, \mathbf{Z})$ and are elements γ such that

$$f(\gamma\tau) = \delta(\gamma, \tau)^k f(\tau), \tag{8.17}$$

then (8.16a) says that (8.17) holds also for $\gamma = g_1 g_2$. If g is an element γ such that (8.17) holds, then (8.16b) says that (8.17) holds also for $\gamma = g^{-1}$. Thus the subset of $SL(2, \mathbf{Z})$ for which (8.17) holds is a subgroup. Equations (8.14) say that this subgroup contains $\begin{pmatrix} 1 & 1 \\ 0 & 1 \end{pmatrix}$ and $\begin{pmatrix} 0 & 1 \\ -1 & 0 \end{pmatrix}$. By Proposition 8.3 the subgroup is all of $SL(2, \mathbf{Z})$.

Theorem 8.5.
(a) Each point of $\mathcal{H}$ can be mapped into R by some element of $\Gamma = SL(2, \mathbf{Z}) / \left\{ \pm \begin{pmatrix} 1 & 0 \\ 0 & 1 \end{pmatrix} \right\}$.

(b) The only points of R that are equivalent with one another under Γ are the points τ and $\tau + 1$ of the vertical sides, and the points τ and $-1/\tau$ of the circular arc.

(c) The only points of R that are fixed by a member $\gamma \neq 1$ of Γ are $\tau = i$ (fixed exactly by the subgroup $\{1, S\}$), $\tau = \rho = e^{2\pi i/3}$ (fixed exactly by the subgroup $\{1, ST, (ST)^2\}$), and $\tau = -\bar{\rho} = e^{\pi i/3}$ (fixed exactly by the subgroup $\{1, TS, (TS)^2\}$).

PROOF. (a) If $\tau = \rho + i\sigma$ is given, then we can compute that

$$\operatorname{Im}\left(\frac{a\tau + b}{c\tau + d}\right) = \frac{\sigma}{|c\tau + d|^2}. \tag{8.18}$$

Since c and d are integers, some choice of $\begin{pmatrix} a & b \\ c & d \end{pmatrix}$ makes $|c\tau + d|^2$ a minimum. Applying this element, we obtain τ' such that $\operatorname{Im} \tau' \geq \operatorname{Im} \gamma\tau'$ for all $\gamma \in \Gamma$. For a suitable choice of n, the translate $\tau'' = T^n \tau'$ will have $|\operatorname{Re} \tau''| \leq \frac{1}{2}$, and it will still be true that $\operatorname{Im} \tau'' \geq \operatorname{Im} \gamma\tau''$ for all $\gamma \in \Gamma$. If τ'' were strictly below the unit circle (at the bottom edge of R), we would have $\operatorname{Im} \tau'' < \operatorname{Im} S\tau''$, contradiction. We conclude that τ'' is in R.

(b,c) Suppose τ is in R and $g = \begin{pmatrix} a & b \\ c & d \end{pmatrix}$ is an element of $SL(2, \mathbf{Z})$ with $g\tau$ in R. Let γ be the image of g in Γ. Since $\gamma\tau$ and $\gamma^{-1}(\gamma\tau)$ are in R, there is no loss of generality in assuming that $\operatorname{Im}(\gamma\tau) \geq \operatorname{Im} \tau$. By (8.18), $|c\tau + d| \leq 1$. Since $\operatorname{Im} \tau > \frac{1}{2}$, we cannot have $|c| \geq 2$. Thus $c = 0, 1,$ or -1. If $c = 0$, then γ must be a power of T, and the points in question are τ and $\tau + 1$ as in (b).

Since $\begin{pmatrix} a & b \\ c & d \end{pmatrix}$ and $\begin{pmatrix} -a & -b \\ -c & -d \end{pmatrix}$ have the same image in Γ, it remains to consider $c = 1$. Then $|\tau + d| \leq 1$. Since $|\operatorname{Re} \tau| \leq \frac{1}{2}$, we must have

$|d| \leq 1$. Thus $d = 0, 1$, or -1. Suppose $d = 0$. Then $|\tau| = 1$. Since $ad - bc = 1$, $b = -1$. Hence $g = \begin{pmatrix} a & -1 \\ 1 & 0 \end{pmatrix}$, and $\gamma\tau = a - \dfrac{1}{\tau}$. This is an integer translate of the point $-\dfrac{1}{\tau}$, which is on the bottom edge of R since $|\tau| = 1$. The possibilities are that $a = -1$ and $\tau = \rho$ as in (c), or $a = 1$ and $\tau = -\bar{\rho}$ as in (c), or $a = 0$ as in (b) and the case $\tau = i$ of (c).

Suppose $d = 1$. Then $\tau \in R$ and $|\tau + 1| \leq 1$ imply $\tau = \rho$. In this case $g = \begin{pmatrix} a & b \\ 1 & 1 \end{pmatrix}$ with $a - b = 1$. So $\gamma\rho = \dfrac{a\rho + a - 1}{\rho + 1} = a - \dfrac{1}{\rho + 1} = a + \rho$. Hence $a = 0$ or 1. If $a = 0$, we have $g = \begin{pmatrix} 0 & -1 \\ 1 & 1 \end{pmatrix}$ as in (c). If $a = 1$, then the points are ρ and $\rho + 1$ as in (b).

Suppose $d = -1$. Then similarly $\tau = -\bar{\rho}$ and $\gamma(-\bar{\rho}) = a + (-\bar{\rho})$ with $a = 0$ or -1. If $a = 0$, we obtain $g = \begin{pmatrix} 0 & -1 \\ 1 & -1 \end{pmatrix}$ as in (c). If $a = 1$, then the points are $-\bar{\rho}$ and $(-\bar{\rho}) + 1$ as in (b).

The fundamental domain allows us to interpret the q expansion of a modular form as an expansion at ∞. In fact, we see from Figure 8.1 that the only way to tend to ∞ in R is vertically. Thus the point $\tau = \rho + i\sigma$ has $\sigma \to \infty$ while ρ remains bounded. The effect on $q = e^{2\pi i \tau}$ is for q to tend to 0. So a power series in q may be regarded as an expansion about $\tau = \infty$. If the q expansion does not have infinitely many negative powers of q, the condition that an unrestricted modular form be modular is a growth condition: There is to be no pole, and the form is thus to be bounded as $\sigma \to \infty$. The additional condition for a cusp form is that the form vanish as $\sigma \to \infty$. This condition can be checked by evaluating the integral defining c_0 in (8.7b).

4. Dimensions of Spaces of Modular Forms

Let $f \not\equiv 0$ be a modular form of weight k. For τ in $\mathcal{H}$, let $v_\tau(f)$ be the order of vanishing of f at τ. If γ is in $SL(2, \mathbb{Z})$, then $v_{\gamma\tau}(f) = v_\tau(f)$ since $f(\gamma\tau) = (c\tau + d)^k f(\tau)$. Let $v_\infty(f)$ be the order of vanishing of the q expansion of f (the expansion at ∞).

Theorem 8.6. If $f \not\equiv 0$ is a modular form of weight k, then

$$v_\infty(f) + \tfrac{1}{2} v_i(f) + \tfrac{1}{3} v_\rho(f) + \sum_\tau{}' v_\tau(f) = \dfrac{k}{12},$$

where $\sum'$ refers to a sum over inequivalent points in the fundamental domain R other than i and ρ.

REMARK. The left side is indeed finite. There can be no zeros of f in a deleted neighborhood of ∞, i.e., beyond some height in R. The remainder of R is compact, and f can have only finitely many zeros there.

PROOF. Essentially we shall apply the Argument Principle, computing $\dfrac{1}{2\pi i} \oint \dfrac{f'(\tau)}{f(\tau)}\, d\tau$ over the boundary of R with positive orientation. Actually we have to adjust our contour, introducing detours to avoid zeros. First assume that there are no zeros on the boundary of R except possibly at i and ρ. We introduce the modified contour in Figure 8.2. It is assumed that all the zeros of f within R, other than those at i and ρ, are within the contour in the figure. The arcs BB', CC', and DD' are circular.

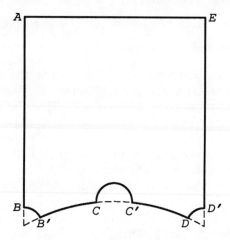

FIGURE 8.2. Contour for calculating number of zeros

We shall make repeated use of the following change of variables formula. If $\tau = \phi(\tau')$ is an analytic change of variables, then

$$\frac{(f\circ\phi)'(\tau')\, d\tau'}{(f\circ\phi)(\tau')} = \frac{f'(\tau)\, d\tau}{f(\tau)}. \tag{8.19}$$

For a first application of this formula, we take $\tau' = q$ and write $\tilde{f}(q) = f(\tau)$. As τ traverses EA, q goes once around a circle with negative orientation. Hence

$$\frac{1}{2\pi i}\int_{EA}\frac{f'(\tau)\, d\tau}{f(\tau)} = -\frac{1}{2\pi i}\oint \frac{\tilde{f}'(q)\, dq}{\tilde{f}(q)} = -v_\infty(f). \tag{8.20}$$

4. DIMENSIONS OF SPACES OF MODULAR FORMS

Next the points of AB are paired with those of ED' by $\tau \to \tau+1$, and $f(\tau+1) = f(\tau)$. Hence

$$\frac{1}{2\pi i}\int_{AB}\frac{f'(\tau)\,d\tau}{f(\tau)} + \frac{1}{2\pi i}\int_{D'E}\frac{f'(\tau)\,d\tau}{f(\tau)} = 0. \tag{8.21}$$

The arc $B'C$ is paired with the arc DC' by $\tau \to \tau' = -1/\tau$. We take $\phi = S$ in (8.19) and obtain

$$\int_{C'D}\frac{f'(\tau)\,d\tau}{f(\tau)} = -\int_{DC'}\frac{f'(\tau)\,d\tau}{f(\tau)} = -\int_{B'C}\frac{(f\circ S)'(\tau)\,d\tau}{(f\circ S)(\tau)}$$

$$= -\int_{B'C}\frac{\frac{d}{d\tau}[(-\tau)^k f(\tau)]\,d\tau}{(-\tau)^k f(\tau)}$$

$$= -\int_{B'C}\frac{f'(\tau)\,d\tau}{f(\tau)} - \int_{B'C}\frac{k\,d\tau}{\tau}.$$

Hence

$$\frac{1}{2\pi i}\int_{B'C}\frac{f'(\tau)\,d\tau}{f(\tau)} + \frac{1}{2\pi i}\int_{C'D}\frac{f'(\tau)\,d\tau}{f(\tau)} = -\frac{k}{2\pi i}\int_{B'C}\frac{d\tau}{\tau} = \frac{k}{12} + o(1), \tag{8.22}$$

where $o(1)$ is a term that tends to 0 as the radii of the arcs BB', CC', DD' tend to 0.

To handle the integrals over the small arcs BB', CC', and DD', we note the following fact from complex variable theory. Let h be a meromorphic function with a simple pole at z_0 and with residue h_{-1} there. Over a small positively oriented circular arc centered at z_0 and consuming a fraction θ of the circumference of the circle, we have

$$\frac{1}{2\pi i}\int_{\text{arc}} h(z)\,dz = \theta h_{-1} + o(1). \tag{8.23}$$

[In fact, we have only to expand $h(z)$ in Laurent series and compute the integral term by term.]

Applying (8.23) to the small arcs in Figure 8.2 and taking into account orientations, we have

$$\frac{1}{2\pi i}\int_{BB'}\frac{f'(\tau)\,d\tau}{f(\tau)} = -\frac{1}{6}v_\rho(f) + o(1)$$

$$\frac{1}{2\pi i}\int_{CC'}\frac{f'(\tau)\,d\tau}{f(\tau)} = -\frac{1}{2}v_i(f) + o(1)$$

$$\frac{1}{2\pi i}\int_{DD'}\frac{f'(\tau)\,d\tau}{f(\tau)} = -\frac{1}{6}v_\rho(f) + o(1).$$

We add these three formulas to the sum of (8.20), (8.21), and (8.22), and we apply the Argument Principle. The result is

$$\sum_{\tau}' v_\tau(f) = -v_\infty(f) + 0 + \frac{k}{12} - \frac{1}{6}v_\rho(f) - \frac{1}{2}v_i(f) - \frac{1}{6}v_\rho(f) + o(1).$$

Letting the radii of the small arcs tend to 0, we obtain the desired result.

We still have to treat the case that f has some zeros on the boundary of R other than at ρ and i. In this case we introduce a circular bubble at each such point, with S or T of that bubble at the congruent point on the boundary of R. The bubbles on the vertical sides of R do not affect the above argument, and the bubbles on the circular bottom side of R introduce error terms that are $o(1)$. This completes the proof.

Let M_k be the vector space of modular forms of weight k, and let S_k be the subspace of cusp forms. These spaces are 0 for k odd, since

$$f(\tau) = f\left(\begin{pmatrix} -1 & 0 \\ 0 & -1 \end{pmatrix}\tau\right) = (-1)^k f(\tau).$$

For even $k \geq 4$, Proposition 8.1 shows that G_k is in M_k but not S_k. On the other hand, S_k has codimension at most 1 in M_k, being given by a single condition $c_0 = 0$. Thus

$$M_k = S_k \oplus \mathbf{C}G_k \qquad \text{for } k \text{ even } \geq 4. \tag{8.24}$$

Corollary 8.7. Let k be even.
(a) $M_k = 0$ for $k < 0$ and $k = 2$.
(b) $M_0 = \mathbf{C}1$, and $M_k = \mathbf{C}G_k$ for $4 \leq k \leq 10$.
(c) Multiplication by $\Delta(\tau)$ defines an isomorphism of M_{k-12} onto S_k.

PROOF. (a) In Theorem 8.6, all terms on the left side are ≥ 0. Thus the right side is ≥ 0, and $k \geq 0$. For $k = 2$, the right side is $\frac{1}{6}$. No nonnegative integer sum $n_1 + \frac{1}{2}n_2 + \frac{1}{3}n_3$ can equal $\frac{1}{6}$, and so no $f \not\equiv 0$ can exist.

(b) When $k < 12$, the right side is < 1. Thus $v_\infty(f) = 0$. Hence $S_k = 0$. If $4 \leq k \leq 10$ and k is even, Proposition 8.1 shows that $M_k = \mathbf{C}G_k$. If $k = 0$, 1 is in M_k. Since S_0 has codimension at most 1 in M_0, $M_0 = \mathbf{C}1$.

(c) Since $\Delta(\tau)$ is a cusp form of weight 12, multiplication by Δ carries M_{k-12} into S_k. On the other hand, $\Delta(\tau)$ is nonvanishing on $\mathcal{H}$ by Corollary 8.2 (or Theorem 6.15) and has a simple zero at ∞, by (8.11). Thus multiplication by $\Delta(\tau)^{-1}$ is a well defined linear map of S_k into M_{k-12}.

4. DIMENSIONS OF SPACES OF MODULAR FORMS

Corollary 8.8. If $k \geq 0$ is even, then

$$\dim M_k = \begin{cases} \left[\dfrac{k}{12}\right] + 1 & \text{if } k \not\equiv 2 \mod 12 \\ \left[\dfrac{k}{12}\right] & \text{if } k \equiv 2 \mod 12. \end{cases}$$

Also $S_k = 0$ for $k < 12$; if $k \geq 12$ is even, then

$$\dim S_k = \begin{cases} \left[\dfrac{k}{12}\right] & \text{if } k \not\equiv 2 \mod 12 \\ \left[\dfrac{k}{12}\right] - 1 & \text{if } k \equiv 2 \mod 12. \end{cases}$$

Corollary 8.9. The q expansion of the cusp form $\Delta(\tau)$ of weight 12 can be regrouped as

$$\Delta(\tau) = (2\pi)^{12} q \prod_{n=1}^{\infty} (1 - q^n)^{24}.$$

More specifically let

$$\eta(\tau) = e^{\pi i \tau/12} \prod_{n=1}^{\infty} (1 - q^n),$$

so that $\Delta(\tau) = (2\pi)^{12} \eta(\tau)^{24}$. Then

$$\eta(\tau + 1) = e^{\pi i/12} \eta(\tau) \quad \text{and} \quad \eta(-1/\tau) = (-i\tau)^{1/2} \eta(\tau).$$

REMARK. This formula for $\Delta(\tau)$ lends itself much better to calculation than the one in (8.2).

PROOF. Once we have proved the two transformation laws for $\eta(\tau)$, it follows that $\eta(\tau)^{24}$ satisfies the transformation laws (8.14) with $k = 12$. Corollary 8.4 then shows that $\eta(\tau)^{24}$ is an unrestricted modular form of weight 12, and hence $\eta(\tau)^{24}$ is a cusp form of weight 12. By Corollary 8.8, $\Delta(\tau) = c\eta(\tau)^{24}$ for some constant c. The coefficient of q for $\eta(\tau)^{24}$ is 1 and for $\Delta(\tau)$ is $(2\pi)^{12}$, by (8.11). Thus $c = (2\pi)^{12}$.

The identity $\eta(\tau + 1) = e^{\pi i/12} \eta(\tau)$ being obvious, we are left with proving

$$\eta(-1/\tau) = (-i\tau)^{1/2} \eta(\tau). \tag{8.25}$$

Let log denote the principal branch of the logarithm cut on the negative reals. Since $|q| < 1$,

$$-\log(1 - q^l) = \sum_{k=1}^{\infty} \frac{1}{k} q^{kl}.$$

Then

$$\eta(\tau)e^{-\pi i \tau/12} = \prod_{l=1}^{\infty} \exp \log(1 - q^l)$$

$$= \exp \sum_{l=1}^{\infty} \log(1 - q^l)$$

$$= \exp\left\{-\sum_{k,l=1}^{\infty} \frac{1}{k} q^{kl}\right\}$$

$$= \exp \sum_{k=1}^{\infty} -\frac{1}{k} \frac{q^k}{1 - q^k}$$

$$= \exp \sum_{k=1}^{\infty} -\frac{1}{k} \left(\frac{1}{e^{-2\pi i k \tau} - 1}\right).$$

Replacing τ by $-1/\tau$ gives

$$\eta(-1/\tau)e^{\pi i/(12\tau)} = \exp \sum_{k=1}^{\infty} -\frac{1}{k} \left(\frac{1}{e^{2\pi i k/\tau} - 1}\right).$$

So

$$\eta(\tau)/\eta(-1/\tau)$$
$$= e^{\frac{\pi i}{12}(\tau + \tau^{-1})} \exp \sum_{k=1}^{\infty} -\frac{1}{k} \left(\frac{1}{e^{-2\pi i k \tau} - 1} - \frac{1}{e^{2\pi i k/\tau} - 1}\right).$$

In order to prove (8.25), it is therefore enough to prove that

$$\sum_{k=1}^{\infty} \frac{1}{k} \left(\frac{1}{e^{-2\pi i k \tau} - 1} - \frac{1}{e^{2\pi i k/\tau} - 1}\right) = \frac{\pi i}{12}(\tau + \tau^{-1}) + \frac{1}{2}\log(-i\tau). \quad (8.26)$$

Fix τ. Put $f(z) = (\cot z)(\cot \frac{z}{\tau})$ and $\nu = (n + \frac{1}{2})\pi$ with $n = 0, 1, 2, \ldots$. The two factors comprising f are together singular only at $z = 0$. Thus

4. DIMENSIONS OF SPACES OF MODULAR FORMS

$z^{-1}f(\nu z)$ has simple poles at $z = \pm\dfrac{\pi k}{\nu}$ and $\pm\dfrac{\pi k \tau}{\nu}$ for $k \neq 0$, and the respective residues are $\dfrac{1}{\pi k}\cot\left(\dfrac{\pi k}{\tau}\right)$ and $\dfrac{1}{\pi k}\cot(\pi k\tau)$ for $k = 1, 2, \ldots$. Also there is a triple pole at $z = 0$, and the residue of $z^{-1}f(\nu z)$ at $z = 0$ can be seen to be $-\frac{1}{3}(\tau + \tau^{-1})$. Let C be the curve in the z plane in the shape of a parallelogram, passing from 1 to τ to -1 to $-\tau$ to 1. Those poles mentioned above that lie within the parallelogram are actually on the diagonals of the parallelogram. Thus the Residue Theorem yields

$$\oint_C f(\nu z)\frac{dz}{z} = -\frac{2\pi i}{3}(\tau + \tau^{-1}) + \frac{2 \cdot 2\pi i}{\pi}\sum_{k=1}^n \frac{1}{k}\left(\cot\left(\frac{\pi k}{\tau}\right) + \cot(\pi k\tau)\right). \tag{8.27}$$

Let us consider the limiting behavior of $f(\nu z)$ on each edge of C as $n \to +\infty$. If we parametrize the edge from 1 to τ by $z = t\tau + (1-t)$ with $0 \leq t \leq 1$, then we can write

$$\cot \nu z = \cot(n + \tfrac{1}{2})\pi z = \frac{i\left(e^{2i(n+\frac{1}{2})\pi(t\tau+(1-t))} + 1\right)}{e^{2i(n+\frac{1}{2})\pi(t\tau+(1-t))} - 1}. \tag{8.28}$$

Let $\tau = \rho + i\sigma$. The exponential here has magnitude

$$e^{-2(n+\frac{1}{2})\pi t \sigma} \tag{8.29}$$

and tends to 0 pointwise (except at $t = 0$) as $n \to +\infty$, since $\sigma > 0$. Thus $\cot \nu z \to -i$ on this edge. Let us show that the convergence occurs boundedly. Let $\epsilon > 0$ be a number to be specified. When $t \geq \dfrac{\epsilon}{n+\frac{1}{2}}$, (8.29) is $\leq e^{-2\pi\epsilon\sigma}$, which can be assumed to be < 1; then (8.28) is bounded. When $t < \dfrac{\epsilon}{n+\frac{1}{2}}$, the $e^{i\theta}$ part of the exponential is

$$e^{2i(n+\frac{1}{2})\pi(t\rho+(1-t))} = e^{i\pi[(2n+1)+t(2n+1)(\rho-1)]} = -e^{i\pi t(2n+1)(\rho-1)}.$$

The exponent on the right is controlled. For a suitably small ϵ, we can make it $< \pi/2$ in absolute value, and we see that the denominator of (8.28) is bounded away from 0. Therefore (8.28) is bounded, and the convergence occurs boundedly. Similarly we find that $\cot \frac{\nu z}{\tau}$ is bounded on this edge and tends pointwise to i. By dominated convergence

$$\lim_{n\to\infty}\int_1^\tau f(\nu z)\frac{dz}{z} = \int_1^\tau \frac{dz}{z} = \log \tau.$$

Similar computations give

$$\lim_{n\to\infty}\int_\tau^{-1} f(\nu z)\frac{dz}{z} = -\int_\tau^{-1}\frac{dz}{z} = -\pi i + \log\tau$$

$$\lim_{n\to\infty}\int_{-1}^{-\tau} f(\nu z)\frac{dz}{z} = \int_{-1}^{-\tau}\frac{dz}{z} = \log(-\tau) + \pi i$$

$$\lim_{n\to\infty}\int_{-\tau}^{1} f(\nu z)\frac{dz}{z} = -\int_{-\tau}^{1}\frac{dz}{z} = \log(-\tau).$$

Since $\log\tau + \log(-\tau) = 2\log\left(\frac{\tau}{i}\right)$, passage to the limit in (8.27) gives

$$4\log\left(\frac{\tau}{i}\right) = -\frac{2\pi i}{3}(\tau + \tau^{-1}) + 4i\sum_{k=1}^\infty \frac{1}{k}\left(\cot(\pi k\tau) + \cot\left(\frac{\pi k}{\tau}\right)\right)$$

$$= -\frac{2\pi i}{3}(\tau + \tau^{-1}) + 4\sum_{k=1}^\infty \frac{1}{k}\left(\frac{e^{-2\pi ik\tau}+1}{e^{-2\pi ik\tau}-1} - \frac{e^{2\pi ik/\tau}+1}{e^{2\pi ik/\tau}-1}\right)$$

$$= -\frac{2\pi i}{3}(\tau + \tau^{-1}) + 8\sum_{k=1}^\infty \frac{1}{k}\left(\frac{1}{e^{-2\pi ik\tau}-1} - \frac{1}{e^{2\pi ik/\tau}-1}\right).$$

This proves (8.26), and we have seen that (8.25) follows.

5. L Function of a Cusp Form

In this section we shall associate to each cusp form an L function given by a Dirichlet series. We shall see that the L function extends to be entire and satisfies a functional equation. The prototype for this study is Theorem 7.19, where we passed from certain θ functions to L functions and proved similar results about the L functions.

Let $f \in S_k$ be a cusp form, and let $f(\tau) = \sum_{n=1}^\infty c_n q^n$ be its q expansion. The **L function** of f is the Dirichlet series

$$L(s,f) = \sum_{n=1}^\infty \frac{c_n}{n^s}. \tag{8.30}$$

The L function can be obtained from f by applying a Mellin transform. If $F : (0,\infty) \to \mathbf{C}$ is given, the **Mellin transform** of F is the function $g(s)$ defined by

$$g(s) = \int_0^\infty F(t) t^s \frac{dt}{t}$$

5. L FUNCTION OF A CUSP FORM

for all values of s for which the integral converges. (For s imaginary, the Mellin transform is the version of the Fourier transform appropriate for the multiplicative group $\mathbf{R}^+$, but we shall not use this fact.)

For our cusp form f, let us write $\tau = \rho + i\sigma$ and compute the Mellin transform of $f(i\sigma)$, with ρ fixed at 0. We proceed formally for now, disregarding convergence. We have

$$g(s) = \int_0^\infty f(i\sigma)\sigma^s \frac{d\sigma}{\sigma} = \int_0^\infty \sum_{n=1}^\infty c_n e^{-2\pi n\sigma} \sigma^s \frac{d\sigma}{\sigma}$$

$$= \sum_{n=1}^\infty c_n \int_0^\infty e^{-t}(2\pi n)^{-s} t^s \frac{dt}{t}$$

$$= (2\pi)^{-s}\Gamma(s) \sum_{n=1}^\infty \frac{c_n}{n^s}$$

$$= (2\pi)^{-s}\Gamma(s)L(s,f). \tag{8.31}$$

We shall show below that the coefficients c_n of a cusp form satisfy $|c_n| \leq Cn^{k/2}$. Then the Dirichlet series (8.30) converges absolutely for $\operatorname{Re} s > \frac{k}{2}+1$, and $L(s,f)$ is analytic in this region. Going over the steps of (8.31), we see that our computation was rigorous in this same region.

Lemma 8.10. Let $f \in S_k$ have q expansion $f(\tau) = \sum_{n=1}^\infty c_n q^n$. Then

(a) the function $\varphi(\tau) = |f(\tau)|\sigma^{k/2}$ is bounded on $\mathcal{H}$ and invariant under $SL(2,\mathbf{Z})$

(b) $|c_n| \leq Cn^{k/2}$.

PROOF. (a) From $f(\tau) = \sum_{n=1}^\infty c_n q^n$ for $|q| < 1$, we have $|f(\tau)| \leq C|q|$ for $|q| \leq \frac{1}{2}$, i.e.,

$$|f(\tau)| \leq Ce^{-2\pi\sigma} \quad \text{for } \sigma \geq \tfrac{1}{2\pi} \log 2. \tag{8.32}$$

Consequently $\varphi(\tau) = |f(\tau)|\sigma^{k/2}$ tends to 0 as τ tends to ∞ through the fundamental domain R for $SL(2,\mathbf{Z})$. Since φ is continuous and the part of R with $\sigma \leq \frac{1}{2\pi}\log 2$ is compact, φ is bounded on R. On $\mathcal{H}$, φ satisfies $\varphi(\tau+1) = \varphi(\tau)$ and

$$\varphi\left(-\frac{1}{\tau}\right) = \left|f\left(-\frac{1}{\tau}\right)\right|\left(\frac{\sigma}{|\tau|^2}\right)^{k/2} = |\tau|^k|f(\tau)|\left(\sigma|\tau|^{-2}\right)^{k/2}$$

$$= |f(\tau)|\sigma^{k/2} = \varphi(\tau).$$

By Proposition 8.3, φ is invariant under $SL(2, \mathbf{Z})$. Being bounded on R, φ must be bounded on $\mathcal{H}$.

(b) We have $c_n = \int_{-\frac{1}{2}}^{\frac{1}{2}} f(\tau) e^{-2\pi i n \tau} d\rho$, and we have just seen that $|f(\tau)| \leq C\sigma^{-k/2}$. Hence

$$|c_n| \leq C\sigma^{-k/2} e^{2\pi n \sigma} \qquad \text{for all } \sigma > 0.$$

Taking $\sigma = \frac{1}{n}$, we get $|c_n| \leq Ce^{2\pi} n^{k/2}$ as required.

REMARK. For fixed σ, the function $f(\rho + i\sigma)$ is smooth and periodic in ρ. Thus its Fourier coefficients decrease faster than any negative power of n. The numbers c_n are not the Fourier coefficients of $f(\rho + i\sigma)$, however, but are $e^{2\pi n \sigma}$ times those Fourier coefficients.

Theorem 8.11 (Hecke). If $f \in S_k$ is a cusp form for $SL(2, \mathbf{Z})$, then the L function $L(s, f)$, initially defined for Re $s > \frac{k}{2} + 1$, extends to be entire in s. Moreover, the function

$$\Lambda(s, f) = (2\pi)^{-s} \Gamma(s) L(s, f) \tag{8.33}$$

satisfies the functional equation

$$\Lambda(s, f) = (-1)^{k/2} \Lambda(k - s, f). \tag{8.34}$$

PROOF. The transformation law for $\begin{pmatrix} 0 & -1 \\ 1 & 0 \end{pmatrix}$ is $f(-1/\tau) = \tau^k f(\tau)$, and we specialize this to $\tau = \rho + i\sigma$ with $\rho = 0$ to obtain

$$f(i/\sigma) = i^k \sigma^k f(i\sigma). \tag{8.35}$$

From (8.31) we have

$$\Lambda(s, f) = \int_0^\infty f(i\sigma) \sigma^{s-1} d\sigma \tag{8.36}$$

for Re $s > \frac{k}{2} + 1$. The argument now proceeds in the same way as for Theorem 7.19c. The formal argument, disregarding convergence, is to make the change of variables $\sigma \to \sigma^{-1}$ in (8.36) and apply (8.35). To make this argument precise, we use the estimate (8.32) to see that

$$\int_1^\infty f(i\sigma) \sigma^{s-1} d\sigma$$

converges for all $s \in \mathbb{C}$ and defines an entire function. For Re $s > \frac{k}{2}+1$, we rewrite (8.36) as

$$\Lambda(s,f) = \int_0^1 f(i\sigma)\sigma^{s-1}\,d\sigma + \int_1^\infty f(i\sigma)\sigma^{s-1}\,d\sigma$$

and transform just the first term, replacing σ by σ^{-1} and applying (8.35). The result is

$$\Lambda(s,f) = i^k \int_1^\infty f(i\sigma)\sigma^{k-s-1}\,d\sigma + \int_1^\infty f(i\sigma)\sigma^{s-1}\,d\sigma. \tag{8.37}$$

The first term is of the same form as the second and thus extends to an entire function of s. Hence $\Lambda(s,f)$ extends to an entire function of s. Since $\Gamma(s)$ is nowhere 0, $L(s,f)$ is entire.

Replacing s by $k-s$ in (8.37) gives, upon multiplication by i^k,

$$i^k \Lambda(k-s,f) = (-1)^k \int_1^\infty f(i\sigma)\sigma^{s-1}\,d\sigma + i^k \int_1^\infty f(i\sigma)\sigma^{k-s-1}\,d\sigma.$$

The right side here matches the right side of (8.37) since $S_k \neq 0$ only when k is even, and (8.34) follows.

6. Petersson Inner Product

In this section we shall introduce a Hermitian inner product on the vector space S_k of cusp forms for $SL(2,\mathbb{Z})$.

Lemma 8.12. The measure $\dfrac{d\rho\,d\sigma}{\sigma^2}$ on $\mathcal{H}$ is invariant under $SL(2,\mathbb{R})$.

PROOF. In terms of τ and $\bar\tau$, we have $d\rho \wedge d\sigma = \dfrac{i}{2}(d\tau \wedge d\bar\tau)$. So it is enough to prove that $\dfrac{d\tau \wedge d\bar\tau}{(\mathrm{Im}\,\tau)^2}$ is an invariant 2-form. For $g = \begin{pmatrix} a & b \\ c & d \end{pmatrix}$, the Jacobian determinant $\dfrac{\partial(g\tau, g\bar\tau)}{\partial(\tau, \bar\tau)}$ is the determinant of a diagonal matrix with diagonal entries $(c\tau+d)^{-2}$ and $(c\bar\tau+d)^{-2}$. Thus

$$d(g\tau) \wedge d(g\bar\tau) = \frac{\partial(g\tau, g\bar\tau)}{\partial(\tau, \bar\tau)} d\tau \wedge d\bar\tau = |c\tau+d|^{-4} d\tau \wedge d\bar\tau.$$

From (8.18) we have

$$\mathrm{Im}(g\tau) = \frac{\mathrm{Im}\,\tau}{|c\tau+d|^2},$$

and the lemma follows.

For f and h in S_k, we define

$$\langle f, h \rangle = \int_R f(\tau)\overline{h(\tau)}\sigma^k \frac{d\rho\, d\sigma}{\sigma^2}, \tag{8.38}$$

where R is the usual fundamental domain in $\mathcal{H}$ for $SL(2,\mathbf{Z})$. By Lemma 8.10, $f(\tau)\overline{h(\tau)}\sigma^k$ is bounded, and therefore the integral is convergent. The result is a well defined Hermitian inner product called the **Petersson inner product** on S_k.

By Lemma 8.10a, $|f(\tau)|^2\sigma^k$ is invariant under $SL(2,\mathbf{Z})$. In view of Lemma 8.12, the measure $|f(\tau)|^2\sigma^k \dfrac{d\rho\, d\sigma}{\sigma^2}$ on $\mathcal{H}$ is invariant under $SL(2,\mathbf{Z})$. It follows that if we cut R into countably many pieces,* map each piece by a member of $SL(2,\mathbf{Z})$, and reassemble the result as a new fundamental domain R', then

$$\int_{R'} |f(\tau)|^2 \sigma^k \frac{d\rho\, d\sigma}{\sigma^2} = \int_R |f(\tau)|^2 \sigma^k \frac{d\rho\, d\sigma}{\sigma^2}.$$

Any measurable fundamental domain for $SL(2,\mathbf{Z})$ can be obtained in this way, and thus $\langle f, f \rangle$ is independent of the choice of fundamental domain. By polarization, the same conclusion holds for $\langle f, h \rangle$. We summarize as follows.

Proposition 8.13. On S_k, the Hermitian inner product

$$\langle f, h \rangle = \int_R f(\tau)\overline{h(\tau)}\sigma^k \frac{d\rho\, d\sigma}{\sigma^2}$$

is independent of the fundamental domain for $SL(2,\mathbf{Z})$.

7. Hecke Operators

Hecke operators are a certain kind of linear operator from M_k to M_k, with S_k mapping to S_k. We shall see that these operators commute and are self adjoint relative to the Petersson inner product. Consequently S_k has an orthogonal basis of simultaneous eigenvectors for the Hecke operators. The end result in §8 will be that the L function of each such eigenvector (normalized to have $c_1 = 1$) has a quadratic Euler product expansion.

---

*Technically we should first remove the part of the boundary of R where the imaginary part is positive, so that R is an exact fundamental domain.

7. HECKE OPERATORS

A **lattice** in $\mathbf{C}$ is all integer combinations of a pair of complex numbers linearly independent over $\mathbf{R}$. Recall from §2 that our first examples of modular forms came from homogeneous functions of lattices: $G_{2k}(\Lambda)$, $g_2(\Lambda)$, $g_3(\Lambda)$, $\Delta(\Lambda)$. Effectively what was shown at the beginning of §2 is the following: Let $\tilde{f}$ be a complex-valued function whose domain is the set of all lattices and which is **homogeneous** of degree $-k$ in the sense that

$$\tilde{f}(\alpha \Lambda) = \alpha^{-k} \tilde{f}(\Lambda) \qquad \text{for } \alpha \in \mathbf{C}^\times. \tag{8.39}$$

Define f on the upper half plane $\mathcal{H}$ by

$$f(\tau) = \tilde{f}(\Lambda_\tau). \tag{8.40}$$

As in the special cases of §2, f satisfies

$$f\left(\begin{pmatrix} a & b \\ c & d \end{pmatrix} \tau\right) = (c\tau + d)^k f(\tau) \qquad \text{for } \begin{pmatrix} a & b \\ c & d \end{pmatrix} \in SL(2, \mathbf{Z}). \tag{8.41}$$

Conversely to any $f : \mathcal{H} \to \mathbf{C}$ satisfying (8.41), we can associate a function $\tilde{f}$ of the lattice variable Λ, homogeneous of degree $-k$, by the definition

$$\tilde{f}(\mathbf{Z}\omega_1 \oplus \mathbf{Z}\omega_2) = \omega_1^{-k} f(\omega_2/\omega_1) \qquad \text{if } \operatorname{Im}(\omega_2/\omega_1) > 0. \tag{8.42}$$

Let us check that this function depends only on Λ, not on the basis. If we have

$$\begin{pmatrix} \omega_2' \\ \omega_1' \end{pmatrix} = \begin{pmatrix} a & b \\ c & d \end{pmatrix} \begin{pmatrix} \omega_2 \\ \omega_1 \end{pmatrix},$$

then

$$\begin{aligned}
\tilde{f}(\mathbf{Z}\omega_1' \oplus \mathbf{Z}\omega_2') &= \omega_1'^{-k} f(\omega_2'/\omega_1') \\
&= (c\omega_2 + d\omega_1)^{-k} (c(\omega_2/\omega_1) + d)^k f(\omega_2/\omega_1) \\
&= \omega_1^{-k} f(\omega_2/\omega_1) \\
&= \tilde{f}(\mathbf{Z}\omega_1 \oplus \mathbf{Z}\omega_2),
\end{aligned}$$

and the independence follows. As a consequence of (8.42) and this independence, $\tilde{f}(\Lambda)$ has the homogeneity property

$$\tilde{f}(\alpha \Lambda) = \alpha^{-k} \tilde{f}(\Lambda) \qquad \text{for } \alpha \in \mathbf{C}^\times.$$

Our initial definition of Hecke operators will be in terms of homogeneous functions of lattices. Then we shall translate the definition into the language of modular forms. Let $\mathcal{L}$ be the free abelian group freely generated by the lattices in $\mathbf{C}$. To avoid confusion, we shall write $n \cdot \Lambda$ or $n(\Lambda)$ for n times the generator in $\mathcal{L}$, and $n\Lambda$ or $(n\Lambda)$ for the dilate of Λ by the factor n. The **Hecke operator** $T(n)$ on lattices, for $n = 1, 2, 3, \ldots$, is defined to be the map $T(n) : \mathcal{L} \to \mathcal{L}$ given by

$$T(n)\Lambda = \sum_{[\Lambda:\Lambda']=n} \Lambda', \tag{8.43}$$

where $[\Lambda : \Lambda']$ is the index of Λ' in Λ. This is a finite sum, since any such Λ' satisfies $n\Lambda \subseteq \Lambda' \subseteq \Lambda$ and so corresponds to a subgroup of $\Lambda/n\Lambda \cong \mathbf{Z}_n \oplus \mathbf{Z}_n$.

To define $T_k(n)$ on an element f of the space M_k of modular forms of weight k, we let $\tilde{f}(\Lambda)$ be the function of lattices given by (8.42), so that $\tilde{f}(\Lambda)$ satisfies (8.39). Then $T_k(n)\tilde{f}$, as a function of lattices, is given by

$$(T_k(n)\tilde{f})(\Lambda) = n^{k-1} \sum_{[\Lambda:\Lambda']=n} \tilde{f}(\Lambda'). \tag{8.44}$$

It is clear that $T_k(n)\tilde{f}$ is another function of lattices homogeneous of degree $-k$. Shortly we shall exhibit the corresponding function of the complex variable τ, denoting it by $T_k(n)f$, and we shall verify that it is a modular form. The resulting operator $T_k(n)$ on modular forms is also called a **Hecke operator**.

To compute $T_k(n)$ explicitly on functions on lattices, let $\Lambda = \mathbf{Z}\omega_1 \oplus \mathbf{Z}\omega_2$ with $\operatorname{Im} \omega_2/\omega_1 > 0$, and let $\Lambda' = \mathbf{Z}\omega_1' \oplus \mathbf{Z}\omega_2'$ with $\operatorname{Im} \omega_2'/\omega_1' > 0$ be a sublattice of index n. If we write

$$\begin{pmatrix} \omega_2' \\ \omega_1' \end{pmatrix} = \begin{pmatrix} a & b \\ c & d \end{pmatrix} \begin{pmatrix} \omega_2 \\ \omega_1 \end{pmatrix}, \tag{8.45}$$

then $\begin{pmatrix} a & b \\ c & d \end{pmatrix}$ will be an integral matrix of determinant n. The most general properly ordered basis of Λ' is obtained by left multiplying the left side of (8.45) by a member of $SL(2, \mathbf{Z})$, and thus the right coset $SL(2, \mathbf{Z}) \begin{pmatrix} a & b \\ c & d \end{pmatrix}$ describes all the matrices leading from $\{\omega_1, \omega_2\}$ to Λ'.

Let $M(n)$ be the set of all integral matrices of determinant n. We have seen that there are only finitely many lattices Λ' of index n in Λ. Consequently there are only finitely many right cosets $SL(2, \mathbf{Z})\alpha$ in $M(n)$ with $\det \alpha = n$. Thus we have a (disjoint) right coset decomposition

$$M(n) = \bigcup_{i=1}^{\nu(n)} SL(2, \mathbf{Z})\alpha_i \tag{8.46}$$

with $\nu(n) < \infty$.

To write $T_k(n)f$ as a function of τ, we make the definition

$$(f \circ [\alpha]_k)(\tau) = f(\alpha\tau)(c\tau + d)^{-k}(\det \alpha)^{k/2} \tag{8.47}$$

for $\alpha = \begin{pmatrix} a & b \\ c & d \end{pmatrix}$ a matrix of positive determinant (so that $\tau \in \mathcal{H}$ implies $\alpha \in \mathcal{H}$). For an analytic function f on $\mathcal{H}$, the condition that f be an unrestricted modular form of weight k is that $f \circ [\gamma]_k = f$ for all γ in $SL(2, \mathbb{Z})$. The formula for $T_k(n)f$ involves this definition for matrices of determinant n.

Proposition 8.14. Let $\{\alpha_i\}_{i=1}^{\nu(n)}$ be a complete set of representatives for the right cosets $SL(2, \mathbb{Z})\alpha$ of $SL(2, \mathbb{Z})$ on $M(n)$. If f is in M_k, then $T_k(n)f$ is given as a function of τ by

$$T_k(n)f = n^{\frac{k}{2}-1} \sum_{i=1}^{\nu(n)} f \circ [\alpha_i]_k. \tag{8.48}$$

Hence $T_k(n)f$ is an unrestricted modular form of weight k.

PROOF. The left side of (8.44) at Λ_τ is just $T_k(n)f(\tau)$. Let $[\Lambda_\tau : \Lambda'] = n$ and let $\alpha_i = \begin{pmatrix} a & b \\ c & d \end{pmatrix}$ be the coset representative corresponding to Λ'. This means that

$$\begin{pmatrix} \omega'_2 \\ \omega'_1 \end{pmatrix} = \begin{pmatrix} a & b \\ c & d \end{pmatrix} \begin{pmatrix} \tau \\ 1 \end{pmatrix}$$

is a basis of Λ'. Then

$$\begin{aligned}
\tilde{f}(\Lambda') &= \tilde{f}(\omega'_1(\mathbb{Z} \oplus (\omega'_2/\omega'_1)\mathbb{Z})) \\
&= \omega'^{-k}_1 f(\omega'_2/\omega'_1) \\
&= (c\tau + d)^{-k} f(\tfrac{a\tau+b}{c\tau+d}) \\
&= (\det \alpha_i)^{-k/2}(f \circ [\alpha_i]_k)(\tau) \\
&= n^{-k/2}(f \circ [\alpha_i]_k)(\tau).
\end{aligned}$$

Substituting into (8.44), we obtain (8.48). This proves the analyticity of $(T_k(n)f)(\tau)$ in $\mathcal{H}$, and the transformation law follows from the homogeneity of $T_k(n)\tilde{f}$ as a function of lattices.

In order to examine the q expansion of $T_k(n)f$, we shall use an explicit set of coset representatives.

Lemma 8.15. The matrices $\begin{pmatrix} a & b \\ 0 & d \end{pmatrix}$ with $ad = n$, $d > 0$, and $0 \leq b < d$ are a complete set of coset representatives for the right cosets of $SL(2, \mathbf{Z})$ on $M(n)$.

PROOF. Let $\begin{pmatrix} a' & b' \\ c' & d' \end{pmatrix}$ be given. Choose relatively prime integers x and y with $xa' + yc' = 0$, and then choose relatively prime integers u and v with $uy + v(-x) = 1$. Then $\begin{pmatrix} u & v \\ x & y \end{pmatrix}$ is in $SL(2, \mathbf{Z})$, and

$$\begin{pmatrix} u & v \\ x & y \end{pmatrix} \begin{pmatrix} a' & b' \\ c' & d' \end{pmatrix}$$

has lower left entry 0. Let us call the resulting matrix $\begin{pmatrix} a'' & b'' \\ 0 & d'' \end{pmatrix}$. Possibly left multiplying by $\begin{pmatrix} -1 & 0 \\ 0 & -1 \end{pmatrix}$, we may assume that $a'' > 0$ and $d'' > 0$. Choose integers q and r with $b'' = d''q + r$ and $0 \leq r < d''$. Then

$$\begin{pmatrix} 1 & -q \\ 0 & 1 \end{pmatrix} \begin{pmatrix} a'' & b'' \\ 0 & d'' \end{pmatrix} = \begin{pmatrix} a'' & r \\ 0 & d'' \end{pmatrix}$$

is a coset representative of the kind described in the lemma.

Now suppose that two of the elements in the lemma are in the same right coset. Say

$$\begin{pmatrix} u & v \\ x & y \end{pmatrix} \begin{pmatrix} a & b \\ 0 & d \end{pmatrix} = \begin{pmatrix} a' & b' \\ 0 & d' \end{pmatrix}$$

with $uy - vx = 1$. The lower left entry forces $x = 0$. The determinant forces $uy = 1$. Consideration of signs forces $u = y = 1$. Finally the inequalities on b and b' force $v = 0$. Thus we have a complete set of representatives.

Proposition 8.16. Let f in M_k have q expansion $f(\tau) = \sum_{n=0}^{\infty} c_n q^n$. Then $T_k(m)f$ has q expansion

$$T_k(m)f(\tau) = \sum_{n=0}^{\infty} b_n q^n,$$

where

$$b_n = \begin{cases} c_0 \sigma_{k-1}(m) & \text{if } n = 0 \\ c_m & \text{if } n = 1 \\ \sum_{a | \text{GCD}(n,m)} a^{k-1} c_{nm/a^2} & \text{if } n > 1, \end{cases}$$

with σ_{k-1} as in Proposition 8.1. Consequently $T_k(m)$ carries M_k to M_k and carries S_k to S_k.

7. HECKE OPERATORS

REMARK. We shall make serious use of the formula for b_n only in the case $n = 1$.

PROOF. We apply (8.48), using the representatives in Lemma 8.15. If $\alpha = \begin{pmatrix} a & b \\ 0 & d \end{pmatrix}$ is such a representative with $ad = m$, then

$$f \circ [\alpha]_k(\tau) = f(\tfrac{a\tau+b}{d})d^{-k}m^{k/2}$$
$$= \sum_{n=0}^{\infty} c_n e^{2\pi i n(a\tau+b)/d} d^{-k} m^{k/2}.$$

Hence

$$T_k(m)f(\tau) = m^{k-1} \sum_{n=0}^{\infty} \sum_{a,b,d} d^{-k} c_n e^{2\pi i n(a\tau+b)/d}, \qquad (8.49)$$

with the inner sum over a, b, d as in the lemma. The sum

$$\sum_{b=0}^{d-1} e^{2\pi i n b/d}$$

equals d if $d \mid n$ and equals 0 otherwise. So we can drop all n's except those of the form $n = ld$, and then (8.49) is

$$= m^{k-1} \sum_{l=0}^{\infty} \sum_{\substack{ad=m \\ d>0}} c_{ld} d^{-k+1} q^{la} = \sum_{l=0}^{\infty} \sum_{\substack{a \mid m \\ a>0}} c_{lm/a} a^{k-1} q^{la}.$$

The coefficient of q^0 comes from $l = 0$ and is

$$= c_0 \sum_{\substack{a \mid m \\ a>0}} a^{k-1} = c_0 \sigma_{k-1}(m).$$

The coefficient of q^1 comes from $a = l = 1$; it is just c_m. For $n \geq 2$, the coefficient of q^n comes from triples (l, a, d) with $la = n$ and $a \mid m$. The factor $c_{lm/a}$ is c_{nm/a^2} with $a \mid n$ and $a \mid m$. Thus the coefficient of q^n is

$$\sum_{a \mid \text{GCD}(n,m)} c_{nm/a^2} a^{k-1}$$

as asserted. From the formula for b_n, we see that $T_k(m)f$ is a modular form; hence $T_k(m)$ carries M_k to M_k. If $c_0 = 0$, then $b_0 = 0$; hence $T_k(m)$ carries S_k to S_k.

Corollary 8.17. Let f in M_k have q expansion $f(\tau) = \sum_{n=0}^{\infty} c_n q^n$. For p prime, $T_k(p)f$ has q expansion $T_k(p)f(\tau) = \sum_{n=0}^{\infty} b_n q^n$ with

$$b_n = \begin{cases} c_{pn} & \text{if } p \nmid n \\ c_{pn} + p^{k-1} c_{n/p} & \text{if } p \mid n. \end{cases}$$

Now we examine how the $T_k(n)$ interact with one another. We begin with the operators $T(n)$ on $\mathcal{L}$. Let $R(n) : \mathcal{L} \to \mathcal{L}$ be given by

$$R(n)(\Lambda) = (n\Lambda).$$

It is clear that $R(n)T(m) = T(m)R(n)$ for all n and m.

Lemma 8.18.
(a) For a prime power p^r with $r \geq 1$,

$$T(p^r)T(p) = T(p^{r+1}) + p \cdot R(p)T(p^{r-1}). \tag{8.50}$$

(b) $T(m)T(n) = T(mn)$ if m and n are relatively prime.

PROOF. (a) Both sides associate to Λ some sublattices Λ' of index p^{r+1}, and we have to check that the multiplicity of a sublattice Λ' is the same for both sides. Fix such a Λ'.

First suppose $\Lambda' \subseteq p\Lambda$. Then $R(p)T(p^{r-1}) = T(p^{r-1})(p\Lambda)$ contains Λ' with multiplicity one. Therefore Λ' occurs on the right side of (8.50) with multiplicity $p+1$. On the left side, $T(p)\Lambda$ is the sum of the $p+1$ lattices strictly between $p\Lambda$ and Λ, and so Λ' occurs on the left side with multiplicity $p+1$.

Now suppose Λ' is not contained in $p\Lambda$. Then it occurs on the right side with multiplicity one, coming only from $T(p^{r+1})$. On the left side it occurs with multiplicity at least one. If it occurs with multiplicity greater than one, then there are two distinct sublattices Λ_1 and Λ_2 of Λ of index p with $\Lambda' \subseteq \Lambda_1$ and $\Lambda' \subseteq \Lambda_2$. Then $\Lambda' \subseteq \Lambda_1 \cap \Lambda_2 = p\Lambda$, contradiction. This proves (a).

(b) If Λ' is a sublattice of index mn in Λ, then Λ/Λ' is finite abelian of order mn. If m and n are relatively prime, we can write

$$\Lambda/\Lambda' = G_1 \oplus G_2$$

uniquely with $|G_1| = n$ and $|G_2| = m$. If Λ'' denotes the pullback to Λ of the members of G_1, then Λ'' is the unique lattice with $\Lambda' \subseteq \Lambda'' \subseteq \Lambda$ and $[\Lambda : \Lambda''] = m$. Hence Λ' occurs on both sides of $T(m)T(n)(\Lambda)$ and $T(mn)(\Lambda)$ with multiplicity one.

We shall translate Lemma 8.18 into a statement about the Hecke operators on modular forms. The lattices that occur in the definition of $T(n)(\Lambda)$ in (8.43) have index n in Λ, and we shall say that $T(n) : \mathcal{L} \to \mathcal{L}$ is an operator of **degree** n. More generally a linear operator $N : \mathcal{L} \to \mathcal{L}$ of the form

$$N(\Lambda) = \sum_{[\Lambda:\Lambda']=n} m_{\Lambda'}(\Lambda') \tag{8.51}$$

will be said to be of **degree** n. In this sense, $R(n)$ has degree n^2.

If $N : \mathcal{L} \to \mathcal{L}$ has degree n and is given as in (8.51), we define N_k on functions $\tilde{f}$ on lattices of homogeneity $-k$ by

$$(N_k \tilde{f})(\Lambda) = n^{k-1} \sum_{[\Lambda:\Lambda']=n} m_{\Lambda'} \tilde{f}(\Lambda'). \tag{8.52}$$

This definition is consistent with (8.44), and it satisfies the properties

$$(S + N)_k = S_k + N_k \tag{8.53}$$

if S and N both have degree n, and

$$(MN)_k = N_k M_k \tag{8.54}$$

if M has degree m and N has degree n (so that MN has degree $m+n$). Property (8.54) argues for putting $\tilde{N}_k$ on the right side of $\tilde{f}$ in (8.52), but the operators of interest to us will commute, and (8.54) will not be a problem. Notice that our definition (8.52) makes

$$(R(n)_k \tilde{f})(\Lambda) = (n^2)^{k-1} \tilde{f}(n\Lambda) = n^{k-2} \tilde{f}(\Lambda). \tag{8.55}$$

Theorem 8.19 (Hecke). On the space M_k, the Hecke operators satisfy

(a) For a prime power p^r with $r \geq 1$,

$$T_k(p^r) T_k(p) = T_k(p^{r+1}) + p^{k-1} T_k(p^{r-1}).$$

Hence $T_k(p^r)$ is a polynomial in $T_k(p)$ with integer coefficients.

(b) $T_k(m) T_k(n) = T_k(mn)$ if m and n are relatively prime.

(c) The algebra generated by the $T_k(n)$ for $n = 1, 2, 3, \ldots$ is generated by the $T_k(p)$ with p prime and is commutative.

PROOF. (a) Let f be in M_k. By (8.53), (8.54), and (8.55), Lemma 8.18a gives

$$T_k(p)T_k(p^r)\tilde{f} = T_k(p^{r+1})\tilde{f} + pp^{k-2}T_k(p^{r-1})\tilde{f}.$$

Hence

$$T_k(p)T_k(p^r)f = T_k(p^{r+1})f + p^{k-1}T_k(p^{r-1})f.$$

Since $T_k(1)$ is the identity, this equation shows recursively that $T_k(p^r)$ is a polynomial in $T_k(p)$ and hence commutes with $T_k(p)$. Then (a) follows.

(b) This follows from Lemma 8.18b and (8.54).

(c) By (b), the algebra generated by the $T_k(n)$ is the same as that generated by the $T_k(p^r)$. By (a) it is the same as that generated by the $T_k(p)$. In turn, these commute by (b).

8. Interaction with Petersson Inner Product

We have proved that the Hecke operators on S_k commute with one another, and we shall next prove that they are self adjoint operators relative to the Petersson inner product. The proof is a little subtle, and we begin with two lemmas.

For any integer $N \geq 1$, we define the **principal congruence subgroup** $\Gamma(N)$ of $SL(2, \mathbf{Z})$ by

$$\Gamma(N) = \left\{ \begin{pmatrix} a & b \\ c & d \end{pmatrix} \in SL(2, \mathbf{Z}) \,\bigg|\, \begin{pmatrix} a & b \\ c & d \end{pmatrix} \equiv \begin{pmatrix} 1 & 0 \\ 0 & 1 \end{pmatrix} \mod N \right\}.$$

The group $\Gamma(N)$ is the kernel of the reduction-modulo-N homomorphism carrying $SL(2, \mathbf{Z})$ to $SL(2, \mathbf{Z}_N)$. Since $SL(2, \mathbf{Z}_N)$ is finite, $\Gamma(N)$ has finite index in $SL(2, \mathbf{Z})$. Observe that $\Gamma(1) = SL(2, \mathbf{Z})$.

For $\alpha = \begin{pmatrix} a & b \\ c & d \end{pmatrix}$ in $M(n)$, we let $\alpha' = n\alpha^{-1} = \begin{pmatrix} d & -b \\ -c & a \end{pmatrix}$. The matrix α' is again in $M(n)$.

Lemma 8.20. If α is in $M(n)$, then

$$\alpha \Gamma(nm) \alpha^{-1} \subseteq \Gamma(m).$$

Consequently

$$\Gamma(n^2) \subseteq \alpha \Gamma(n) \alpha^{-1} \subseteq SL(2, \mathbf{Z}). \tag{8.56}$$

8. INTERACTION WITH PETERSSON INNER PRODUCT

PROOF. Let $\alpha = \begin{pmatrix} a & b \\ c & d \end{pmatrix}$ and let γ be in $\Gamma(nm)$. Then

$$n\alpha\gamma\alpha^{-1} = \begin{pmatrix} a & b \\ c & d \end{pmatrix} \gamma \begin{pmatrix} d & -b \\ -c & a \end{pmatrix}$$

$$\equiv \begin{pmatrix} a & b \\ c & d \end{pmatrix} \begin{pmatrix} d & -b \\ -c & a \end{pmatrix} \mod nm$$

$$= \begin{pmatrix} n & 0 \\ 0 & n \end{pmatrix} \mod nm,$$

and

$$\alpha\gamma\alpha^{-1} \equiv \begin{pmatrix} 1 & 0 \\ 0 & 1 \end{pmatrix} \mod m.$$

Taking $m = 1$ gives the right hand inclusion of (8.56). Taking $m = n$ and replacing α by α' gives the left hand inclusion.

Lemma 8.21. If α is in $M(n)$, then the group

$$\Gamma(n) \cap \alpha\Gamma(n)\alpha^{-1}$$

has the same finite index in $\Gamma(n)$ as it does in $\alpha\Gamma(n)\alpha^{-1}$.

REMARK. The index is finite since both groups lie between $\Gamma(n^2)$ and $SL(2, \mathbb{Z})$, by Lemma 8.20, and since $\Gamma(n^2)$ has finite index in $SL(2, \mathbb{Z})$.

PROOF. Let $\tilde{R}$ be a fundamental domain in $\mathcal{H}$ for the action of $\Gamma(n)$ (obtained, for example, as the union of the $[SL(2, \mathbb{Z}) : \Gamma(n)]$ translates of R by left coset representatives of $SL(2, \mathbb{Z})/\Gamma(n)$). Let $\tilde{R}'$ be a fundamental domain in $\mathcal{H}$ for the action of $\Gamma(n) \cap \alpha\Gamma(n)\alpha^{-1}$, obtained as the union of

$$[\Gamma(n) : \Gamma(n) \cap \alpha\Gamma(n)\alpha^{-1}]$$

translates of $\tilde{R}$ by elements of $\Gamma(n)$. If μ is the measure $\dfrac{d\rho\,d\sigma}{\sigma^2}$ in $\mathcal{H}$, then

$$\mu(\tilde{R}') = [\Gamma(n) : \Gamma(n) \cap \alpha\Gamma(n)\alpha^{-1}]\mu(\tilde{R}). \tag{8.57}$$

Now $\alpha\tilde{R}$ is a fundamental domain in $\mathcal{H}$ for the action of $\alpha\Gamma(n)\alpha^{-1}$. [In fact, if τ is given, choose $\gamma \in \Gamma(n)$ with $\gamma(\alpha^{-1}\tau) \in \tilde{R}$; then $(\alpha\gamma\alpha^{-1})\tau$ is in $\alpha\tilde{R}$. Uniqueness follows similarly.] Moreover,

$$\mu(\alpha\tilde{R}) = \mu(\tilde{R}) \tag{8.58}$$

by Lemma 8.12 applied to the element $n^{-1/2}\alpha$ of $SL(2,\mathbf{R})$. Next let $(\alpha\tilde{R})'$ be a fundamental domain in $\mathcal{H}$ for the action of $\Gamma(n)\cap\alpha\Gamma(n)\alpha^{-1}$, obtained as the union of

$$[\alpha\Gamma(n)\alpha^{-1}:\Gamma(n)\cap\alpha\Gamma(n)\alpha^{-1}]$$

translates of $\alpha\tilde{R}$ by elements of $\alpha\Gamma(n)\alpha^{-1}$. Then

$$\mu((\alpha\tilde{R})')=[\alpha\Gamma(n)\alpha^{-1}:\Gamma(n)\cap\alpha\Gamma(n)\alpha^{-1}]\mu(\alpha\tilde{R}). \quad (8.59)$$

Since $\tilde{R}'$ and $(\alpha\tilde{R})'$ are fundamental domains for the intersection, the same argument as at the end of §6 shows that $\mu(\tilde{R}')=\mu((\alpha\tilde{R})')$. Comparing (8.57) and (8.59) and using (8.58), we obtain the result of the lemma.

Theorem 8.22 (Petersson). The Hecke operators $T_k(n)$ on the space of cusp forms S_k are self adjoint relative to the Petersson inner product.

PROOF. By Theorem 8.19, it is enough to prove the result for n equal to a prime p. We shall use Lemmas 8.20 and 8.21 and the notation in the proof of Lemma 8.21. Also, as in the proof of Corollary 8.4, we let $\delta(g,\tau)=c\tau+d$ if $g=\begin{pmatrix}a&b\\c&d\end{pmatrix}$ has determinant one. If f and h are in S_k, we are to prove that

$$\int_R T_k(p)f(\tau)\overline{h(\tau)}\sigma^k\frac{d\rho\,d\sigma}{\sigma^2}=\int_R f(\tau)\overline{T_k(p)h(\tau)}\sigma^k\frac{d\rho\,d\sigma}{\sigma^2}. \quad (8.60)$$

By Proposition 8.13 it is enough to prove that

$$\int_{\tilde{R}} T_k(p)f(\tau)\overline{h(\tau)}\sigma^k\frac{d\rho\,d\sigma}{\sigma^2}=\int_{\tilde{R}} f(\tau)\overline{T_k(p)h(\tau)}\sigma^k\frac{d\rho\,d\sigma}{\sigma^2}, \quad (8.61)$$

since each side of (8.61) is the same multiple of the corresponding side of (8.60). Let $\{\alpha_i\}$ be right coset representatives of $M(p)$ relative to $SL(2,\mathbf{Z})$, as in (8.46). For each choice of α_i as in Lemma 8.15, the element $\alpha_i'=p\alpha_i^{-1}$ has $\alpha_i'=\gamma_i\alpha_i\gamma_i'$ for some γ_i and γ_i' in $SL(2,\mathbf{Z})$. Namely

$$\alpha_i=\begin{pmatrix}1&b\\0&p\end{pmatrix} \quad \text{has} \quad \alpha_i'=\begin{pmatrix}-b&-1\\1&0\end{pmatrix}\alpha_i\begin{pmatrix}b&1\\-1&0\end{pmatrix}$$

and

$$\alpha_i=\begin{pmatrix}p&0\\0&1\end{pmatrix} \quad \text{has} \quad \alpha_i'=\begin{pmatrix}0&-1\\1&0\end{pmatrix}\alpha_i\begin{pmatrix}0&1\\-1&0\end{pmatrix}.$$

8. INTERACTION WITH PETERSSON INNER PRODUCT

Since $f \circ [\gamma_i'^{-1}]_k = f$ and $h \circ [\gamma_i]_k = h$, Lemma 8.12 shows that

$$\int_{\tilde{R}} f(\tau)\overline{h \circ [\alpha_i]_k(\tau)}\sigma^k \frac{d\rho\,d\sigma}{\sigma^2} = \int_{\gamma_i'\tilde{R}} f(\tau)\overline{h \circ [\gamma_i\alpha_i\gamma_i']_k(\tau)}\sigma^k \frac{d\rho\,d\sigma}{\sigma^2}$$

$$= \int_{\tilde{R}} f(\tau)\overline{h \circ [\gamma_i\alpha_i\gamma_i']_k(\tau)}\sigma^k \frac{d\rho\,d\sigma}{\sigma^2}$$

$$= \int_{\tilde{R}} f(\tau)\overline{h \circ [\alpha_i']_k(\tau)}\sigma^k \frac{d\rho\,d\sigma}{\sigma^2}.$$

In view of Proposition 8.14, proving (8.61) therefore reduces to proving

$$\int_{\tilde{R}} f \circ [\alpha]_k(\tau)\overline{h(\tau)}\sigma^k \frac{d\rho\,d\sigma}{\sigma^2} = \int_{\tilde{R}} f(\tau)\overline{h \circ [\alpha']_k(\tau)}\sigma^k \frac{d\rho\,d\sigma}{\sigma^2} \qquad (8.62)$$

for each $\alpha = \alpha_i$.

We shall change variables on the left side of (8.62), replacing $\alpha\tau$ by a new variable τ'. Note that $\alpha^{\#} = n^{-1/2}\alpha$ has determinant 1 and satisfies

$$f \circ [\alpha]_k = f \circ [\alpha^{\#}]_k.$$

Thus (8.16) and (8.18) give

$$f \circ [\alpha]_k(\tau)\overline{h(\tau)}\sigma^k = f \circ [\alpha^{\#}]_k(\tau)\overline{h(\tau)}(\operatorname{Im}\tau)^k$$

$$= f(\alpha^{\#}\tau)\overline{h(\alpha^{\#-1}(\alpha^{\#}\tau))}\delta(\alpha^{\#},\tau)^{-k}(\operatorname{Im}\tau)^k$$

$$= f(\alpha^{\#}\tau)\overline{h(\alpha^{\#-1}(\alpha^{\#}\tau))}\,\overline{\delta(\alpha^{\#},\tau)}^k \left(\frac{\operatorname{Im}\tau}{|\delta(\alpha^{\#},\tau)|^2}\right)^k$$

$$= f(\alpha^{\#}\tau)\overline{h(\alpha^{\#-1}(\alpha^{\#}\tau))}\,\overline{\delta(\alpha^{\#-1},\alpha^{\#}\tau)}^{-k}(\operatorname{Im}\alpha^{\#}\tau)^k$$

$$= f(\tau')\overline{h \circ [\alpha']_k(\tau')}(\operatorname{Im}\tau')^k.$$

Taking Lemma 8.12 into account, we obtain

$$\int_{\tilde{R}} f \circ [\alpha]_k(\tau)\overline{h(\tau)}\sigma^k \frac{d\rho\,d\sigma}{\sigma^2} = \int_{\alpha\tilde{R}} f(\tau)\overline{h \circ [\alpha']_k(\tau)}\sigma^k \frac{d\rho\,d\sigma}{\sigma^2} \qquad (8.63)$$

Now f satisfies

$$f(\gamma\tau) = \delta(\gamma,\tau)^k f(\tau) \qquad \text{for } \gamma \in \alpha\Gamma(p)\alpha^{-1}$$

since $\alpha\Gamma(p)\alpha^{-1} \subseteq SL(2,\mathbf{Z})$ by Lemma 8.20. Also we readily check that $h \circ [\alpha']_k$ satisfies

$$h \circ [\alpha']_k(\gamma\tau) = \delta(\gamma,\tau)^k h \circ [\alpha']_k(\tau) \qquad \text{for } \gamma \in \alpha\Gamma(p)\alpha^{-1},$$

since $\alpha'(\alpha\Gamma(p)\alpha^{-1})\alpha'^{-1} = \Gamma(p) \subseteq SL(2,\mathbf{Z})$. By (8.18), the integrand on the right side of (8.63) is invariant under $\alpha\Gamma(p)\alpha^{-1}$. Since $\alpha\tilde{R}$ is a fundamental domain for $\alpha\Gamma(p)\alpha^{-1}$, the right side of (8.63) is unaffected by replacing $\alpha\tilde{R}$ by any other such fundamental domain. Thus we have

$$\int_{\tilde{R}} f \circ [\alpha]_k(\tau)\overline{h(\tau)}\sigma^k \frac{d\rho\, d\sigma}{\sigma^2} =$$
$$\frac{1}{[\alpha\Gamma(p)\alpha^{-1} : \Gamma(p) \cap \alpha\Gamma(p)\alpha^{-1}]} \int_{(\alpha\tilde{R})'} f(\tau)\overline{h \circ [\alpha']_k(\tau)}\sigma^k \frac{d\rho\, d\sigma}{\sigma^2}$$
(8.64)

Meanwhile, on the right side of (8.62), f satisfies

$$f(\gamma\tau) = \delta(\gamma,\tau)^k f(\tau) \qquad \text{for } \gamma \in \Gamma(p),$$

and we readily check that $h \circ [\alpha']_k$ satisfies

$$h \circ [\alpha']_k(\gamma\tau) = \delta(\gamma,\tau)^k h \circ [\alpha']_k(\tau) \qquad \text{for } \gamma \in \Gamma(p)$$

since $\alpha'\Gamma(p)\alpha'^{-1} \subseteq SL(2,\mathbf{Z})$ by Lemma 8.20. Thus the integrand on the right side of (8.62) is invariant under $\Gamma(p)$. Since $\tilde{R}$ is a fundamental domain for $\Gamma(p)$, the right side of (8.62) is unaffected by replacing $\tilde{R}$ by any other such fundamental domain. Thus we have

$$\int_{\tilde{R}} f(\tau)\overline{h \circ [\alpha']_k(\tau)}\sigma^k \frac{d\rho\, d\sigma}{\sigma^2} =$$
$$\frac{1}{[\Gamma(p) : \Gamma(p) \cap \alpha\Gamma(p)\alpha^{-1}]} \int_{\tilde{R}'} f(\tau)\overline{h \circ [\alpha']_k(\tau)}\sigma^k \frac{d\rho\, d\sigma}{\sigma^2}. \quad (8.65)$$

The numerical coefficients on the right sides of (8.64) and (8.65) are equal, by Lemma 8.21. Also the common integrand is invariant under $\Gamma(p) \cap \alpha\Gamma(p)\alpha^{-1}$, and $\tilde{R}'$ and $(\alpha\tilde{R})'$ are two fundamental domains for this group. Therefore the two integrals are equal. Tracing back, we see that (8.62) is a valid equality, and the theorem follows.

We have now shown that the Hecke operators $T_k(n)$ are a commuting family of self adjoint operators on the space S_k of cusp forms for $SL(2,\mathbf{Z})$ of weight k. Consequently S_k has an orthogonal basis of simultaneous eigenvectors.

8. INTERACTION WITH PETERSSON INNER PRODUCT

Proposition 8.23. Let $f \in S_k$ be a simultaneous eigenvector of $T_k(n)$ with $T_k(n)f = \lambda(n)f$. If f has q expansion $f(\tau) = \sum_{n=1}^{\infty} c_n q^n$, then
$$c_n = \lambda(n)c_1. \tag{8.66}$$

Consequently

(a) $f \not\equiv 0$ implies $c_1 \neq 0$

(b) the system of eigenvalues $\{\lambda(n)\}$ determines f up to a scalar.

PROOF. In the q expansion of $T_k(n)f$, the coefficient of q has to be $\lambda(n)c_1$. But Proposition 8.16 gives it as c_n. Thus (8.66) follows. Then (a) and (b) are immediate consequences.

Suppose now that $f \in S_k$ is a simultaneous eigenvector of $T_k(n)$. Proposition 8.23 allows us to normalize f so that the q expansion $f(\tau) = \sum_{n=1}^{\infty} c_n q^n$ has $c_1 = 1$. Then (8.66) says that c_n is the eigenvalue of $T_k(n)$. From Theorem 8.19 we see that

$$c_{p^r} c_p = c_{p^{r+1}} + p^{k-1} c_{p^{r-1}} \quad \text{for } p \text{ prime} \tag{8.67a}$$

$$c_m c_n = c_{mn} \quad \text{if } \mathrm{GCD}(m,n) = 1. \tag{8.67b}$$

Let us return to the L function of f, defined in §5 by

$$L(s,f) = \sum_{n=1}^{\infty} \frac{c_n}{n^s}.$$

According to §VII.2, the condition (8.67b) means that $L(s,f)$ has an Euler product. By Corollary 7.8, (8.67a) means that the Euler product is quadratic, the pth factor being

$$\frac{1}{1 - c_p p^{-s} + p^{k-1-2s}}.$$

(Here we are taking $d_p = -p^{k-1}$ in (7.23).) We summarize as follows.

Theorem 8.24 (Hecke-Petersson). The space S_k of cusp forms has an orthogonal basis of simultaneous eigenvectors under the Hecke operators $T_k(n)$. Each such eigenvector f can be normalized so that its q expansion $f(\tau) = \sum_{n=1}^{\infty} c_n q^n$ has $c_1 = 1$. With such a normalization the coefficients c_n satisfy (8.67). Moreover, the L function $L(s,f)$ has an Euler product expansion

$$L(s,f) = \prod_{p \text{ prime}} \left[\frac{1}{1 - c_p p^{-s} + p^{k-1-2s}} \right] \tag{8.68}$$

convergent for $\mathrm{Re}\, s > \frac{k}{2} + 1$.

CHAPTER IX

MODULAR FORMS FOR HECKE SUBGROUPS

1. Hecke Subgroups

In §VIII.8 we already defined the **principal congruence subgroup** $\Gamma(N)$ of $SL(2,\mathbb{Z})$, for any integer $N \geq 1$, by

$$\Gamma(N) = \left\{ \begin{pmatrix} a & b \\ c & d \end{pmatrix} \in SL(2,\mathbb{Z}) \,\bigg|\, \begin{pmatrix} a & b \\ c & d \end{pmatrix} \equiv \begin{pmatrix} 1 & 0 \\ 0 & 1 \end{pmatrix} \mod N \right\}. \tag{9.1}$$

This is the kernel of the reduction-modulo-N homomorphism carrying $SL(2,\mathbb{Z})$ to $SL(2,\mathbb{Z}_N)$ and therefore is a normal subgroup of finite index in $SL(2,\mathbb{Z})$. The **Hecke subgroups** are

$$\Gamma_0(N) = \left\{ \begin{pmatrix} a & b \\ c & d \end{pmatrix} \in SL(2,\mathbb{Z}) \,\bigg|\, c \equiv 0 \mod N \right\}. \tag{9.2}$$

These satisfy

$$\Gamma(N) \subseteq \Gamma_0(N) \subseteq SL(2,\mathbb{Z}) \tag{9.3}$$

and hence have finite index in $SL(2,\mathbb{Z})$.

We can compute the index of $\Gamma_0(N)$ by relating this subgroup to the set $M(N)$ of integer matrices of determinant N. Let $M^*(N)$ be the subset of **primitive** such matrices $\begin{pmatrix} a & b \\ c & d \end{pmatrix}$, those with $\gcd(a,b,c,d) = 1$.

Lemma 9.1. With $\Gamma = SL(2,\mathbb{Z})$,

$$M^*(n) = \Gamma \begin{pmatrix} n & 0 \\ 0 & 1 \end{pmatrix} \Gamma = \bigcup_\alpha \Gamma\alpha \tag{9.4}$$

disjointly, where α runs over all integer matrices $\begin{pmatrix} a & b \\ 0 & d \end{pmatrix}$ with $ad = n$, $d > 0$, $0 \leq b < d$, and $\gcd(a,b,d)=1$. The number of such right cosets $\Gamma\alpha$ is

$$[M^*(n) : \Gamma] = n \prod_{\substack{p \text{ prime} \\ p|n}} (1 + \tfrac{1}{p}). \tag{9.5}$$

1. HECKE SUBGROUPS

PROOF. If we multiply a nonprimitive member of $M(n)$ on either side by a member of Γ, the result is still nonprimitive. Taking complements, we see that $M^*(n)$ is preserved under these operations.

The conclusion that $M^*(n) = \Gamma \begin{pmatrix} n & 0 \\ 0 & 1 \end{pmatrix} \Gamma$ is elementary divisor theory: In the proof that a doubly generated subgroup of a free abelian group of rank 2 is free abelian, one does integer row and column operations on a 2-by-2 integer matrix and arrives at a diagonal matrix. Starting from $\begin{pmatrix} a' & b' \\ c' & d' \end{pmatrix}$, we can do those operations here and arrive at $\begin{pmatrix} a & 0 \\ 0 & d \end{pmatrix}$ with $\text{GCD}(a, d) = 1$. This answer corresponds to having $\mathbf{Z}_a \oplus \mathbf{Z}_d$ as the quotient of the free abelian groups. Since $\text{GCD}(a, d) = 1$, $\mathbf{Z}_a \oplus \mathbf{Z}_d \cong \mathbf{Z}_n$, and the theory shows we could have arrived at $\begin{pmatrix} n & 0 \\ 0 & 1 \end{pmatrix}$.

For the conclusion $M^*(n) = \bigcup_\alpha \Gamma \alpha$, we can apply Lemma 8.15 to $\begin{pmatrix} a' & b' \\ c' & d' \end{pmatrix} \in M^*(n)$ and see that $\begin{pmatrix} a' & b' \\ c' & d' \end{pmatrix}$ is in a unique $\Gamma \alpha$ such that $\alpha = \begin{pmatrix} a & b \\ 0 & d \end{pmatrix}$, $ad = n$, $d > 0$, and $0 \leq b < d$. Since α must be primitive, we have also $\text{GCD}(a, b, d) = 1$. Hence $M^*(n) = \bigcup_\alpha \Gamma \alpha$ in the asserted fashion.

For the index formula (9.5), we first show that the index is multiplicative. In fact, let $\text{GCD}(n, m) = 1$. Write $M^*(n) = \bigcup_i \Gamma \alpha_i$ and $M^*(m) = \bigcup_i \Gamma \beta_i$ as in (9.4). It is clear that $M^*(nm) \supseteq \bigcup_{i,j} \Gamma \alpha_i \beta_j$. We shall show that equality holds and that the $\Gamma \alpha_i \beta_j$ are disjoint. In fact, if δ is in $M^*(nm)$, then

$$\delta = \gamma_1 \begin{pmatrix} nm & 0 \\ 0 & 1 \end{pmatrix} \gamma_2 \quad \text{by (9.4)}$$

$$= \gamma_1 \begin{pmatrix} n & 0 \\ 0 & 1 \end{pmatrix} \begin{pmatrix} m & 0 \\ 0 & 1 \end{pmatrix} \gamma_2$$

$$= \gamma_1 \begin{pmatrix} n & 0 \\ 0 & 1 \end{pmatrix} \gamma_3 \beta_j \quad \text{for some } \gamma_3 \in \Gamma \text{ and some } j$$

$$= \gamma_4 \alpha_i \beta_j \quad \text{for some } \gamma_4 \in \Gamma \text{ and some } i.$$

Thus $M^*(nm) = \bigcup_{i,j} \Gamma \alpha_i \beta_j$. If $\Gamma \alpha_i \beta_j = \Gamma \alpha_{i'} \beta_{j'}$, then $\alpha_i \beta_j = \gamma \alpha_{i'} \beta_{j'}$ and $\beta_{j'} \beta_j^{-1} \in \alpha_i^{-1} \Gamma \alpha_{i'}$. From this relation we see that $\beta_{j'} \beta_j^{-1}$ has entries in $m^{-1}\mathbf{Z}$ and also in $n^{-1}\mathbf{Z}$. Since $\text{GCD}(m, n) = 1$, $\beta_{j'} \beta_j^{-1}$ is in Γ. Thus $\beta_j = \beta_{j'}$. Then $\alpha_i = \gamma \alpha_{i'}$ and $\alpha_i = \alpha_{i'}$.

To complete the proof of the index formula, we may take $n = p^r$ with p prime. We can simply count the number of α's in (9.4). The number

is
$$p^r + (p^{r-1} - p^{r-2}) + (p^{r-2} - p^{r-3}) + \cdots + (p-1) + 1 = p^r + p^{r-1},$$
as required.

Lemma 9.2. For $\Gamma = SL(2, \mathbf{Z})$,
$$\Gamma \cap \begin{pmatrix} N & 0 \\ 0 & 1 \end{pmatrix}^{-1} \Gamma \begin{pmatrix} N & 0 \\ 0 & 1 \end{pmatrix} = \Gamma_0(N).$$

PROOF. The left side consists of all members of Γ of the form
$$\begin{pmatrix} N^{-1} & 0 \\ 0 & 1 \end{pmatrix} \begin{pmatrix} a & b \\ c & d \end{pmatrix} \begin{pmatrix} N & 0 \\ 0 & 1 \end{pmatrix} = \begin{pmatrix} a & bN^{-1} \\ cN & d \end{pmatrix}$$
with $\begin{pmatrix} a & b \\ c & d \end{pmatrix} \in \Gamma$. Hence the left side is contained in the right side. If $\begin{pmatrix} a & b \\ c & d \end{pmatrix}$ is in $\Gamma_0(N)$, then $\begin{pmatrix} a & b \\ c & d \end{pmatrix}$ is in Γ and
$$\begin{pmatrix} a & b \\ c & d \end{pmatrix} = \begin{pmatrix} N & 0 \\ 0 & 1 \end{pmatrix}^{-1} \begin{pmatrix} a & bN \\ cN^{-1} & d \end{pmatrix} \begin{pmatrix} N & 0 \\ 0 & 1 \end{pmatrix}$$
is in $\begin{pmatrix} N & 0 \\ 0 & 1 \end{pmatrix}^{-1} \Gamma \begin{pmatrix} N & 0 \\ 0 & 1 \end{pmatrix}$.

Proposition 9.3. $\Gamma = SL(2, \mathbf{Z})$ acts transitively by right multiplication on $\Gamma \backslash M^*(N)$ with isotropy subgroup $\Gamma_0(N)$ at the coset $\Gamma \begin{pmatrix} N & 0 \\ 0 & 1 \end{pmatrix}$. Therefore
$$[\Gamma : \Gamma_0(N)] = N \prod_{\substack{p \text{ prime} \\ p|N}} (1 + \tfrac{1}{p}). \tag{9.6}$$

PROOF. The action is transitive by (9.4). By Lemma 9.2, we are to show that the isotropy subgroup at $\Gamma \begin{pmatrix} N & 0 \\ 0 & 1 \end{pmatrix}$ is $\Gamma \cap \begin{pmatrix} N & 0 \\ 0 & 1 \end{pmatrix}^{-1} \Gamma \begin{pmatrix} N & 0 \\ 0 & 1 \end{pmatrix}$. If $\gamma \in \Gamma$ is in the intersection, then $\gamma = \begin{pmatrix} N & 0 \\ 0 & 1 \end{pmatrix}^{-1} \gamma' \begin{pmatrix} N & 0 \\ 0 & 1 \end{pmatrix}$ and
$$\Gamma \begin{pmatrix} N & 0 \\ 0 & 1 \end{pmatrix} \gamma = \Gamma \gamma' \begin{pmatrix} N & 0 \\ 0 & 1 \end{pmatrix} = \Gamma \begin{pmatrix} N & 0 \\ 0 & 1 \end{pmatrix}.$$
Hence the intersection is contained in the isotropy subgroup. In the reverse direction, if γ is in the isotropy subgroup, then γ is a member of Γ with
$$\Gamma \begin{pmatrix} N & 0 \\ 0 & 1 \end{pmatrix} \gamma = \Gamma \begin{pmatrix} N & 0 \\ 0 & 1 \end{pmatrix},$$

1. HECKE SUBGROUPS

i.e.,
$$\Gamma \begin{pmatrix} N & 0 \\ 0 & 1 \end{pmatrix} \gamma \begin{pmatrix} N & 0 \\ 0 & 1 \end{pmatrix}^{-1} = \Gamma.$$

This says $\begin{pmatrix} N & 0 \\ 0 & 1 \end{pmatrix} \gamma \begin{pmatrix} N & 0 \\ 0 & 1 \end{pmatrix}^{-1}$ is in Γ. So γ is in the intersection.
It follows that
$$|\Gamma \backslash M^*(N)| = |\Gamma_0(N) \backslash \Gamma|,$$
and the index formula (9.6) is then a consequence of (9.5).

EXAMPLE 1. Let $N = p$ be prime. Then Proposition 9.3 gives $|\Gamma/\Gamma_0(p)| = p + 1$, and coset representatives for the left coset decomposition $\Gamma = \bigcup_{j=0}^{p} \beta_j \Gamma_0(p)$ may be taken as

$$\beta_p = \begin{pmatrix} 1 & 0 \\ 0 & 1 \end{pmatrix} \quad \text{and} \quad \beta_j = \begin{pmatrix} j & 1 \\ -1 & 0 \end{pmatrix} \quad \text{for } 0 \leq j < p. \quad (9.7)$$

To see that these elements represent distinct cosets, we have only to observe that $\beta_p^{-1} \beta_j = \beta_j$ is not in $\Gamma_0(p)$ and that

$$\beta_i^{-1} \beta_j = \begin{pmatrix} 0 & -1 \\ 1 & i \end{pmatrix} \begin{pmatrix} j & 1 \\ -1 & 0 \end{pmatrix} = \begin{pmatrix} 1 & 0 \\ j-i & 1 \end{pmatrix}$$

is not in $\Gamma_0(p)$ unless $i = j$.

EXAMPLE 2. Let $N = p^r$ with p prime. One easily checks that coset representatives for the left coset decomposition $\Gamma = \bigcup \beta \Gamma_0(p)$ may be taken as

$$\begin{pmatrix} 1 & 0 \\ pl & 1 \end{pmatrix} \quad \text{with } 0 \leq l < p^{r-1} \quad (9.8a)$$

and

$$\begin{pmatrix} j & 1 \\ -1 & 0 \end{pmatrix} \quad \text{with } 0 \leq j < p^r. \quad (9.8b)$$

Let R be the usual fundamental domain in $\mathcal{H}$ for $SL(2,\mathbb{Z})$. (If there is a need to be quite precise, we exclude the part of the boundary where $\text{Re}\,\tau > 0$, so that no two distinct points of R are congruent.) Write the left coset decomposition of $SL(2,\mathbb{Z})/\Gamma_0(N)$ as

$$SL(2,\mathbb{Z}) = \bigcup_j \beta_j \Gamma_0(N),$$

and put
$$R_N = \bigcup_j \beta_j^{-1} R.$$

Then R_N is a fundamental domain for $\Gamma_0(N)$. [In fact, if $\tau \in \mathcal{H}$ is given, we choose $\beta \in SL(2,\mathbf{Z})$ with $\beta\tau \in R$. If we write $\beta = \beta_j \gamma$ with $\gamma \in \Gamma_0(N)$, then $\gamma\tau$ is in $\beta_j^{-1} R \subseteq R_N$. So every point of $\mathcal{H}$ is congruent to a point of R_N. Uniqueness is proved similarly.]

A picture of this fundamental domain for $\Gamma_0(2)$ appears in Figure 9.1. We use the β_j's of Example 1 above, namely

$$\beta_2 = \begin{pmatrix} 1 & 0 \\ 0 & 1 \end{pmatrix}, \qquad \beta_0 = \begin{pmatrix} 0 & 1 \\ -1 & 0 \end{pmatrix}, \qquad \beta_1 = \begin{pmatrix} 1 & 1 \\ -1 & 0 \end{pmatrix}.$$

In terms of $S = \begin{pmatrix} 0 & 1 \\ -1 & 0 \end{pmatrix}$ and $T = \begin{pmatrix} 1 & 1 \\ 0 & 1 \end{pmatrix}$, R_2 is comprised of

$$R, \qquad \beta_0^{-1} R = S(R), \qquad \beta_1^{-1} R = ST(R).$$

The image under each nontrivial β^{-1} of the part of R where Im τ is large contributes to a set in the form of a cusp above the real line. Also we regard R itself as contributing a cusp. A feature of this diagram is that $S(R)$ and $ST(R)$ contribute to the same cusp. Thus there are only two cusps, at ∞ and 0, even though $[SL(2,\mathbf{Z}) : \Gamma_0(2)] = 3$.

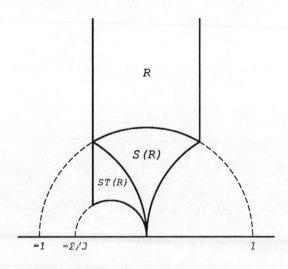

FIGURE 9.1. Fundamental domain for $\Gamma_0(2)$

2. Modular and Cusp Forms

It will help now to make systematic use of the notation

$$\delta(g, \tau) = c\tau + d \qquad \text{for } g = \begin{pmatrix} a & b \\ c & d \end{pmatrix} \text{ in } SL(2, \mathbf{R}). \tag{9.9}$$

This notation was introduced in the proof of Corollary 8.4. We recall that $\delta(g, \tau)$ satisfies

$$\begin{aligned} \delta(g_1 g_2, \tau) &= \delta(g_1, g_2 \tau) \delta(g_2, \tau) \\ \delta(g, \tau)^{-1} &= \delta(g^{-1}, g\tau). \end{aligned} \tag{9.10}$$

As in (8.47), we shall write

$$f \circ [\beta]_k(\tau) = \delta(\beta, \tau)^{-k} f(\beta \tau) \qquad \text{for } \beta \in SL(2, \mathbf{R}).$$

For a matrix α of positive determinant, we let $\alpha^\# = (\det \alpha)^{-1/2} \alpha$. Then $\alpha^\#$ has determinant one, and we define

$$f \circ [\alpha]_k = f \circ [\alpha^\#]_k. \tag{9.11}$$

This construction is a group operation, in that

$$f \circ [\alpha_1 \alpha_2]_k = (f \circ [\alpha_1]_k) \circ [\alpha_2]_k. \tag{9.12}$$

An **unrestricted modular form** of weight $k \in \mathbf{Z}$ and level $N \geq 1$ is an analytic function f on $\mathcal{H}$ with

$$f(\gamma \tau) = \delta(\gamma, \tau)^k f(\tau) \qquad \text{for all } \gamma \in \Gamma_0(N). \tag{9.13}$$

Since $\begin{pmatrix} -1 & 0 \\ 0 & -1 \end{pmatrix}$ is in $\Gamma_0(N)$, we may as well assume k is an even integer. Such a function f is a **modular form** if it is "holomorphic at the cusps," and it is a **cusp form** if it "vanishes at the cusps." We need to explain these conditions on f at the "cusps."

If R is the standard fundamental domain for $SL(2, \mathbf{Z})$, then we can regard ∞ as a boundary point of $\mathcal{H}$ that is also a limit point of R. Each set $\beta^{-1} R$, with $\beta \in SL(2, \mathbf{Z})$, is also a fundamental domain for $SL(2, \mathbf{Z})$. If $\beta^{-1} = \begin{pmatrix} a & b \\ c & d \end{pmatrix}$, then $\beta^{-1}(\infty) = \frac{a}{c}$ is a boundary point of $\mathcal{H}$ and a limit point of $\beta^{-1} R$. Because of the geometry of $\beta^{-1} R$ (as in Figure

9.1), we say that $\frac{a}{c}$ is a **cusp** of β^{-1}. To eliminate the dependence on β in this definition of "cusp," it is customary to define "cusps" by taking $\mathbf{Q} \cup \{\infty\}$ and identifying those equivalent under the group in question. With $SL(2, \mathbf{Z})$, there is only one equivalence class, hence only one cusp.

Now let us pass to $\Gamma_0(N)$. With the above conventions, a **cusp** is an equivalence class of $\mathbf{Q} \cup \{\infty\}$ under the action of $\Gamma_0(N)$. If we write the coset space $SL(2, \mathbf{Z})/\Gamma_0(N)$ as

$$SL(2, \mathbf{Z}) = \bigcup_j \beta_j \Gamma_0(N), \tag{9.14}$$

then each $\beta_j^{-1}(\infty)$ represents a cusp. [In fact, let $\frac{a}{c}$ be in $\mathbf{Q}$ with $\mathrm{GCD}(a, c) = 1$. Choose b and d with $ad - bc = 1$, so that $\beta^{-1} = \begin{pmatrix} a & b \\ c & d \end{pmatrix}$ is in $SL(2, \mathbf{Z})$ and has $\beta^{-1}(\infty) = \frac{a}{c}$. Choose $\gamma \in \Gamma_0(N)$ and β_j as above with $\beta = \beta_j \gamma$. Then $\gamma^{-1}(\beta_j^{-1}(\infty)) = \beta^{-1}(\infty) = \frac{a}{c}$, so that $\frac{a}{c}$ and $\beta_j^{-1}(\infty)$ represent the same cusp.]

There can be duplication among the cusps $\beta_j^{-1}(\infty)$. This is what happens in Figure 9.1. The condition that $\beta_i^{-1}(\infty)$ and $\beta_j^{-1}(\infty)$ represent the same cusp is that $\beta_i^{-1}(\infty) = \gamma \beta_j^{-1}(\infty)$ for some $\gamma \in \Gamma_0(N)$. This means that $\beta_i \gamma \beta_j^{-1}$ fixes ∞, hence is of the form $\pm \begin{pmatrix} 1 & l \\ 0 & 1 \end{pmatrix}$ for some sign and some l. A sufficient condition is that $\beta_i \beta_j^{-1}$ is of the form $\pm \begin{pmatrix} 1 & l \\ 0 & 1 \end{pmatrix}$. In $\Gamma_0(p)$ this condition is satisfied by all pairs $\beta_1, \ldots, \beta_p$ in (9.7). But the condition is not necessary. In $\Gamma_0(8)$, which is discussed in Example 2, $\beta_i = \begin{pmatrix} 1 & 0 \\ 2 & 1 \end{pmatrix}$ and $\beta_j = \begin{pmatrix} 1 & 0 \\ 6 & 1 \end{pmatrix}$ yield the same cusp (take $\gamma = \begin{pmatrix} 7 & 1 \\ -8 & -1 \end{pmatrix}$), but $\beta_i \beta_j^{-1}$ is not upper triangular.

Suppose f is an unrestricted modular form of weight k and level N. Let us examine the behavior of f near a cusp $\beta^{-1}(\infty)$. We consider the function $f \circ [\beta^{-1}]_k(\tau)$, whose behavior near $\tau = \infty$ (i.e., for Im τ large) reflects the behavior of f near $\beta^{-1}(\infty)$. Since

$$\beta^{-1} \begin{pmatrix} 1 & N \\ 0 & 1 \end{pmatrix} \beta \in \beta^{-1} \Gamma(N) \beta \cap \Gamma(N) \subseteq \Gamma_0(N), \tag{9.15}$$

we have

$$f \circ [\beta^{-1}]_k(\tau + N) = (f \circ [\beta^{-1}]_k) \circ \left[\begin{pmatrix} 1 & N \\ 0 & 1 \end{pmatrix}\right]_k(\tau)$$

$$= (f \circ [\beta^{-1} \begin{pmatrix} 1 & N \\ 0 & 1 \end{pmatrix} \beta]_k) \circ [\beta^{-1}]_k(\tau) \quad \text{by (9.12)}$$

$$= f \circ [\beta^{-1}]_k(\tau) \quad \text{by (9.13)}$$

2. MODULAR AND CUSP FORMS

Arguing as for (8.7), we see that

$$f \circ [\beta^{-1}]_k(\tau) = \sum_{n=-\infty}^{\infty} c_n^{(\beta)} q_N^n \quad \text{with } q_N = e^{2\pi i \tau/N}. \tag{9.16}$$

We say that f is **holomorphic at the cusp** $\beta^{-1}(\infty)$ if $c_n^{(\beta)}$ occurs only for $n \geq 0$ and f **vanishes at the cusp** $\beta^{-1}(\infty)$ if $c_n^{(\beta)}$ occurs only for $n \geq 1$.

If β is replaced by a different coset representative $\beta\gamma$ with $\gamma \in \Gamma_0(N)$, then

$$f \circ [(\beta\gamma)^{-1}]_k = (f \circ [\gamma^{-1}]_k) \circ [\beta^{-1}]_k = f \circ [\beta^{-1}]_k,$$

so that the expansion (9.16) is completely unchanged. Therefore our definitions do not depend on the choice of coset representatives.

But more is true. Suppose $\beta_i^{-1}(\infty)$ and $\beta_j^{-1}(\infty)$ represent the same cusp, i.e., $\beta_i \gamma \beta_j^{-1} = \pm \begin{pmatrix} 1 & l \\ 0 & 1 \end{pmatrix}$ for some $\gamma \in \Gamma_0(N)$ and $l \in \mathbf{Z}$. Then we have

$$f \circ [\beta_i^{-1}]_k(\begin{pmatrix} 1 & l \\ 0 & 1 \end{pmatrix} \tau) = (f \circ [\beta_i^{-1}]_k) \circ [\pm \begin{pmatrix} 1 & l \\ 0 & 1 \end{pmatrix}]_k(\tau) \quad \text{since } k \text{ is even}$$

$$= f \circ [\pm \beta_i^{-1} \begin{pmatrix} 1 & l \\ 0 & 1 \end{pmatrix}]_k(\tau)$$

$$= f \circ [\gamma \beta_j^{-1}]_k(\tau)$$

$$= (f \circ [\gamma]_k) \circ [\beta_j^{-1}]_k(\tau)$$

$$= f \circ [\beta_j^{-1}]_k(\tau).$$

Thus changing from $\beta = \beta_j$ to $\beta = \beta_i$ makes $c_n^{(\beta)}$ in (9.16) get multiplied by $e^{2\pi i l n/N}$. The validity of "holomorphic at the cusp" and "vanishes at the cusp" are unaffected.

The expansion (9.16) involves $q_N = e^{2\pi i \tau/N}$, which is periodic in τ of period N. It can happen that a smaller period will work for all f. Namely if $\beta^{-1}(\infty)$ is a cusp, then

$$\{l \in \mathbf{Z} \mid \beta^{-1} \begin{pmatrix} 1 & l \\ 0 & 1 \end{pmatrix} \beta \in \Gamma_0(N)\} \tag{9.17}$$

is a subgroup of $\mathbf{Z}$ that contains N, according to (9.15). Let h be its positive generator. Going over the above argument shows that

$$f \circ [\beta^{-1}]_k(\tau + h) = f \circ [\beta^{-1}]_k(\tau).$$

Thus we are led to an expansion in powers of $q_h = e^{2\pi i \tau/h}$. We call h the **width** of the cusp.

EXAMPLES.
(1) The width of the cusp at ∞ is 1.
(2) For $\Gamma_0(p)$ with p prime, the width of the cusp at 0 is p. (See Figure 9.1.)
(3) For $\Gamma_0(4)$, there are three cusps—the one at ∞ of width 1 corresponding to one left coset in (9.14), one at 0 of width 4 corresponding to four left cosets, and one at $-\frac{1}{2}$ of width 1 corresponding to one left coset.

Proposition 9.4. Let $\beta^{-1}(\infty)$ be a cusp of $\Gamma_0(N)$, where β is in $SL(2,\mathbb{Z})$. Then there exists a unique $\alpha_i = \begin{pmatrix} a & b \\ 0 & d \end{pmatrix}$ in $M^*(N)$ with $ad = N$, $d > 0$, $0 \leq b < d$, and $\text{GCD}(a,b,d) = 1$ such that

$$\begin{pmatrix} N & 0 \\ 0 & 1 \end{pmatrix} \beta^{-1} = \gamma \alpha_i \qquad (9.18)$$

for some γ in $SL(2,\mathbb{Z})$. Moreover, the width of the cusp equals $d/\text{GCD}(a,d)$.

PROOF. The first conclusion follows from Lemma 9.1. For the second conclusion, we conjugate $\begin{pmatrix} 1 & l \\ 0 & 1 \end{pmatrix}$ by (9.18), obtaining

$$\begin{pmatrix} N & 0 \\ 0 & 1 \end{pmatrix} \beta^{-1} \begin{pmatrix} 1 & l \\ 0 & 1 \end{pmatrix} \beta \begin{pmatrix} N & 0 \\ 0 & 1 \end{pmatrix}^{-1}$$
$$= \frac{1}{N} \gamma \begin{pmatrix} a & b \\ 0 & d \end{pmatrix} \begin{pmatrix} 1 & l \\ 0 & 1 \end{pmatrix} \begin{pmatrix} d & -b \\ 0 & a \end{pmatrix} \gamma^{-1} = \gamma \begin{pmatrix} 1 & al/d \\ 0 & 1 \end{pmatrix} \gamma^{-1}.$$

Thus

$$\beta^{-1} \begin{pmatrix} 1 & l \\ 0 & 1 \end{pmatrix} \beta = \begin{pmatrix} N & 0 \\ 0 & 1 \end{pmatrix}^{-1} \left[\gamma \begin{pmatrix} 1 & al/d \\ 0 & 1 \end{pmatrix} \gamma^{-1} \right] \begin{pmatrix} N & 0 \\ 0 & 1 \end{pmatrix}.$$

The left side is in Γ, and this equality shows it is in

$$\begin{pmatrix} N & 0 \\ 0 & 1 \end{pmatrix}^{-1} SL(2,\mathbb{Z}) \begin{pmatrix} N & 0 \\ 0 & 1 \end{pmatrix}$$

if and only if al/d is an integer. By Lemma 9.2, it is in $\Gamma_0(N)$ if and only if al/d is an integer. Hence l is in the subgroup (9.17) defining width if and only if al/d is an integer, and it follows that $l = d/\text{GCD}(a,d)$.

The spaces of modular and cusp forms of weight k and level N will be denoted $M_k(\Gamma_0(N))$ and $S_k(\Gamma_0(N))$, respectively. These spaces are 0 for k odd.

3. Examples of Modular Forms

There are several constructions of modular forms in one situation from modular forms in another. Typically the new transformation law (9.13) is a consequence of some computation with matrices. Behavior at the cusps is handled by the following result.

Proposition 9.5. Suppose that f is analytic in the upper half plane and, for each β in $SL(2,\mathbb{Z})$, $f \circ [\beta^{-1}]_k(\tau)$ has a holomorphic q_N expansion at ∞, where $q_N = e^{2\pi i \tau / N}$. If α is in $M(m)$, then $(f \circ [\alpha]_k) \circ [\beta^{-1}]_k(\tau)$ has a holomorphic q_{mN} expansion at ∞ for each β in $SL(2,\mathbb{Z})$. If all of these expansions for f have 0 constant term, then all of these expansions for $f \circ [\alpha]_k$ have 0 constant term.

PROOF. Scaling α so that its entries are relatively prime, we can apply Lemma 9.1 we and rescale to write $\alpha \beta^{-1} = \gamma^{-1} \begin{pmatrix} a & b \\ 0 & d \end{pmatrix}$ with γ in $SL(2,\mathbb{Z})$ and $\begin{pmatrix} a & b \\ 0 & d \end{pmatrix}$ in $M(m)$. Then

$$(f \circ [\alpha]_k) \circ [\beta^{-1}]_k(\tau) = (f \circ [\gamma^{-1}]_k) \circ \left[\begin{pmatrix} a & b \\ 0 & d \end{pmatrix}\right]_k(\tau). \qquad (9.19)$$

By assumption we can write $f \circ [\gamma^{-1}]_k(\tau) = \sum_{n=0}^{\infty} c_n e^{2\pi i n \tau / N}$. Hence the right side of (9.19) is

$$= (m^{-1/2} d)^{-k} f \circ [\gamma^{-1}]_k \left(\tfrac{a\tau+b}{d}\right)$$

$$= (m^{-1/2} d)^{-k} \sum_{n=0}^{\infty} c_n e^{2\pi i n b / (Nd)} e^{2\pi i n a^2 \tau / (mN)}, \qquad (9.20)$$

and the proposition follows.

The first two examples give methods for obtaining modular forms of one level from those at a lower level, with the weight remaining the same.

EXAMPLE 1. If f is in $M_k(\Gamma_0(N))$, then f is in $M_k(\Gamma_0(rN))$ for $r \geq 1$, and similarly for $S_k(\Gamma_0(N))$ and $S_k(\Gamma_0(rN))$. The new transformation law (9.13) is valid because a subset of γ's is involved, and the cusp conditions are already built into the properties relative to $\Gamma_0(N)$.

EXAMPLE 2. If $f(\tau)$ is in $M_k(\Gamma_0(N))$, then $f(r\tau)$ is in $M_k(\Gamma_0(rN))$ for $r \geq 1$. The transformation law (9.13) follows from Lemma 8.20:

$$(f \circ [\begin{pmatrix} r & 0 \\ 0 & 1 \end{pmatrix}]_k) \circ [\gamma]_k(\tau) = (f \circ [\begin{pmatrix} r & 0 \\ 0 & 1 \end{pmatrix} \gamma \begin{pmatrix} r & 0 \\ 0 & 1 \end{pmatrix}^{-1}]_k) \circ [\begin{pmatrix} r & 0 \\ 0 & 1 \end{pmatrix}]_k(\tau)$$
$$= f \circ [\begin{pmatrix} r & 0 \\ 0 & 1 \end{pmatrix}]_k(\tau)$$

for γ in $\Gamma_0(rN)$. The conditions at the cusps are satisfied because of Proposition 9.4 with $\alpha = \begin{pmatrix} r & 0 \\ 0 & 1 \end{pmatrix}$.

There is one more simple construction to obtain new modular forms from old ones. This one changes the weight.

EXAMPLE 3. The product of two modular forms for $\Gamma_0(N)$, one of weight k and one of weight l, is a modular form of weight $k + l$. If one is a cusp form, so is the product.

Now we give two completely different constructions, largely without proof.

EXAMPLE 4. Define

$$G_2(\tau) = \sum_n \sum_m{}' \frac{1}{(m + n\tau)} \quad \text{for } \tau \in \mathcal{H}, \tag{9.21}$$

where the inner sum is taken over m such that $(m,n) \neq (0,0)$. One can prove that this series converges nicely when summed in the order indicated. Moreover, we have $G_2(\tau + 1) = G_2(\tau)$, and one can prove that

$$G_2(-1/\tau) = \tau^2 G_2(\tau) - 2\pi i \tau.$$

Then it follows that

$$G_2(\tfrac{a\tau+b}{c\tau+d}) = (c\tau + d)^2 G_2(\tau) - 2\pi i c(c\tau + d) \tag{9.22}$$

for $\begin{pmatrix} a & b \\ c & d \end{pmatrix}$ in $SL(2, \mathbf{Z})$. The argument of Proposition 8.1 shows that

$$G_2(\tau) = 2\zeta(2) - 8\pi^2 \sum_{n=1}^{\infty} \sigma_1(n) q^n. \tag{9.23}$$

Put

$$G_2^{(2)}(\tau) = G_2(\tau) - 2G_2(2\tau). \tag{9.24}$$

Combining (9.22) and the identity

$$\begin{pmatrix} 2 & 0 \\ 0 & 1 \end{pmatrix} \begin{pmatrix} a & b \\ c & d \end{pmatrix} \begin{pmatrix} 2 & 0 \\ 0 & 1 \end{pmatrix}^{-1} = \begin{pmatrix} a & 2b \\ c/2 & d \end{pmatrix},$$

we see that $G_2^{(2)} \circ [\gamma]_2 = G_2^{(2)}$ for all $\gamma \in \Gamma_0(2)$. Also it is not hard to see from (9.23) that $G_2^{(2)}$ is holomorphic at the cusps. Therefore $G_2^{(2)}$ is in $M_2(\Gamma_0(2))$.

EXAMPLE 5 (Hecke). If $p > 3$ is prime and $\eta(\tau)$ is as in Corollary 8.9, then $(\eta^p(\tau)/\eta(p\tau))^2$ is in $M_{p-1}(\Gamma_0(p))$. The difficult thing to prove is the transformation law. To handle the behavior at the cusps, we consider the 12th power, which is $\Delta^p(\tau)/\Delta(p\tau)$.

Since p is prime, the only cusps are at 0 and ∞. For $\Delta(\tau)$, the q expansion at ∞ begins with a multiple of q; the same thing is true for the q expansion at 0 since $\Delta \circ [\begin{pmatrix} 0 & -1 \\ 1 & 0 \end{pmatrix}]_{12} = \Delta$. For $\Delta(p\tau)$, the q expansion at ∞ clearly begins with a multiple of q^p. For the q expansion at 0, we follow the steps of the proof of Proposition 9.5, using $\alpha = \begin{pmatrix} p & 0 \\ 0 & 1 \end{pmatrix}$, $\beta^{-1} = \gamma^{-1} = \begin{pmatrix} 0 & -1 \\ 1 & 0 \end{pmatrix}$, and $\begin{pmatrix} a & b \\ 0 & d \end{pmatrix} = \begin{pmatrix} 1 & 0 \\ 0 & p \end{pmatrix}$. Then (9.20) shows that $(\Delta \circ [\alpha]_k) \circ [\beta^{-1}]_k(\tau)$ begins with $q^{1/p}$. We conclude that $\Delta^p(\tau)/\Delta(p\tau)$ has q expnasion at ∞ holomorphic nonvanishing, while at 0 the first term involves $q^{p-\frac{1}{p}}$.

Hence $(\eta^p(\tau)/\eta(p\tau))^2$ is holomorphic at the cusps. Modifications to it will yield cusp forms. For example, let $p = 12n-1$. Then $(\eta(p\tau)/\eta^p(\tau))^2$ has weight $-(p-1)$, and its q expansions begin with q^0 at ∞ and $q^{-\frac{1}{12}(p-\frac{1}{p})}$ at 0. Multiplying by

$$\Delta^n = \Delta^{\frac{1}{12}(p+1)} = \eta^{2(p+1)},$$

we see that $\eta(p\tau)^2 \eta(\tau)^2$ has weight 2 and has q expansions beginning with q^n at ∞ and with $q^{n - \frac{1}{12}(p - \frac{1}{p})} = q^{(p+1)/(12p)}$ at 0. Hence

$$\eta(p\tau)^2 \eta(\tau)^2$$

is a cusp form of weight 2 when $p \equiv -1 \mod 12$.

4. L Function of a Cusp Form

Let $f \in S_k(\Gamma_0(N))$ be a cusp form, and let $f(\tau) = \sum_{n=1}^{\infty} c_n q^n$ be its q expansion at the cusp ∞. The L function of f is the Dirichlet series

$$L(s, f) = \sum_{n=1}^{\infty} \frac{c_n}{n^s}, \tag{9.25}$$

just as in (8.30), and we have

$$\int_0^\infty f(i\sigma)\sigma^s \frac{d\sigma}{\sigma} = (2\pi)^{-s}\Gamma(s)L(s,f) \qquad (9.26)$$

for all values of s for which the integral converges.

Lemma 9.6. Let $f \in S_k(\Gamma_0(N))$ have q expansion $f(\tau) = \sum_{n=1}^\infty c_n q^n$ at the cusp ∞. Then
(a) the function $\varphi(\tau) = |f(\tau)|\sigma^{k/2}$ is bounded on $\mathcal{H}$ and invariant under $\Gamma_0(N)$
(b) $|c_n| \leq C n^{k/2}$.

PROOF. Write $SL(2,\mathbf{Z})/\Gamma_0(N)$ as in (9.14). First we prove that φ is bounded on R_N. The cusps of R_N are the points $\beta_j^{-1}(\infty)$, and it is enough to prove boundedness near each, i.e., boundedness of $\varphi(\beta_j^{-1}\tau)$ for Im $\tau \geq 2$ and all j. Since f vanishes at the cusp $\beta_j^{-1}(\infty)$, we have

$$f \circ [\beta_j^{-1}]_k(\tau) = \sum_{n=1}^\infty c_n^{(\beta)} e^{2\pi i n \tau/N}$$

and hence

$$|f \circ [\beta_j^{-1}]_k(\tau)| \leq C_j e^{-2\pi(\text{Im }\tau)/N} \qquad \text{for Im } \tau \geq 2. \qquad (9.27)$$

Thus

$$|f \circ [\beta_j^{-1}]_k(\tau)|(\text{Im }\tau)^{k/2} \leq C_j' \qquad \text{for Im } \tau \geq 2. \qquad (9.28)$$

Since

$$\text{Im}(\beta\tau) = \frac{\text{Im }\tau}{|\delta(\beta,\tau)|^2} \qquad \text{for } \beta \in SL(2,\mathbf{R}), \qquad (9.29)$$

we have

$$\varphi(\beta_j^{-1}\tau) = |f(\beta_j^{-1}\tau)|(\text{Im }\beta_j^{-1}\tau)^{k/2}$$
$$= |f \circ [\beta_j^{-1}]_k(\tau)\delta(\beta_j^{-1},\tau)^k| \, (\text{Im }\tau)^{k/2} \, |\delta(\beta_j^{-1},\tau)|^{-k}$$
$$= |f \circ [\beta_j^{-1}]_k(\tau)|(\text{Im }\tau)^{k/2}. \qquad (9.30)$$

In combination with (9.28), (9.30) shows that φ is bounded on R_N. Replacing β_j^{-1} in the calculation (9.30) by a member γ of $\Gamma_0(N)$, we see that φ is invariant under $\Gamma_0(N)$. Hence φ is bounded on $\mathcal{H}$, and (a) is proved. To see that (a) implies (b), we argue just as in Lemma 8.10, starting from (9.28) with $\beta_j = 1$.

4. L FUNCTION OF A CUSP FORM

For $S_k = S_k(\Gamma_0(1))$, we obtained a functional equation for $L(s,f)$ as Theorem 8.11, using the transformation law for f under $\begin{pmatrix} 0 & -1 \\ 1 & 0 \end{pmatrix}$. For $\Gamma_0(N)$ with $N > 1$, the element $\begin{pmatrix} 0 & -1 \\ 1 & 0 \end{pmatrix}$ is not in $\Gamma_0(N)$, and the functional equation needs an extra hypothesis. This hypothesis will be given in terms of the effect of the matrix $\begin{pmatrix} 0 & -1 \\ N & 0 \end{pmatrix}$.

Let $\alpha_N = \begin{pmatrix} 0 & -1 \\ N & 0 \end{pmatrix}$, so that $\alpha_N^\# = \frac{1}{\sqrt{N}} \begin{pmatrix} 0 & -1 \\ N & 0 \end{pmatrix}$. If $\gamma = \begin{pmatrix} a & b \\ c & d \end{pmatrix}$, then $\alpha_N \gamma \alpha_N^{-1} = \begin{pmatrix} d & -c/N \\ -Nb & a \end{pmatrix}$. Therefore

$$\alpha_N \Gamma_0(N) \alpha_N^{-1} \subseteq \Gamma_0(N). \tag{9.31}$$

For $f \in M_k(\Gamma_0(N))$, we consider the map $w_N f = f \circ [\alpha_N]_k$. For $\gamma \in \Gamma_0(N)$, we have

$$(f \circ [\alpha_N]_k) \circ [\gamma]_k = (f \circ [\alpha_N \gamma \alpha_N^{-1}]_k) \circ [\alpha_N]_k = f \circ [\alpha_N]_k,$$

so that $f \circ [\alpha_N]_k$ is an unrestricted modular form of the same weight and level as f.

Let us record the fact that w_N maps $M_k(\Gamma_0(N))$ to itself and $S_k(\Gamma_0(N))$ to itself. This result is a special case of Proposition 9.5.

Proposition 9.7. The map w_N carries $M_k(\Gamma_0(N))$ to itself and $S_k(\Gamma_0(N))$ to itself.

PROOF. If f is in $M_k(\Gamma_0(N))$, we apply Proposition 9.5 with $m = N$ and $\alpha = \alpha_N$. The proposition says that

$$(w_N f) \circ [\beta^{-1}]_k(\tau) = \sum_{n=0}^{\infty} c_n e^{2\pi i n \tau / N^2}. \tag{9.32}$$

But $w_N f$ is an unrestricted modular form for $\Gamma_0(N)$, and hence $(w_N f) \circ [\beta^{-1}]_k(\tau)$ is periodic with period N. Thus the coefficients c_n in (9.32) are 0 unless $N \mid n$, and f is holomorphic at the cusp $\beta^{-1}(\infty)$. Similarly if f is in $S_k(\Gamma_0(N))$, then $w_N f$ is in $S_k(\Gamma_0(N))$.

Taking into account Proposition 9.7 and the identity $(\alpha_N^\#)^2 = -I$, we see that w_N is an involution on $M_k(\Gamma_0(N))$ and on $S_k(\Gamma_0(N))$. These spaces therefore each split as the sum of the eigenspaces for eigenvalues $+1$ and -1. In the case of $S_k(\Gamma_0(N))$, we denote these eigenspaces by $S_k^\pm(\Gamma_0(N))$.

Theorem 9.8 (Hecke). Let $f \in S_k(\Gamma_0(N))$ be a cusp form in one of the eigenspaces $S_k^\varepsilon(\Gamma_0(N))$ of w_N, where $\varepsilon = \pm$. Then $L(s,f)$ is initially defined for $\operatorname{Re} s > \frac{k}{2} + 1$ and extends to be entire in s. Moreover, the function
$$\Lambda(s,f) = N^{s/2}(2\pi)^{-s}\Gamma(s)L(s,f) \tag{9.33a}$$
satisfies the functional equation
$$\Lambda(s,f) = \varepsilon(-1)^{k/2}\Lambda(k-s,f). \tag{9.33b}$$

REMARKS. The proof is similar to that of Theorem 8.11, except that $\begin{pmatrix} 0 & -1 \\ 1 & 0 \end{pmatrix}$ gets replaced by $\alpha^\#$. The integration from 0 to ∞ is broken at $1/\sqrt{N}$ instead of 1 because $i/\sqrt{N}$ is the fixed point of w_N on the imaginary axis.

PROOF. The initial region of convergence for $L(s,f)$ is $\operatorname{Re} s > \frac{k}{2} + 1$, by Lemma 9.6b. Since $w_N f = \varepsilon f$, the inversion law for f under the action of w_N is
$$f\left(\frac{i}{N\sigma}\right) = \varepsilon N^{k/2} i^k \sigma^k f(i\sigma). \tag{9.34}$$
By (9.26) we have
$$\Lambda(s,f) = N^{s/2} \int_0^\infty f(i\sigma)\sigma^{s-1}\, d\sigma. \tag{9.35}$$
From (9.27) we see that
$$\int_{1/\sqrt{N}}^\infty f(i\sigma)\sigma^{s-1}\, d\sigma \tag{9.36}$$
converges for all $s \in \mathbb{C}$ and defines an entire function. We rewrite (9.35) for $\operatorname{Re} s > \frac{k}{2} + 1$ as
$$\Lambda(s,f) = N^{s/2}\int_0^{1/\sqrt{N}} f(i\sigma)\sigma^{s-1}\, d\sigma + N^{s/2}\int_{1/\sqrt{N}}^\infty f(i\sigma)\sigma^{s-1}\, d\sigma.$$
In the first term, we replace σ by $(N\sigma)^{-1}$ and then use (9.34). The result is
$$\Lambda(s,f) = \varepsilon N^{\frac{1}{2}(k-s)} i^k \int_{1/\sqrt{N}}^\infty f(i\sigma)\sigma^{k-s-1}\, d\sigma + N^{s/2}\int_{1/\sqrt{N}}^\infty f(i\sigma)\sigma^{s-1}\, d\sigma. \tag{9.37}$$

In view of (9.36) the first term extends to be entire in s, and hence so does $\Lambda(s, f)$ itself. Since $\Gamma(s)$ is nowhere 0, $L(s, f)$ is entire. Replacing s by $k - s$ in (9.37) and multiplying by $\varepsilon i^k = \varepsilon(-1)^{k/2}$, we obtain

$$\varepsilon(-1)^{k/2}\Lambda(k - s, f)$$
$$= N^{\frac{1}{2}s}\int_{1/\sqrt{N}}^{\infty} f(i\sigma)\sigma^{s-1}\, d\sigma + \varepsilon N^{\frac{1}{2}(k-s)} i^k \int_{1/\sqrt{N}}^{\infty} f(i\sigma)\sigma^{k-s-1}\, d\sigma.$$

Comparison with (9.37) yields (9.33b).

5. Dimensions of Spaces of Cusp Forms

The kind of explicit argument used to prove Theorem 8.6 is not available for $\Gamma_0(N)$. But we can get at the finite-dimensionality of $S_k(\Gamma_0(N))$ rather painlessly, and this finite-dimensionality will be relevant in obtaining simultaneous eigenvectors for the Hecke operators.

Theorem 9.9. The vector space $M_k(\Gamma_0(N))$ is finite-dimensional.

SKETCH OF PROOF. (Some of the steps will be treated in more detail in Chapter XI.) The first step is to make $\mathcal{R} = \Gamma_0(N)\backslash\mathcal{H}$ into a Riemann surface by introducing appropriate local parameters at the fixed points of elements of $\Gamma_0(N)$ of finite order. Then we form a compactification $\mathcal{R}^*$ by adjoining one point for each cusp and using local parameters built from q in the case of the cusp at ∞, or a transformed q_h (h being the width) in the case of the other cusps. The result is a compact Riemann surface.

Let f be in $M_k(\Gamma_0(N))$. Then f^{12} is in $M_{12k}(\Gamma_0(N))$, and so is Δ^k. Hence $F = f^{12}/\Delta^k$ is an analytic function on $\mathcal{H}$ invariant under $\Gamma_0(N)$. The function F descends to a holomorphic function on $\mathcal{R}$ that extends to be meromorphic on $\mathcal{R}^*$. The only possible poles are at worst of order k, coming from the zeros of Δ^k. These poles occur at worst at each cusp, and the number of cusps is $\leq \mu = [SL(2, \mathbb{Z}) : \Gamma_0(N)]$. Hence the number of poles of F, counting multiplicities, is $\leq k\mu$.

It follows that the number of zeros of F, counting multiplicities, is $\leq k\mu$. In fact, the meromorphic differential form $\omega = \dfrac{F'(\tau)d\tau}{F(\tau)}$ is globally well defined on $\mathcal{R}^*$, as a consequence of (8.19). Let us triangulate $\mathcal{R}^*$ so that each singularity of ω is in the interior of one triangle. Let $\Delta_1, \ldots, \Delta_l$ be the 2-simplices in the triangulation. By the Argument Principle,

$$\#\{\text{zeros of } F\} - \#\{\text{poles of } F\} = \frac{1}{2\pi i}\sum_{j=1}^{l}\int_{\partial\Delta_j}\omega,$$

where $\partial\Delta_j$ is the boundary of Δ_j with positive orientation. In the sum on the right, each 1-simplex appears twice, with opposite orientations, and the right side is therefore 0. We conclude that F has $\leq k\mu$ zeros, counting multiplicities.

Now let f and therefore F vary. Select $k\mu + 1$ distinct points $p_1, \ldots, p_{k\mu+1}$ in $\mathcal{R}$ and consider the linear map $M_k(\Gamma_0(N)) \to \mathbf{C}^{k\mu+1}$ given by $f \to (F(p_1), \ldots, F(p_{k\mu+1}))$. By the above, this map has 0 kernel. Hence $\dim M_k(\Gamma_0(N)) \leq k\mu + 1$.

Our interest ultimately will be in $S_2(\Gamma_0(N))$. With more machinery, one can calculate the dimension of $S_2(\Gamma_0(N))$ exactly. We shall say what is involved without giving the details.

The first step is to compute the genus of the compact Riemann surface $\mathcal{R}^*$ in the proof of Theorem 9.9. This is possible because we have a natural map
$$\mathcal{R}^* \longrightarrow (SL(2,\mathbf{Z})\backslash\mathcal{H})^* \tag{9.38}$$
as a consequence of the canonical quotient map
$$\Gamma_0(N)\backslash SL(2,\mathbf{R}) \longrightarrow SL(2,\mathbf{Z})\backslash SL(2,\mathbf{R}).$$
Applying the Riemann-Hurwitz formula to this map gives a formula for the genus of $\mathcal{R}^*$. The space $S_2(\Gamma_0(N))$ is easily seen to be isomorphic to the space of holomorphic differentials on $\mathcal{R}^*$, and the latter space can be shown to have dimension equal to the genus of $\mathcal{R}^*$. We summarize as follows.

Theorem 9.10. The dimension of $S_2(\Gamma_0(N))$ is
$$g = 1 + \frac{\mu(N)}{12} - \frac{\mu_2(N)}{4} - \frac{\mu_3(N)}{3} - \frac{\mu_\infty(N)}{2},$$
where
$$\mu(N) = [SL(2,\mathbf{Z})/\Gamma_0(N)] = N\prod_{p|N}(1+\tfrac{1}{p})$$

$$\mu_2(N) = \begin{cases} 0 & \text{if } 4 \mid N \\ \prod_{p|N}(1+(\tfrac{-1}{p})) & \text{otherwise} \end{cases}$$

$$\mu_3(N) = \begin{cases} 0 & \text{if } 2 \mid N \text{ or } 9 \mid N \\ \prod_{p|N}(1+(\tfrac{-3}{p})) & \text{otherwise} \end{cases}$$

$$\mu_\infty(N) = \sum_{d|N}\varphi(\text{GCD}(d,N/d)).$$

Here $(\tfrac{-1}{p})$ and $(\tfrac{-3}{p})$ are Legendre symbols (referring to quadratic residues), and φ is the Euler φ function.

Corollary 9.11. If $p > 3$ is prime, then

$$\dim S_2(\Gamma_0(p)) = \begin{cases} n-1 & \text{if } p = 12n+1 \\ n & \text{if } p = 12n+5 \\ n & \text{if } p = 12n+7 \\ n+1 & \text{if } p = 12n+11. \end{cases}$$

6. Hecke Operators

As with $\Gamma_0(1) = SL(2, \mathbb{Z})$, Hecke operators for $\Gamma_0(N)$ are built from lattices. However, additional structure plays a role. Instead of working with lattices Λ, we shall work with pairs (Λ, C), where Λ is a lattice in $\mathbb{C}$ and C is a cyclic subgroup of $\mathbb{C}/\Lambda$ of exact order N. Such a pair we shall call a **modular pair**.

If Λ is a lattice, we write P_Λ for the quotient map of $\mathbb{C}$ into $\mathbb{C}/\Lambda$. To a modular pair (Λ, C) and a nonzero complex number α, we can associate another modular pair $(\alpha\Lambda, \alpha C)$ as follows: The lattice $\alpha\Lambda$ is as earlier, and the cyclic subgroup αC is given by

$$\alpha C = P_{\alpha\Lambda}(\alpha P_\Lambda^{-1}(C)). \tag{9.39}$$

A complex-valued function $\tilde{f}$ on modular pairs is said to be **homogeneous** of degree $-k$ if

$$\tilde{f}(\alpha\Lambda, \alpha C) = \alpha^{-k}\tilde{f}(\Lambda, C) \qquad \text{for } \alpha \in \mathbb{C}^\times. \tag{9.40}$$

To such a function $\tilde{f}$, we can associate a function f on $\mathcal{H}$ by the definition

$$f(\tau) = \tilde{f}(\Lambda_\tau, P_{\Lambda_\tau}(\tfrac{1}{N}\mathbb{Z})).$$

Let us check that this function f satisfies

$$f(\gamma\tau) = \delta(\gamma, \tau)^k f(\tau) \qquad \text{for } \gamma \in \Gamma_0(N). \tag{9.41}$$

In fact, we recall from §VIII.2 that $\Lambda_{\gamma\tau} = \delta(\gamma, \tau)^{-1}\Lambda_\tau$. Thus we have

$$f(\gamma\tau) = \tilde{f}(\Lambda_{\gamma\tau}, P_{\Lambda_{\gamma\tau}}(\tfrac{1}{N}\mathbb{Z}))$$
$$= \tilde{f}(\delta(\gamma,\tau)^{-1}\Lambda_\tau, P_{\delta(\gamma,\tau)^{-1}\Lambda_\tau}(\tfrac{1}{N}\mathbb{Z}))$$
$$= \delta(\gamma,\tau)^k \tilde{f}(\Lambda_\tau, \delta(\gamma,\tau) P_{\delta(\gamma,\tau)^{-1}\Lambda_\tau}(\tfrac{1}{N}\mathbb{Z})) \qquad \text{by (9.40)}$$
$$= \delta(\gamma,\tau)^k \tilde{f}(\Lambda_\tau, P_{\Lambda_\tau}(\delta(\gamma,\tau) P_{\delta(\gamma,\tau)^{-1}\Lambda_\tau}^{-1} P_{\delta(\gamma,\tau)^{-1}\Lambda_\tau}(\tfrac{1}{N}\mathbb{Z})))$$

by (9.39), and this is

$$= \delta(\gamma,\tau)^k \tilde{f}(\Lambda_\tau, P_{\Lambda_\tau}(\delta(\gamma,\tau)(\tfrac{1}{N}\mathbf{Z} + \delta(\gamma,\tau)^{-1}\Lambda_\tau)))$$
$$= \delta(\gamma,\tau)^k \tilde{f}(\Lambda_\tau, P_{\Lambda_\tau}(\delta(\gamma,\tau)\tfrac{1}{N}\mathbf{Z} + \Lambda_\tau)).$$

In turn this is

$$= \delta(\gamma,\tau)^k \tilde{f}(\Lambda_\tau, P_{\Lambda_\tau}(\tfrac{1}{N}\mathbf{Z})) = \delta(\gamma,\tau)^k f(\tau),$$

since easy computation shows that

$$(c\tau + d)\tfrac{1}{N}\mathbf{Z} + \mathbf{Z} + \tau\mathbf{Z} = \tfrac{1}{N}\mathbf{Z} + \mathbf{Z} + \tau\mathbf{Z}$$

if $\begin{pmatrix} a & b \\ c & d \end{pmatrix}$ is in $\Gamma_0(N)$. This proves (9.41).

Conversely suppose that f is a complex-valued function on $\mathcal{H}$, We shall define a function $\tilde{f}$ on modular pairs such that (9.40) holds. Let (Λ, C) be given. Then Λ is a sublattice of index N in $P_\Lambda^{-1}(C)$ with $P_\Lambda^{-1}(C)/\Lambda$ cyclic. It follows that we can find a $\mathbf{Z}$ basis $\{\omega_1, \omega_2\}$ of Λ such that $\{\tfrac{1}{N}\omega_1, \omega_2\}$ is a $\mathbf{Z}$ basis of $P_\Lambda^{-1}(C)$. We may assume moreover that $\text{Im}(\omega_2/\omega_1) > 0$. In terms of this basis, we define

$$\tilde{f}(\Lambda, C) = \omega_1^{-k} f(\omega_2/\omega_1). \tag{9.42}$$

To see that $\tilde{f}(\Lambda, C)$ is well defined, let $\{\omega_1', \omega_2'\}$ be another basis of Λ such that $\{\tfrac{1}{N}\omega_1', \omega_2'\}$ is a basis of $P_\Lambda^{-1}(C)$ and $\text{Im}(\omega_2'/\omega_1')$ is positive. If we write

$$\begin{pmatrix} \omega_2' \\ \omega_1' \end{pmatrix} = \begin{pmatrix} a & b \\ c & d \end{pmatrix} \begin{pmatrix} \omega_2 \\ \omega_1 \end{pmatrix}, \tag{9.43a}$$

then $\begin{pmatrix} a & b \\ c & d \end{pmatrix}$ is in $SL(2,\mathbf{Z})$ since $\{\omega_1, \omega_2\}$ and $\{\omega_1', \omega_2'\}$ are both properly ordered bases for Λ. Since $\{\tfrac{1}{N}\omega_1, \omega_2\}$ and $\{\tfrac{1}{N}\omega_1', \omega_2'\}$ are both bases for $P_\Lambda^{-1}(C)$, the equality

$$\begin{pmatrix} \omega_2' \\ \tfrac{1}{N}\omega_1' \end{pmatrix} = \begin{pmatrix} a & Nb \\ \tfrac{1}{N}c & d \end{pmatrix} \begin{pmatrix} \omega_2 \\ \tfrac{1}{N}\omega_1 \end{pmatrix} \tag{9.43b}$$

forces the coefficient matrix to be integral. Thus $N \mid c$, and $\gamma = \begin{pmatrix} a & b \\ c & d \end{pmatrix}$ is in $\Gamma_0(N)$. By (9.41) we have

$$\omega_1'^{-k} f(\omega_2'/\omega_1') = \omega_1'^{-k} f(\gamma(\omega_2/\omega_1))$$
$$= \omega_1'^{-k}(c(\omega_2/\omega_1) + d)^k f(\omega_2/\omega_1) = \omega_1^{-k} f(\omega_2/\omega_1),$$

6. HECKE OPERATORS

and thus (9.42) is well defined.

Finally we check the homogeneity of $\tilde{f}$. If $\{\omega_1, \omega_2\}$ is a properly ordered basis for Λ as above, then $\{\alpha\omega_1, \alpha\omega_2\}$ is a properly ordered basis for $\alpha\Lambda$. Hence

$$\tilde{f}(\alpha\Lambda, \alpha C) = (\alpha\omega_1)^{-k} f((\alpha\omega_2)/(\alpha\omega_1)) = \alpha^{-k}\tilde{f}(\Lambda, C),$$

and $\tilde{f}$ satisfies (9.40).

The two constructions $\tilde{f} \to f$ and $f \to \tilde{f}$ are inverse to one another, and thus we have a one-one correspondence between functions $\tilde{f}$ on modular pairs homogeneous of degree $-k$ and functions f on $\mathcal{H}$ satisfying (9.41).

Let $\mathcal{L}$ be the free abelian group freely generated by the modular pairs (Λ, C). In analogy with (8.43), the **Hecke operator** $T(n)$ on modular pairs, for $n = 1, 2, 3, \ldots$, is defined to be the map $T(n) : \mathcal{L} \to \mathcal{L}$ given by

$$T(n)(\Lambda, C) = \sum_{\substack{[\Lambda:\Lambda']=n \\ nC \twoheadrightarrow C'}} (\Lambda', C'). \qquad (9.44)$$

The notation $nC \twoheadrightarrow C'$ means the following: We have an inclusion $n\Lambda \subseteq \Lambda'$ and an induced map $\mathbf{C}/(n\Lambda) \to \mathbf{C}/\Lambda'$. Under the induced map the cyclic group nC is to map onto the cyclic group C'.

The sum (9.44) is a finite sum since $n\Lambda \subseteq \Lambda' \subseteq \Lambda$ and since (Λ, C) and Λ' uniquely determine C'. It is easy to check that the condition $nC \twoheadrightarrow C'$ is automatically satisfied if $\gcd(n, N) = 1$. One might expect that $nC \twoheadrightarrow C'$ could be replaced by $C' \twoheadrightarrow C$ and that an equivalent or perhaps better theory would result. But this expectation is misplaced. For one thing, (Λ, C) and Λ' would no longer determine C'; for another, the analog of Proposition 8.14 would break down in a serious way.

We define $T_k(n)$ on the space of functions on modular pairs, homogeneous of degree $-k$, just as in (8.44):

$$(T_k(n)\tilde{f})(\Lambda, C) = n^{k-1} \sum_{\substack{[\Lambda:\Lambda']=n \\ nC \twoheadrightarrow C'}} \tilde{f}(\Lambda', C'). \qquad (9.45)$$

It is clear that $T_k(n)\tilde{f}$ is another function on modular pairs, homogeneous of degree $-k$. Shortly we shall exhibit the corresponding operator on functions of the variable $\tau \in \mathcal{H}$, and we shall see that it carries modular forms for $\Gamma_0(N)$ to modular forms for $\Gamma_0(N)$, and cusp forms to cusp forms. This operator too will be denoted $T_k(n)$ and is called a **Hecke operator**.

Let us compute the relationship between (Λ, C) and (Λ', C') in (9.45) in terms of matrices. Choose a basis $\{\omega_1, \omega_2\}$ of Λ such that $\{\frac{1}{N}\omega_1, \omega_2\}$ is a basis of $P_\Lambda^{-1}(C)$ and $\text{Im}(\omega_2/\omega_1) > 0$, and let $\{\omega_1', \omega_2'\}$ be a similar sort of basis for Λ'. Then we can write

$$\begin{pmatrix} \omega_2' \\ \omega_1' \end{pmatrix} = \begin{pmatrix} a & b \\ c & d \end{pmatrix} \begin{pmatrix} \omega_2 \\ \omega_1 \end{pmatrix}, \tag{9.46}$$

with $\begin{pmatrix} a & b \\ c & d \end{pmatrix}$ in $M(n)$. The inverse images of nC and C' are given by

$$P_{n\Lambda}^{-1}(nC) = \frac{n}{N}\mathbf{Z}\omega_1 + n\mathbf{Z}\omega_2 \tag{9.47a}$$

$$P_{\Lambda'}^{-1}(C') = \frac{1}{N}\mathbf{Z}\omega_1' + \mathbf{Z}\omega_2'. \tag{9.47b}$$

Since

$$\frac{n}{N}\omega_1 = \frac{1}{N}(a)(c\omega_2 + d\omega_1) - \left(\frac{c}{N}\right)(a\omega_2 + b\omega_1) = \frac{1}{N}(a)\omega_1' - \left(\frac{c}{N}\right)\omega_2'$$

and

$$n\omega_2 = (-bn)(c\omega_2 + d\omega_1) + (nd)(a\omega_2 + b\omega_1) = \frac{1}{N}(-bnN)\omega_1' + (nd)\omega_2',$$

(9.47a) is contained in (9.47b) if and only if $\dfrac{c}{N}$ is an integer. Thus nC maps into C' if and only if $\dfrac{c}{N}$ is an integer. Suppose this happens. For nC to map onto C', $\dfrac{n}{N}\omega_1$ must have exact order N relative to Λ'. The order is the least $m \geq 1$ such that

$$\frac{nm}{N}\omega_1 = r\omega_1' + s\omega_2'$$

for some integers r and s. Inversion of (9.46) allows us to substitute $\omega_1 = \dfrac{1}{n}(-c\omega_2' + a\omega_1')$ and rewrite this condition as

$$\frac{m}{N}(-c\omega_2' + a\omega_1') = r\omega_1' + s\omega_2'.$$

Since $N \mid c$, the condition is just that $ma/N = r$. The order is therefore $N/\text{GCD}(a, N)$, and order N occurs if and only if $\text{GCD}(a, N) = 1$. We conclude that $\{\omega_1', \omega_2'\}$ and $\{\omega_1, \omega_2\}$ are related by a matrix in the set

$$M(n, N) = \{\begin{pmatrix} a & b \\ c & d \end{pmatrix} \in M(n) \mid c \equiv 0 \bmod N \text{ and } \text{GCD}(a, N) = 1\}.$$

The most general similar sort of basis for Λ' is related to $\{\omega_1', \omega_2'\}$ by a member of $\Gamma_0(N)$, according to (9.43). Thus the modular pairs (Λ', C') in the sum (9.44) are parametrized by the right cosets $\Gamma_0(N)\backslash M(n, N)$.

6. HECKE OPERATORS   277

Proposition 9.12. Let $\{\alpha_i\}$ be a complete set of representatives for the right cosets $\Gamma_0(N)\alpha$ of $\Gamma_0(N)$ on $M(n, N)$. If f is in $M_k(\Gamma_0(N))$, then $T_k(n)f$ is given as a function of τ by

$$T_k(n)f = n^{\frac{k}{2}-1} \sum_i f \circ [\alpha_i]_k. \tag{9.48}$$

Hence $T_k(n)f$ is an unrestricted modular form of weight k and level N.

PROOF. The argument is the same as for Proposition 8.14.

Corollary 9.13. The Hecke operator $T_k(n)$ carries $M_k(\Gamma_0(N))$ to itself and $S_k(\Gamma_0(N))$ to itself.

PROOF. If f is in $M_k(\Gamma_0(N))$, we apply Proposition 9.5 with $m = n$, $\alpha = \alpha_i$, and $\beta \in SL(2, \mathbb{Z})$ to see that $(f \circ [\alpha_i]_k) \circ [\beta^{-1}]_k(\tau)$ has a holomorphic q_{nN} expansion at ∞. By (9.48) the same thing is true of $(T_k(n)f) \circ [\beta^{-1}]_k(\tau)$. But $T_k(n)f$ is an unrestricted modular form for $\Gamma_0(N)$, and hence $(T_k(n)f) \circ [\beta^{-1}]_k(\tau)$ is periodic with period N. Hence the terms in the q_{nN} expansion whose index is not a multiple of n vanish, and $T_k(n)f$ is holomorphic at the cusp $\beta^{-1}(\infty)$. A similar argument works if f is in $S_k(\Gamma_0(N))$.

To see concretely the effect of $T_k(n)$ on q expansions, we use explicit coset representatives α_i in (9.48). Arguing as in the proof of Lemma 8.15 and noting that N will have to divide x at the start of the proof, we arrive at the following.

Lemma 9.14. The matrices $\begin{pmatrix} a & b \\ 0 & d \end{pmatrix}$ with $ad = n$, $d > 0$, $\mathrm{GCD}(a, N) = 1$, and $0 \leq b < d$ are a complete set of coset representatives for the right cosets of $\Gamma_0(N)$ on $M(n, N)$.

REMARK. The set of representatives here is a subset of the set in Lemma 8.15. The subset is proper if and only if $\mathrm{GCD}(n, N) = 1$.

Proposition 9.15. Let f in $M_k(\Gamma_0(N))$ have q expansion $f(\tau) = \sum_{n=0}^{\infty} c_n q^n$. Then $T_k(m)f$ has q expansion

$$T_k(m)f(\tau) = \sum_{n=0}^{\infty} b_n q^n,$$

where

$$b_n = \begin{cases} c_0 \sum_{\substack{a|m,\, a>0 \\ \text{GCD}(a,N)=1}} a^{k-1} & \text{if } n=0 \\ c_m & \text{if } n=1 \\ \sum_{\substack{a|\text{GCD}(n,m) \\ \text{GCD}(a,N)=1}} a^{k-1} c_{nm/a^2} & \text{if } n>1. \end{cases}$$

PROOF. The argument is substantially the same as for Proposition 8.16. The condition $\text{GCD}(a, N) = 1$ needs to be carried along throughout.

Next we examine how the $T_k(n)$ interact with one another. First we formalize the definition of dilation on modular pairs made in (9.39). We let

$$R(n)(\Lambda, C) = (n\Lambda, nC).$$

This definition is consistent with the one in §VIII.7, and $R(n)T(m) = T(m)R(n)$ for all m and n. The following lemma generalizes Lemma 8.18.

Lemma 9.16.
(a) For a prime power p^r with $r \geq 1$ such that $p \nmid N$,

$$T(p^r)T(p) = T(p^{r+1}) + p \cdot R(p)T(p^{r-1}). \tag{9.49}$$

(b) For a prime power p^r with $r \geq 1$ such that $p \mid N$,

$$T(p^r) = T(p)^r. \tag{9.50}$$

(c) $T(m)T(n) = T(mn)$ if m and n are relatively prime.

PROOF. Parts (a) and (c) are proved in the same way as for Lemma 8.18, the cyclic group being of no significance in the argument. For (b), we shall show that

$$T(p^r) = T(p^{r-1})T(p), \tag{9.51}$$

and then the result follows by induction. If $T(p)(\Lambda, C)$ contains a term (Λ'', C'') and $T(p^{r-1})(\Lambda'', C'')$ contains a term (Λ', C'), then

$$[\Lambda : \Lambda''] = p, \quad pC \twoheadrightarrow C'', \quad [\Lambda'', \Lambda'] = p^{r-1}, \quad p^{r-1}C'' \twoheadrightarrow C'. \tag{9.52}$$

Hence $[\Lambda : \Lambda'] = p^r$ and $p^r C \twoheadrightarrow p^{r-1}C'' \twoheadrightarrow C'$. Thus $T(p^r)(\Lambda, C)$ contains the term (Λ', C').

6. HECKE OPERATORS

In the reverse direction, if $T(p^r)(\Lambda, C)$ contains (Λ', C'), it contains it only once. So we want to see that there is only one Λ'' such that (9.52) holds. Fix such a Λ''. Choose a basis $\{\omega_1, \omega_2\}$ of Λ so that $\{\frac{1}{N}\omega_1, \omega_2\}$ is a basis of $P_\Lambda^{-1}(C)$ and $\text{Im}(\omega_2/\omega_1) > 0$. Fix a similar basis $\{\omega_1'', \omega_2''\}$ of Λ''. Changing basis for Λ'' by a member of $\Gamma_0(N)$, we may assume that

$$\begin{pmatrix} \omega_2'' \\ \omega_1'' \end{pmatrix} = \begin{pmatrix} a & b \\ 0 & d \end{pmatrix} \begin{pmatrix} \omega_2 \\ \omega_1 \end{pmatrix}$$

with the matrix as in Lemma 9.14. Since $ad = p$ and $\text{GCD}(a, N) = 1$ and $p \mid N$, we must have $a = 1$. Thus

$$\begin{pmatrix} \omega_2'' \\ \omega_1'' \end{pmatrix} = \begin{pmatrix} 1 & b \\ 0 & p \end{pmatrix} \begin{pmatrix} \omega_2 \\ \omega_1 \end{pmatrix}, \quad 0 \leq b < p.$$

Similarly we can find a basis $\{\omega_1', \omega_2'\}$ of Λ' such that

$$\begin{pmatrix} \omega_2' \\ \omega_1' \end{pmatrix} = \begin{pmatrix} 1 & b' \\ 0 & p^{r-1} \end{pmatrix} \begin{pmatrix} \omega_2'' \\ \omega_1'' \end{pmatrix}, \quad 0 \leq b' < p^{r-1}.$$

Then

$$\begin{pmatrix} \omega_2' \\ \omega_1' \end{pmatrix} = \begin{pmatrix} 1 & b + b'p \\ 0 & p^r \end{pmatrix} \begin{pmatrix} \omega_2 \\ \omega_1 \end{pmatrix}.$$

The uniqueness in Lemma 9.14 shows that $b + b'p$ here is determined by Λ and Λ'. But $b + b'p$ determines b and b'. Hence Λ'' is completely determined by Λ and Λ'. This proves (9.51) and the lemma.

To get from operators on modular pairs to operators on functions $\tilde{f}$ on modular pairs, we use the definitions of (8.51) and (8.52). Lemma 9.16 then yields the following analog of Theorem 8.19.

Theorem 9.17 (Hecke). On the space $M_k(\Gamma_0(N))$, the Hecke operators satisfy

(a) For a prime power p^r with $r \geq 1$ such that $p \nmid N$,

$$T_k(p^r)T_k(p) = T_k(p^{r+1}) + p^{k-1}T_k(p^{r-1}).$$

Hence $T_k(p^r)$ is a polynomial in $T_k(p)$ with integer coefficients.

(b) For a prime power p^r with $r \geq 1$ such that $p \mid N$,

$$T_k(p^r) = T_k(p)^r.$$

(c) $T_k(m)T_k(n) = T_k(mn)$ if m and n are relatively prime.

(d) The algebra generated by the $T_k(n)$ for $n = 1, 2, 3, \ldots$ is generated by the $T_k(p)$ with p prime and is commutative.

As a consequence of Lemma 9.6, we can define a **Petersson inner product** on $S_k(\Gamma_0(N))$. The definition is

$$\langle f, h\rangle = \int_{R_N} f(\tau)\overline{h(\tau)}\sigma^k \frac{d\rho\, d\sigma}{\sigma^2},$$

where R_N is a fundamental domain for $\Gamma_0(N)$. The inner product does not depend on the choice of fundamental domain. In an effort to prove an analog of Theorem 8.22 (that the $T_k(n)$ are self adjoint), we again take n to be a prime p, by Theorem 9.17d. Let $\tilde{R}_N$ be a fundamental domain for $\Gamma(pN)$ obtained as the union of translates of R_N. Then the argument in the proof of Theorem 8.22 still shows that

$$\int_{\tilde{R}_N} f \circ [\alpha]_k(\tau)\overline{h(\tau)}\,\frac{d\rho\, d\sigma}{\sigma^2} = \int_{\tilde{R}_N} f(\tau)\overline{h \circ [\alpha']_k(\tau)}\sigma^k \frac{d\rho\, d\sigma}{\sigma^2} \qquad (9.53)$$

for $\alpha \in M(p)$, just as in (8.62).

We now assume that $p \nmid N$. For the element $\alpha_i = \begin{pmatrix} 1 & b \\ 0 & p \end{pmatrix}$, choose integers x and y so that

$$p^2 x - Ny = 1 + pbN.$$

Then

$$\begin{pmatrix} p & -b \\ 0 & 1 \end{pmatrix} = \begin{pmatrix} px - bN & y \\ N & p \end{pmatrix} \begin{pmatrix} 1 & b \\ 0 & p \end{pmatrix} \begin{pmatrix} x & xb + y + pb \\ N & Nb + p^2 \end{pmatrix}^{-1}$$

shows that $\alpha_i' = \gamma_i \alpha_i \gamma_i'$ for some γ_i and γ_i' in $\Gamma_0(N)$. Inverting this equation for $b = 0$, we see that we have a similar relation $\alpha_i' = \gamma_i \alpha_i \gamma_i'$ for $\alpha_i = \begin{pmatrix} p & 0 \\ 0 & 1 \end{pmatrix}$. Thus the same argument as the one in Theorem 8.22 that obtained that theorem from (8.62) is applicable here, but only under the assumption $p \nmid N$. We summarize as follows.

Theorem 9.18 (Petersson). The Hecke operators $T_k(n)$ with $\mathrm{GCD}(n, N) = 1$, on the space of cusp forms $S_k(\Gamma_0(N))$, are self adjoint relative to the Petersson inner product.

It is not always true that the operators $T_k(n)$ with $\mathrm{GCD}(n, N) > 1$ are self adjoint. Thus Theorem 9.18 does not give us quite as good a result as we had with $SL(2, \mathbf{Z})$. Since the Hecke operators commute, we can conclude that $S_k(\Gamma_0(N))$ splits into the orthogonal sum of simultaneous eigenspaces for the operators $T_k(n)$ with $\mathrm{GCD}(n, N) = 1$. An eigenvector cusp form under the $T_k(n)$ with $\mathrm{GCD}(n, N) = 1$ is called an **eigenform**; eigenforms in the same such eigenspace are said to be **equivalent**.

Proposition 9.19. The involution w_N of $S_k(\Gamma_0(N))$ given in §4 is self adjoint and commutes with all $T_k(n)$ such that $GCD(n, N) = 1$.

PROOF. We have $w_N(f) = f \circ [\alpha_N]_k$. The argument of Theorem 8.22 establishes (9.53) for $\alpha = \alpha_N$ if we regard $\tilde{R}_N$ as a fundamental domain for $\Gamma(N^2)$. But $[\alpha'_N] = [\alpha_N]$, and it follows that w_N is self adjoint.

To prove that $w_N T_k(n) = T_k(n) w_N$, we may assume that n is a prime p with $p \nmid N$, by Theorem 9.17. The matrices α_i to use in the formula (9.48) for $T_k(p)$ are

$$\begin{pmatrix} p & 0 \\ 0 & 1 \end{pmatrix}, \quad \begin{pmatrix} 1 & 0 \\ 0 & p \end{pmatrix}, \quad \begin{pmatrix} 1 & b \\ 0 & p \end{pmatrix} \text{ for } 1 \le b \le p-1. \quad (9.54)$$

What is to be checked is that

$$\sum_i f \circ [\alpha_N \alpha_i]_k = \sum_i f \circ [\alpha_i \alpha_N]_k. \quad (9.55)$$

The sum of the contributions from the first two matrices in (9.54) to each side of (9.55) is the same since

$$\alpha_N \begin{pmatrix} p & 0 \\ 0 & 1 \end{pmatrix} = \begin{pmatrix} 1 & 0 \\ 0 & p \end{pmatrix} \alpha_N \quad \text{and} \quad \alpha_N \begin{pmatrix} 1 & 0 \\ 0 & p \end{pmatrix} = \begin{pmatrix} p & 0 \\ 0 & 1 \end{pmatrix} \alpha_N.$$

We shall show that the sum of the contributions from the other matrices in (9.54) to each side of (9.55) is also the same. Specifically we shall show that the matrix with b contributes to the left side what the matrix with e contributes to the right side, where $1 \le e \le p-1$ and $e \equiv (-Nb)^{-1}$ mod p. In fact, we can calculate that

$$\begin{pmatrix} 0 & -1 \\ N & 0 \end{pmatrix} \begin{pmatrix} 1 & b \\ 0 & p \end{pmatrix} = \begin{pmatrix} p & -e \\ -Nb & p^{-1}(1+ebN) \end{pmatrix} \begin{pmatrix} 1 & e \\ 0 & p \end{pmatrix} \begin{pmatrix} 0 & -1 \\ N & 0 \end{pmatrix}.$$

The first matrix γ on the right side is in $\Gamma_0(N)$, and $f \circ [\gamma]_k = f$. Thus

$$f \circ \left[\alpha_N \begin{pmatrix} 1 & b \\ 0 & p \end{pmatrix} \right]_k = f \circ \left[\begin{pmatrix} 1 & e \\ 0 & p \end{pmatrix} \alpha_N \right]_k$$

as asserted.

Consequently the decomposition of $S_k(\Gamma_0(N))$ into spaces of equivalent eigenforms is compatible with the decomposition of $S_k(\Gamma_0(N))$ into $S_k^+(\Gamma_0(N))$ and $S_k^-(\Gamma_0(N))$.

The Hecke operators $T_k(n)$ with $GCD(n, N) \ne 1$ commute with the other $T_k(n)$ and hence map the spaces of equivalent eigenforms into themselves. In each such space there will be at least one eigenvector of all the $T_k(n)$, and its eigenvalues will be as in Proposition 9.20 below. We cannot assert anything at this stage about a relationship between w_N and the operators $T_k(n)$ for $GCD(n, N) \ne 1$.

Proposition 9.20. Suppose $f \in S_k(\Gamma_0(N))$ is an eigenform that is an eigenvector of all $T_k(n)$, say with $T_k(n)f = \lambda(n)f$. If the q expansion of f at ∞ is $f(\tau) = \sum_{n=1}^{\infty} c_n q^n$, then

$$c_n = \lambda(n) c_1. \tag{9.56}$$

Consequently
(a) $f \not\equiv 0$ implies $c_1 \neq 0$
(b) the system of eigenvalues $\{\lambda(n)\}$ determines f up to a scalar.

PROOF. This follows from Proposition 9.15 in the same way that Proposition 8.23 follows from Proposition 8.16.

Under the assumptions of Proposition 9.20, we can normalize f so that the q expansion $f(\tau) = \sum_{n=1}^{\infty} c_n q^n$ has $c_1 = 1$. Then (9.56) says that c_n is the eigenvalue of $T_k(n)$. From Theorem 9.17 we see that

$$c_{p^r} c_p = c_{p^{r+1}} + p^{k-1} c_{p^{r-1}} \quad \text{for } p \text{ prime, } p \nmid N \tag{9.57a}$$
$$c_{p^r} = (c_p)^r \quad \text{for } p \text{ prime, } p \mid N \tag{9.57b}$$
$$c_m c_n = c_{mn} \quad \text{if GCD}(m,n) = 1. \tag{9.57c}$$

As in §VIII.8 it follows from (9.57) that the L function $L(s,f) = \sum_{n=1}^{\infty} \frac{c_n}{n^s}$ has an Euler product expansion. In this case the Euler factor when $p \nmid N$ is quadratic, as before. But when $p \mid N$, it is first order of the form

$$\frac{1}{1 - c_p p^{-s}}.$$

We summarize part of this discussion as follows.

Theorem 9.21 (Hecke-Petersson). The whole space $S_k(\Gamma_0(N))$ of cusp forms is the orthogonal sum of the spaces of equivalent eigenforms. Each space of equivalent eigenforms has a member that is an eigenvector for all $T_k(n)$. Any eigenform f in $S_k(\Gamma_0(N))$ that is an eigenvector for all $T_k(n)$ can be normalized so that its q expansion $f(\tau) = \sum_{n=1}^{\infty} c_n q^n$ has $c_1 = 1$. With such a normalization the coefficients satisfy (9.57). Moreover, the L function $L(s,f)$ has an Euler product expansion

$$L(s,f) = \prod_{\substack{p \text{ prime} \\ p \mid N}} \left[\frac{1}{1 - c_p p^{-s}} \right] \prod_{\substack{p \text{ prime} \\ p \nmid N}} \left[\frac{1}{1 - c_p p^{-s} + p^{k-1-2s}} \right] \tag{9.58}$$

convergent for $\operatorname{Re} s > \frac{k}{2} + 1$.

7. Oldforms and Newforms

Theorem 9.21 is not the same kind of definitive result that we obtained in Chapter VIII. Since we were unable in Theorem 9.21 to relate w_N to the $T_k(p)$ for which $p \mid N$, we ended up with no correlation betweeen L functions having Euler products and L functions having functional equations.

It turns out that the difficulty is the presence of cusp forms that come trivially from lower levels. These are the cusp forms considered in Examples 1 and 2 in §3. A specific example is the pair of members $\Delta(\tau)$ and $\Delta(2\tau)$ of $S_{12}(\Gamma_0(2))$. We shall see after (9.61) that both are eigenforms with the same eigenvalues under all $T_{12}(n)$ with n odd, yet they are linearly independent.

More generally the first kind of example was $f(\tau)$ in $S_k(\Gamma_0(N))$ when $f(\tau)$ was given in $S_k(\Gamma_0(N/r))$ and $r \mid N$. When $\mathrm{GCD}(n,N) = 1$, the formula for $T_k(n)f$ is the same relative to $\Gamma_0(N)$ as relative to $\Gamma_0(N/r)$. Hence an eigenform for $\Gamma_0(N/r)$ becomes an eigenform for $\Gamma_0(N)$ with the same eigenvalues.

The second kind of example was $f(r\tau)$ in $S_k(\Gamma_0(N))$ when $f(\tau)$ was given in $S_k(\Gamma_0(N/r))$ and $r \mid N$. Let $A_k(r)$ be the operator

$$A_k(r)f = f \circ \left[\begin{pmatrix} r & 0 \\ 0 & 1 \end{pmatrix}\right]_k. \tag{9.59}$$

Since $f(\tau) = \sum_{n=1}^{\infty} c_n q^n$ has

$$A_k(r)f(\tau) = r^{k/2} f(r\tau), \tag{9.60}$$

we know from §3 that $A_k(r)$ carries $S_k(\Gamma_0(N/r))$ to $S_k(\Gamma_0(N))$. We shall see in Lemma 9.23 below that

$$A_k(r)T_k(n) = T_k(n)A_k(r) \quad \text{if } \mathrm{GCD}(n,N) = 1. \tag{9.61}$$

Consequently if $f(\tau)$ is an eigenform for $\Gamma_0(N/r)$, then $f(r\tau)$ is an eigenform for $\Gamma_0(N)$ with the same eigenvalues.

We can combine the two examples in sequence as follows: If $r_1 r_2 \mid N$ and if $f(\tau)$ is an eigenform for $\Gamma_0(N/(r_1 r_2))$, then $f(r_2 \tau)$ is an eigenform for $\Gamma_0(N)$ with the same eigenvalues. Such an eigenform we call an **oldform**. The linear span of the oldforms is denoted $S_k^{\mathrm{old}}(\Gamma_0(N))$, and its orthogonal complement is denoted $S_k^{\mathrm{new}}(\Gamma_0(N))$. The eigenforms in $S_k^{\mathrm{new}}(\Gamma_0(N))$ are called **newforms** for $\Gamma_0(N)$. Since $T_k(n)$ is self adjoint when $\mathrm{GCD}(n,N) = 1$, $S_k^{\mathrm{new}}(\Gamma_0(N))$ is spanned by newforms. The key result, whose proof we omit, is the following Multiplicity One Theorem.

Theorem 9.22 (Atkin-Lehner). If $f \in S_k(\Gamma_0(N))$ is a newform, then its equivalence class is one-dimensional, i.e., consists of the multiples of f.

Before deriving some of the consequences for newforms of Theorem 9.22, we supply a proof of (9.61).

Lemma 9.23. If $\text{GCD}(n, N) = 1$ and $r \mid N$, then

$$A_k(r)T_k(n) = T_k(n)A_k(r)$$

on $S_k(\Gamma_0(N/r))$.

PROOF. Let $\begin{pmatrix} a & b \\ 0 & d \end{pmatrix}$ range over the usual matrices with $ad = n$, $d > 0$, $\text{GCD}(a, N) = 1$, and b in a complete residue system modulo d. (Since $\text{GCD}(n, N) = 1$, the condition $\text{GCD}(a, N) = 1$ is vacuous.) By Lemma 9.14 and (9.48)

$$T_k(n)A_k(r)f = n^{\frac{k}{2}-1} \sum f \circ \left[\begin{pmatrix} r & 0 \\ 0 & 1 \end{pmatrix} \begin{pmatrix} a & b \\ 0 & d \end{pmatrix} \right]_k$$
$$= n^{\frac{k}{2}-1} \sum f \circ \left[\begin{pmatrix} a & br \\ 0 & d \end{pmatrix} \begin{pmatrix} r & 0 \\ 0 & 1 \end{pmatrix} \right]_k.$$

Since br goes through a complete residue system modulo d, the right side is

$$= A_k(r)T_k(n)f,$$

and the lemma follows.

We return to the consequences of Theorem 9.22. The operators w_N and $T_k(p)$ with $p \mid N$ commute with the $T_k(p)$ having $p \nmid N$, and thus they map each equivalence class into itself. If f is a newform, then the theorem says that $\mathbb{C}f$ is an equivalence class. Consequently f is an eigenvector for w_N and the $T_k(n)$ with $p \mid N$. The first of these conclusions means that $L(s, f)$ has a functional equation, according to Theorem 9.8. The second of these conclusions means that $L(s, f)$ has an Euler product expansion (after f is normalized), according to Theorem 9.21.

But there are further operators of this kind if N is composite, and they give some insight into the eigenvalue ± 1 of w_N. Fix a prime p

7. OLDFORMS AND NEWFORMS

dividing N, and let $Q = p^l$ be the exact power of p dividing N. Then we can choose integers α_0 and γ_0 with $Q\alpha_0 - (N/Q)\gamma_0 = 1$, and the matrix

$$\begin{pmatrix} Q\alpha_0 & 1 \\ N\gamma_0 & Q \end{pmatrix}$$

will have determinant Q. More generally we consider any matrix

$$w(Q) = \begin{pmatrix} Q\alpha & \beta \\ N\gamma & Q\delta \end{pmatrix} \tag{9.62}$$

of determinant Q. For f in $S_k(\Gamma_0(N))$, let $w_Q f = f \circ [w(Q)]_k$.

Lemma 9.24. The operator w_Q carries $S_k(\Gamma_0(N))$ into itself. It is independent of the choice of defining matrix $w(Q)$, and $w_Q^2 = 1$. It commutes with all $T_k(n)$ with $\mathrm{GCD}(n, N) = 1$. If p' is another prime dividing N and if Q' is the corresponding Q, then w_Q and $w_{Q'}$ commute. As p varies over all primes dividing N, the product of the various w_Q's is w_N.

PROOF. Let $\begin{pmatrix} a & b \\ c & d \end{pmatrix}$ be in $\Gamma_0(N)$. Carrying out the multiplications, we see that $w(Q) \begin{pmatrix} a & b \\ c & d \end{pmatrix} w(Q)^{-1}$ is in $\Gamma_0(N)$. The same argument as with w_N in §4 then shows that w_Q carries $S_k(\Gamma_0(N))$ into itself.

If we have two representatives $w(Q)$ and $w'(Q)$, we calculate that $w(Q)w'(Q)^{-1}$ is an element γ' of $\Gamma_0(N)$. For f in $S_k(\Gamma_0(N))$, we therefore have

$$f \circ [w(Q)]_k = (f \circ [\gamma']_k) \circ [w'(Q)]_k = f \circ [w'(Q)]_k.$$

Thus w_Q is independent of the representative. Similarly we calculate that $Q^{-1} w(Q)^2$ is in $\Gamma_0(N)$, and it follows that $w_Q^2 = 1$.

For the commutativity with $T_k(n)$, it is enough to prove commutativity with $T_k(p')$, where p' is a prime with $p' \nmid N$. Using the facts that $p' \nmid N$ and $p' \neq p$, we shall choose the representative $w(Q)$ in (9.62) to be congruent to $\begin{pmatrix} Q & 0 \\ 0 & 1 \end{pmatrix}$ modulo p'. Namely choose an integer Q_1 with $QQ_1 \equiv 1 \mod p'$. Let $Q_0 = Q_1 + mp'$, where m is the product of all primes dividing N but not Q_1; then $QQ_0 \equiv 1 \mod p'$ and $\mathrm{GCD}(Q_0, N) = 1$. Since $\mathrm{GCD}(Q^2 Q_0, Np'^2) = Q$, we can choose α' and γ' with

$$Q^2 Q_0 \alpha' - Np'^2 \gamma' = Q.$$

Then

$$w(Q) = \begin{pmatrix} Q\alpha' & p' \\ Np'\gamma' & QQ_0 \end{pmatrix}$$

has the required properties.

If $m(b)$ denotes $\begin{pmatrix} 1 & b \\ 0 & p' \end{pmatrix}$, then we can calculate that

$$m(b)w(Q)m(Q_0 b)^{-1} = w'(Q)$$

with $w'(Q)$ of the form (9.62). Hence

$$\sum_{b \bmod p'} (f \circ [m(b)]_k) \circ [w(Q)]_k = \sum_{b \bmod p'} (w_Q f) \circ [m(Q_0 b)]_k$$

$$= \sum_{b \bmod p'} (w_Q f) \circ [m(b)]_k. \quad (9.63a)$$

With the same $w'(Q)$ we have

$$\begin{pmatrix} 1 & 0 \\ 0 & p \end{pmatrix} w(Q) \begin{pmatrix} 1 & 0 \\ 0 & p \end{pmatrix}^{-1} = w'(Q),$$

and with another $w''(Q)$ we have

$$\begin{pmatrix} p & 0 \\ 0 & 1 \end{pmatrix} w(Q) \begin{pmatrix} p & 0 \\ 0 & 1 \end{pmatrix}^{-1} = w''(Q).$$

Thus

$$\left(f \circ \left[\begin{pmatrix} 1 & 0 \\ 0 & p \end{pmatrix} \right]_k \right) \circ [w(Q)]_k = (w_Q f) \circ \left[\begin{pmatrix} 1 & 0 \\ 0 & p \end{pmatrix} \right]_k \quad (9.63b)$$

$$\left(f \circ \left[\begin{pmatrix} p & 0 \\ 0 & 1 \end{pmatrix} \right]_k \right) \circ [w(Q)]_k = (w_Q f) \circ \left[\begin{pmatrix} p & 0 \\ 0 & 1 \end{pmatrix} \right]_k. \quad (9.63c)$$

Adding the equations (9.63), we obtain

$$w_Q T_k(p) f = T_k(p) w_Q f$$

as required.

If p' is another prime, we calculate that

$$(w(Q)w(Q'))(w(Q')w(Q))^{-1}$$

is in $\Gamma_0(N)$. Then it follows that $w_Q w_{Q'} = w_{Q'} w_Q$ on $S_k(\Gamma_0(N))$.

For the product formula for w_N, we extend the definition of $w(Q)$ in (9.62) to any divisor Q of N with $\gcd(Q, N/Q) = 1$. If Q and Q' are relatively prime such divisors, we find that $w(Q)w(Q')$ is a version of $w(QQ')$. Hence the product of all $w(Q)$ with Q a prime power is a version of $w(N)$. But $\begin{pmatrix} 0 & 1 \\ N & 0 \end{pmatrix}$ is another version of $w(N)$. Then it follows that w_N is the product of all w_Q with Q a prime power dividing N such that $\gcd(Q, N/Q) = 1$. This completes the proof of the lemma.

7. OLDFORMS AND NEWFORMS

If p is prime, let $\Gamma_0(r,p)$ be the subgroup of $\Gamma_0(r)$ given by

$$\Gamma_0(r,p) = \left\{ \begin{pmatrix} a & b \\ c & d \end{pmatrix} \in \Gamma_0(r) \,\Big|\, b \equiv 0 \mod p \right\}.$$

Lemma 9.25. If p is prime, then a complete set of right coset representatives for $\Gamma_0(r,p)$ in $\Gamma_0(r)$ is

$$\begin{pmatrix} 1 & j \\ 0 & 1 \end{pmatrix} \text{ with } 0 \leq j \leq p-1 \qquad\qquad \text{if } p \mid r \qquad (9.64a)$$

$$\begin{pmatrix} 1 & j \\ 0 & 1 \end{pmatrix} \text{ with } 0 \leq j \leq p-1, \text{ and } \begin{pmatrix} p\alpha & 1 \\ r\gamma & 1 \end{pmatrix} \text{ if } p \nmid r, \quad (9.64b)$$

where α and γ are integers such that $p\alpha - r\gamma = 1$.

PROOF. Let $\begin{pmatrix} a & b \\ cr & d \end{pmatrix}$ be given in $\Gamma_0(r,p)$. If $p \nmid a$, we can choose j so that $p \mid (b - aj)$, and then

$$\begin{pmatrix} a & b \\ cr & d \end{pmatrix} \text{ is in } \Gamma_0(r,p) \begin{pmatrix} 1 & j \\ 0 & 1 \end{pmatrix}.$$

If $p \mid r$, then $p \mid a$ is impossible, and the elements (9.64a) represent all cosets. If $p \nmid r$ and $p \mid a$, then

$$\begin{pmatrix} a & b \\ cr & d \end{pmatrix} \text{ is in } \Gamma_0(r,p) \begin{pmatrix} p\alpha & 1 \\ r\gamma & 1 \end{pmatrix}.$$

Hence the elements (9.64b) represent all cosets in this case. We readily check that the elements in question represent distinct cosets, and the proof is complete.

Lemma 9.26. Let f be in $S_k(\Gamma_0(N))$, and let p be a prime dividing N. If $p^2 \mid N$, then $T_k(p)f$ is in $S_k(\Gamma_0(N/p))$. If $p \mid N$ and $p^2 \nmid N$, then $T_k(p)f + p^{\frac{k}{2}-1}w_p f$ is in $S_k(\Gamma_0(N/p))$.

PROOF. Let $\gamma = \begin{pmatrix} a & bp \\ cN/p & d \end{pmatrix}$ be in $\Gamma_0(N/p,p)$. Then

$$\begin{pmatrix} p & 0 \\ 0 & 1 \end{pmatrix}^{-1} \gamma \begin{pmatrix} p & 0 \\ 0 & 1 \end{pmatrix} = \begin{pmatrix} a & b \\ cN & d \end{pmatrix}$$

is in $\Gamma_0(N)$. Hence

$$f \circ \left[\begin{pmatrix} p & 0 \\ 0 & 1 \end{pmatrix}^{-1}\right]_k = f \circ \left[\begin{pmatrix} a & b \\ cN & d \end{pmatrix}\begin{pmatrix} p & 0 \\ 0 & 1 \end{pmatrix}^{-1}\right]_k$$

$$= \left(f \circ \left[\begin{pmatrix} p & 0 \\ 0 & 1 \end{pmatrix}^{-1}\right]_k\right) \circ [\gamma]_k,$$

and $f \circ \left[\begin{pmatrix} p & 0 \\ 0 & 1 \end{pmatrix}^{-1}\right]_k$ transforms according to $\Gamma_0(N/p, p)$.

Let $\{R_i\}$ be a system of right coset representatives for $\Gamma_0(N/p, p)$ in $\Gamma_0(N/p)$. If γ is in $\Gamma_0(N/p)$, then $\{R_i\gamma\}$ is another system of right coset representatives. Hence

$$\sum_i \left(f \circ \left[\begin{pmatrix} p & 0 \\ 0 & 1 \end{pmatrix}^{-1}\right]_k\right) \circ [R_i]_k = \sum_i \left(f \circ \left[\begin{pmatrix} p & 0 \\ 0 & 1 \end{pmatrix}^{-1}\right]_k\right) \circ [R_i\gamma]_k, \quad (9.65)$$

and the left side of (9.65) transforms according to $\Gamma_0(N/p)$.

If we use the representatives in Lemma 9.25, the desired result now follows. In fact,

$$\begin{pmatrix} p & 0 \\ 0 & 1 \end{pmatrix}^{-1}\begin{pmatrix} 1 & j \\ 0 & 1 \end{pmatrix} = \frac{1}{p}\begin{pmatrix} 1 & 0 \\ 0 & p \end{pmatrix}\begin{pmatrix} 1 & j \\ 0 & 1 \end{pmatrix} = \frac{1}{p}\begin{pmatrix} 1 & j \\ 0 & p \end{pmatrix}$$

shows that the contribution to the left side of (9.65) from all $\begin{pmatrix} 1 & j \\ 0 & 1 \end{pmatrix}$ is just $p^{-\frac{k}{2}+1}T_k(p)f$. If $p \mid N$ and $p^2 \nmid N$, then $p \nmid (N/p)$, and there is one more coset representative, namely $\begin{pmatrix} p\alpha & 1 \\ N\gamma/p & 1 \end{pmatrix}$. It satisfies

$$\begin{pmatrix} p & 0 \\ 0 & 1 \end{pmatrix}^{-1}\begin{pmatrix} p\alpha & 1 \\ N\gamma/p & 1 \end{pmatrix} = \frac{1}{p}\begin{pmatrix} 1 & 0 \\ 0 & p \end{pmatrix}\begin{pmatrix} p\alpha & 1 \\ N\gamma/p & 1 \end{pmatrix}$$

$$= \frac{1}{p}\begin{pmatrix} p\alpha & 1 \\ N\gamma & p \end{pmatrix} = \frac{1}{p}w(p),$$

and the contribution to the left side of (9.65) is $w_p f$.

7. OLDFORMS AND NEWFORMS

Theorem 9.27 (Atkin-Lehner). The space $S_k^{\text{new}}(\Gamma_0(N))$ of newforms is the orthogonal sum of one-dimensional equivalence classes of eigenforms. If f is such an eigenform, then f can be normalized so that its q expansion $f(\tau) = \sum_{n=1}^{\infty} c_n q^n$ has $c_1 = 1$. In this case the eigenform f is an eigenvector of all $T_k(n)$ for all n, of all w_Q for primes dividing N, and of w_N, and the eigenvalues are as follows:

(a) $T_k(n)f = c_n f$ for all n
(b) $w_Q f = \lambda(Q) f$ with $\lambda(Q) = \pm 1$, if $p \mid N$ and Q corresponds to p
(c) $w_N f = \prod_{p \mid N} \lambda(Q) f$.

Moreover, $c_p = 0$ if $p^2 \mid N$, while $c_p = -p^{\frac{k}{2}-1} \lambda(p)$ if $p \mid N$ and $p^2 \nmid N$. Consequently the L function $L(s, f)$ has an Euler product expansion

$$L(s,f) = \prod_{\substack{p \text{ prime} \\ p \mid N, p^2 \nmid N}} \left[\frac{1}{1 + \lambda(p) p^{\frac{k}{2}-1-s}} \right] \prod_{\substack{p \text{ prime} \\ p \nmid N}} \left[\frac{1}{1 - c_p p^{-s} + p^{k-1-2s}} \right],$$

(9.66)

and $L(s, f)$ satisfies a functional equation (9.33) with $\varepsilon = \prod_{p \mid N} \lambda(Q)$.

PROOF. The first sentence is by Theorem 9.22 and the self adjointness of the $T_k(n)$ with $\gcd(n, N) = 1$. Since the $T_k(n)$ commute, any such eigenform must then be an eigenvector of all $T_k(n)$. By Theorem 9.21, $c_1 \neq 0$; also if f is normalized to make $c_1 = 1$, then (a) holds.

Lemma 9.24 shows that f is an eigenvector of all w_Q. Since $w_Q^2 = 1$, (b) holds. Lemma 9.24 also proves (c).

Let $p^2 \mid N$. By Lemma 9.26, $T_k(p)f$ is an oldform equivalent to f. Since f is orthogonal to $S_k^{\text{old}}(\Gamma_0(N))$, $T_k(p)f = 0$. By (a), $c_p = 0$.

Let $p \mid N$ but $p^2 \nmid N$. Lemma 9.26 and the same argument show that $T_k(p)f + p^{\frac{k}{2}-1} w_p f = 0$. By (a) and (b),

$$c_p f = T_k(p)f = -p^{\frac{k}{2}-1} w_p f = -p^{\frac{k}{2}-1} \lambda(p) f.$$

Hence $c_p = -p^{\frac{k}{2}-1} \lambda(p)$.

The Euler product expansion follows from Theorem 9.21 and these evaluations of c_p when $p \mid N$. The functional equation is by Theorem 9.8.

CHAPTER X

L FUNCTION OF AN ELLIPTIC CURVE

1. Global Minimal Weierstrass Equations

Let E be an elliptic curve over $\mathbf{Q}$. The L function of E is a certain Euler product that takes into account information about the reduction of E modulo each prime p. This section will deal with some preliminaries that make the definition invariant under admissible changes of variables over $\mathbf{Q}$.

From the start we may assume that the equation is as in (3.23) with integer coefficients. The discriminant Δ will then be an integer, and the p-adic norm will satisfy $|\Delta|_p \leq 1$ with equality if and only if $p \nmid \Delta$. An equation (3.23) is called **minimal** for the prime p if the power of p dividing Δ cannot be decreased by making an admissible change of variables over $\mathbf{Q}$ with the property that the new coefficients are p-integral. It is the same to say that $|\Delta|_p$ cannot be increased by such a change of variables. The equation (3.23) is called a **global minimal Weierstrass equation** if it is minimal for all primes and if its coefficients are integers.

Before considering existence and uniqueness questions for these notions, it will be helpful to have close at hand detailed formulas for an admissible change of variables. Such a change of variables is given as in (3.43a) by

$$x = u^2 x' + r \qquad \text{and} \qquad y = u^3 y' + su^2 x' + t. \tag{10.1}$$

The effect on the coefficients a_i of the Weierstrass equation (3.23) and of the related coefficients b_i, c_i, and Δ is given in Table 10.1. The new coefficients are denoted by primes.

1. GLOBAL MINIMAL WEIERSTRASS EQUATIONS

$$ua'_1 = a_1 + 2s$$
$$u^2 a'_2 = a_2 - sa_1 + 3r - s^2$$
$$u^3 a'_3 = a_3 + ra_1 + 2t$$
$$u^4 a'_4 = a_4 - sa_3 + 2ra_2 - (t+rs)a_1 + 3r^2 - 2st$$
$$u^6 a'_6 = a_6 + ra_4 + r^2 a_2 + r^3 - ta_3 - t^2 - rta_1$$

$$u^2 b'_2 = b_2 + 12r$$
$$u^4 b'_4 = b_4 + rb_2 + 6r^2$$
$$u^6 b'_6 = b_6 + 2rb_4 + r^2 b_2 + 4r^3$$
$$u^8 b'_8 = b_8 + 3rb_6 + 3r^2 b_4 + r^3 b_2 + 3r^4$$

$$u^4 c'_4 = c_4$$
$$u^6 c'_6 = c_6$$
$$u^{12} \Delta' = \Delta$$

TABLE 10.1 Effect of an admissible change of variables

Lemma 10.1. Suppose p is a prime and all the coefficients a_i in (3.23) are p-integral. If $|\Delta|_p > p^{-12}$ or $|c_4|_p > p^{-4}$ or $|c_6|_p > p^{-6}$, then the equation is minimal for the prime p. Conversely if $p > 3$ and $|\Delta|_p \leq p^{-12}$ and $|c_4|_p \leq p^{-4}$, then the equation is not minimal for the prime p.

REMARK. The proof will show how constructively to achieve minimality simultaneously for all primes $p > 3$.

PROOF. Suppose a change of variables (10.1) leads to a system of p-integral coefficients $\{a'_i\}$ with $1 \geq |\Delta'|_p > |\Delta|_p$. Since $u^{12} \Delta' = \Delta$, we have $|u|_p^{12} |\Delta'|_p = |\Delta|_p$, so that $|u|_p < 1$. Then $|u|_p \leq p^{-1}$, and

$$|\Delta|_p = |u|_p^{12} |\Delta'|_p \leq p^{-12} \cdot 1 = p^{-12}.$$

The arguments for c_4 and c_6 are similar.

Conversely let $p > 3$ and $|\Delta|_p \leq p^{-12}$ and $|c_4|_p \leq p^{-4}$. Then (3.31) gives $1728\Delta = c_4^3 - c_6^2$. Since $|1728|_p = 1$, we see that $|c_6|_p \leq p^{-6}$. From §III.2, there is an admissible change of variables leading from (3.23) to

$$y^2 = x^3 - 27c_4 x - 54c_6$$

with discriminant $\Delta' = 2^{12}3^{12}\Delta$. If we make a change of variables (10.1) with $u = p$ and $r = s = t = 0$, we are led to

$$y^2 = x^3 - 27(c_4 p^{-4})x - 54(c_6 p^{-6}).$$

This has p-integral coefficients, since $|c_4 p^{-4}|_p \leq 1$ and $|c_6 p^{-6}|_p \leq 1$, and the discriminant $\Delta'' = p^{-12}\Delta'$ has $|\Delta''|_p = p^{12}|\Delta'|_p = p^{12}|\Delta|_p$. Hence the given equation was not minimal for the prime p.

Proposition 10.2. Fix a prime p and an elliptic curve E over $\mathbf{Q}$.

(a) There exists an admissible change of variables for E over $\mathbf{Q}$ such that the resulting equation is minimal for the prime p.

(b) If E has p-integral coefficients, then the change of variables in (a) has u, r, s, t all p-integral.

(c) Two equations that are minimal for the prime p and that come from E are related by an admissible change of variables in which $|u|_p = 1$ and r, s, t are p-integral.

PROOF. (a) Without loss of generality we may assume E has p-integral coefficients (or actually integral coefficients). Then $|\Delta|_p \leq 1$. Since the range of $|\cdot|_p$ is discrete away from 0, $|\Delta|_p$ can be increased only finitely many times if we are to maintain $|\Delta|_p \leq 1$. Hence in finitely many steps, we can pass to an equation minimal for the prime p.

(b) Let E have coefficients $\{a_i\}$, and let the minimal equation have coefficients $\{a_i'\}$. Since $|\Delta'|_p \geq |\Delta|_p$, we must have $|u|_p \leq 1$. From (3.24), we see that all $\{b_i\}$ and $\{b_i'\}$ are p-integral. Suppose $p \neq 3$. If $|r|_p > 1$, then the equation for $u^8 b_8'$ in Table 10.1 has $3r^4$ as strictly the largest term in p-norm on the right side, contradiction; we conclude $|r|_p \leq 1$. If $p = 3$, we can argue similarly with $u^6 b_6'$ and the term $4r^3$ to see that $|r|_p \leq 1$. Similar arguments with $u^2 a_2'$ and $-s^2$, and then with $u^6 a_6'$ and $-t^2$ give $|s|_p \leq 1$ and $|t|_p \leq 1$.

(c) We apply (b) to the change of variables relating two minimal equations, finding that $|u|_p \leq 1$ and that r, s, t are p-integral. Applying (b) to the inverse change of variables, which involves u^{-1}, we see that $|u^{-1}|_p \leq 1$. Thus $|u|_p = 1$.

Theorem 10.3 (Néron). If E is an elliptic curve over $\mathbf{Q}$, then there exists an admissible change of variables over $\mathbf{Q}$ such that the resulting equation is a global minimal Weierstrass equation. Two such resulting global minimal Weierstrass equations are related by an admissible change of variables with $u = \pm 1$ and with r, s, t in $\mathbf{Z}$.

1. GLOBAL MINIMAL WEIERSTRASS EQUATIONS

PROOF. The uniqueness is immediate from Proposition 10.2c, and we are to prove existence. For existence we may assume that E has integer coefficients a_i. For each p dividing Δ, choose an admissible change of variables $\{u_p, r_p, s_p, t_p\}$ over $\mathbf{Q}$ such that the resulting equation has coefficients $a_{i,p}$ and is minimal for the prime p. By Proposition 10.2b,

$$\text{the rationals } u_p, r_p, s_p, t_p \text{ are } p\text{-integral.} \tag{10.2}$$

If the new discriminant is denoted Δ_p, then Table 10.1 gives

$$|u_p|_p^{12} |\Delta_p|_p = |\Delta|_p. \tag{10.3}$$

Let us write
$$u_p = p^{d_p} v_p \quad \text{with } |v_p|_p = 1. \tag{10.4a}$$

Define
$$u = \prod_{p | \Delta} p^{d_p}. \tag{10.4b}$$

We shall make an admissible change of variables $\{u, r, s, t\}$ in the original equation that leads to an equation with integer coefficients a_i' and discriminant Δ'. Since $u^{12} \Delta' = \Delta$, we have

$$|\Delta'|_p = |u|_p^{-12} |\Delta|_p = |u_p|_p^{-12} |\Delta|_p = |\Delta_p|_p \tag{10.5}$$

by (10.3). Thus the new equation is minimal for all p, hence is globally minimal.

For each p with $p \mid \Delta$, let us write $r_p = p^{\rho_p} m_p / n_p$ with m_p and n_p in $\mathbf{Z}$ and with $|m_p|_p = |n_p|_p = 1$. Let n_p^{-1} be an inverse to n_p modulo p^{6d_p}. We set up the congruence

$$r \equiv p^{\rho_p} m_p n_p^{-1} \mod p^{6d_p}. \tag{10.6}$$

By the Chinese Remainder Theorem, we can find an integer r such that (10.6) is satisfied for all p with $p \mid \Delta$. Then $|n_p r - p^{\rho_p} m_p|_p \leq p^{-6d_p}$ and

$$|r - r_p|_p \leq p^{-6d_p}$$

for all p. Similarly we can find integers s and t such that

$$|s - s_p|_p \leq p^{-6d_p} \quad \text{and} \quad |t - t_p|_p \leq p^{-6d_p} \quad \text{for all } p.$$

Our admissible change of variables $\{u, r, s, t\}$ is now defined, and we are left with showing that the new coefficients $\{a_i'\}$ are integers. We check for all primes p that $|a'_1|_p \leq 1, \ldots, |a'_6|_p \leq 1$, using the formulas of Table 10.1. For $p \nmid \Delta$, there is no problem: Since $|u|_p = 1$ and r, s, t are integers, we have $|a'_i|_p \leq 1$. For $p \mid \Delta$, we estimate each $|a'_i|_p$. The estimates are similar, and we illustrate with a'_2 only. We have

$$u^2 a'_2 = a_2 - sa_1 + 3r - s^2$$
$$= (a_2 - s_p a_1 + 3r_p - s_p^2) - (s - s_p)a_1 + 3(r - r_p) - (s^2 - s_p^2)$$
$$= u_p^2 a'_{2,p} - (s - s_p)a_1 + 3(r - r_p) - (s - s_p)(s + s_p)$$

$|u|_p^2 |a'_2|_p$

$\leq \max\{|u_p^2|_p |a'_{2,p}|_p, |(s - s_p)a_1|_p, |3(r - r_p)|_p, |(s - s_p)(s + s_p)|_p\}$
$\leq \max\{|u_p^2|_p, |s - s_p|_p, |r - r_p|_p\}$ by (10.2)
$\leq \max\{|u_p^2|_p, p^{-6d_p}\} \leq |u_p^2|_p$ by (10.4a).

By (10.4), $|u|_p^2 = |u_p^2|_p$. Thus $|a'_2|_p \leq 1$, and the proof is complete.

The argument in Theorem 10.3 is constructive, provided we know how to produce, for each individual p, an equation that is minimal for the prime p. The proof of Lemma 10.1 shows how to produce such an equation for primes > 3, and an algorithm of Tate, which we do not discuss here, handles the cases $p = 2$ and $p = 3$.

2. Zeta Functions and L Functions

To define the L function of an elliptic curve E over $\mathbf{Q}$ we assume that E is given by a globally minimal Weierstrass equation. This condition is no loss of generality in view of Theorem 10.3.

For each prime p, we consider the reduction E_p of E modulo p. This curve was introduced in §V.2 and is defined over $\mathbf{Z}_p$. It is singular if and only if $p \mid \Delta$. The singular cases were discussed in §III.5. In both the nonsingular and the singular cases we define

$$a_p = p + 1 - \#E_p(\mathbf{Z}_p), \tag{10.7}$$

where $E_p(\mathbf{Z}_p)$ is as usual the set of projective solutions. The **local L factor** for the prime p is the formal power series given by

$$L_p(u) = \begin{cases} \dfrac{1}{1 - a_p u + pu^2} & \text{if } p \nmid \Delta \\ \dfrac{1}{1 - a_p u} & \text{if } p \mid \Delta. \end{cases} \tag{10.8}$$

2. ZETA FUNCTIONS AND L FUNCTIONS

The **L function** of E is the product of the local L factors, with u replaced in the pth factor by p^{-s}:

$$L(s, E) = \prod_{p \mid \Delta} \left[\frac{1}{1 - a_p p^{-s}} \right] \prod_{p \nmid \Delta} \left[\frac{1}{1 - a_p p^{-s} + p^{1-2s}} \right]. \tag{10.9}$$

An elementary convergence result for this Euler product is given in the next proposition. This result will be improved in the next section.

Proposition 10.4. (a) For every prime p, $|a_p| \leq p$.
(b) For $p \nmid \Delta$, the reciprocal roots of $1 - a_p u + p u^2$ are $\leq \sqrt{p}$ in absolute value.
(c) The Euler product defining $L(s, E)$ converges for $\operatorname{Re} s > 2$ and is given there by an absolutely convergent Dirichlet series.

PROOF. The members of $E_p(\mathbf{Z}_p)$ include ∞ and cannot consist of more than two other points for each x in $\mathbf{Z}_p$. Thus $1 \leq \#E_p(\mathbf{Z}_p) \leq 2p+1$ and $|a_p| \leq p$. This proves (a). The reciprocal roots are $\frac{1}{2}(a_p \pm \sqrt{a_p^2 - 4p})$, which is $\leq |a_p|$ in absolute value; thus (a) implies (b). By Proposition 7.6, (b) implies (c).

When $p \mid \Delta$, we can calculate a_p exactly. According to §III.5, when there is a singularity, there is only one, and it is classified as a cusp, a split case of a node, or a nonsplit case of a node. Proposition 3.11 counted the nonsingular points in $\#E_p(\mathbf{Z}_p)$ in each case. Adding one for the singularity, we arrive at the following formula for a_p when $p \mid \Delta$:

$$a_p = \begin{cases} 0 & \text{for the case of a cusp} \\ +1 & \text{for a split case of a node} \\ -1 & \text{for a nonsplit case of a node.} \end{cases} \tag{10.10}$$

Although it is not necessary for the logical development, we shall give some indication of how the local L factors $L_p(u)$ arise. An arithmetically defined L function typically is part of a more naturally defined zeta function, or a variant of such a function. In the case at hand, the zeta function $Z(u, E_p)$ is a generating function that encodes how many points are on the curve in each finite extension of $\mathbf{Z}_p$. If $\mathbf{F}_{p^n}$ denotes the field of p^n elements, the definition is

$$Z(u, E_p) = \exp\left(\sum_{n=1}^{\infty} \frac{\#E_p(\mathbf{F}_{p^n}) u^n}{n} \right).$$

The definition is arranged so that additive formulas for $\#E_p(\mathbf{F}_{p^n})$ make multiplicative contributions to $Z(u, E_p)$: Operationally one calculates with the formula

$$u \frac{d}{du} \log Z(u, E_p) = \sum_{n=1}^{\infty} \#E_p(\mathbf{F}_{p^n}) u^n.$$

For our elliptic curve, calculation of $Z(u, E_p)$ leads to a combination of three polynomials, two appearing in the denominator and one in the numerator:

$$Z(u, E_p) = \begin{cases} \dfrac{1 - a_p u + p u^2}{(1-u)(1-pu)} & \text{if } p \nmid \Delta \\ \dfrac{1 - a_p u}{(1-u)(1-pu)} & \text{if } p \mid \Delta. \end{cases}$$

These three polynomials have separate significance, and the product over p of the (reciprocal of) any of them (after substitution of $u = p^{-s}$) is in principle a meaningful object. However, $\prod \dfrac{1}{1-p^{-s}}$ and $\prod \dfrac{1}{1-p^{1-s}}$ are just $\zeta(s)$ and $\zeta(s-1)$ and give no useful information about E. The remaining polynomial leads to $L(s, E)$, which encodes a great deal of information.

3. Hasse's Theorem

The goal of this section is to establish the following improvement of Proposition 10.4a.

Theorem 10.5 (Hasse). Let E be an elliptic curve over $\mathbf{Q}$ with integer coefficients. For each prime $p \nmid \Delta$, let E_p be the reduction modulo p. Then

$$|p + 1 - \#E_p(\mathbf{Z}_p)| < 2\sqrt{p}. \tag{10.11}$$

Corollary 10.6. The Euler product defining $L(s, E)$ converges for $\operatorname{Re} s > \frac{3}{2}$ and is given there by an absolutely convergent Dirichlet series.

PROOF OF COROLLARY. Let $p \nmid \Delta$. If $a_p = p + 1 - \#E_p(\mathbf{Z}_p)$, the reciprocal roots r of $1 - a_p u + p u^2$ are $r = \frac{1}{2}(a_p \pm \sqrt{a_p^2 - 4p})$. By Theorem 10.5, the square root in this expression is imaginary. Hence $|r|^2 = \frac{1}{4}(a_p^2 + (4p - a_p^2)) = p$ and $|r| = \sqrt{p}$. The corollary therefore follows from Proposition 7.6.

3. HASSE'S THEOREM

The proof of the theorem that we give is due to Manin. In order to be able to normalize E, we first dispose of the cases $p = 2$ and $p = 3$. For these values of p, we have $p < 2\sqrt{p}$. In these cases (10.11) therefore follows from Proposition 10.4a.

For $p > 3$, we can make an admissible change of variables that does not affect the condition $p \nmid \Delta$, does not change $\#E_p(\mathbf{Z}_p)$, and brings the equation of E into the form

$$y^2 = x^3 + ax + b. \tag{10.12}$$

We may therefore assume from the outset that E is given by (10.12).

We shall work with the nonsingular cubic

$$Y^2 = \frac{X^3 + aX + b}{x^3 + ax + b} \tag{10.13}$$

defined over the field $\mathbf{Z}_p(x)$ of rational functions with coefficients in $\mathbf{Z}_p$. Two solutions of (10.13) are

$$(X, Y) = (x, 1) \quad \text{and} \quad (X, Y) = (x^p, (x^3 + ax + b)^{\frac{1}{2}(p-1)}).$$

If we specify ∞ as identity, the projective solutions of (10.13) over $\mathbf{Z}_p(x)$ form a group, by Theorem 3.8, and we form the group element

$$Z_n = (x^p, (x^3 + ax + b)^{\frac{1}{2}(p-1)})) + n(x, 1) \tag{10.14}$$

for each integer n, $-\infty < n < \infty$. We define a corresponding sequence of integers $d_n \geq 0$ as follows: If $Z_n = \infty$, then $d_n = 0$. Otherwise Z_n is of the form (X_n, Y_n); in this case we reduce X_n to lowest terms in $\mathbf{Z}_p(x)$, and we let d_n be the larger of the degree of the numerator and the degree of the denominator of X_n. Let us isolate the statements of the two main steps.

Lemma 10.7. $d_{-1} - d_0 - 1 = \#E_p(\mathbf{Z}_p) - p - 1$.

PROOF. By (10.14) with $n = 0$, $d_0 = p$. Let $N_p = \#E_p(\mathbf{Z}_p) - 1$ be the number of affine solutions. We are to prove that

$$d_{-1} = N_p + 1. \tag{10.15}$$

When the addition formulas of (3.73) are recomputed to take into account the form (10.13), we obtain

$$X_{-1} = -x - x^p + \frac{[1 + (x^3 + ax + b)^{\frac{1}{2}(p-1)}]^2[x^3 + ax + b]}{(x - x^p)^2}. \tag{10.16}$$

Clearing fractions, we see that

$$X_{-1} = \frac{x^{2p+1} + R(x)}{(x - x^p)^2}$$

with $\deg(R) \leq 2p$. Therefore the degree of the numerator is 1 greater than the degree of the denominator, and the degree of the denominator is $d_{-1} - 1$. We shall compute the degree of the denominator after reducing X_{-1} as much as possible.

The denominator splits over $\mathbf{Z}_p$ as $\prod_{j \in \mathbf{Z}_p}(x - j)^2$. The linear factors $x - k$ of the numerator of the fraction in (10.16) are those such that the numerator vanishes at k. In terms of the Legendre symbol (referring to quadratic residues), we have

$$(j^3 + aj + b)^{\frac{1}{2}(p-1)} = \left(\frac{j^3 + aj + b}{p}\right).$$

The factor $[1+(j^3+aj+b)^{\frac{1}{2}(p-1)}]$ vanishes if and only if $\left(\frac{j^3 + aj + b}{p}\right) = -1$, and in this case $x - j$ occurs as a factor of the numerator exactly twice. The other factor $[j^3 + aj + b]$ vanishes if and only if $x - j$ is a factor of $x^3 + ax + b$, and in this case $x - j$ occurs as a factor of the numerator exactly once. The factors that remain in the denominator are $(x-j)^2$ when $\left(\frac{j^3 + aj + b}{p}\right) = +1$ and $x - j$ when $\left(\frac{j^3 + aj + b}{p}\right) = 0$. They correspond exactly to the affine $\mathbf{Z}_p$ solutions of E_p, the first ones giving two values of y and the second ones giving just $y = 0$.

Hence the number of surviving factors in the denominator, which we saw earlier is $d_{-1} - 1$, is exactly N_p. This equality is (10.15), and the lemma is proved.

Lemma 10.8. For $-\infty < n < \infty$,

$$d_{n-1} + d_{n+1} = 2d_n + 2. \tag{10.17}$$

PROOF. First suppose one of Z_{n-1}, Z_n, Z_{n+1} is ∞. Then neither of the other two is ∞, in view of (10.14). Say $Z_n = \infty$. Then $d_n = 0$ and

$$Z_{n+1} = (x, 1), \qquad Z_{n-1} = -(x, 1) = (x, -1).$$

Hence $d_{n+1} = d_{n-1} = 1$, and (10.17) is verified.

Say $Z_{n-1} = \infty$. Then $d_{n-1} = 0$ and

$$Z_n = (x, 1), \qquad Z_{n+1} = \left(\frac{(x^2 - a)^2 - 8bx}{4(x^3 + ax + b)}, Y_{n+1}\right)$$

by a recomputation of (3.74). Hence $d_n = 1$ and $d_{n+1} = 4$, and (10.17) is verified. The remaining case, with $Z_{n+1} = \infty$, is similar.

Now suppose that none of Z_{n-1}, Z_n, Z_{n+1} is ∞. Write X_{n-1}, X_n, X_{n+1} in lowest terms as

$$X_{n-1} = \frac{A}{B}, \quad X_n = \frac{P}{Q}, \quad X_{n+1} = \frac{C}{D}.$$

The addition formulas allow us to express X_{n-1} and X_{n+1} in terms of X_n as follows:

$$X_{n-1} = \frac{-(Qx+P)(Qx-P)^2 + (1+Y_n)^2(x^3+ax+b)Q^3}{Q(Qx-P)^2}$$
$$X_{n+1} = \frac{-(Qx+P)(Qx-P)^2 + (1-Y_n)^2(x^3+ax+b)Q^3}{Q(Qx-P)^2}.$$
(10.18)

Addition of the two formulas (10.18) and use of (10.13) give

$$\tfrac{1}{2}(X_{n-1}+X_{n+1})$$
$$= \frac{-(Qx+P)(Qx-P)^2 + (x^3+ax+b)Q^3 + ((\tfrac{P}{Q})^3 + a(\tfrac{P}{Q}) + b)Q^3}{Q(Qx-P)^2}$$
$$= \frac{PQx^2 + P^2x + axQ^2 + 2bQ^2 + aPQ}{(Qx-P)^2}.$$
(10.19)

Multiplication of the two formulas (10.18) and some manipulations give

$$X_{n-1}X_{n+1} = \frac{(Px-aQ)^2 - 4bQ(Qx+P)}{(Qx-P)^2}$$
(10.20)

Now

$$X_{n-1}X_{n+1} = \frac{AC}{BD}$$

also, and the claim is that

$$BD = (Qx-P)^2,$$
(10.21)

up to a $\mathbf{Z}_p$ factor. If S is the greatest common divisor of AC and BD, (10.20) gives

$$AC = S[(Px-aQ)^2 - 4bQ(Qx+P)] \quad (10.22a)$$
$$BD = S(Qx-P)^2, \quad (10.22b)$$

while the numerator of (10.19) gives

$$AD + BC = 2S[PQx^2 + P^2x + axQ^2 + 2bQ^2 + aPQ]. \qquad (10.22c)$$

Let F be a prime factor of S. Then (10.22b) shows that $F \mid BD$. Without loss of generality, suppose $F \mid B$. Then $F \nmid A$ since $\text{GCD}(A, B) = 1$. Since (10.22a) shows that $F \mid AC$, $F \mid C$. Thus $F \mid BC$. By (10.22c), $F \mid (AD + BC)$. So $F \mid AD$. Since $F \nmid A$, $F \mid D$. Then $F \mid C$ and $F \mid D$, in contradiction to $\text{GCD}(C, D) = 1$. We conclude that S is a scalar, and (10.21) is proved.

The integers in (10.15) are

$$d_{n-1} = \max\{\deg A, \deg B\}$$
$$d_{n+1} = \max\{\deg C, \deg D\}$$
$$d_n = \max\{\deg P, \deg Q\}.$$

We divide the proof of (10.17) into the following cases:
(a) $d_{n-1} = \deg A$ and $d_{n+1} = \deg C$
(b) $d_{n-1} = \deg B$ and $d_{n+1} = \deg D$
(c) $d_{n-1} = \deg A$ and $d_{n+1} = \deg D$, but not Case (a) or (b)
(d) $d_{n-1} = \deg B$ and $d_{n+1} = \deg C$, but not Case (a) or (b).

Case (a). By (10.21) and (10.22), we have

$$d_{n-1} + d_{n+1} = \deg(AC) = \deg[(Px - aQ)^2 - 4bQ(Qx + P)].$$

If $\deg P \geq \deg Q$, then the unique term on the right of highest degree is P^2x^2; so the right side is $d_n + 2$, and (10.17) follows. If $\deg P < \deg Q$, then (10.21) gives $\deg BD = 2\deg Q + 2$. Then

$$\deg(AC) \leq \max\{2\deg P + 2, 2\deg Q, 2\deg Q + 1, \deg P + \deg Q\}$$
$$\leq 2\deg Q + 1 < \deg(BD).$$

But this inequality contradicts the hypothesis of Case (a). Hence $\deg P < \deg Q$ is impossible.

Case (b). By (10.21), we have

$$d_{n-1} + d_{n+1} = \deg(BD) = \deg[(Qx - P)^2].$$

If $\deg Q \geq \deg P$, then the unique term on the right of highest degree is Q^2x^2; so the right side is $d_n + 2$, and (10.17) follows. If $\deg Q < \deg P$, then (10.22a) gives

$$\deg(AC) = \deg(P^2x^2) > \deg(P^2) \geq \deg(BD).$$

But this inequality contradicts the hypothesis of Case (b). Hence $\deg Q < \deg P$ is impossible.

Case (c). Since we are not in Case (a) or (b), $\deg A > \deg B$ and $\deg D > \deg C$. Thus

$$\deg AD > \deg AC, \qquad \deg AD > \deg BD, \qquad \deg AD > \deg BC. \tag{10.23}$$

From the third of these and from (10.22c), we have

$$\begin{aligned}\deg(AD) &= \deg(AD + BC) \\ &= \deg[PQx^2 + P^2x + axQ^2 + 2bQ^2 + aPQ].\end{aligned} \tag{10.24}$$

If $\deg P \geq \deg Q$, (10.24) is

$$\leq \deg(P^2x^2) = \deg(AC),$$

in contradiction to the first inequality of (10.23). If $\deg P < \deg Q$, then (10.24) is

$$\leq \deg(Q^2x^2) = \deg(BD),$$

in contradiction to the second inequality of (10.23). Thus Case (c) is impossible.

Case (d). This case is symmetric with Case (c) and similarly is impossible.

This completes the proof of the lemma.

It is an easy matter to combine the two lemmas to prove the theorem. Induction forwards and backwards by means of Lemma 10.8 gives the formula

$$d_n = n^2 - (d_{-1} - d_0 - 1)n + d_0. \tag{10.25}$$

(The base cases for the induction are $n = 0$ and $n = -1$, for which (10.25) is trivial.) Substitution from Lemma 10.7 and use of $d_0 = p$ gives

$$d_n = n^2 + a_p n + p,$$

where $a_p = p + 1 - \#E_p(\mathbf{Z}_p)$ as in (10.7). The d_n's are degrees of polynomials and are therefore ≥ 0. Moreover, two consecutive d_n's cannot both be 0. Since a_p is an integer, it follows that $r^2 + a_p r + p \geq 0$ for all real r. The discriminant of this polynomial must be ≤ 0, and thus $|a_p| \leq 2\sqrt{p}$. This completes the proof of Theorem 10.5.

CHAPTER XI

EICHLER-SHIMURA THEORY

1. Overview

We return to the theme announced in the preface and discussed a little in §§VII.5 and VIII.1. We saw in Chapter VII that the Riemann zeta function $\zeta(s)$ and the Dirichlet L functions $L(s,\chi)$, which encode subtle arithmetic information about primes, can be obtained as Mellin transforms of certain θ functions that have transformation laws akin to those of modular forms. In other words, $\zeta(s)$ and $L(s,\chi)$ arise as automorphic L functions. Consequently they have analytic continuations and functional equations, and their analytic properties are more manageable.

Our objective in this chapter and the next is to address the corresponding problem for the L function $L(s, E)$ of an elliptic curve over $\mathbf{Q}$. We would like $L(s, E)$ to have an analytic continuation and satisfy a functional equation, and the likely way to obtain these conclusions is to identify $L(s, E)$ as an automorphic L function. A special case of a theory of Eichler and Shimura gives a clue to the nature of the automorphic objects to use. Their theory takes certain cusp forms f in $S_2(\Gamma_0(N))$ and gives a geometric construction of elliptic curves over $\mathbf{Q}$ such that $L(s, E) = L(s, f)$.

We shall discuss this theory in the present chapter. The Taniyama-Weil Conjecture anticipates that every elliptic curve over $\mathbf{Q}$ is obtained in this way from $S_2(\Gamma_0(N))$ if the Eichler-Shimura map is followed by a relatively simple kind of map called an "isogeny." This conjecture will be the subject of Chapter XII.

By way of introduction we begin with a construction that passes from members f of $S_2(\Gamma_0(N))$ to homomorphisms from $\Gamma_0(N)$ into the additive complex numbers. For f in $S_2(\Gamma_0(N))$, $f(\zeta)\,d\zeta$ is invariant under $\Gamma_0(N)$. Fix τ_0 in $\mathcal{H}$ and define

$$F(\tau) = \int_{\tau_0}^{\tau} f(\zeta)\,d\zeta. \tag{11.1}$$

1. OVERVIEW

Since f is analytic, $F(\tau)$ is independent of the path of the integral. For $\gamma \in \Gamma_0(N)$, the invariance of $f(\zeta)\, d\zeta$ gives

$$F(\gamma(\tau)) = \int_{\tau_0}^{\gamma(\tau)} f(\zeta)\, d\zeta = \int_{\gamma(\tau_0)}^{\gamma(\tau)} f(\zeta)\, d\zeta + \int_{\tau_0}^{\gamma(\tau_0)} f(\zeta)\, d\zeta$$

$$= \int_{\tau_0}^{\tau} f(\zeta)\, d\zeta + \int_{\tau_0}^{\gamma(\tau_0)} f(\zeta)\, d\zeta = F(\tau) + \int_{\tau_0}^{\gamma(\tau_0)} f(\zeta)\, d\zeta. \tag{11.2}$$

Put

$$\Phi_f(\gamma) = \int_{\tau_0}^{\gamma(\tau_0)} f(\zeta)\, d\zeta.$$

We claim that $\Phi_f(\gamma)$ is independent of τ_0. In fact, if we define $F_1(\tau)$ by (11.1) with τ_1 in place of τ_0, then (11.2) gives

$$F_1(\gamma(\tau)) = F_1(\tau) + \int_{\tau_1}^{\gamma(\tau_1)} f(\zeta)\, d\zeta. \tag{11.3}$$

Since $F_1 = F + \int_{\tau_1}^{\tau_0} f(\zeta)\, d\zeta$, comparison of (11.2) and (11.3) gives

$$\int_{\tau_0}^{\gamma(\tau_0)} f(\zeta)\, d\zeta = \int_{\tau_1}^{\gamma(\tau_1)} f(\zeta)\, d\zeta,$$

as asserted.

Proposition 11.1. For f in $S_2(\Gamma_0(N))$, Φ_f is a homomorphism of $\Gamma_0(N)$ into the additive complex numbers. If $\gamma_0 \in \Gamma_0(N)$ is **elliptic** (i.e., $|\text{Tr } \gamma_0| < 2$) or **parabolic** (i.e., $|\text{Tr } \gamma_0| = 2$), then $\Phi_f(\gamma_0) = 0$.

PROOF. The invariance of $f(\zeta)\, d\zeta$ under $\Gamma_0(N)$ gives

$$\Phi_f(\gamma_1 \gamma_2) = \int_{\tau_0}^{\gamma_1 \gamma_2 \tau_0} = \int_{\tau_0}^{\gamma_1 \tau_0} + \int_{\gamma_1 \tau_0}^{\gamma_1 \gamma_2 \tau_0}$$

$$= \int_{\tau_0}^{\gamma_1 \tau_0} + \int_{\tau_0}^{\gamma_2 \tau_0} = \Phi_f(\gamma_1) + \Phi_f(\gamma_2),$$

and Φ_f is a homomorphism. If $\gamma_0 \in \Gamma_0(N)$ is elliptic, it has finite order and hence so does its image in $\mathbb{C}$. Thus $\Phi_f(\gamma_0) = 0$.

Suppose $\gamma_0 = \begin{pmatrix} a & b \\ c & d \end{pmatrix}$ is parabolic in $\Gamma_0(N)$. Since $|\text{Tr } \gamma_0| = 2$ and $\det \gamma_0 = 1$, we readily find that the equation $\dfrac{az+b}{cz+d} = z$ has a double root, namely $z = \frac{a-1}{c}$. Thus $\gamma_0(\frac{a-1}{c}) = \frac{a-1}{c}$. Choose β in $SL(2,\mathbf{Z})$ with $\frac{a-1}{c} = \beta^{-1}(\infty)$. Then $\gamma_1 = \beta\gamma_0\beta^{-1}$ has

$$\gamma_1(\infty) = \beta\gamma_0\beta^{-1}(\infty) = \beta\gamma_0(\tfrac{a-1}{c}) = \beta(\tfrac{a-1}{c}) = \infty.$$

Thus $\gamma_1 = \pm \begin{pmatrix} 1 & l \\ 0 & 1 \end{pmatrix}$ for some l in $\mathbf{Z}$. By (9.17), the integer l is a multiple of the width h of the cusp $\beta^{-1}(\infty) = \frac{a-1}{c}$. The Fourier expansion

$$(f \circ [\beta^{-1}]_2)(\tau) = \sum_{n=1}^{\infty} c_n e^{2\pi i n \tau / h}$$

has zero constant term since f is a cusp form, and the analog for Chapter IX of (8.7b) therefore gives

$$\int_{\tau_1}^{\tau_1 + h} (f \circ [\beta^{-1}]_2)(\zeta) \, d\zeta = 0$$

for any τ_1. Hence also

$$\int_{\tau_1}^{\tau_1 + l} (f \circ [\beta^{-1}]_2)(\zeta) \, d\zeta = 0. \tag{11.4}$$

Now the change of variables $\zeta' = \beta^{-1}(\zeta)$ gives

$$\Phi_f(\gamma_0) = \int_{\tau_0}^{\beta^{-1}\gamma_1\beta\tau_0} f(\zeta') \, d\zeta' = \int_{\beta\tau_0}^{\gamma_1\beta\tau_0} f(\beta^{-1}\zeta)\delta(\beta^{-1},\zeta)^{-2} \, d\zeta$$
$$= \int_{\beta\tau_0}^{\beta\tau_0 + l} f \circ [\beta^{-1}]_2(\zeta) \, d\zeta,$$

and this is 0 by (11.4). This completes the proof.

Let us consider the relatively simple case that $N = 11$. According to Theorem 9.10, $S_2(\Gamma_0(11))$ is one-dimensional. A nonzero element is

$$f(\tau) = \eta(\tau)^2 \eta(11\tau)^2 = \sum_{n=1}^{\infty} c_n q^n \tag{11.5}$$

$$- q - 2q^2 - q^3 + 2q^4 + q^5 + 2q^6 - 2q^7$$
$$- 2q^9 - 2q^{10} + q^{11} - 2q^{12} + 4q^{13} + 4q^{14}$$
$$- q^{15} - 4q^{16} - 2q^{17} + 4q^{18} + 2q^{20} + \cdots,$$

1. OVERVIEW

according to Example 5 in §IX.3. The group $\Gamma_0(11)$ turns out to be generated by

$$T = \begin{pmatrix} 1 & 1 \\ 0 & 1 \end{pmatrix}, \quad V_4 = \begin{pmatrix} 8 & 1 \\ -33 & -4 \end{pmatrix}, \quad V_6 = \begin{pmatrix} 9 & 1 \\ -55 & -6 \end{pmatrix}.$$

Since T is parabolic, Proposition 11.1 shows that the image of Φ_f is generated by $\Phi_f(V_4)$ and $\Phi_f(V_6)$. We have

$$\Phi_f(\gamma) = \int_{\tau_0}^{\gamma(\tau_0)} f(\zeta)\,d\zeta$$

$$= \frac{1}{2\pi i} \int_{e^{2\pi i \tau_0}}^{e^{2\pi i \gamma(\tau_0)}} \sum_{n=1}^{\infty} c_n q^{n-1}\,dq = \frac{1}{2\pi i}\left(H(e^{2\pi i \gamma(\tau_0)}) - H(e^{2\pi i \tau_0})\right),$$

where

$$H(q) = \sum_{n=1}^{\infty} \frac{c_n q^n}{n}.$$

Judicious choice of τ_0 cuts down considerably on the number of terms needed. For $\tau_0 = i$, computation with 10000 terms gives 11 digits accuracy. For $\tau_0 = .125 + .025i$ with 600 terms, we obtain

$$H(e^{2\pi i \tau_0}) \doteq .2628106125215867 + .5230455366717368i$$
$$H(e^{2\pi i V_4(\tau_0)}) \doteq .897415264661364 - .93577108802667581i$$
$$H(e^{2\pi i V_6(\tau_0)}) \doteq 1.1532019916801141 + .523045536671738i$$

with 14 digits accuracy. (With 300 terms, we would have had 10 digits accuracy.) Thus

$$\begin{aligned} \omega_1 &= \Phi_f(V_4) \doteq -.232177875650357 - .101000467297158i \\ \omega_2 &= \Phi_f(V_6) \doteq -.202000934594317i. \end{aligned} \qquad (11.6)$$

The complex numbers ω_1 and ω_2 are independent over $\mathbf{R}$ and therefore define a lattice Λ in $\mathbf{C}$. For any γ in $\Gamma_0(N)$, (11.2) shows that $F(\gamma(\tau)) - F(\tau)$ is in Λ, hence projects to the 0 element of $\mathbf{C}/\Lambda$. Thus the map F of (11.1), which initially carries $\mathcal{H}$ into $\mathbf{C}$ (and hence also $\mathbf{C}/\Lambda$), descends to a holomorphic map of $\mathcal{R} = \Gamma_0(11)\backslash\mathcal{H}$ into $\mathbf{C}/\Lambda$. In turn, Theorem 6.14 exhibits a biholomorphic map of $\mathbf{C}/\Lambda$ onto $E(\mathbf{C})$ for an elliptic curve E. If $X_0(11)$ denotes the compactification of $\mathcal{R}$ discussed briefly in the proof of Theorem 9.9 and called $\mathcal{R}^*$ there, then F actually yields a holomorphic map of the compact Riemann surface $X_0(11)$ onto $\mathbf{C}/\Lambda$ and then onto $E(\mathbf{C})$.

Two miracles occur in this construction. The first miracle is that $X_0(11)$, E, and the mapping can be defined compatibly over $\mathbf{Q}$. We have not yet defined rational maps and rational projective curves (other than plane curves), and we shall need to make such definitions in order to make our assertion precise. We return to a discussion of rationality shortly.

The second miracle is that the L function of E matches the L function of the cusp form f in (11.5). This assertion relies on introducing new interpretations of Hecke operators and their eigenvalues.

Let us give a numerical indication of why E can be defined over $\mathbf{Q}$. Substituting from (11.6) into (6.50), we find that

$$g_2(\Lambda) \doteq 64419.8788704867 - .0000000003i$$
$$g_3(\Lambda) \doteq -5699399.99557174 + .00000002i.$$

Then (8.2) and (8.3) give

$$j(\Lambda) \doteq -757.672637860052 + .000000000008i. \qquad (11.7a)$$

The actual value is

$$j(\Lambda) = -\frac{(2^4 \cdot 31)^3}{11^5} \doteq -757.672637860057. \qquad (11.7b)$$

The significance of $j(\Lambda)$ for the present context is explained in Proposition 3.7. With the value of $j(\Lambda)$ as in (11.7b), the proposition shows that E can be defined over $\mathbf{Q}$.

Let us classify the inequivalent $\mathbf{Q}$ structures on E and see that there are infinitely many of them. It will turn out that any two of them are equivalent over a quadratic extension of $\mathbf{Q}$. If E ultimately is given in global minimal Weierstrass form, then the integer tuple (c_4, c_6, Δ) determines the $\mathbf{Q}$ structure of E, and the equation of E is unique up to the kind of admissible change of variables described in Theorem 10.3. Since

$$j = \frac{c_4^3}{\Delta} = 1728 + \frac{c_6^2}{\Delta},$$

we readily find that the only possibilities are

$$\begin{aligned}c_4 &= 2^4 \cdot 31 m^2 \\ c_6 &= -2^3 \cdot 2501 m^3 \\ \Delta &= -11^5 m^6\end{aligned} \qquad (11.8)$$

1. OVERVIEW

for an integer $m \neq 0$. We shall see that there are some restrictions on m. The curve

$$y^2 = x^3 - 4mx^2 - 160m^2x - 1264m^3 \tag{11.9}$$

has

$$c_4 = 2^4(2^4 \cdot 31m^2), \qquad c_6 = 2^6(-2^3 \cdot 2501m^3), \qquad \Delta = 2^{12}(-11^5 m^6)$$

and thus must arise from E by an admissible change of variables with $u = 2$. We claim that m has no odd prime square factor p^2. In fact, otherwise we could use $u = 1/p$ in (11.9) and change variables to find that E was not minimal at the prime p. Similarly 16 does not divide m, since otherwise we could use $u = 1/4$ to find that E was not minimal at the prime 2.

If $m \equiv 1 \bmod 4$, then (11.9) is equivalent with the curve

$$y^2 + y = x^3 - mx^2 - 10m^2 x - \tfrac{1}{4}(79m^3 + 1), \tag{11.10a}$$

which has (c_4, c_6, Δ) given by (11.8) and is therefore global minimal, by Lemma 10.1. If $m \equiv 2 \bmod 4$, then similarly (11.9) is equivalent with

$$y^2 = x^3 - mx^2 - 10m^2 x - 158(\tfrac{1}{2}m)^3, \tag{11.10b}$$

which has (c_4, c_6, Δ) given by (11.8) and hence is global minimal. If $m \equiv 3 \bmod 4$, then we can attempt a reduction at the prime 2 by rewriting (11.9) as

$$(y + ax + b)^2 = (x + c)^3 - 4m(x + c)^2 - 160m^2(x + c) - 1264m^3$$

or

$$\begin{aligned} y^2 + 2axy + 2by = x^3 &+ (3c - 4m - a^2)x^2 \\ &+ (3c^2 - 8mc - 2ab - 160m^2)x \\ &+ (c^3 - 4mc^2 - b^2 - 160m^2 c - 1264m^3). \end{aligned}$$

To have a reduction, we need 2^j to divide the usual coefficient a_j of this equation. From a_3 we obtain $4 \mid b$. From a_4 we obtain $2 \mid c$. Then a_2 gives $2 \mid a$, and a_4 gives $4 \mid c$. Hence a_6 gives $64 \mid (b^2 + 1264m^3)$. Writing $b = 4b'$, we see that $4 \mid (b'^2 + 79m^3)$. But this is impossible since $79m^3 \equiv 1 \bmod 4$. Thus (11.9) is already global minimal if $m \equiv 3 \bmod 4$.

If $4 \mid m$, write $m = 4m'$ with $4 \nmid m'$. Then we can reduce (11.9) to

$$y^2 = x^3 - 4m'x^2 - 160m'^2 x - 1264m'^3, \tag{11.11}$$

and we must have arrived at a global minimal form. Since we have established that $4 \nmid m'$, the preceding paragraph shows that $m' \equiv 3$ mod 4.

Putting everything together, we see that we get a $\mathbf{Q}$ structure whose global minimal equation satisfies (11.8) exactly when m has no odd square factor and m is congruent modulo 16 to one of

$$1, 2, 5, 6, 9, 10, 12, 13, 14. \tag{11.12}$$

The simplest $\mathbf{Q}$ structure comes from $m = 1$, and the global minimal equation can be taken as

$$E: \qquad y^2 + y = x^3 - x^2 - 10x - 20. \tag{11.13}$$

This is the equation of interest. The $\mathbf{Q}$ structures corresponding to other values of m are called **twists** of E.

We have obtained a mapping of $X_0(11)$ onto $E(\mathbf{C})$. There are other elliptic curves that $X_0(11)$ maps onto. Define

$$\begin{pmatrix} \omega_2' \\ \omega_1' \end{pmatrix} = \begin{pmatrix} 1 & 3 \\ 0 & 5 \end{pmatrix} \begin{pmatrix} \omega_2 \\ \omega_1 \end{pmatrix}, \tag{11.14}$$

and let $\Lambda' = \mathbf{Z}\omega_1' \oplus \mathbf{Z}\omega_2'$. Calculating as with Λ, we find that

$$j(\Lambda') \doteq -372.363636363639 - .000000000004i.$$

The actual value is

$$j(\Lambda') = -\frac{16^3}{11} \doteq -372.363636363636.$$

The result is an elliptic curve E' that can be defined over $\mathbf{Q}$. The simplest $\mathbf{Q}$ structure corresponds to the equation

$$E': \qquad y^2 + y = x^3 - x^2 \tag{11.15}$$

with $(c_4, c_6, \Delta) = (16, -152, -11)$. The same analysis as with E yields twists for the same m's as in (11.12), but only (11.15) will be of interest to us.

1. OVERVIEW

The inclusion $\Lambda' \subseteq \Lambda$ yields a holomorphic homomorphism

$$E'(\mathbf{C}) \cong \mathbf{C}/\Lambda' \longrightarrow \mathbf{C}/\Lambda \cong E(\mathbf{C}). \tag{11.16a}$$

Such a map is called an **isogeny**. (Cf. §VI.4.) This particular isogeny can be defined over $\mathbf{Q}$. Namely if (x, y) satisfies (11.15) and if X and Y are defined by

$$X = x + \frac{1}{x^2} + \frac{2}{x-1} + \frac{1}{(x-1)^2}$$

$$Y = y - (2y+1)\left(\frac{1}{x^3} + \frac{1}{(x-1)^3} + \frac{1}{(x-1)^2}\right),$$

then (X, Y) satisfies (11.13).

To any isogeny corresponds a **dual isogeny** in the reverse direction. Namely (11.14) yields

$$\begin{pmatrix} 5\omega_2 \\ 5\omega_1 \end{pmatrix} = \begin{pmatrix} 5 & 0 \\ 0 & 5 \end{pmatrix} \begin{pmatrix} \omega_2 \\ \omega_1 \end{pmatrix}$$

$$= \begin{pmatrix} 5 & -3 \\ 0 & 1 \end{pmatrix} \begin{pmatrix} 1 & 3 \\ 0 & 5 \end{pmatrix} \begin{pmatrix} \omega_2 \\ \omega_1 \end{pmatrix} = \begin{pmatrix} 5 & -3 \\ 0 & 1 \end{pmatrix} \begin{pmatrix} \omega'_2 \\ \omega'_1 \end{pmatrix}.$$

Since $\mathbf{C}/(5\Lambda) \cong \mathbf{C}/\Lambda$, we obtain a holomorphic homomorphism

$$E(\mathbf{C}) \cong \mathbf{C}/(5\Lambda) \longrightarrow \mathbf{C}/\Lambda' \cong E'(\mathbf{C}). \tag{11.16b}$$

Since (11.16a) is defined over $\mathbf{Q}$, so is (11.16b). But we shall not write down explicit formulas. In any event, composition of $X_0(11) \to E(\mathbf{C})$ with (11.16b) yields a rational map of $X_0(11)$ onto E'.

Similarly if we define

$$\begin{pmatrix} \omega''_2 \\ \omega''_1 \end{pmatrix} = \begin{pmatrix} 1 & 0 \\ 0 & 5 \end{pmatrix} \begin{pmatrix} \omega_2 \\ \omega_1 \end{pmatrix}, \tag{11.17}$$

and let $\Lambda'' = \mathbf{Z}\omega''_1 \oplus \mathbf{Z}\omega''_2$, then the actual value of j is

$$j(\Lambda'') = -\frac{(375376)^3}{11},$$

and we are led to the equation

$$E'' : \quad y^2 + y = x^3 - x^2 - 7820x - 263580, \tag{11.18}$$

as well as various twists. The inclusion $\Lambda'' \subseteq \Lambda$ yields an isogeny $E''(\mathbb{C}) \to E(\mathbb{C})$, and this isogeny is defined over $\mathbb{Q}$. The dual isogeny yields by composition a rational map of $X_0(N)$ onto E''.

Thus $X_0(11)$ maps rationally onto the three curves E, E', and E'', all as a result of the one cusp form f in (11.5). Elliptic curves that are isogenous over $\mathbb{Q}$ have the same L function (Theorem 11.67). By the Eichler-Shimura theory the common L function for these three curves is the same as the L function as f. The maps of $X_0(N)$ to E, E', and E'' are called **modular parametrizations** of these curves. These maps are determined by giving their x and y coordinates, which are meromorphic functions on $X_0(N)$ with some additional property to reflect the rationality. The rationality is important: f gives us holomorphic maps of $X_0(11)$ onto the twists of E, E', and E'', but we do not get a match of L functions in these cases.

So far, we have discussed in detail $X_0(11)$ only. The special feature of this case is that $X_0(11)$ has genus one and that correspondingly $S_2(\Gamma_0(11))$ has dimension one. For general $\Gamma_0(N)$, Proposition 11.1 says that any cusp form f in $S_2(\Gamma_0(N))$ yields a homomorphism $\Phi_f : \Gamma_0(N) \to \mathbb{C}$ that annihilates the elliptic and parabolic elements, but the situation is of interest only when the image of Φ_f is a lattice. It turns out that we do get a lattice when f is a newform and all the eigenvalues of the Hecke operators are integers. Then the resulting elliptic curve E is defined over $\mathbb{Q}$, and so are $X_0(N)$ and the map from $X_0(N)$ to E. Moreover the L functions of f and E match.

The approach to these results is a little indirect. The "Jacobian variety" $J(X_0(N))$ of $X_0(N)$ is a certain complex torus with the universal mapping property that any map of $X_0(N)$ to an elliptic curve must factor through $J(X_0(N))$. Because of the universal property, the construction of the map from $X_0(N)$ to E amounts to obtaining E as a quotient of $J(X_0(N))$. The construction does not yield an equation for E directly. Instead, such an equation has to be worked out from side information; we shall take up this matter in §XII.3.

A certain amount of this chapter is about background material. Sections 2-4 deal with compact Riemann surfaces in general and with $X_0(N)$ in particular. Some of the results that we quote about compact Riemann surfaces will not be used in the form as stated but are helpful as motivation for properties of projective curves later in the chapter. Section 5 applies some of the more elementary properties of compact Riemann surfaces to derive properties of Hecke operators on $S_2(\Gamma_0(N))$. We shall see that $S_2(\Gamma_0(N))$ has a basis in which the Hecke operators are given by integer matrices. Consequently the eigenvalues of the Hecke operators are algebraic integers. Sections 6-8 deal with general projective curves and show how $X_0(N)$ can be given a canonical $\mathbb{Q}$ structure. Sections 9 and 10 discuss abstract elliptic curves, isogenies, abelian varieties, and the algebraic Jacobian variety. The main results of Eichler-Shimura theory are in §§11-12.

2. Riemann Surface $X_0(N)$

In this section we shall give a more careful construction of the compactification $X_0(N)$ of $\Gamma_0(N)\backslash \mathcal{H}$ than we gave in §IX.5. In §§3-4 we shall go on to discuss some generalities about Riemann surfaces. We omit many proofs, giving references in the section of Notes at the end.

Let $\mathcal{C} = \mathbb{Q} \cup \{\infty\}$, the set whose equivalence classes form the cusps of §IX.2. Put $\mathcal{H}^* = \mathcal{H} \cup \mathcal{C}$, and topologize $\mathcal{H}^*$ as follows: A basic open set about a point of $\mathcal{H}$ is an open disc wholly within $\mathcal{H}$, and a basic open set about the member ∞ of $\mathcal{C}$ is $\{\text{Im } \tau > r\}$ for each $r > 0$. If $x \in \mathbb{Q}$ is in $\mathcal{C}$, a basic open set about x is of the form $D \cup \{x\}$, where D is an open disc in $\mathcal{H}$ of radius $y > 0$ and center $x + iy$. The resulting topology on $\mathcal{H}^*$ is Hausdorff, $\mathcal{H}$ is an open subset, and $SL(2, \mathbb{Z})$ acts continuously.

Let $X_0(N) = \Gamma_0(N)\backslash \mathcal{H}^*$. It is easy to see that $X_0(N)$ is compact. Moreover, with some effort one can show that $X_0(N)$ is Hausdorff. A key step in the argument is Lemma 11.2 below. For z in $\mathcal{H}^*$, let $\Gamma_z = \{\gamma \in \Gamma_0(N) \mid \gamma(z) = z\}$.

Lemma 11.2. If τ is in $\mathcal{H}$, then there exists an open neighborhood V_τ of τ in $\mathcal{H}$ with $V_\tau \cap \gamma(V_\tau) = \emptyset$ for all γ in $\Gamma_0(N)$ not in Γ_τ. The set V_τ may be taken to be stable under Γ_τ.

We shall introduce a system of charts

$$\{(U_z, \varphi_z) \mid z \in \mathcal{H}^*\} \quad \text{on } X_0(N) \tag{11.19}$$

that makes the compact Hausdorff space $X_0(N)$ into a Riemann surface. Let $\pi : \mathcal{H}^* \to X_0(N)$ be the quotient map. If V is open in $\mathcal{H}^*$, then $\pi^{-1}(\pi(V)) = \bigcup_{\gamma \in \Gamma_0(N)} \gamma(V)$ is open; thus π is an open map. In defining the charts, there are cases and subcases.

Case 1. $z_0 = \tau$ is in $\mathcal{H}$. We choose V_τ as in Lemma 11.2 and let $U_\tau = \pi(V_\tau)$; U_τ is open since π is open.

Case 1a. $\Gamma_\tau = \{\pm I\}$. Then $\pi : V_\tau \to U_\tau$ is a homeomorphism. We let φ_τ be its inverse, and then (U_τ, φ_τ) is the required chart.

Case 1b. $z_0 = \tau$ is in $\mathcal{H}$, and $\Gamma_\tau \neq \{\pm I\}$. Since $\Gamma_\tau \supseteq \{\pm I\}$ and $\Gamma_0(N) \subseteq SL(2, \mathbb{Z})$, the order of Γ_τ is 4 or 6; we write $2n$ for this number.

Let λ be a linear fractional transformation from $\mathcal{H}$ onto the unit disc carrying τ to 0, e.g., $\lambda(z) = \dfrac{z-\tau}{z-\bar{\tau}}$. Then $\varphi_\tau : U_\tau \to \mathbb{C}$ is well defined by the formula $\varphi_\tau(\pi(z)) = \lambda(z)^n$, and (U_τ, φ_τ) is the required chart.

Case 2. $z_0 = x$ is in $\mathcal{C}$. Choose β in $SL(2,\mathbb{Z})$ with $\beta(x) = \infty$. Then $\beta \Gamma_x \beta^{-1} = \{\pm \begin{pmatrix} 1 & mh \\ 0 & 1 \end{pmatrix} \mid m \in \mathbb{Z}\}$, where h is the width of the cusp. Let $V_x = \beta^{-1}(\{\text{Im } \tau > 2\})$. If $\gamma \in \Gamma_0(N)$ has $\gamma(V_x) \cap V_x \neq \emptyset$, then
$$\beta\gamma\beta^{-1}(\{\text{Im } \tau > 2\}) \cap \{\text{Im } \tau > 2\} \neq \emptyset.$$
Writing $\beta\gamma\beta^{-1} = \begin{pmatrix} a & b \\ c & d \end{pmatrix}$ and taking τ in the nonempty intersection, we have
$$2 \leq \text{Im } \beta\gamma\beta^{-1}(\tau) = \frac{\text{Im } \tau}{|c\tau + d|^2} \leq \frac{1}{|c|^2 \text{Im } \tau} \leq \frac{1}{2|c|^2}.$$
Thus $c = 0$, $\beta\gamma\beta^{-1}$ fixes ∞, and γ fixes x. In other words, γ is in Γ_x. Put $U_x = \pi(V_x)$. Then $\varphi_x : U_x \to \mathbb{C}$ is well defined by the formula $\varphi_x(\pi(z)) = e^{2\pi i \beta(z)/h}$, and (U_x, φ_x) is the required chart.

Proposition 11.3. The charts in (11.19) are compatible, and $X_0(N)$ becomes a compact Riemann surface.

3. Meromorphic Differentials

This section treats meromorphic and holomorphic differentials of compact (connected) Riemann surfaces, with particular attention to $X_0(N)$ and elliptic curves. We omit proofs of results that do not specifically address these applications.

Thus let X be a compact Riemann surface, and let $\{(U_i, \varphi_i)\}_{i \in I}$ be an atlas. The meromorphic functions on X form a field, which we denote by $K(X)$. A system $\omega = \{\omega_i\}_{i \in I}$ of scalar-valued meromorphic functions ω_i on U_i is called a **meromorphic differential** if

$$\omega_i \circ \varphi_i^{-1} = (\omega_j \circ \varphi_j^{-1})(\varphi_j \circ \varphi_i^{-1})' \quad \text{on } \varphi_i(U_i \cap U_j) \subseteq \mathbb{C} \quad (11.20a)$$

whenever $U_i \cap U_j \neq \emptyset$. The classical notation that is used for $\omega_i \circ \varphi_i^{-1}$ in order to capture the transformation law is $\omega_i(\varphi_i^{-1}(z))\,dz$, where z is a local parameter. In this notation if $w = \varphi_j \circ \varphi_i^{-1}(z)$, then $dw = (\varphi_j \circ \varphi_i^{-1})'(z)\,dz$, and (11.20a) says that

$$\omega_i(\varphi_i^{-1}(z))\,dz = \omega_j(\varphi_j^{-1}(w))\,dw. \quad (11.20b)$$

The space of meromorphic differentials is denoted $\Omega(X)$. If F is in $K(X)$, then $dF = \{(F \circ \varphi_i^{-1})' \circ \varphi_i\}_{i \in I}$ is a meromorphic differential. In the development of the theory of compact Riemann surfaces, one of the early steps is to prove the following.

3. MEROMORPHIC DIFFERENTIALS

Proposition 11.4. *If X is a compact Riemann surface, then the space $\Omega(X)$ of meromorphic differentials is nonzero.*

If $\omega = \{\omega_i\}_{i \in I}$ is a meromorphic differential, then ω has a meaning in every compatible chart. Namely if (U_0, φ_0) is a compatible chart, then we can adjoin $\omega_0 : U_0 \to \mathbf{C}$ to $\{\omega_i\}_{i \in I}$ if ω_0 is defined by

$$\omega_0 = (\omega_i)((\varphi_i \circ \varphi_0^{-1})' \circ \varphi_0) \qquad \text{on } U_0 \cap U_i.$$

A meromorphic differential is a **holomorphic differential** if all the ω_i are holomorphic functions. The space of holomorphic differentials is denoted $\Omega_{\text{hol}}(X)$. If $\omega = \{\omega_i\}_{i \in I}$ is a holomorphic differential, then ω is locally of the form df. Namely let (U_0, φ_0) be a simply connected compatible chart. Then $\omega_0 \circ \varphi_0^{-1}$ is analytic on $\varphi_0(U_0)$ and is the derivative of an analytic function g. If we define $f = g \circ \varphi_0$ on U_0, then $(f \circ \varphi_0^{-1})' \circ \varphi_0 = \omega_0$, as asserted. We say that ω locally has an indefinite integral.

Let
$$y^2 + a_1 xy + a_3 y = x^3 + a_2 x^2 + a_4 x + a_6$$

be the equation of an elliptic curve E over $\mathbf{C}$. In Corollary 6.32 we saw that $E(\mathbf{C})$ is a compact Riemann surface of genus 1. Formally we have

$$2y\,dy + a_1 x\,dy + a_1 y\,dx + a_3\,dy = (3x^2 + 2a_2 x + a_4)\,dx$$

and therefore

$$\frac{dx}{2y + a_1 x + a_3} = \frac{dy}{3x^2 + 2a_2 x + a_4 - a_1 y}. \tag{11.21a}$$

At points of $E(\mathbf{C})$ where $\dfrac{\partial}{\partial y} \neq 0$, x is a local coordinate of $E(\mathbf{C})$, and the left side defines (locally) a holomorphic differential. At points where $\dfrac{\partial}{\partial x} \neq 0$, y is a local coordinate, and the right side defines (locally) a holomorphic differential. Together these sets of points cover $E(\mathbf{C})$ except for the point at ∞. In terms of affine local coordinates $(X, 1, W)$ for projective space, we obtain a similar relation from (3.23a):

$$\frac{-dW}{3X^2 + 2a_2 XW + a_4 W^2 - a_1 W} = \frac{dX}{3a_6 W^2 + 2a_4 XW + a_2 X^2 - 2a_3 W - a_1 X - 1}. \tag{11.21b}$$

The relationship between the two coordinate systems is $[(X,1,W)] = [(x,y,1)]$. Thus $X = \dfrac{x}{y}$ and $W = \dfrac{1}{y}$. In particular, $dW = -y^{-2}\,dy$. Calculating, we find that the left side of (11.21b) matches the right side of (11.21a) on the overlap. The common value of (11.21a) and (11.21b) is therefore a (globally defined) holomorphic differential on $E(\mathbf{C})$. It is called the **invariant differential** for E, for reasons given below.

In the theory of compact Riemann surfaces, one of the preliminary results used for the Riemann-Roch Theorem is the following.

Proposition 11.5. If X is a compact Riemann surface of genus g, then the vector space $\Omega_{\text{hol}}(X)$ of holomorphic differentials has dimension g.

In the case of $E(\mathbf{C})$, the genus is one, and the invariant differential is the unique holomorphic differential, up to a scalar. We know from Corollary 6.32 that $E(\mathbf{C})$ is biholomorphic with $\mathbf{C}/\Lambda$ for a lattice Λ. Then dz on $\mathbf{C}$ has a meaning on $\mathbf{C}/\Lambda$ and defines a holomorphic differential on $\mathbf{C}/\Lambda$. By Proposition 11.5, it agrees with the invariant differential in (11.21), up to a scalar. The invariance of dz under translations of $\mathbf{C}/\Lambda$, which we know correspond to group translations in $E(\mathbf{C})$, is the reason for the name "invariant differential" for (11.21).

Now consider $X_0(N)$. Let $\pi : \mathcal{H}^* \to X_0(N)$ be the quotient map. If $\omega = \{\omega_i\}_{i \in I}$ is a holomorphic differential on $X_0(N)$, then we define a function $f_\omega : \mathcal{H} \to \mathbf{C}$ by

$$f_\omega(\tau) = \omega_i(\pi(\tau))(\varphi_i \circ \pi)'(\tau) \tag{11.22}$$

if (U_i, φ_i) is a chart about $\pi(\tau)$.

Proposition 11.6. The map $\omega \to f_\omega$ is well defined and is an isomorphism of $\Omega_{\text{hol}}(X_0(N))$ onto $S_2(\Gamma_0(N))$.

PROOF. To see that f_ω is well defined, let (U_i, φ_i) and (U_j, φ_j) be two charts with $\pi(\tau)$ in $U_i \cap U_j$. Let $t_i = \varphi_i(\pi(\tau))$ and $t_j = \varphi_j(\pi(\tau))$. Then $\varphi_j \circ \pi = (\varphi_j \circ \varphi_i^{-1}) \circ (\varphi_i \circ \pi)$ implies

$$(\varphi_j \circ \pi)'(\tau) = (\varphi_j \circ \varphi_i^{-1})'(t_i)(\varphi_i \circ \pi)'(\tau).$$

Hence

$$\begin{aligned}
\omega_j(\pi(\tau))(\varphi_j \circ \pi)'(\tau) &= (\omega_j \circ \varphi_j^{-1})(t_j)(\varphi_j \circ \varphi_i^{-1})'(t_i)(\varphi_i \circ \pi)'(\tau) \\
&= (\omega_i \circ \varphi_i^{-1})(t_i)(\varphi_i \circ \pi)'(\tau) \qquad \text{by (11.20a)} \\
&= \omega_i(\pi(\tau))(\varphi_i \circ \pi)'(\tau),
\end{aligned}$$

3. MEROMORPHIC DIFFERENTIALS   315

and f_ω is well defined. It is clear that f_ω is analytic.

If γ is in $\Gamma_0(N)$, then $\pi(\tau) = \pi(\gamma\tau)$, so that

$$\begin{aligned}f_\omega(\gamma\tau) &= \omega_i(\pi(\gamma\tau))(\varphi_i \circ \pi)'(\gamma\tau)\\ &= \omega_i(\pi(\tau))(\varphi_i \circ \pi \circ \gamma)'(\tau)/\gamma'(\tau)\\ &= \delta(\gamma,\tau)^2 \omega_i(\pi(\tau))(\varphi_i \circ \pi)'(\tau)\\ &= \delta(\gamma,\tau)^2 f_\omega(\tau).\end{aligned}$$

Hence f_ω is an unrestricted modular form of weight 2 and level N.

For the behavior at the cusps, let $x = \beta^{-1}(\infty)$ be a cusp, and let (U_x, φ_x) be the chart about x given by (11.19). Here

$$U_x = \pi(\beta^{-1}\{\operatorname{Im}\tau > 2\})$$

and

$$\varphi_x(\pi(\beta^{-1}\tau)) = e^{2\pi i \tau/h},$$

where h is the width of the cusp. Then we have

$$\begin{aligned}f_\omega \circ [\beta^{-1}]_2(\tau) &= f_\omega(\beta^{-1}\tau)\delta(\beta^{-1},\tau)^{-2}\\ &= \omega_x(\pi(\beta^{-1}\tau))(\varphi_x \circ \pi)'(\beta^{-1}\tau)\delta(\beta^{-1},\tau)^{-2}\\ &= \omega_x \circ \varphi_x^{-1}(e^{2\pi i \tau/h})\frac{(\varphi_x \circ \pi \circ \beta^{-1})'(\tau)}{\beta^{-1\prime}(\tau)}\delta(\beta^{-1},\tau)^{-2}\\ &= \omega_x \circ \varphi_x^{-1}(e^{2\pi i \tau/h})\frac{2\pi i}{h}e^{2\pi i \tau/h}. \qquad (11.23)\end{aligned}$$

Since the differential is holomorphic, the first factor is an analytic function of $q_h = e^{2\pi i \tau/h}$. The remaining factor is a multiple of q_h. Therefore f_ω is holomorphic at the cusp $\beta^{-1}(\infty)$ and vanishes there.

Clearly the map $\omega \to f_\omega$ is one-one. To see that it is onto, let f be in $S_2(\Gamma_0(N))$ and let x be in $\mathcal{H}^*$. Form the chart (U_x, φ_x). For z in U_x, choose τ in $\mathcal{H}^*$ with $z = \varphi_x^{-1}(\pi(\tau))$. Define

$$\omega_x(z) = f(\tau)((\varphi_x \circ \pi)'(\tau))^{-1}$$

if τ is in $\mathcal{H}$. Modular invariance makes ω_x well defined. Also ω_x is holomorphic everywhere if x is not in $\pi(\mathcal{C})$, and it is holomorphic except at x if x is in $\pi(\mathcal{C})$. In the latter case, to define ω_x at x (the only point where τ would be forced to be a cusp), we unravel the argument (11.23) to see that ω_x is bounded in a neighborhood of x. By the Riemann removable singularity theorem, ω_x extends to be holomorphic at x. Then it is easy to see that $\omega = \{\omega_x\}_{x \in \mathcal{H}^*}$ is a holomorphic differential that maps to f.

4. Properties of Compact Riemann Surfaces

This section will summarize, without any proofs, some deeper properties of compact Riemann surfaces. The initial three main theorems about compact Riemann surfaces are the Riemann-Roch Theorem, Abel's Theorem, and the Jacobi Inversion Theorem. All are expressed in the language of divisors.

Let X be a compact Riemann surface. We denote by $\mathrm{Div}(X)$ the free abelian group on the points of X. An element D of $\mathrm{Div}(X)$ is called a **divisor**. Such an element is written as

$$D = \sum_{x \in X} \mathrm{ord}_x(D) x \qquad \text{with } \mathrm{ord}_x(D) \in \mathbf{Z}. \tag{11.24}$$

We write $D_1 \geq D_2$ if $\mathrm{ord}_x(D_1) \geq \mathrm{ord}_x(D_2)$ for all x in X.

If f is in $K(X)^\times$ and x is in X, choose a chart (U_i, φ_i) with $x_i \in U_i$, and put

$$\mathrm{ord}_x(f) = \mathrm{ord}_{\varphi_i(x)}(f \circ \varphi_i^{-1}). \tag{11.25}$$

The right side is > 0 at a zero and < 0 at a pole. The expression (11.25) does not depend on the choice of (U_i, φ_i).

The divisor $[f]$ of $f \in K(X)^\times$ is given by

$$[f] = \sum_{x \in X} \mathrm{ord}_x(f) x.$$

Such a divisor is called a **principal divisor**, and the set of principal divisors is denoted $\mathrm{Div}_0(X)$. The quotient group

$$\mathrm{Pic}(X) = \mathrm{Div}(X)/\mathrm{Div}_0(X)$$

is called the **divisor class group**.

By Proposition 11.4, X possesses nonzero meromorphic differentials ω. Let $[\omega]$ be the divisor of ω, defined in the same way as $[f]$. The divisor class of $[\omega]$ is independent of the choice of ω. In fact, if ω and ω' are two such, then reference to (11.20a) shows that $\omega = f\omega'$ with f in $K(X)^\times$, and thus the assertion follows. The divisor class of $[\omega]$ in $\mathrm{Pic}(X)$ is called the **canonical class**.

For a divisor D, the **linear system** $L(D)$ is the vector space

$$L(D) = \{f \in K(X)^\times \mid [f] + D \geq 0\} \cup \{0\}$$
$$= \{f \in K(X)^\times \mid \mathrm{ord}_x(f) + \mathrm{ord}_x(D) \geq 0 \text{ for all } x \in X\} \cup \{0\}.$$

Then $L(0) = \mathbf{C}$ since the only holomorphic functions on X are the constants. Also it is clear that

$$D \geq D' \quad \text{implies} \quad L(D) \supseteq L(D'). \tag{11.26}$$

The next result limits how much $L(D')$ can fail to be all of $L(D)$.

4. PROPERTIES OF COMPACT RIEMANN SURFACES

Proposition 11.7. Let X be a compact Riemann surface, let D' be a divisor, and let x be in X. Then
$$\dim L(D' + x) \leq \dim L(D') + 1.$$
Consequently $L(D)$ is finite-dimensional for each D in $\text{Div}(X)$.

The **degree** $\deg(D)$ of the divisor D in (11.24) is $\sum_{x \in X} \text{ord}_x(D)$. A proof of the following result was sketched in §IX.5.

Proposition 11.8. Each principal divisor on a compact Riemann surface has degree 0.

Theorem 11.9 (Riemann-Roch Theorem). Let X be a compact Riemann surface of genus g, let D be a divisor, and let W be in the canonical class of $\text{Pic}(X)$. Then
$$\dim L(D) = \deg(D) + \dim L(W - D) - g + 1.$$

Corollary 11.10. Let X be a compact Riemann surface of genus g, and let W be in the canonical class of $\text{Pic}(X)$. Then $\deg W = 2g - 2$ and $\dim L(W) = g$.

Corollary 11.11. Let X be a compact Riemann surface of genus g, and let D be a divisor with $\deg(D) > 2g - 2$. Then
$$\dim L(D) = \deg(D) - g + 1.$$

Corollary 11.12. Let X be a compact Riemann surface of genus $g \geq 1$. Then there is no point of X where all holomorphic differentials vanish.

Until further notice in this section, we assume that our compact Riemann surface X has genus $g \geq 1$. Since X is a closed orientable surface of genus g, the ordinary homology group $H_1(X, \mathbb{Z})$ is free abelian on $2g$ generators. One usually writes a standard homology basis over $\mathbb{Z}$ in the form $a_1, \ldots, a_g, b_1, \ldots, b_g$ with special intersection properties. But we shall be content to fix any $\mathbb{Z}$ basis $c_1, \ldots, c_{2g}$ of $H_1(X, \mathbb{Z})$.

Proposition 11.13. Let X be a compact Riemann surface of genus $g \geq 1$, and let $\omega_1, \ldots, \omega_g$ be a basis of $\Omega_{\text{hol}}(X)$ over $\mathbb{C}$. Then the $2g$ vectors
$$\begin{pmatrix} \int_{c_k} \omega_1 \\ \vdots \\ \int_{c_k} \omega_g \end{pmatrix}$$
in $\mathbb{C}^g$ are linearly independent over $\mathbb{R}$.

Consequently the vectors in the proposition are a $\mathbf{Z}$ basis for a lattice $\Lambda(X)$ in $\mathbf{C}^g$. The lattice $\Lambda(X)$ is unchanged if $\{c_k\}$ is replaced by a different $\mathbf{Z}$ basis of $H_1(X, \mathbf{Z})$. If $\{\omega_j\}$ is replaced by a different $\mathbf{Z}$ basis of $\Omega_{\text{hol}}(X)$, the effect is to transform $\Lambda(X)$ by a member of $GL(g, \mathbf{C})$.

The **Jacobian variety** of X is the g-dimensional complex torus $J(X) = \mathbf{C}^g/\Lambda(X)$. We define a map $\Phi : X \to J(X)$ by fixing a point x_0 in X and setting

$$\Phi(x) = \left\{ \int_{x_0}^{x} \omega_j \right\}_{j=1}^{g}. \tag{11.27}$$

According to the previous paragraph, if the base point is fixed, the pair $(J(X), \Phi)$ is well defined up to the action of a member of $GL(g, \mathbf{C})$ simultaneously on the ω_j's, $\mathbf{C}^g$, and $\Lambda(X)$.

Proposition 11.14. For a compact Riemann surface X of genus ≥ 1, the map Φ of X into $J(X)$ is well defined and holomorphic, and its rank (over $\mathbf{C}$) is everywhere 1.

The first statement of the proposition has already been addressed, and the second statement follows from Corollary 11.12. In the case of $X = X_0(N)$, Proposition 11.6 shows that the map Φ is essentially the same as the tuple of maps Φ_{f_j} of §1 if $\{f_j\}$ is a basis of $S_2(\Gamma_0(N))$. This tuple $\{\Phi_{f_j}\}$ will be used in §§10-11 and will be called $\tilde{\Phi}$ there.

Since $J(X)$ is a group, we can extend $\Phi : X \to J(X)$ to a map $\Phi : \text{Div}(X) \to J(X)$ in the obvious way: If $D = \sum_{x \in X} \text{ord}_x(D) x$, then

$$\Phi(D) = \sum_{x \in X} \text{ord}_x(D) \Phi(x).$$

For general D, this definition depends on the base point x_0 in (11.27), but it is independent of x_0 if D has degree 0.

Recall from Proposition 11.8 that all principal divisors have degree 0.

Theorem 11.15 (Abel's Theorem). Let X be a compact Riemann surface of genus ≥ 1, and let D be a divisor. Then D is principal if and only if $\deg(D) = 0$ and $\Phi(D)$ is the 0 element of $J(X)$.

Corollary 11.16.
(a) If X has genus 1, then $\Phi : X \to J(X)$ is one-one onto and biholomorphic.
(b) If X has genus > 1 then $\Phi : X \to J(X)$ is a one-one holomorphic mapping onto a proper complex submanifold of $J(X)$.

4. PROPERTIES OF COMPACT RIEMANN SURFACES

With g equal to the genus of X, let

$$\mathrm{Div}^{(g)}(X) = \{D \in \mathrm{Div}(X) \mid D \geq 0 \text{ and } \deg(D) = g\}.$$

Then we can restrict Φ from $\mathrm{Div}(X)$ to $\mathrm{Div}^{(g)}(X)$ and obtain a map $\Phi : \mathrm{Div}^{(g)}(X) \to J(X)$. The special case $g = 1$ is what was considered in Corollary 11.16a.

Theorem 11.17 (Jacobi Inversion Theorem). For a compact Riemann surface of genus $g \geq 1$, the map Φ carries $\mathrm{Div}^{(g)}(X)$ onto $J(X)$.

Corollary 11.18. As a group, $J(X)$ is isomorphic to the group of divisors of degree 0 modulo the subgroup of principal divisors.

Theorem 11.19. If $F : X \to T$ is a holomorphic mapping of a compact Riemann surface of genus ≥ 1 into a complex torus, then F factors through the Jacobian variety: $F = f \circ \Phi$ for some holomorphic mapping $f : J(X) \to T$ that is the sum of a translation and a holomorphic homomorphism.

This completes our discussion of the three initial main theorems about compact Riemann surfaces. The final part of this section addresses the nature of the function field $K(X)$, indicating part of the connection between Riemann surface theory and the algebraic geometry of curves.

Theorem 11.20. Let X be a compact Riemann surface of genus $g \geq 0$, and let x be a nonconstant meromorphic function on X (existence by Corollary 11.11). Then there exist a nonconstant $y \in K(X)$ and an irreducible polynomial P in two variables such that $P(x, y) = 0$ and

$$K(X) \cong \mathbb{C}(x)[y]/(P(x, y)).$$

We should think of Theorem 11.20 as giving us a mapping of X into $P_2(\mathbb{C})$ by $z \to (x(z), y(z))$, with the image contained in the plane curve defined by P. In §7, we shall briefly discuss this approach to defining $X_0(N)$ over $\mathbb{Q}$.

Theorem 11.21. Let X and Y be compact Riemann surfaces, and suppose $F : K(Y) \to K(X)$ is a nonzero $\mathbb{C}$ algebra homomorphism. Then F is one-one and is implemented by a holomorphic map of X onto Y.

5. Hecke Operators on Integral Homology

We now return to the subject of modular forms. We shall examine Hecke operators in detail in this section. Hecke operators act on many things, and they do so consistently. Understanding these consistent actions is a key to unlocking the power of the Hecke operators.

One such consistent action is on $H_1(X_0(N), \mathbf{Z})$. Using this action, we shall see that $S_2(\Gamma_0(N))$ has a basis such that all Hecke operators $T_2(n)$ act by integer matrices. One consequence is that their eigenvalues are algebraic integers. This same realization will allow us also to characterize the algebra generated by the Hecke operators $T_2(n)$.

Before considering $H_1(X_0(N), \mathbf{Z})$, we consider points and divisors. Recall from §IX.6 that Hecke operators were introduced as acting on the free abelian group generated by modular pairs (Λ, C), where Λ is a lattice in $\mathbf{C}$ and C is a cyclic subgroup of $\mathbf{C}/\Lambda$ of exact order N. The set of equivalence classes of such pairs, with equivalence defined essentially by $\mathbf{C}^\times$, is really $X_0(N)$ without the cusps, and thus we should expect an action of Hecke operators on the free abelian group generated by the points of $X_0(N)$. This is the divisor group. Rather than try to unwind our original definition, however, we shall start afresh.

In §IX.6 we defined $M(n, N)$ as the set of 2-by-2 integer matrices of determinant n whose upper left entry is prime to N and whose lower left entry is divisible by N. As in Proposition 9.12, we write $M(n, N)$ as a finite disjoint union

$$M(n, N) = \bigcup_{i=1}^{K} \Gamma_0(N)\alpha_i. \tag{11.28}$$

If τ is in $\mathcal{H}^*$, let $[\tau]$ be the corresponding member of $X_0(N)$. For such τ we define

$$T(n)(\tau) = \sum_{i=1}^{K} [\alpha_i \tau] \tag{11.29}$$

as a member of $\text{Div}(X_0(N))$. If the α_i's are replaced by different coset representatives in (11.28), then it is clear that the right side of (11.29) is unchanged.

Let us see that $T(n)(\tau)$ depends only on $[\tau]$. In fact, if γ is in $\Gamma_0(N)$, then $M(n, N)\gamma \subseteq M(n, N)$. Hence we can write

$$\alpha_i \gamma = \gamma_i \alpha_{j(i)} \tag{11.30}$$

5. HECKE OPERATORS ON INTEGRAL HOMOLOGY

with $\gamma_i \in \Gamma_0(N)$, and we readily check that

$$i \to j(i) \quad \text{is a permutation of } \{1, \ldots, K\}. \tag{11.31}$$

Thus

$$\begin{aligned}
T(n)(\gamma\tau) - T(n)(\tau) &= \sum_i [\alpha_i \gamma \tau] - \sum_i [\alpha_i \tau] \\
&= \sum_i [\gamma_i \alpha_{j(i)} \tau] - \sum_i [\alpha_i \tau] \\
&= \sum_i \left([\gamma_i \alpha_{j(i)} \tau] - [\alpha_{j(i)} \tau]\right) \\
&= 0
\end{aligned} \tag{11.32}$$

by (11.31). Equation (11.32) says that (11.29) is a consistent definition of $T(n)[\tau]$.

Thus $T(n)$ has been defined as a $\mathbb{Z}$ linear map of $\text{Div}(X_0(N))$ to itself. This suggests a natural definition of $T(n)$ on 1-cycles on $X_0(N)$. If we think of a 1-cycle as a sum of loops on $X_0(N)$, we just need a definition of $T(n)$ on a loop. We lift the loop to a path from τ_0 to some $\gamma \tau_0$ in $\mathcal{H}^*$, writing the path as $[\tau_0, \gamma \tau_0]$, and we try to define

$$T(n)[\tau_0, \gamma\tau_0] = \sum_{i=1}^{K} [\tau_0, \gamma_i \tau_0] \tag{11.33}$$

with γ_i as in (11.30). Then we hope that this definition descends to homology. But there is much to check—that the definition depends only on the homotopy class of the loop, then that is independent of the lift, and finally that it depends only on the homology class. The motivation for an efficient approach comes from Proposition 11.1.

Proposition 11.22. Let Γ_{ep} be the (normal) subgroup of $\Gamma_0(N)$ generated by all elliptic and parabolic elements, and let $(\cdot)^{\text{ab}}$ refer to a group modulo its commutator. In terms of a specified point τ_0 in $\mathcal{H}^*$, there is a canonical isomorphism

$$H_1(X_0(N), \mathbb{Z}) \cong \Gamma_0(N)^{\text{ab}} / \Gamma_{ep}^{\text{ab}}. \tag{11.34}$$

REMARK. To simplify matters, we shall always assume that τ_0 is neither an elliptic fixed point nor a cusp.

PROOF. We remove from $X_0(N)$ the finite set of images of elliptic fixed points and cusps (parabolic fixed points) under $\Gamma_0(N)$; the remaining set will be called $X_0(N)'$, and its preimage in $\mathcal{H}^*$ will be called $\mathcal{H}'$. Then $e : \mathcal{H}' \to X_0(N)'$ is a covering map. We shall use τ_0 and $e(\tau_0)$ as base points for fundamental groups, but we drop them from the notation.

Let $\bar{\Gamma}_0(N) = \Gamma_0(N)/\{\pm 1\}$. Then $\bar{\Gamma}_0(N)\backslash\mathcal{H}' \cong X_0(N)'$ shows that $\bar{\Gamma}_0(N)$ is the group of deck transformations of $\mathcal{H}'$ over $X_0(N)'$ and that $\bar{\Gamma}_0(N)$ acts transitively on the fibers. Hence $e_*(\pi_1(\mathcal{H}'))$ is normal in $\pi_1(X_0(N)')$ with quotient $\bar{\Gamma}_0(N)$. Let

$$p : \pi_1(X_0(N)') \longrightarrow \pi_1(X_0(N)')/e_*(\pi_1(\mathcal{H}')) \cong \bar{\Gamma}_0(N)$$

be the quotient homomorphism.

Let $Y_0(N) = \bar{\Gamma}_0(N)\backslash\mathcal{H}$, i.e., $Y_0(N)$ is $X_0(N)'$ with the images of the elliptic fixed points restored. We shall use the Van Kampen Theorem to show that

$$\pi_1(Y_0(N)) \cong \bar{\Gamma}_0(N)/\bar{\Gamma}_e,$$

where $\bar{\Gamma}_e$ is the (normal) subgroup generated by the elliptic elements. Visualize restoring one point to $Y_0(N)$ at a time. We apply the Van Kampen Theorem to a small disc about this point and the partially restored $Y_0(N)$. Let l be a simple loop about the point within the small disc. The theorem says that the effect on the fundamental group of restoring the point is to convert l into a relation, i.e., to factor by the normal subgroup generated by l. If $l_1, \ldots, l_r$ are the loops used as all the points are restored, the result is that

$$\pi_1(Y_0(N)) \cong \pi_1(X_0(N)')/\langle l_1, \ldots, l_r \rangle$$

in obvious notation.

Using p, we have a natural homomorphism onto:

$$\pi_1(Y_0(N)) \cong \pi_1(X_0(N)')/\langle l_1, \ldots, l_r \rangle \longrightarrow \bar{\Gamma}_0(N)/p\langle l_1, \ldots, l_r \rangle. \quad (11.35)$$

Suppose an element σ of $\pi_1(X_0(N)')$ maps to the identity coset. Then $p(\sigma)$ is a product of elliptic elements. Under p, however, the elements $l_1, \ldots, l_r$ map to elliptic elements that generate $\bar{\Gamma}_e$. Hence $p(\sigma) = p(\sigma_0)$ for some σ_0 in $\langle l_1, \ldots, l_r \rangle$. Adjusting σ by σ_0^{-1} and changing notation, we may assume that $p(\sigma) = 1$. Then σ is in $e(\pi_1(\mathcal{H}'))$. But the punctures of $\mathcal{H}'$ get filled in during the passage from $X_0(N)'$ to $Y_0(N)$. Thus σ is contractible in $Y_0(N)$ and must be in $\langle l_1, \ldots, l_r \rangle$. Thus (11.35) is one-one and is an isomorphism.

5. HECKE OPERATORS ON INTEGRAL HOMOLOGY

Now we pass from $Y_0(N)$ to $X_0(N)$ in the same way, restoring the images of the cusps one at a time. At the start we have

$$\pi_1(Y_0(N)) \cong \bar{\Gamma}_0(N)/\bar{\Gamma}_e.$$

Each restoration of a point has the effect of adjoining a new relation (by the Van Kampen Theorem), and the relations are easily seen to be parabolic generators corresponding to each cusp. The end result is that

$$\pi_1(X_0(N)) \cong \bar{\Gamma}_0(N)/\bar{\Gamma}_{ep} \cong \Gamma_0(N)/\Gamma_{ep}.$$

Finally we have

$$H_1(X_0(N), \mathbb{Z}) \cong \pi_1(X_0(N))^{\mathrm{ab}} \cong (\Gamma_0(N)/\Gamma_{ep})^{\mathrm{ab}}.$$

We can readily check that

$$(\Gamma_1/\Gamma_2)^{\mathrm{ab}} \cong \Gamma_1^{\mathrm{ab}}/\Gamma_2^{\mathrm{ab}}$$

whenever Γ_2 is normal in Γ_1, and then (11.34) follows.

Now we can make $T(n)$ act on $H_1(X_0(N), \mathbb{Z})$. Guided by (11.34), we regard $H_1(X_0(N), \mathbb{Z})$ as the free abelian group on symbols $[\gamma]$, $\gamma \in \Gamma_0(N)$, subject to the rules

$$\begin{aligned}[\gamma_1 \gamma_2] &= [\gamma_1] + [\gamma_2] \\ [\gamma] &= 0 \quad \text{if } \gamma \text{ is elliptic or parabolic.}\end{aligned} \qquad (11.36)$$

This correspondence of symbols $[\gamma]$ and cycles c respects integration. Namely suppose $f \in S_2(\Gamma_0(N))$ corresponds under Proposition 11.6 to $\omega \in \Omega_{\mathrm{hol}}(X)$ and $[\gamma]$ corresponds to c. Then

$$\int_{\tau_0}^{\gamma(\tau_0)} f(\tau)\, d\tau = \int_c \omega. \qquad (11.37)$$

(Recall from (11.13) that the left side is independent of τ_0.) To see this, let l be a loop in $X_0(N)$ that maps to c in homology, and lift l to a path $\tilde{l}$ in $\mathcal{H}^*$. Then $\tilde{l}$ goes from τ_0 to some point $\gamma(\tau_0)$ in $\mathcal{H}^*$. For this γ and c, the two sides of (11.37) are equal, and γ does correspond to c since $\tilde{l}$ projects to the loop l mapping onto c in homology.

Let the compact Riemann surface $X_0(N)$ have genus g. We know that $H_1(X_0(N), \mathbb{Z})$ is free abelian on $2g$ generators, and hence so is the isomorphic group of $[\gamma]$'s; in particular, it is torsion free.

We can make (11.33) into a rigorous definition by setting

$$T(n)[\gamma] = \sum_{i=1}^K [\gamma_i], \qquad (11.38)$$

with γ_i as in (11.30).

Proposition 11.23. The equation (11.38) makes $T(n)$ into a well defined $\mathbf{Z}$ linear operator on $H_1(X_0(N), \mathbf{Z})$, independent of the coset representatives α_i in (11.28). Moreover, if c is a 1-cycle on $X_0(N)$ and ω is a holomorphic differential (corresponding to a member f of $S_2(\Gamma_0(N))$), then

$$\int_{T(n)c} \omega = \int_c T(n)\omega, \qquad (11.39a)$$

where $T(n)\omega$ is the holomorphic differential corresponding to $T_2(n)f$.

PROOF. To see that $T(n)$ is well defined on the group of $[\gamma]$'s, we have to check that $T(n)[\gamma] = 0$ if γ is elliptic or parabolic and that

$$T(n)[\gamma_1\gamma_2] = T(n)[\gamma_1] + T(n)[\gamma_2];$$

these conclusions are enough, according to (11.36). First we check this identity.

Let γ_1 and γ_2 be in $\Gamma_0(N)$, and write

$$\alpha_i\gamma_1 = \gamma_i\alpha_{j(i)} \qquad \text{and} \qquad \alpha_j\gamma_2 = \gamma'_j\alpha_{k(j)}.$$

Then

$$\alpha_i(\gamma_1\gamma_2) = \gamma_i\alpha_{j(i)}\gamma_2 = (\gamma_i\gamma'_{j(i)})\alpha_{k(j(i))},$$

so that (11.38) gives

$$T(n)[\gamma_1\gamma_2] = \sum_{i=1}^{K}[\gamma_i\gamma'_{j(i)}] = \sum_{i=1}^{K}[\gamma_i] + \sum_{i=1}^{K}[\gamma'_{j(i)}]$$
$$= \sum_{i=1}^{K}[\gamma_i] + \sum_{j=1}^{K}[\gamma'_j] = T(n)[\gamma_1] + T(n)[\gamma_2].$$

Thus $T(n)$ is compatible with the first relation of (11.36).

Suppose γ is parabolic with $\alpha_i\gamma = \gamma_i\alpha_{j(i)}$. Let $j^{(k)}(\cdot)$ be the kth iterate of the permutation j. Then

$$\alpha_i\gamma^K = \gamma_i\alpha_{j(i)}\gamma^{K-1} = \gamma_i\gamma_{j(i)}\alpha_{j^{(2)}(i)}\gamma^{K-2}$$
$$= \cdots = \gamma_i\gamma_{j(i)}\cdots\gamma_{j^{(K-1)}(i)}\alpha_{j^{(K)}(i)}$$
$$= \gamma_i\gamma_{j(i)}\cdots\gamma_{j^{(K-1)}(i)}\alpha_i.$$

Since γ is parabolic, so is

$$\alpha_i\gamma^K\alpha_i^{-1} = \gamma_i\gamma_{j(i)}\cdots\gamma_{j^{(K-1)}(i)},$$

5. HECKE OPERATORS ON INTEGRAL HOMOLOGY

and the right side shows that this element is in $\Gamma_0(N)$. Since

$$\alpha_i \gamma^K = (\alpha_i \gamma^K \alpha_i^{-1}) \alpha_i,$$

(11.38) gives

$$K \cdot T(n)[\gamma] = T(n)[\gamma^K] = \sum_{i=1}^{K} [\alpha_i \gamma^K \alpha_i^{-1}] = 0.$$

Since the group of $[\gamma]$'s is torsion free, $T(n)[\gamma] = 0$.

Suppose γ is elliptic with $\gamma^p = 1$. It is clear from (11.38) that

$$T(n)[1] = \sum_{i=1}^{K} [1] = 0.$$

Then

$$pT(n)[\gamma] = T(n)[\gamma^p] = T(n)[1] = 0,$$

and $T(n)[\gamma] = 0$ since the group of $[\gamma]$'s is torsion free. Thus $T(n)$ is well defined on the group of $[\gamma]$'s.

The thrust of Proposition 11.22 is that no further relations need to be checked in order to transfer $T(n)$ to a well defined operator on $H_1(X_0(N), \mathbf{Z})$.

Suppose α_i is changed to $\alpha'_i = \varepsilon_i \alpha_i$ with ε_i in $\Gamma_0(N)$. From $\alpha_i \gamma = \gamma_i \alpha_{j(i)}$, we obtain

$$\alpha'_i \gamma = \varepsilon_i \alpha_i \gamma = \varepsilon_i \gamma_i \alpha_{j(i)} = \varepsilon_i \gamma_i \varepsilon_{j(i)}^{-1} \alpha'_{j(i)}.$$

Then (11.36) gives

$$\sum_{i=1}^{K} [\varepsilon_i \gamma_i \varepsilon_{j(i)}^{-1}] = \sum_i [\varepsilon_i] + \sum_i [\gamma_i] - \sum_i [\varepsilon_{j(i)}]$$
$$= \sum_i [\varepsilon_i] + \sum_i [\gamma_i] - \sum_j [\varepsilon_j] = \sum_i [\gamma_i].$$

By (11.38), $T(n)$ gives the same result from the α'_i's as from the α_i's.

Now let c be a 1-cycle on $X_0(N)$. With τ_0 as base point in $\mathcal{H}$, we can choose a path $[\tau_0, \gamma \tau_0]$ whose projection to $X_0(N)$ is homologous to c. By (11.38), we can write

$$T(n)[\gamma] = \sum_{i=1}^{K} [\gamma_i]$$

with γ_i as in (11.30). By (11.38), equation (11.39a) is the assertion that

$$\sum_{i=1}^{K} \int_{\tau_0}^{\gamma_i(\tau_0)} f(\tau) \, d\tau = \int_{\tau_0}^{\gamma(\tau_0)} T_2(n) f(\tau) \, d\tau. \tag{11.39b}$$

To prove (11.39b), we calculate the right side as

$$= \sum_i \int_{\tau_0}^{\gamma(\tau_0)} f \circ [\alpha_i]_2(\tau) \, d\tau \quad \text{by Proposition 9.12}$$

$$= \sum_i \int_{\tau_0}^{\gamma(\tau_0)} f(\alpha_i(\tau)) \delta(\alpha_i, \tau)^{-2} \, d\tau$$

$$= \sum_i \int_{\alpha_i(\tau_0)}^{\alpha_i \gamma(\tau_0)} f(\tau') \, d\tau' \qquad \text{under } \tau' = \alpha_i(\tau)$$

$$= \sum_i \int_{\alpha_i(\tau_0)}^{\gamma_i \alpha_{j(i)}(\tau_0)} f(\tau) \, d\tau$$

$$= \sum_i \int_{\alpha_i(\tau_0)}^{\alpha_{j(i)}(\tau_0)} f(\tau) \, d\tau + \sum_i \int_{\alpha_{j(i)}(\tau_0)}^{\gamma_i \alpha_{j(i)}(\tau_0)} f(\tau) \, d\tau. \tag{11.40}$$

Since $j(\cdot)$ is a permutation, the sum of paths

$$\sum_i [\alpha_i(\tau_0), \alpha_{j(i)}(\tau_0)]$$

is a 1-cycle on $\mathcal{H}$. Hence the first term on the right side of (11.40) is 0 by the Cauchy Integral Theorem. The second term is equal to the left side of (11.39b) since $\Phi_f(\gamma_i)$ in §1 is independent of base point. Thus (11.39b) is proved, and the proposition follows.

In the setting of Proposition 11.23, let us write

$$\langle c, \omega \rangle = \int_c \omega \tag{11.41}$$

for c in $H_1(\mathbb{Z}) = H_1(X_0(N), \mathbb{Z})$, so that the proposition says

$$\langle T(n)c, \omega \rangle = \langle c, T(n)\omega \rangle.$$

We can extend (11.41) by linearity to be defined for c in $H_1(\mathbb{R}) = H_1(X_0(N), \mathbb{R})$ or even for c in $H_1(\mathbb{C})$, with ω in $\Omega_{\text{hol}}(X_0(N))$. Let us observe that

$$(\langle c, \omega \rangle = 0 \text{ for some } c \text{ in } H_1(\mathbb{R}) \text{ and all } \omega) \implies (c = 0). \tag{11.42}$$

In fact, let $c_1, \ldots, c_{2g}$ be a $\mathbf{Z}$ basis of $H_1(\mathbf{R})$, and write

$$c = \sum_{k=1}^{2g} r_k c_k \quad \text{with } r_k \in \mathbf{R}.$$

If $\omega_1, \ldots, \omega_g$ is a basis of $\Omega_{\text{hol}}(X_0(N))$ over $\mathbf{C}$, then we have

$$\sum_{k=1}^{2g} r_k \langle c_k, \omega_j \rangle = 0$$

for all j. By Proposition 11.13, all the r_k are 0. Thus $c = 0$. This proves (11.42).

We would like to say that the c's and ω's are in dual vector spaces, with the c's in $H_1(\mathbf{Z})$ constrained to the integer points. But the complex vector space of c's, namely $H_1(\mathbf{C})$, is $2g$-dimensional, while the vector space of ω's, namely $\Omega_{\text{hol}}(X_0(N))$, is g-dimensional. So we shall first split $H_1(\mathbf{C})$ into two pieces of dimension g.

The mapping $(\cdot)^* : \tau \to -\bar{\tau}$ of $\mathcal{H}$ into itself induces an involution of $X_0(N)$. Namely if $\tau_2 = \gamma \tau_1$ with $\gamma = \begin{pmatrix} a & b \\ c & d \end{pmatrix}$ in $\Gamma_0(N)$, then the element $\gamma^* = \begin{pmatrix} a & -b \\ -c & d \end{pmatrix}$ of $\Gamma_0(N)$ has $\tau_2^* = \gamma^* \tau_1^*$. This involution induces an involution on $H_1(\mathbf{Z})$, which we extend by linearity to $H_1(\mathbf{R})$ and $H_1(\mathbf{C})$. The vector spaces have eigenspace decompositions under $*$ that we write as

$$\begin{aligned} H_1(\mathbf{R}) &= H_1^+(\mathbf{R}) \oplus H_1^-(\mathbf{R}) \\ H_1(\mathbf{C}) &= H_1^+(\mathbf{C}) \oplus H_1^-(\mathbf{C}). \end{aligned} \tag{11.43}$$

Proposition 11.24.

(a) The spaces H_1^+ and H_1^- each have dimension g, and the pairing $\langle c, \omega \rangle$ exhibits the spaces $H_1^+(\mathbf{C})$ and $\Omega_{\text{hol}}(X_0(N))$ as dual to one another.

(b) The linear extension of $T(n)$ to $H_1(\mathbf{C})$ maps $H_1^+(\mathbf{C})$ into itself, and the restriction of $T(n)$ to $H_1^+(\mathbf{C})$ is the transpose of $T_2(n)$ on $\Omega_{\text{hol}}(X_0(N))$.

(c) The intersection $H_1(\mathbf{Z}) \cap H_1^+(\mathbf{R})$ is a lattice in $H_1^+(\mathbf{R})$, and a $\mathbf{Z}$ basis for this intersection is a basis for $H_1^+(\mathbf{C})$.

PROOF. We write the involution on 1-cycles as $c \to c^*$. If we lift the 1-cycles to paths in $\mathcal{H}^*$ without changing their names and if c is given by $[\tau_0, \tau_1]$, then c^* is given by $[\tau_0^*, \tau_1^*]$. Also if we realize members ω of

$\Omega_{\text{hol}}(X_0(N))$ as functions in $S_2(\Gamma_0(N))$, then (11.37) allows us to write the pairing as

$$\langle c, f \rangle = \int_c f(\tau)\, d\tau.$$

The space $S_2(\Gamma_0(N))$ has a conjugate-linear involution $f \to f^*$ given by

$$f^*(\tau) = -\overline{f(\tau^*)}.$$

We readily check that f^* is again in $S_2(\Gamma_0(N))$ and that

$$\langle c^*, f \rangle = \overline{\langle c, f^* \rangle} \qquad \text{for } c \text{ in } H_1(\mathbf{C}).$$

If f has $f = f^*$, then if has $(if)^* = -if$. Hence the real subspace $S_{2,+}$ of f's in $S_2(\Gamma_0(N))$ with $f^* = f$ has real dimension g, and so does the real subspace $S_{2,-}$ with $f^* = -f$. If c is in $H_1^+(\mathbf{R})$, then

$$\langle c, f \rangle = \langle c^*, f \rangle = \overline{\langle c, f^* \rangle}$$

shows that $\langle c, f \rangle$ is real for f in $S_{2,+}$. If $\dim_{\mathbf{R}} H_1^+(\mathbf{R}) > g$, we can therefore find some $c_0 \neq 0$ such that $\langle c_0, f \rangle = 0$ for all $f \in S_{2,+}$. Since $\langle \cdot, \cdot \rangle$ is complex bilinear, $\langle c_0, if \rangle = 0$ also. But if is the most general member of $S_{2,-}$. So $\langle c_0, S_2(\Gamma_0(N)) \rangle = 0$. From (11.42) we see that $c_0 = 0$, contradiction. Thus $\dim_{\mathbf{R}} H_1^+(\mathbf{R}) \leq g$. A similar argument with c in $H_1^-(\mathbf{R})$ and f in $S_{2,-}$ shows that $\dim_{\mathbf{R}} H_1^-(\mathbf{R}) \leq g$. By (11.43), equality must hold in both cases.

(b) In Proposition 11.23 we choose τ_0 imaginary, so that if $c \leftrightarrow [\tau_0, \gamma\tau_0]$, then $c^* \leftrightarrow [\tau_0, \gamma^*\tau_0]$. The formula for $T(n)$ is

$$T(n)[\gamma] = \sum_{i=1}^{K} [\gamma_i]$$

with $\alpha_i \gamma = \gamma_i \alpha_{j(i)}$. On matrices the operation $(\cdot)^*$ is multiplicative. Thus

$$\alpha_i^* \gamma^* = \gamma_i^* \alpha_{j(i)}^*.$$

If we use the standard α's as in Lemma 9.14, then α_i^* and $\alpha_{j(i)}^*$ are different members of the same set. Therefore

$$T(n)[\gamma^*] = \sum_{i=1}^{K} [\gamma_i^*] = (T(n)[\gamma])^*.$$

By linearity, $T(n)c^* = (T(n)c)^*$ on $H_1(\mathbf{C})$, and it follows that $T(n)$ maps $H_1^+(\mathbf{C})$ into itself. By (a) and (11.39a), the action of $T(n)$ on $H_1^+(\mathbf{C})$ is the transpose of the action of $T_2(n)$ on $\Omega_{\text{hol}}(X_0(N))$.

5. HECKE OPERATORS ON INTEGRAL HOMOLOGY

(c) If c is in $H_1(\mathbf{Z})$, then

$$c = \tfrac{1}{2}(c+c^*) + \tfrac{1}{2}(c-c^*) \qquad \text{in } H_1^+(\mathbf{R}) \oplus H_1^-(\mathbf{R}).$$

Hence $\Lambda_1 = \{\tfrac{1}{2}(c+c^*) \mid c \in H_1(\mathbf{Z})\}$ spans $H_1^+(\mathbf{R})$. On the other hand, $\Lambda_2 = \{c+c^* \mid c \in H_1(\mathbf{Z})\}$ is contained in $H_1(\mathbf{Z})$ and hence is discrete. Since Λ_2 has finite index in Λ_1 and since $H_1(\mathbf{Z}) \cap H_1^+(\mathbf{R})$ lies between Λ_2 and Λ_1, $H_1(\mathbf{Z}) \cap H_1^+(\mathbf{R})$ is a lattice in $H_1^+(\mathbf{R})$. This completes the proof.

Theorem 11.25. Fix a $\mathbf{Z}$ basis $c_1, \ldots, c_g$ of $H_1(\mathbf{Z}) \cap H_1^+(\mathbf{R})$, so that $c_1, \ldots, c_g$ is a basis over $\mathbf{C}$ of $H_1^+(\mathbf{C})$. Let $f_1, \ldots, f_g$ be the dual basis of $S_2(\Gamma_0(N))$. Then each Hecke operator $T_2(n)$ is given in the basis $f_1, \ldots, f_g$ by a matrix with integer entries.

REMARK. The existence of the bases $c_1, \ldots, c_g$ and $f_1, \ldots, f_g$ in the theorem is assured by Proposition 11.24.

PROOF. $T(n)$ acts on $H_1(\mathbf{Z})$, with linear extensions to $H_1(\mathbf{R})$ and $H_1(\mathbf{C})$. By Proposition 11.24b, it maps $H_1^+(\mathbf{C})$ to itself, hence $H_1^+(\mathbf{R})$ to itself. Thus $T(n)$ maps $H_1(\mathbf{Z}) \cap H_1^+(\mathbf{R})$ to itself. Hence it is given in the basis $c_1, \ldots, c_g$ by a matrix with integer entries. Proposition 11.24 shows that $T_2(n)$ is the transpose of $T(n)$, and the theorem follows.

Corollary 11.26. The eigenvalues of $T_2(n)$ on $S_2(\Gamma_0(N))$ are algebraic integers.

PROOF. By Theorem 11.24, $T_2(n)$ is given in a suitable basis by an integer matrix. The characteristic polynomial is then a monic polynomial with integer coefficients, and the roots must be algebraic integers.

Theorem 11.27. Let $c_1 \in S_2(\Gamma_0(N))'$ be the linear functional that extracts the first Fourier coefficient, and let $\mathcal{T}$ be the commutative subalgebra of $\operatorname{End}_\mathbf{C}(S_2(\Gamma_0(N)))$ generated by the identity and the Hecke operators. Then the map $L : \mathcal{T} \to S_2(\Gamma_0(N))'$ given by

$$L(T) = c_1 \circ T \tag{11.44}$$

is one-one onto and exhibits the left regular representation of $\mathcal{T}$ on itself as equivalent with the representation of $\mathcal{T}$ on $S_2(\Gamma_0(N))' \cong H_1^+(\mathbf{C})$. In particular, $\dim_\mathbf{C} \mathcal{T} = g$.

REMARKS. We continue to write g for the genus of $X_0(N)$. It will be convenient in this proof to drop the subscript 2 on Hecke operators $T_2(n)$.

PROOF. We can write $L(T) = \langle c_1 \circ T, \cdot \rangle$, where $\langle \cdot, \cdot \rangle$ is the pairing of $S_2(\Gamma_0(N))'$ with $S_2(\Gamma_0(N))$. For T_0 and T in $\mathcal{T}$, we have

$$\begin{aligned} T_0(L(T))(f) &= \langle T_0 \langle c_1 \circ T, \cdot \rangle, f \rangle \\ &= \langle \langle c_1 \circ T, \cdot \rangle, T_0 f \rangle \\ &= \langle c_1 \circ T, T_0 f \rangle \\ &= c_1(TT_0 f) \end{aligned}$$

and

$$L(T_0 T)(f) = (c_1 \circ T_0 T)(f) = c_1(T_0 T f).$$

These two are equal since $\mathcal{T}$ is commutative. Hence L is equivariant.

Before proving that L is one-one onto, we make a construction. Let $f_1, \ldots, f_g$ be a basis of $S_2(\Gamma_0(N))$, and write

$$\mathbf{f}(\tau) = \begin{pmatrix} f_1(\tau) \\ \vdots \\ f_g(\tau) \end{pmatrix} = \sum_{n=1}^{\infty} \mathbf{c}_n e^{2\pi i n \tau}$$

with $\mathbf{c}_n \in \mathbf{C}^g$. Our first observation is that

$$\{\mathbf{c}_n\}_{n=1}^{\infty} \text{ spans } \mathbf{C}^g. \tag{11.45}$$

In fact, let ξ be any linear functional on $\mathbf{C}^g$ with $\xi(\mathbf{c}_n) = 0$ for all n. Since ξ is continuous,

$$\xi(\mathbf{f}(\tau)) = 0 \qquad \text{for all } \tau. \tag{11.46}$$

Let $\tau_1, \ldots, \tau_r$ be such that the set of r vectors $\begin{pmatrix} f_1(\tau_j) \\ \vdots \\ f_g(\tau_j) \end{pmatrix}$ with $1 \leq j \leq r$ is a maximal linearly independent set among the infinite set of all $\begin{pmatrix} f_1(\tau) \\ \vdots \\ f_g(\tau) \end{pmatrix}$. For each τ we can then find scalars $a_1(\tau), \ldots, a_r(\tau)$ such that

$$\begin{pmatrix} f_1(\tau) \\ \vdots \\ f_g(\tau) \end{pmatrix} = a_1(\tau) \begin{pmatrix} f_1(\tau_1) \\ \vdots \\ f_g(\tau_1) \end{pmatrix} + \cdots + a_r(\tau) \begin{pmatrix} f_1(\tau_r) \\ \vdots \\ f_g(\tau_r) \end{pmatrix}.$$

5. HECKE OPERATORS ON INTEGRAL HOMOLOGY

It follows that the functions $f_1, \ldots, f_g$ lie in the span of $a_1, \ldots, a_r$. Hence $g \leq r$, and the vectors $\begin{pmatrix} f_1(\tau_j) \\ \vdots \\ f_g(\tau_j) \end{pmatrix}$ span $\mathbf{C}^g$. By (11.46), $\xi(\mathbf{C}^g) = 0$. Thus $\xi = 0$, and (11.45) follows.

For each T in $\mathcal{T}$, we define a g-by-g matrix $A(T)$ by

$$\begin{pmatrix} Tf_1 \\ \vdots \\ Tf_g \end{pmatrix} = A(T) \begin{pmatrix} f_1 \\ \vdots \\ f_g \end{pmatrix}. \tag{11.47}$$

If we write

$$\begin{pmatrix} T(m)f_1 \\ \vdots \\ T(m)f_g \end{pmatrix} = \sum_{n=1}^{\infty} \mathbf{b}_n e^{2\pi i n \tau}$$

with $\mathbf{b}_n \in \mathbf{C}^g$, we have

$$\sum_{n=1}^{\infty} \mathbf{b}_n e^{2\pi i n \tau} = \begin{pmatrix} T(m)f_1 \\ \vdots \\ T(m)f_g \end{pmatrix}$$

$$= A(T(m)) \begin{pmatrix} f_1 \\ \vdots \\ f_g \end{pmatrix} = A(T(m)) \sum_{n=1}^{\infty} \mathbf{c}_n e^{2\pi i n \tau}.$$

Equating coefficients, we obtain

$$\mathbf{b}_n = A(T(m))\mathbf{c}_n.$$

Referring to Proposition 9.15, we see that $\mathbf{b}_1 = \mathbf{c}_m$. Hence

$$\mathbf{c}_m = A(T(m))\mathbf{c}_1. \tag{11.48}$$

We can now prove that L is one-one onto. Suppose that $L(T) = 0$. Since L is equivariant and $\mathcal{T}$ is commutative, we have

$$0 = T(m)L(T) = L(T(m)T) = L(TT(m)) = c_1 \circ TT(m) \tag{11.49}$$

for all m. Let e_j be the jth standard basis vector of $\mathbf{C}^g$, and apply (11.49) to f_j. Then

$$0 = c_1(TT(m)f_j) = c_1\left(\begin{pmatrix} TT(m)f_1 \\ \vdots \\ TT(m)f_g \end{pmatrix} \cdot e_j\right)$$

$$= c_1(A(T)A(T(m))\begin{pmatrix} f_1 \\ \vdots \\ f_g \end{pmatrix} \cdot e_j) \qquad \text{by (11.47)}$$

$$= A(T)A(T(m))\begin{pmatrix} c_1(f_1) \\ \vdots \\ c_1(f_g) \end{pmatrix} \cdot e_j \qquad \text{by linearity of } c_1(\cdot)$$

$$= A(T)A(T(m))\mathbf{c}_1 \cdot e_j$$

$$= A(T)\mathbf{c}_m \cdot e_j \qquad \text{by (11.48).}$$

Since j is arbitrary, $A(T)\mathbf{c}_m = 0$. By (11.45), $A(T) = 0$. Substituting into (11.47), we have $Tf_1 = \cdots = Tf_g = 0$. Thus $T = 0$, and L is one-one.

Suppose $L(\mathcal{T})$ does not span $S_2(\Gamma_0(N))'$. Choose $f \neq 0$ with

$$L(T)(f) = 0 \qquad \text{for all } T \in \mathcal{T}. \qquad (11.50)$$

Write $f = \sum_{j=1}^g z_j f_j$. We have

$$L(T(m))f_j = c_1(T(m)f_j) = c_1\left(\begin{pmatrix} T(m)f_1 \\ \vdots \\ T(m)f_g \end{pmatrix} \cdot e_j\right)$$

$$= c_1(A(m)\begin{pmatrix} f_1 \\ \vdots \\ f_g \end{pmatrix} \cdot e_j) \qquad \text{by (11.47)}$$

$$= A(m)\begin{pmatrix} c_1(f_1) \\ \vdots \\ c_1(f_g) \end{pmatrix} \cdot e_j \qquad \text{by linearity of } c_1(\cdot)$$

$$= A(m)\mathbf{c}_1 \cdot e_j$$

$$= \mathbf{c}_m \cdot e_j \qquad \text{by (11.48).}$$

Multiplying by z_j and summing on j, we have

$$L(T(m))f = \mathbf{c}_m \cdot \begin{pmatrix} z_1 \\ \vdots \\ z_g \end{pmatrix} \qquad (11.51)$$

for all m. The left side of (11.51) is 0 by (11.50). By (11.45), $\begin{pmatrix} z_1 \\ \vdots \\ z_g \end{pmatrix}$ is the 0 vector. Thus $f = 0$, contradiction. We conclude that L is onto.

6. Modular Function $j(\tau)$

Our objective in this section is to address the special properties of the compact Riemann surface $X_0(N)$ that allow us to realize it in a particular way as the set of $\mathbf{C}$ points of a projective curve defined over $\mathbf{Q}$. The function $j(\tau)$ introduced in §VIII.2 will play a critical role in our construction.

A meromorphic function f on $\mathcal{H}$ is said to be **automorphic of weight** k **and level** N if $f \circ [\gamma]_k = f$ for all $\gamma \in \Gamma_0(N)$ and such that the following condition holds: For every cusp $\beta^{-1}(\infty)$ relative to $\Gamma_0(N)$, there is some C_β such that $f \circ [\beta^{-1}]_k(\tau)$ is analytic for Im $\tau > C_\beta$ with an expansion

$$f \circ [\beta^{-1}]_k(\tau) = \sum_{n=-\infty}^{\infty} c_n^{(\beta)} q_N^n$$

that has only finitely many $c_n^{(\beta)} \neq 0$ for $n < 0$. The space of such functions is denoted $A_k(\Gamma_0(N))$. It is an enlargement of the space $M_k(\Gamma_0(N))$ of modular forms so as to allow poles in $\mathcal{H}$ and at the cusps.

Our interest will be in the space $A_0(\Gamma_0(N))$, which is a field. This space corresponds to the field $K(X_0(N))$ of meromorphic functions on $X_0(N)$ in the same way that $S_2(\Gamma_0(N))$ corresponds to the space of holomorphic differentials. In more detail, let $\{(U_i, \varphi_i)\}_{i \in I}$ be an atlas for $X_0(N)$. A system $\{f_i\}_{i \in I}$ of scalar-valued meromorphic functions f_i on U_i yields a meromorphic function on $X_0(N)$ if $f_i \circ \varphi_i^{-1} = f_j \circ \varphi_j^{-1}$ on $\varphi(U_i \cap U_j)$ for all i and j. Let $\pi : \mathcal{H}^* \to X_0(N)$ be the quotient map. If $\{f_i\}_{i \in I}$ is a meromorphic function on $X_0(N)$, then we define a corresponding scalar-valued function f on $\mathcal{H}$ by

$$f(\tau) = f_i(\pi(\tau)) \tag{11.52}$$

if (U_i, φ_i) is a chart about $\pi(\tau)$. The same kind of argument as for Proposition 11.6 yields the following result: The map $\{f_i\} \to f$ of (11.52) is well defined and is a field isomorphism of $K(X_0(N))$ onto $A_0(\Gamma_0(N))$.

Typically we shall use the same name for a member of $A_0(\Gamma_0(N))$ and its counterpart in $K(X_0(N))$.

Let us take $N = 1$ for the moment. The function $j(\tau)$ in $A_0(\Gamma_0(N))$ comes from a member of $K_0(X_0(1))$, according to the proposition, and we are going to denote this member of $K(X_0(N))$ by j also. For the case of $\Gamma_0(1) = SL(2,\mathbf{Z})$, there is just one cusp, and $j(\tau)$ has a simple pole there, according to Proposition 8.2. Moreover, j is analytic on $\mathcal{H}$ by (8.3).

Proposition 11.28. $j : SL(2,\mathbf{Z})\backslash\mathcal{H} \to \mathbf{C}$ is one-one onto.

PROOF. Fix z_0 in $\mathbf{C}$. As a meromorphic function on $X_0(1)$, $j - z_0$ has divisor of degree 0, by Proposition 11.8. Since the only pole is at $\pi(\infty)$ and is simple, the divisor must be

$$[j - z_0] = p - \pi(\infty)$$

for some $p \neq \pi(\infty)$ in $SL(2,\mathbf{Z})\backslash\mathcal{H}^*$. Then j takes the value z_0 at $p \in SL(2,\mathbf{Z})\backslash\mathcal{H}$ and only there.

Lemma 11.29. Let f in $A_0(\Gamma_0(1))$ be analytic on $\mathcal{H}$ and have q expansion $f(\tau) = \sum_{n=-M}^{\infty} c_n q^n$ at ∞. Then f is a polynomial in j with coefficients in the module over $\mathbf{Z}$ generated by the coefficients c_n.

PROOF. We induct on M for $M \geq 0$. For $M = 0$, f corresponds to an analytic function on $X_0(1)$ and must therefore be constant. Hence f is the polynomial c_0 in j of degree 0. Assuming the result for $M - 1$, we observe from Proposition 8.2 that $f - c_{-M} j^M$ has a q expansion $\sum_{n=-M+1}^{\infty} d_n q^n$. Since the coefficients of j are in $\mathbf{Z}$, each d_n is in $c_n + c_{-M}\mathbf{Z}$. Hence the lemma follows by induction.

Theorem 11.30. $K(X_0(1)) = \mathbf{C}(j)$.

PROOF. If f is in $K(X_0(1))$, then f has only a finite number of poles. If f has a pole at a point $p \neq \pi(\infty)$, then Proposition 11.28 shows that $(j - j(p))^{-1}$ has a simple pole at p and no other pole. Arguing as in the proof of Lemma 11.29, we can subtract from f a linear combination of powers of $(j - j(p))^{-1}$ and obtain a function with no pole at p and with no new pole. Continuing in this way, we can reduce to the case that f has its only pole at $\pi(\infty)$. Then Lemma 11.29 completes the proof.

Now we consider general N. We have denoted by $M^*(N)$ the set of 2-by-2 primitive integer matrices of determinant N. Let us write $\Gamma = \Gamma_0(1) = SL(2,\mathbf{Z})$. Lemma 9.1 says that

$$M^*(N) = \Gamma \begin{pmatrix} N & 0 \\ 0 & 1 \end{pmatrix} \Gamma = \bigcup_{i=1}^{\psi(N)} \Gamma\alpha_i \qquad (11.53)$$

6. MODULAR FUNCTION $j(\tau)$

and gives an explicit set of coset representatives. The number $\psi(N)$ is given by (9.5). The **modular polynomial** of order N is the polynomial $\Phi_N(X) = \Phi_N(j, X)$ of degree $\psi(N)$ defined by

$$\Phi_N(X) = \prod_{i=1}^{\psi(N)} (X - j \circ \alpha_i). \tag{11.54}$$

For $\gamma \in \Gamma$, we have $j \circ (\gamma \alpha_i) = j \circ [\gamma]_0 \circ \alpha_i = j \circ \alpha_i$; hence $\Phi_N(X)$ does not depend on the choice of coset representatives.

The first thing to notice about Φ_N is that one of its roots is $j_N = j \circ \begin{pmatrix} N & 0 \\ 0 & 1 \end{pmatrix}$. The same argument as with Examples 1 and 2 of §IX.3 shows that both j and j_N are in the field $K(X_0(N))$. In fact, Theorem 11.32 below will show that j_N is algebraic over $\mathbf{C}(j)$ and that Φ_N is its minimal polynomial. The theorem says even more, restricting the nature of the coefficients.

Lemma 11.31. With $\alpha_1, \ldots, \alpha_{\psi(N)}$ as in Lemma 9.1, the functions $j \circ \alpha_1, \ldots, j \circ \alpha_{\psi(N)}$ are distinct on $\mathcal{H}$.

PROOF. Let us write

$$j(\tau) = \frac{1}{q} + P(q), \tag{11.55}$$

where P is a power series with integer coefficients. If $\alpha_i = \begin{pmatrix} a & b \\ 0 & d \end{pmatrix}$ is as in Lemma 9.1 and if $q_d = e^{2\pi i \tau / d}$ and $\zeta_d = e^{2\pi i / d}$, then

$$j \circ \alpha_i(\tau) = j\left(\frac{a\tau + b}{d}\right) = \frac{1}{q_d^a \zeta_d^b} + P(q_d^a \zeta_d^b). \tag{11.56}$$

Suppose $\alpha_{i'} = \begin{pmatrix} a' & b' \\ 0 & d' \end{pmatrix}$ has $j \circ \alpha_i = j \circ \alpha_{i'}$. Taking the quotient and letting q tend to 0 (i.e., letting $\text{Im}(\tau)$ tend to $+\infty$), we obtain

$$q_d^a \zeta_d^b = q_{d'}^{a'} \zeta_{d'}^{b'}. \tag{11.57}$$

It follows first that $\dfrac{a}{d} = \dfrac{a'}{d'}$. Since $ad = a'd' = N$ and all are positive, we obtain $a = a'$ and $d = d'$. Substituting into (11.57), we obtain $b = b'$. Thus $\alpha_i = \alpha_{i'}$.

Theorem 11.32. The polynomial $\Phi_N(X)$ has coefficients in $\mathbf{Z}[j]$ and is irreducible over $\mathbf{C}(j)$.

PROOF. Any γ in Γ acts on $j \circ \alpha_i$, sending it to $j \circ \alpha_i \circ \gamma$. By (11.53) the result is some $j \circ \alpha_k$, and hence the effect of γ is to permute these functions. Since all of the functions contribute once to Φ_N, the coefficients of Φ_N are invariant under Γ. The coefficients are elementary symmetric polynomials in the functions $j \circ \alpha_i$ and thus are analytic on $\mathcal{H}$. The same argument as in Proposition 9.5 shows that each $j \circ \alpha_i$ has a meromorphic q_N expansion at ∞, and hence the same thing is true of the coefficients of Φ_N. It follows from the Γ invariance that the q expansions of the coefficients of Φ_N are meromorphic at ∞. By Theorem 11.30, the coefficients of Φ_N are in $\mathbf{C}(j)$.

The polynomial Φ_N splits in the field $\mathbf{C}(j \circ \alpha_1, \ldots, j \circ \alpha_{\psi(N)})$, and Γ acts as automorphisms of this field, fixing $\mathbf{C}(j)$ and permuting the roots of Φ_N. Since α_i is in $\Gamma\alpha_1\Gamma$, we have $\alpha_i = \gamma\alpha_1\gamma'$ or $\gamma^{-1}\alpha_i = \alpha_1\gamma'$. Thus Γ acts transitively on the roots. By Lemma 11.31 the roots are distinct. Consequently Φ_N is irreducible over $\mathbf{C}(j)$.

We are left with showing that the coefficients are actually in $\mathbf{Z}[j]$. Let us write $j(\tau)$ and $j \circ \alpha_i(\tau)$ as in (11.55) and (11.56). When we form the elementary symmetric polynomials in the functions (11.56), we know that the result involves only integral powers of q, and we can see that the coefficients of this resulting series are in $\mathbf{Z}[\zeta_N]$ since $d \mid N$.

Regard the coefficients of the series as in $\mathbf{Q}(\zeta_N)$, and consider the automorphism of this field given by $\zeta_N \to \zeta_N^r$, where $\mathrm{GCD}(r, N) = 1$. The expression (11.56) is mapped to

$$\frac{1}{q_d^a \zeta_d^{b'}} + P(q_d^a \zeta_d^{b'}),$$

where b' is the least residue ≥ 0 of rb modulo d. This expression is just $j \circ \alpha_{i'}(\tau)$, where $\alpha_{i'} = \begin{pmatrix} a & b' \\ 0 & d \end{pmatrix}$. The map $(a, b, d) \to (a, b', d)$ represents a permutation of the representatives α_i. Thus the effect of $\zeta_N \to \zeta_N^r$ on the coefficients of the q series expansion of a coefficient of Φ_N is the same as the effect of a certain permutation of the roots, i.e., to leave things fixed. Since $\mathbf{Q}(\zeta_N)$ is Galois over $\mathbf{Q}$, the coefficients of the q series expansion of a coefficient of Φ_N are in $\mathbf{Q} \cap \mathbf{Z}[\zeta_N] = \mathbf{Z}$. By Lemma 11.29 the coefficients of Φ_N are in $\mathbf{Z}[j]$.

Theorem 11.33. $K(X_0(N)) = \mathbf{C}(j, j_N)$.

The proof will use the following two lemmas.

Lemma 11.34. If $g \in \Gamma$ has $j(N(g\tau)) = j(N\tau)$ for all $\tau \in \mathcal{H}$, then g is in $\Gamma_0(N)$.

PROOF. Let $\alpha = \begin{pmatrix} N & 0 \\ 0 & 1 \end{pmatrix}$. Since $N\tau = \alpha(\tau)$, we have

$$j(\alpha g \alpha^{-1} N\tau) = j(N\tau)$$

for all τ in $\mathcal{H}$. Hence

$$j(\alpha g \alpha^{-1} \tau) = j(\tau)$$

for all τ in $\mathcal{H}$. By Proposition 11.28, there exists β_τ in Γ with

$$\alpha g \alpha^{-1}(\tau) = \beta_\tau(\tau)$$

for all τ in $\mathcal{H}$. Fix τ_0 in the interior of the standard fundamental domain R of Γ. The point $\alpha g \alpha^{-1}(\tau_0)$ is in the interior of $\beta_{\tau_0}(R)$ and hence so is $\alpha g \alpha^{-1}(\tau)$ for τ near τ_0. Consequently $\beta_\tau = \pm \beta_{\tau_0}$ for τ near τ_0. By analyticity $\alpha g \alpha^{-1} = \pm \beta_{\tau_0}$. Consequently g is in

$$SL(2,\mathbb{Z}) \cap \alpha^{-1} SL(2,\mathbb{Z})\alpha,$$

and this is $\Gamma_0(N)$ by Lemma 9.2.

Lemma 11.35. Let F be a field of characteristic 0, and let G be a finite subgroup of $\text{Aut}_\mathbb{Q}(F)$. Then F is a finite Galois extension of

$$F^G = \{x \in F \mid \varphi(x) = x \text{ for all } \varphi \in G\},$$

and $\text{Aut}_{F^G}(F) = G$.

PROOF. For $x \in F$, let $f_x(X)$ be the member of $F[X]$ given by

$$f_x(X) = \prod_{\varphi \in G}(X - \varphi(x)). \tag{11.58}$$

Then f_x is in $F^G[X]$. Since $f_x(x) = 0$, x is algebraic over F^G and $[F^G(x) : F^G] \leq |G|$. Choose y in F such that $[F^G(y) : F^G]$ is maximal. For any z in F, the Theorem of the Primitive Element yields some w in F such that $F^G(z,y) = F^G(w)$. By maximality

$$[F^G(z,y) : F^G] = [F^G(w) : F^G] \leq [F^G(y) : F^G],$$

and thus z is in $F^G(y)$. Therefore $F = F^G(y)$ and $[F : F^G] \leq |G|$. In particular, F is a finite algebraic extension of F^G.

Fix an algebraic closure $\bar{F}$ of F. If ψ is in $\text{Hom}_{F^G}(F, \bar{F})$, then $f_y(\psi(y)) = 0$ since f_y has coefficients in F^G. Referring to (11.58), we see that $\psi(y) = \varphi(y)$ for some φ in G. Since y generates F over F^G, we have $\psi = \varphi$ on F. We conclude that every member of $\text{Hom}_{F^G}(F, \bar{F})$ carries F into itself and coincides with a member of G. The first of these conclusions implies that F is a Galois extension of F^G, and the second implies that G is the Galois group.

PROOF OF THEOREM 11.33. Let $\Gamma(N)$ be the principal congruence subgroup of level N. Let $K = A_0(\Gamma(N))$ be the field of meromorphic functions f on $\mathcal{H}$ such that $f \circ [\gamma]_0 = f$ for all $\gamma \in \Gamma(N)$ and such that the following condition holds: For each β in $\gamma = SL(2, \mathbf{Z})$, there is some C_β such that $f \circ [\beta^{-1}]_0(\tau)$ is analytic for Im $\tau > C_\beta$ with an expansion

$$f \circ [\beta^{-1}]_0(\tau) = \sum_{n=-\infty}^{\infty} c_n^{(\beta)} q_N^n$$

that has only finitely many $c_n^{(\beta)} \neq 0$ for $n < 0$.

Then $A_0(\Gamma_0(N)) \cong K(X_0(N))$ is contained in K. The group Γ acts on K by automorphisms under $f \to f \circ \alpha = f \circ [\alpha]_0$, and $\Gamma(N)$ fixes K. Hence $G = \Gamma/\Gamma(N)$ acts on K. By Lemma 11.35, K is a finite Galois extension of $K^G = K^\Gamma$, and the Galois group is G/G_0, where G_0 is the subgroup of G that fixes K. A member f of K^Γ has $f = f \circ [\alpha]_0$ for all $\alpha \in \Gamma$, hence is in $K(X_0(1))$, which Theorem 11.30 identifies as $\mathbf{C}(j)$. Thus K is a Galois extension of $\mathbf{C}(j)$ with Galois group G/G_0.

Let us determine the Galois group G' of K over the intermediate field $\mathbf{C}(j, j_N)$. If g is in G', we can regard g as a member of Γ satisfying $j_0 N \circ g = j_0 N$. By Lemma 11.34, g is in $\Gamma_0(N)$. Conversely $\Gamma_0(N)/\Gamma(N)$ does fix $\mathbf{C}(j, j_N)$. Hence $G' = (\Gamma_0(N)/\Gamma(N))/G_0$.

By the Fundamental Theorem of Galois Theory,

$$\mathbf{C}(j, j_N) = K^{G'} = K^{\Gamma_0(N)}.$$

Since $K(X_0(N)) \subseteq K$, we have $K(X_0(N)) \subseteq K^{\Gamma_0(N)}$. On the other hand, comparison of the definitions of $K(X_0(N))$ and K shows that $K(X_0(N)) \supseteq K^{\Gamma_0(N)}$. Thus $\mathbf{C}(j, j_N) = K^{\Gamma_0(N)} = K(X_0(N))$.

The $\mathbf{Q}$ structure on $X_0(N)$ will result from a relationship between $K(X_0(N)) = \mathbf{C}(j, j_N)$ and $\mathbf{Q}(j, j_N)$ that is given by the equivalent conditions in the next theorem, in which we eventually will take $x = j$, $y = j_N$, and $k_0 = \mathbf{Q}$. The conditions are satisfied as a consequence of Theorem 11.32.

Theorem 11.36. If $\mathbf{C}(x, y)$ is a field with x transcendental over $\mathbf{C}$ and y algebraic over $\mathbf{C}(x)$, then the following conditions on a subfield k_0 of $\mathbf{C}$ are equivalent:

(a) y is algebraic over $k_0(x)$, and the minimal polynomial of y over $k_0(x)$ remains irreducible over $\mathbf{C}(x)$,
(b) $[k_0(x, y) : k_0(x)] = [\mathbf{C}(x, y) : \mathbf{C}(x)]$,
(c) $\mathbf{C} \cap k_0(x, y) = k_0$,
(d) $\bar{k}_0 \cap k_0(x, y) = k_0$.

REMARK. Condition (d) may be restated as: k_0 is algebraically closed in $k_0(x,y)$.

PROOF. (a) $\Leftrightarrow$ (b). This is clear.
(b) $\Rightarrow$ (c). Let ω be in $\mathbb{C}$ and in $k_0(x,y)$. From the inclusions

$$k_0(x,y) \supseteq k_0(x,\omega) \supseteq k_0(x)$$
$$\mathbb{C}(x,y) \supseteq \mathbb{C}(x,\omega) \supseteq \mathbb{C}(x),$$

we have

$$[k_0(x,y) : k_0(x)] = [k_0(x,y) : k_0(x,\omega)][k_0(x,\omega) : k_0(x)]$$
$$[\mathbb{C}(x,y) : \mathbb{C}(x)] = [\mathbb{C}(x,y) : \mathbb{C}(x,\omega)][\mathbb{C}(x,\omega) : \mathbb{C}(x)], \qquad (11.59)$$

with the left sides equal, by (b). Since y is algebraic over $k_0(x,\omega)$ and $k_0(x)$, we have

$$[k_0(x,y) : k_0(x,\omega)] \geq [\mathbb{C}(x,y) : \mathbb{C}(x,\omega)]$$
$$[k_0(x,\omega) : k_0(x)] \geq [\mathbb{C}(x,\omega) : \mathbb{C}(x)]. \qquad (11.60)$$

Comparing (11.59) and (11.60), we see that actually the inequalities (11.60) are equalities. Since ω is in $\mathbb{C}$, we have

$$[k_0(x,\omega) : k_0(x)] = [\mathbb{C}(x,\omega) : \mathbb{C}(x)] = 1,$$

and we conclude that ω is in $k_0(x)$. Let $\omega = P_1(x)/P_2(x)$ with P_1 and P_2 in $k[x]$ and $P_2 \neq 0$. Put $P(X) = P_1 - \omega P_2(X)$ in $\mathbb{C}[X]$. If P is not the 0 polynomial, then $P(x) = 0$ shows that x is algebraic over $\mathbb{C}$, contradiction. So $P = 0$ and every coefficient of $P_1 - \omega P_2$ is 0. Since P_1 and P_2 have coefficients in k_0 and P_2 is not 0, it follows that ω is in k_0.
(c) $\Rightarrow$ (d). This is clear.
(d) $\Rightarrow$ (b). First let us show that (d) implies

$$[k_0(x,\omega) : k_0(x)] = [\bar{k}_0(x,\omega) : \bar{k}_0(x)]. \qquad (11.61)$$

In fact, otherwise we can find a finite Galois extension k_1 of k_0 for which

$$[k_0(x,\omega) : k_0(x)] > [k_1(x,\omega) : k_1(x)].$$

Hence

$$[k_1(x) : k_0(x)] > [k_1(x,y) : k_0(x,y)]. \qquad (11.62)$$

The left side here equals $[k_1 : k_0]$ because any element of $k_1(x)$ can have its denominator "rationalized" by $\mathrm{Aut}_{k_0}(k_1)$ so as to be written in the form $P_1(x)/P_2(x)$ with P_1 in $k_1[x]$ and P_2 in $k_0[x]$. Thus $k_1(x)$ is Galois over $k_0(x)$, and $k_1(x,y)$ is Galois over $k_0(x,y)$. Since k_1 is Galois over k_0, any k_0 isomorphism of k_1 into a field containing k_1 sends k_1 into k_1. Hence every member of $\mathrm{Aut}_{k_0(x,y)}(k_1(x,y))$ sends k_1 into k_1 and fixes x, thus sends $k_1(x)$ into itself while fixing $k_0(x)$. Thus we have a homomorphism

$$\mathrm{Aut}_{k_0(x,y)}(k_1(x,y)) \longrightarrow \mathrm{Aut}_{k_0(x)}(k_1(x))$$

that is clearly one-one. Let H be the image in $\mathrm{Aut}_{k_0(x)}(k_1(x))$. The assumed inequality (11.62) means that H is not all of $\mathrm{Aut}_{k_0(x)}(k_1(x))$. Let k_1' be the field with $k_1 \supseteq k_1' \supseteq k_0$ and $k_1'(x) = k_1(x)^H$; here $k_1'(x) \supsetneq k_0(x)$. Form $k_1'(x,y)$. Every element of $\mathrm{Aut}_{k_0(x,y)}(k_1(x,y))$ fixes this, by construction. Hence $k_1'(x,y) = k_0(x,y)$. Thus $k_1' \subseteq k_0(x,y)$. By (d), $k_1' = k_0$. Since $k_1'(x) \supsetneq k_0(x)$, (11.62) has led to a contradiction. Thus (11.61) holds.

To complete the proof of (b), we shall show that

$$[\bar{k}_0(x,y) : \bar{k}_0(x)] = [\mathbb{C}(x,y) : \mathbb{C}(x)], \tag{11.63}$$

even without assuming (d). Let $P(Y) \in \bar{k}_0(x)[Y]$ be an irreducible polynomial with $P(y) = 0$. Multiplying the coefficients by a suitable member of $\bar{k}_0(x)$, we may assume that $P(Y)$ is in $\bar{k}_0[x][Y]$ and is primitive over $\bar{k}_0[x]$. We are to prove that $P(Y)$ remains irreducible over $\mathbb{C}(x)$. Suppose on the contrary that P factors nontrivially over $\mathbb{C}(x)$. By Gauss's Lemma it factors over $\mathbb{C}[x]$. We can thus write

$$P(Y) = Q(Y)R(Y) \tag{11.64}$$

with Q and R primitive over $\mathbb{C}[x]$. Let us make x explicit by writing (11.64) as

$$P(x,Y) = Q(x,Y)R(x,Y) \tag{11.65}$$

with $P(x,Y)$ in $\bar{k}_0[x,Y]$ and with $Q(x,Y)$ and $R(x,Y)$ in $\mathbb{C}[x,Y]$. Applying the evaluation homomorphism $x \to c \in \bar{k}_0$, we obtain a factorization of the polynomial $P(c,Y)$ of one variable, and we see that $Q(c,Y)$ and $R(c,Y)$ are in $\bar{k}_0[Y]$.

Write $Q(X,Y) = \sum Q_i(X)Y^i$. We have seen that $Q_i(c)$ is in $\bar{k}_0$ for every c in $\bar{k}_0$. If $\deg Q_i = n$, we can construct $Q_i^{\#}(X)$ in $\bar{k}_0[X]$ of degree n with $Q_i^{\#}(l) = Q_i(l)$ for $1 \leq l \leq n$. Then it follows that

$$Q_i^{\#}(X) - Q_i(X) = c \prod_{l=1}^{n}(X - l)$$

with $c \neq 0$ in $\mathbf{C}$. Evaluating at $n+1$, we see that c is in $\bar{k}_0$. Hence Q_i is in $\bar{k}_0[X]$, and $Q(X,Y)$ is in $\bar{k}_0[X,Y]$. Similarly $R(X,Y)$ is in $\bar{k}_0[X,Y]$. Thus (11.65) contradicts the irreducibility of P over $\bar{k}_0(x)$. This proves (11.63), and (b) follows by combining (11.61) and (11.63).

7. Varieties and Curves

In this section and the next we shall complete the task of realizing $X_0(N)$ in a particular way as the set of $\mathbf{C}$ points of a projective curve defined over $\mathbf{Q}$. So far our discussion of curves has been limited to plane curves. In order to treat $X_0(N)$ adequately, we shall need to carry out two additional steps. First, in this section, we extend our discussion from plane curves to varieties and curves in higher dimensional projective spaces, and we define types of mappings between such varieties and curves. Second, in §8, we characterize curves in terms of their function fields.

In the case of $X_0(N)$, we can regard the irreducible polynomial $\Phi_N(X)$ of Theorem 11.32 as a polynomial $\Phi_N(X_1, X_2)$ over $\mathbf{Q}$ in two variables such that $\Phi_N(j, j_N) = 0$. Theorem 11.33 shows that $\tau \to (j(\tau), j_N(\tau))$ gives a one-one parametric mapping of $X_0(N)$ into the $\mathbf{C}$ points of Φ. We could then apply a desingularizing process to obtain a nonsingular curve over $\mathbf{Q}$ in a higher dimensional projective space whose $\mathbf{C}$ points are biholomorphic with $X_0(N)$.

But as we shall see, the theory of algebraic curves provides a faster procedure. By working directly with the function field $K(X_0(N))$ and using Theorems 11.32, 11.33, and 11.36, we can arrive at an appropriate nonsingular curve over $\mathbf{Q}$ without working further with explicit equations. The speed of our procedure is achieved at the price of not easily being able to make explicit computations. This disadvantage was already noted in §1, where we mentioned that the equations for the elliptic curves that arise in Eichler-Shimura theory had to be deduced from side information.

We begin by assembling background information about varieties and curves in general. We omit many of the proofs. We shall not return to $X_0(N)$ until §8.

Let k_0 be a field, and let $\bar{k}$ be an algebraically closed field with $k_0 \subseteq \bar{k}$. We shall assume that k_0 has characteristic 0 or is a finite field and that the fixed field of $\mathrm{Aut}_{k_0}(\bar{k})$ is k_0. In many situations that will be clear from the context, a definition or result valid for k_0 is also valid for $\bar{k}$.

To any ideal I in $\bar{k}[X_1, \ldots, X_n]$, we associate its zero set in $\bar{k}^n$:

$$V_I = \{x = (x_1, \ldots, x_n) \in \bar{k}^n \mid f(x) = 0 \text{ for all } f \in I\}.$$

Any such set V_I is called an **affine algebraic set**. If $V \subseteq \bar{k}^n$ is an affine algebraic set, its **ideal** (of polynomials vanishing on it) is

$$I(V) = \{f \in \bar{k}[X_1, \ldots, X_n] \mid f(x) = 0 \text{ for all } x \in V\}.$$

Then $I(V_I) \supseteq I$.

An ideal in a polynomial ring over a field is finitely generated (i.e., the ring is **Noetherian**), by the Hilbert Basis Theorem. We say that an affine algebraic set V is **defined over** k_0 if $I(V)$ can be generated by members of $k_0[X_1, \ldots, X_n]$. In this case we often abbreviate "V defined over k_0" by V/k_0, we define $I(V/k_0) = I(V) \cap k_0[X_1, \ldots, X_n]$, and we define the set of k_0 **points** or k_0 **rational points** of V to be $V(k_0) = V \cap k_0^n$.

The group action by $\text{Aut}_{k_0}(\bar{k})$ on $\bar{k}$ extends to $\bar{k}[X_1, \ldots, X_n]$ with all X_j fixed by all of $\text{Aut}_{k_0}(\bar{k})$. Then $k_0[X_1, \ldots, X_n]$ is the full fixed ring under $\text{Aut}_{k_0}(\bar{k})$. In the case of V/k_0, $I(V)$ is mapped into itself by $\text{Aut}_{k_0}(\bar{k})$, and $I(V/k_0)$ is the subset of $I(V)$ fixed by $\text{Aut}_{k_0}(\bar{k})$. Also $V(k_0)$ is the subset of V fixed by $\text{Aut}_{k_0}(\bar{k})$.

Theorem 11.37 (Hilbert Nullstellensatz). If I is an ideal in $\bar{k}[X_1, \ldots, X_n]$ and f is in $I(V_I)$ then f^r is in I for some integer $r > 0$.

If I is a prime ideal, then it follows from Theorem 11.37 that $I = I(V_I)$, and the situation is more manageable. Accordingly we shall systematically work with this situation (even though we did not do so in Chapter II). An affine algebraic set V is **irreducible** if it is not the union of two proper affine algebraic sets, or equivalently if $I(V)$ is prime. An irreducible affine algebraic set is called an **affine variety**. For example, if $P(X, Y)$ is an irreducible polynomial in $\bar{k}[X, Y]$, then its zero locus in $\bar{k}^2$ is an affine variety, being V_I for I equal to the prime ideal (P).

The **affine coordinate ring** $\bar{k}[V]$ of an affine variety V is the Noetherian ring defined by

$$\bar{k}[V] = \bar{k}[X_1, \ldots, X_n]/I(V).$$

Since $I(V)$ is prime, $\bar{k}[V]$ is an integral domain. Its quotient field $\bar{k}(V)$ is called the **function field** of V; it is finitely generated over $\bar{k}$ since $\bar{k}[V]$ has this same property. In the case of V/k_0, the **affine coordinate ring** is

$$k_0[V] = k_0[X_1, \ldots, X_n]/I(V/k_0).$$

Since $I(V)$ is prime, so is $I(V/k_0)$, and therefore $k_0[V]$ is an integral domain. Its quotient field $k_0(V)$ is the **function field** of V/k_0 and is finitely generated over k_0.

Also in the case of V/k_0, $\text{Aut}_{k_0}(\bar{k})$ acts on $\bar{k}[X_1,\ldots,X_n]$ and carries $I(V)$ into itself; therefore it acts on $\bar{k}[V]$. One can prove that the fixed ring is naturally identified with $k_0[V]$. Similar remarks apply to $\bar{k}(V)$ and $k_0(V)$.

The **dimension** of an affine variety V is the transcendence degree of $\bar{k}(V)$. An **affine curve** is an affine variety of dimension one. In other words, there is to exist some f in $\bar{k}(V)$ not in $\bar{k}$, so that $\bar{k}(f)$ is transcendental over $\bar{k}$, and any g in $\bar{k}(V)$ is to have the property that $\bar{k}(f,g)$ is algebraic over $\bar{k}(f)$. In the case of characteristic 0, it follows from the Theorem of the Primitive Element that $\bar{k}(V) = \bar{k}(f,h)$ for some suitable h. (Cf. Theorem 11.20.)

Let $V \subseteq \bar{k}^n$ be an affine variety, and let $f_1,\ldots,f_l$ be a set of generators for $I(V)$. A point x of V is a **nonsingular** point if the matrix $\left(\dfrac{\partial f_i}{\partial x_j}(x)\right)$ has rank $n - d$, where d is the dimension of V. The variety V is **nonsingular** if every point of it is a nonsingular point. These definitions do not depend on the set of generators. There are two different constructs with which we can see this independence. The first one is the easier to use in computations. For it, let x be in V, and let m_x be the ideal in $\bar{k}[V]$ given by

$$m_x = \{f \in \bar{k}[V] \mid f(x) = 0\}.$$

This is a maximal ideal in $\bar{k}[V]$, since evaluation at x is an isomorphism of $\bar{k}[V]/m_x$ with $\bar{k}$, and the quotient m_x/m_x^2 is a finite-dimensional vector space over $\bar{k}$. The following proposition shows that the definition of nonsingularity does not depend on the system of generators.

Proposition 11.38. A point x of an affine variety V is nonsingular if and only if the vector space dimension of m_x/m_x^2 equals the dimension of V.

The second construct for seeing that nonsingularity is intrinsic uses quotients of elements of $\bar{k}[V]$. Let $\bar{k}[V]_x$ be the subring of $\bar{k}(V)$ given by

$$\bar{k}[V]_x = \left\{ F \in \bar{k}(V) \,\middle|\, \begin{array}{l} F = g/h, \\ g \text{ and } h \text{ are in } \bar{k}[V], \\ h(x) \neq 0 \end{array} \right\}.$$

This is a **local ring** in the sense that it has a unique maximal ideal M_x, namely the members vanishing at x. It is called the **local ring** of V at x. It is Noetherian since $\bar{k}[V]$ is Noetherian, and it is an integral domain since it is contained in a field. The members of $\bar{k}[V]_x$ are the elements of

$\bar{k}(V)$ with a definite value at x, i.e., $F = g/h$ has $F(x) = g(x)/h(x)$ well defined and finite. Such elements are said to be **regular** or **defined** at x. Nonsingularity can be decided by M_x/M_x^2, according to Proposition 11.38 and the following result.

Proposition 11.39. The inclusion $m_x \subseteq M_x$ yields an isomorphism $m_x/m_x^2 \cong M_x/M_x^2$.

PROOF. We get a ring homomorphism $m_x/m_x^2 \to M_x/M_x^2$. If g/h is given in M_x, then

$$h(x)^{-1}g = \frac{g}{h} + \left(\frac{g}{h}\right)\left(\frac{h(x)^{-1}h - 1}{1}\right)$$

exhibits $h(x)^{-1}g$ in M_x as mapping to $\frac{g}{h} + M_x^2$. Hence the map is onto. If g in m_x maps to $\sum \left(\frac{g_i}{h_i}\right)\left(\frac{g_i'}{h_i'}\right)$ in M_x^2, we can clear fractions and write $hg = \sum g_i g_i' h_i''$ for an element h of $\bar{k}[V]$ with $h(x) \neq 0$. Here $\sum g_i g_i' h_i''$ is in m_x^2. The set $\{f \in \bar{k}[V] \mid fg \in m_x^2\}$ is an ideal in $\bar{k}[V]$ containing m_x and also h, which is not in m_x. Since m_x is maximal, 1 is in the set and g is in m_x^2. Thus the mapping $m_x/m_x^2 \to M_x/M_x^2$ is one-one.

Projective n-space $P_n(\bar{k})$ is defined in the same way as with $P_2(\bar{k})$, namely as the quotient of $\{(x_0, \ldots, x_n) \in \bar{k}^{n+1} - \{(0, \ldots, 0)\}\}$ by an equivalence relation, two points being equivalent if one is a $\bar{k}^\times$ multiple of the other. Our notation for writing a point of $P_n(\bar{k})$ with care is $[(x_0, \ldots, x_n)]$. We let $P_n(k_0)$ be the subset of $P_n(\bar{k})$ for which $x_0, \ldots, x_n$ can be taken in k_0.

To any *homogeneous* ideal I in $\bar{k}[X_0, \ldots, X_n]$, we associate its zero set in $P_n(\bar{k})$:

$$W_I = \{[(x_0, \ldots, x_n)] \in P_n(\bar{k}) \mid f(x_0, \ldots, x_n) = 0 \text{ for all } f \in I\}.$$

Any such set W_I is called a **projective algebraic set**. If $W \subseteq P_n(\bar{k})$ is a projective algebraic set, its **homogeneous ideal** (of polynomials vanishing on it) is the ideal $I(W)$ generated by all homogeneous f in $\bar{k}[X_0, \ldots, X_n]$ such that $f(x_0, \ldots, x_n) = 0$ for all $[(x_0, \ldots, x_n)] \in W$. It is always true that $I(W_I) \supseteq I$.

We say that a projective algebraic set is **defined over** k_0 if $I(W)$ can be generated by homogeneous members of $k_0[X_0, \ldots, X_n]$. In this case we abbreviate "W defined over k_0" by W/k_0, and we define the set of k_0 **points** or k_0 **rational points** of W to be $W(k_0) = W \cap P_n(k_0)$.

7. VARIETIES AND CURVES

A projective algebraic set W is **irreducible** if it is not the union of two proper projective algebraic sets, or equivalently if $I(W)$ is prime. An irreducible projective algebraic set is called a **projective variety**.

Projective transformations of $P_n(\bar{k})$ onto $P_n(\bar{k})$ are defined in analogy with the case of $n = 2$ that was discussed in §II.1. About any point $[(x_0, \ldots, x_n)]$ of $P_n(\bar{k})$ we can introduce various systems of **affine local coordinates**. Namely choose Φ in $GL(n+1, \bar{k})$ with $\Phi(x_0, \ldots, x_n) = (1, 0, \ldots, 0)$. (For many purposes we ignore $(x_0, \ldots, x_n)$, and it suffices to take Φ to be the identity or a permutation matrix that interchanges 0 and i.) Then we can define affine local coordinates on $[\Phi^{-1}(\bar{k}^\times \times \bar{k}^n)]$ to $\bar{k}^n$ by the one-one map

$$\varphi([\Phi^{-1}(y_0, \ldots, y_n)]) = \left(\frac{y_1}{y_0}, \ldots, \frac{y_n}{y_0}\right). \tag{11.66}$$

If W is a projective algebraic set with homogeneous ideal $I(W)$ and if Φ is given as above, we shall write $W \cap \bar{k}^n$ for $\varphi([\Phi^{-1}(\bar{k}^\times \times \bar{k}^n)] \cap W)$. This is an affine algebraic set with ideal $I(W \cap \bar{k}^n) \subseteq \bar{k}[X_1, \ldots, X_n]$ given by

$$I(W \cap \bar{k}^n) = \left\{ f \;\middle|\; \begin{array}{l} f(X_1, \ldots, X_n) = F(\Phi^{-1}(1, X_1, \ldots, X_n)) \\ \text{with } F \in I(W) \end{array} \right\}. \tag{11.67}$$

Conversely if f is given in $\bar{k}[X_1, \ldots, X_n]$ of degree d and if Φ is given as above, then we can define f_H as a homogeneous element of $\bar{k}[X_0, \ldots, X_n]$ in two steps by

$$g_H(X_0, \ldots, X_n) = X_0^d f\left(\frac{X_1}{X_0}, \ldots, \frac{X_n}{X_0}\right) \tag{11.68}$$

$$f_H = g_H \circ \Phi.$$

If V is an affine algebraic set with ideal $I(V)$, then the **projective closure** of V relative to Φ is the projective algebraic set $W = \bar{V}$ whose homogeneous ideal is generated by $\{f_H \in \bar{k}[X_0, \ldots, X_n] \mid f \in I(V)\}$.

If we combine these two constructions, passing from F to f as in (11.67) and then from f to f_H as in (11.68), we find that

$$F = (X_0 \circ \Phi)^r f_H \tag{11.69}$$

for an integer $r \geq 0$. As a consequence we can prove the following.

Proposition 11.40. Fix Φ in $GL(n+1,\bar{k})$ to determine affine local coordinates.

(a) If V is an affine variety, then $W = \bar{V}$ is a projective variety and $V = W \cap \bar{k}^n$.

(b) If W is a projective variety, then $V = W \cap \bar{k}^n$ is an affine variety, and $\bar{V} \subseteq W$. The inclusion is strict only if $X_0 \circ \Phi$ is in $I(W)$, and in this case $V = \bar{V} = \emptyset$.

(c) If Φ is in $GL(n+1,k_0)$ and W is a projective variety with $W \cap \bar{k}^n$ nonempty, then W is defined over k_0 if and only if $W \cap \bar{k}^n$ is defined over k_0.

PROOF. Let us check that $I(W)$ prime implies $I(W \cap \bar{k}^n)$ prime. If homogeneous F and G lead to f and g with fg in $I(W \cap \bar{k}^n)$, then FG vanishes on $W \cap \bar{k}^n$ and $(X_0 \circ \Phi)FG$ vanishes on W. Hence $(X_0 \circ \Phi)FG$ is in $I(W)$. Then $(X_0 \circ \Phi)F$ or G is in $I(W)$, and it follows that f or g is in $I(W \cap \bar{k}^n)$.

Next let us check that $I(V)$ prime implies $I(\bar{V})$ prime. If f and g lead to f_H and g_H with $f_H g_H = h_H$ in $I(\bar{V})$, choose h in $I(V)$ leading to h_H. Then $fg = h$, f or g is in $I(V)$, and f_H or g_H is in $I(\bar{V})$.

Then (a) is clear, and (11.69) implies that $I(W) \subseteq I(\bar{V})$ in (b). Hence $\bar{V} \subseteq W$. If the inclusion is strict, let f_H be in $I(\bar{V})$ but not in $I(W)$. Then f_H comes from some f in $I(V)$, which comes from some F in $I(W)$. By (11.69), $F = (X_0 \circ \Phi)^r f_H$. Since $I(W)$ is prime and f_H is not in $I(W)$, $X_0 \circ \Phi$ is in $I(W)$. Then 1 is in $I(V)$, and $V = \emptyset$. This proves (b), and (c) follows from the definitions.

If W is a projective variety, we define the **function field** $\bar{k}(W)$ of W to be the field of quotients $F(X) = g(X)/h(X)$ in $\bar{k}(X_0,\ldots,X_n)$ such that g and h are polynomials homogeneous of the same degree, $h(X)$ is not in $I(W)$, and quotients $\dfrac{g}{h}$ and $\dfrac{g'}{h'}$ are identified if $gh' - g'h$ is in $I(W)$. In this way, $\bar{k}(W)$ has a canonical definition independent of Φ. We can describe $\bar{k}(W)$ alternatively using affine local coordinates. Namely (b) shows that we can choose Φ in $GL(n+1,\bar{k})$ so that $\overline{W \cap \bar{k}^n} = W$. Fix such a choice of Φ. Using Φ, we can identify the function field $\bar{k}(W)$ with $\bar{k}(W \cap \bar{k}^n)$. Consequently $\bar{k}(W)$ is finitely generated over $\bar{k}$. If W is defined over k_0, we can treat $k_0(W)$ similarly, as a consequence of Proposition 11.40c; here we must take Φ in $GL(n+1,k_0)$.

The **dimension** of W is again the transcendence degree of $\bar{k}(W)$. If x is in W, we can choose Φ so that x is mapped by φ in (11.66) to a point of $W \cap \bar{k}^n$. Then it is meaningful to speak of the **local ring** $\bar{k}[W]_x$ of W at x. We regard $\bar{k}[W]_x$ as a subring of $\bar{k}(W)$; it is the subring of elements that are **regular** or **defined** at x. The ring $\bar{k}[W]_x$ is Noetherian.

7. VARIETIES AND CURVES

Finally we can consistently define x to be **nonsingular** if $\dim M_x/M_x^2 = \dim W$, by virtue of Propositions 11.38 and 11.39. The projective variety W is **nonsingular** if it is nonsingular at every point.

The group $\text{Aut}_{k_0}(\bar{k})$ plays the same role for projective varieties that it does for affine varieties. We shall not list the details.

We shall encounter products of projective varieties and shall want to regard them as projective varieties. First we consider the **product** of two projective spaces. If $P_m(\bar{k})$ has points $[(x_0, \ldots, x_m)]$ and $P_n(\bar{k})$ has points $[(y_0, \ldots, y_n)]$, then the **Segre embedding**, which maps

$$[(x_0, \ldots, x_m)] \times [(y_0, \ldots, y_n)] \longrightarrow [(\ldots, x_i y_j, \ldots)] = [(\ldots, z_{ij}, \ldots)]$$

in lexicographic order, exhibits $P_m(\bar{k}) \times P_n(\bar{k})$ as contained in $P_M(\bar{k})$ with $M = mn + m + n$. The image of the Segre embedding is the variety of the ideal generated by all $z_{ij} z_{i'j'} - z_{ij'} z_{i'j}$. It is a projective variety in $P_M(\bar{k})$. If $W \subseteq P_m(\bar{k})$ and $W' \subseteq P_n(\bar{k})$ are projective varieties defined by homogeneous polynomials $\{f(X)\}$ and $\{g(Y)\}$, then the **product** $W \times W' \subseteq P_M(\bar{k})$ is the variety of the ideal generated by all $z_{ij} z_{i'j'} - z_{ij'} z_{i'j}$, all $f(X) Y_j^{\deg f}$, and all $g(Y) X_i^{\deg g}$, in obvious notation. It too is a projective variety in $P_M(\bar{k})$. If W and W' are nonsingular, so is $W \times W'$. If W and W' are defined over k_0, so is $W \times W'$.

Let $V_1 \subseteq P_m(\bar{k})$ and $V_2 \subseteq P_n(\bar{k})$ be projective varieties. A **rational map** F from V_1 to V_2 is a tuple $F = (f_0, \ldots, f_n)$ of members of $\bar{k}(V_1)$ such that for every point $x \in V_1$ where $f_0, \ldots, f_m$ are all defined,

$$F(x) = [(f_0(x), \ldots, f_n(x))]$$

is in V_2. If V_1 and V_2 are defined over k_0 and if $f_0, \ldots, f_n$ can be multiplied by a common member of $\bar{k}^\times$ so as to be in $k_0(V_1)$, then F is said to be **defined over** k_0.

EXAMPLE 1. Let V_1 be $P_1(\mathbb{C}) = \{[(x_0, x_1)]\}$ with $I(V_1) = 0$, and let V_2 be the curve in $P_2(\mathbb{C}) = \{[(y_0, y_1, y_2)]\}$ given by $y_0 y_2 = y_1^2$, i.e., having homogeneous ideal $(y_0 y_2 - y_1^2)$ in $\mathbb{C}[y_0, y_1, y_2]$. Then $F = (x_0^2, x_0 x_1, x_1^2)$ is a rational map. It is defined over $\mathbb{Q}$.

Let $G : V_2 \to V_1$ be given by $G = \left(1, \dfrac{y_1}{y_0}\right)$. This too is a rational map defined over $\mathbb{Q}$.

EXAMPLE 2. Let V_1 be the elliptic curve (11.15) with $I(V_1)$ generated within $\mathbb{C}[x_0, x_1, x_2]$ (in the current notation) by

$$x_0 x_2^2 + x_0^2 x_2 - x_1^3 + x_0 x_1^2.$$

Let V_2 be the elliptic curve (11.13) with $I(V_2)$ generated by
$$x_0 x_2^2 + x_0^2 x_2 - x_1^3 + x_0 x_1^2 + 10 x_0^2 x_1 + 20 x_0^3.$$
The map (11.16a) is $F = (f_0, f_1, f_2)$, where $f_0 = 1$ and
$$f_1(x_0, x_1, x_2) = \frac{x_1}{x_0} + \frac{x_0^2}{x_1^2} + \frac{2x_0}{x_1 - x_0} + \frac{x_0^2}{(x_1 - x_0)^2}$$
$$f_2(x_0, x_1, x_2) = \frac{x_2}{x_0} - \frac{2x_2 + x_0}{x_0} \left(\frac{x_0^3}{x_1^3} + \frac{x_0^3}{(x_1 - x_0)^3} + \frac{x_0^2}{(x_1 - x_0)^2} \right).$$
Then F is a rational map, and it is defined over $\mathbb{Q}$.

EXAMPLE 3. The affine curves $v^2 = 2u^4 - 1$ and $y^2 = x^3 + 8x$ are related by the transformations (3.19) and (3.21). If the curves and the transformations are recast in projective form, then the transformations are rational maps.

A rational map $F = (f_0, \ldots, f_n)$ between projective varieties V_1 and V_2 is **regular** or **defined** at x in V_1 if there is some g in $\bar{k}(V_1)$ such that $g f_i$ is regular at x and
$$((g f_0)(x), \ldots, (g f_n)(x))$$
is not the 0 tuple. A rational map that is regular at every point of V_1 is **a morphism**.

EXAMPLES. In Example 1 above, both F and G are morphisms. This is clear for F. In the case of G, G is given also by $G = \left(\frac{y_0}{y_1}, 1 \right)$. Hence the only question is about points of V_2 where $y_0 = y_1 = 0$. But $y_1^2 / y_0 = y_2$ on V_2, and hence
$$G = \left(1, \frac{y_1}{y_0} \right) = \left(y_1, \frac{y_1^2}{y_0} \right) = (y_1, y_2) = \left(\frac{y_1}{y_2}, 1 \right)$$
is regular at $[(0, 0, 1)]$.

Example 2 is a morphism, as a consequence of Theorem 11.42 below. It fixes ∞ and hence is a homomorphism of the group structures, by Proposition 11.61 below. Its kernel is the 5-element torsion subgroup of $V_1/\mathbb{Q}$.

In Example 3, the two rational maps (3.19) and (3.21) are inverse to each other, and thus we say that they are **birational**. The first curve has a singularity at $u = w = 0$ if the curve is written projectively in variables (u, v, w), while the second curve is nonsingular. It will follow that the two transformations cannot both be morphisms.

Two projective varieties V_1 and V_2 are **isomorphic** if there are morphisms $F : V_1 \to V_2$ and $G : V_2 \to V_1$ such that $F \circ G$ and $G \circ F$ are the respective identity maps. The two varieties of Example 1 are isomorphic. In the case of elliptic curves as in (3.23), Proposition 11.58 will note that an isomorphism must be given by an admissible change of variables over $\bar{k}$. Thus the current definition of "isomorphism" for elliptic curves agrees with the earlier one.

Two projective varieties V_1/k_0 and V_2/k_0 are **isomorphic over** k_0 if the above F and G can be defined over k_0.

Let us address how maps between varieties affect function fields. If $F : V_1 \to V_2$ is a rational map between projective varieties, we can try to define a $\bar{k}$ algebra map $F^* : \bar{k}(V_2) \to \bar{k}(V_1)$ between their function fields by composition: $F^*(f) = f \circ F$. Some condition is needed, however, for this formula to be meaningful. For example, the image of F should not completely be contained in the set where f is not defined. It turns out that it is enough that F have dense image in a suitable topology.

We shall not pursue this point in full generality, however. Suppose instead that F is a morphism and carries V_1 onto V_2. Then F is called a **dominant morphism**. In this case F^* is well defined and is one-one. Moreover, it carries local rings to local rings:

$$F^*(\bar{k}[V_2]_{F(x)}) \subseteq \bar{k}[V_1]_x.$$

Consequently if F is an isomorphism, then F^* is an isomorphism of function fields and an isomorphism of local rings at each point. In particular, F carries nonsingular points to nonsingular points.

8. Canonical Model of $X_0(N)$

In this section we shall relate curves to their function fields and then show how to introduce a canonical $\mathbf{Q}$ structure on $X_0(N)$. We continue with fields $k_0 \subseteq \bar{k}$ and assumptions on them as in §7. By a **function field of dimension** 1, we shall mean any finitely generated extension of $\bar{k}$ of transcendence degree one.

Let C be a projective curve, and let x be a nonsingular point. The key to analyzing C about x lies in the fact that every rational function on C has a well defined order at x. To define the order, we use the following lemma.

Lemma 11.41. Let R be a Noetherian local ring that is an integral domain and a $\bar{k}$ algebra, and let M be the maximal ideal. Suppose that

(a) $\dim_{\bar{k}}(M/M^2) = 1$, and
(b) every element of R not in M is a unit.

Then M is principal, being given by $M = (t)$ for any element t of M not in M^2. Also $\bigcap_{l=1}^{\infty} M^l = \{0\}$.

PROOF. Let t be in M but not M^2. Since R is Noetherian, M has a finite set of generators $t, s_1, \ldots, s_r$. Without loss of generality, suppose r is as small as possible. By (a), $s_j - c_j t$ is in M^2 for some $c_j \in \bar{k}$. Changing notation, we may assume that our generators $t, s_1, \ldots, s_r$ of M have the property that each s_j is in M^2. Since the products of pairs of generators of M generate M^2, we can write

$$s_r = at^2 + \sum_j b_j t s_j + \sum_{j \leq l} c_{jl} s_j s_l.$$

Thus

$$s_r(1 - b_r t - \sum_j c_{jr} s_j) = at^2 + \sum_{j<r} b_j t s_j + \sum_{j \leq l < r} c_{jl} s_j s_l. \qquad (11.70)$$

The right side of (11.70) is in $(t, s_1, \ldots, s_{r-1})$, and the coefficient of s_r on the left side is a unit, by (b). Hence s_r is in $(t, s_1, \ldots, s_{r-1})$, in contradiction to the minimal choice of r (unless $r = 0$). We conclude that $M = (t)$.

Let f be in $\bigcap_{l=1}^{\infty} M^l = \bigcap_{l=1}^{\infty} (t^l)$, and write $f = t^l f_l$. Since $f_l = t f_{l+1}$, we have $(f_1) \subseteq (f_2) \subseteq (f_3) \subseteq \ldots$. Since R is Noetherian, $(f_m) = (f_{m+1})$ for some m. Then $f_{m+1} = a f_m = at f_{m+1}$, and $at = 1$ if $f_{m+1} \neq 0$. Since t is in M, $at = 1$ is impossible. Thus $f_{m+1} = 0$ and $f = 0$.

Let us return to the projective curve C and the nonsingular point x. The local ring $\bar{k}[C]_x$ of C at x satisfies the hypotheses of the lemma. If f is in $\bar{k}[C]_x$, we define $\text{ord}_x(f)$ to be the least $l \geq 0$ such that f is in M_x^l but not M_x^{l+1}. This is an integer ≥ 0 if $f \neq 0$ and is $+\infty$ if $f = 0$. Let t be any member of M_x but not M_x^2, so that $M_x = (t)$; t is called a **uniformizer** of C at x. If $f \neq 0$ and if $l = \text{ord}_x(f)$, then $f = at^l$ for some unit a in $\bar{k}[C]_x$.

The function field $\bar{k}(C)$ is the quotient field of $\bar{k}[C]_x$, and we can extend $\text{ord}_x(\cdot)$ to all of $\bar{k}(C)$. Namely any F in $\bar{k}(C)$ is of the form $F = f/g$ with f and g in $\bar{k}[C]_x$ and $g \neq 0$. We define

$$\text{ord}_x(F) = \text{ord}_x(f) - \text{ord}_x(g).$$

To see that this is well defined, we let $l = \text{ord}_x(f)$ and $m = \text{ord}_x(g)$. Then $f = at^l$ and $g = bt^m$ with a and b units in $\bar{k}[C]_x$. Hence $F = ab^{-1} t^{l-m}$. So $l - m = \text{ord}_x(F)$ is characterized as the unique power of

8. CANONICAL MODEL OF $X_0(N)$

t such that F/t^{l-m} is a unit in $\bar{k}[C]_x$. Then $\text{ord}_x(\cdot)$ has the following properties:

(a) $\text{ord}_x(FG) = \text{ord}_x(F) + \text{ord}_x(G)$,
(b) $\text{ord}_x(F + G) \geq \min\{\text{ord}_x(F), \text{ord}_x(G)\}$,
(c) $\bar{k}[C]_x = \{F \in \bar{k}(C) \mid \text{ord}_x(F) \geq 0\}$,
(d) $M_x = \{F \in \bar{k}(C) \mid \text{ord}_x(F) > 0\}$,
(e) $\text{ord}_x(F) = +\infty$ if and only if $F = 0$.

The function $\text{ord}_x(\cdot)$ allows us to prove a theorem mentioned in connection with Example 2 in the preceding section.

Theorem 11.42. Let C be a projective curve, let x be a nonsingular point of C, let V be a projective variety, and let $F : C \to V$ be a rational map. Then F is regular at x. Consequently if C is nonsingular, then F is a morphism.

PROOF. Let $F = (f_0, \ldots, f_n)$ with $f_j \in \bar{k}(C)$. Let t be a uniformizer at x, and let
$$m = \min_{0 \leq j \leq n} \{\text{ord}_x(f_j)\}.$$

Then $\text{ord}_x(t^{-m} f_j) \geq 0$ for all j with equality for some $j = j_0$. By (c), $t^{-m} f_j$ is regular at x, and $(t^{-m} f_{j_0})(x) \neq 0$ by (d). Hence $g = t^{-m}$ exhibits F as regular at x.

To each nonsingular point x of a projective curve C, we have associated the local subring $\bar{k}[C]_x \subseteq \bar{k}(C)$ and the function $\text{ord}_x(\cdot) : \bar{k}(C) \to \mathbb{Z} \cup \{\infty\}$ satisfying (a) through (e) above. Let us abstract this situation. A proper subring $R \subseteq \bar{k}(C)$ containing $\bar{k}$ is called a **discrete valuation ring** of $\bar{k}(C)$ over $\bar{k}$ if there exists a function $v : \bar{k}(C) \to \mathbb{Z} \cup \{\infty\}$ (called the **valuation**) such that

(a) $v(FG) = v(F) + v(G)$,
(b) $v(F + G) \geq \min\{v(F), v(G)\}$,
(c) $R = \{F \in \bar{k}(C) \mid v(F) \geq 0\}$.

Then R is a local ring with maximal ideal
$$M_R = \{F \in \bar{k}(C) \mid v(F) > 0\}.$$

For a nonsingular curve C, the next proposition says that this abstraction captures the functions $\text{ord}_x(\cdot)$ completely and gives nothing additional.

Proposition 11.43. Let C be a nonsingular projective curve, and let R be a discrete valuation ring of $\bar{k}(C)$ over $\bar{k}$, with valuation v. Then there exists a point x of C such that v is a positive multiple of $\operatorname{ord}_x(\cdot)$. For this x, $R = \bar{k}[C]_x$ and $M_R = M_x$.

PROOF. Let $(x_0, \ldots, x_n)$ be coordinates for the projective space in which C lies. The x_i's with $x_i \in I_C$ (so that $x_i = 0$ everywhere on C) will not play a role, and we discard them. For the remaining indices we can regard each x_j/x_0 as in $\bar{k}(C)$. If $v(x_j/x_0) \geq 0$ for all j, we use the standard system of affine local coordinates (with $\Phi = 1$) and pass to the affine curve $C \cap \bar{k}^n$, which is not empty since $x_0 \neq 0$ somewhere on C. The members of the affine coordinate ring $\bar{k}[C \cap \bar{k}^n]$ are polynomials in the x_j/x_0, and thus v is ≥ 0 on the whole ring. Hence $\bar{k}[C \cap \bar{k}^n] \subseteq R$. On the other hand, if $v(x_j/x_0) < 0$ for some j, let $j = j_0$ be an index for which it is smallest. Using affine local coordinates in which Φ interchanges 0 and j_0, we readily see similarly that $C \cap \bar{k}^n$ is nonempty and that $\bar{k}[C \cap \bar{k}^n] \subseteq R$.

Consequently $I = M_R \cap \bar{k}[C \cap \bar{k}^n]$ is a proper ideal in $\bar{k}[C \cap \bar{k}^n] = \bar{k}[\bar{k}^n]/I_C$. Let $\tilde{I}$ be the inverse image of I in $\bar{k}[\bar{k}^n]$. By Theorem 11.37, all members of $\tilde{I}$ vanish at some x in $\bar{k}^n$. Since $I_C \subseteq \tilde{I}$, x is in C. Thus every member f of I has the property that $\operatorname{ord}_x(f) > 0$.

Since $C \cap \bar{k}^n$ is nonempty, we can identify $\bar{k}(C)$ with $\bar{k}(C \cap \bar{k}^n)$ and $\bar{k}[C]_x$ with $\bar{k}[C \cap \bar{k}^n]_x$. Suppose F is in $M_R \cap \bar{k}[C \cap \bar{k}^n]_x$. We can write $F = f/g$ with f and g in $\bar{k}[C \cap \bar{k}^n]$ and $\operatorname{ord}_x(g) = 0$. Since g is in $\bar{k}[C \cap \bar{k}^n] \subseteq R$, $f = Fg$ is in I; thus the previous paragraph gives $\operatorname{ord}_x(f) > 0$. Since $\operatorname{ord}_x(g) = 0$, we have $\operatorname{ord}_x(F) > 0$. This proves that

$$\bigl(v(F) > 0 \text{ and } \operatorname{ord}_x(F) \geq 0\bigr) \implies \operatorname{ord}_x(F) > 0 \tag{11.71}$$

for general F in $\bar{k}(C)$. The contrapositive of (11.71) is

$$\operatorname{ord}_x(F) = 0 \implies v(F) \leq 0.$$

Applying this to $1/F$, we obtain

$$\operatorname{ord}_x(F) = 0 \implies v(F) = 0. \tag{11.72}$$

By Proposition 11.39 we can choose t in $\bar{k}[C \cap \bar{k}^n]$ that is a uniformizing parameter at x. Since $\bar{k}[C \cap \bar{k}^n] \subseteq R$, we have $v(t) \geq 0$. A general member F of $\bar{k}(C)$ is of the form $F = t^l F_0$ with $\operatorname{ord}_x(F_0) = 0$. By (11.72) this F has

$$v(F) = lv(t) + v(F_0) = lv(t) = v(t)\operatorname{ord}_x(F).$$

Now v is not identically 0 since R is a proper subring of $k(C)$. Since $v(t) \geq 0$, we conclude that $v(t) > 0$ and that v is a positive multiple of $\operatorname{ord}_x(\cdot)$. This completes the proof.

8. CANONICAL MODEL OF $X_0(N)$

Proposition 11.43 gives a clue how to reconstruct a nonsingular projective curve from its function field. The point is that the same idea can be used to associate a nonsingular projective curve to any function field of dimension 1.

Theorem 11.44. Let K be a function field of dimension 1 over $\bar{k}$, and let C_K be the set of discrete valuation rings of K over $\bar{k}$. Then there exists a nonsingular projective curve C such that $\bar{k}(C) \cong K$ and such that C_K gets canonically identified with the set of points of C.

REMARK. The curve C is unique up to isomorphism, by Theorem 11.48 below.

We shall give a brief indication of some of the steps. First, if B is any integral domain that is finitely generated as a $\bar{k}$ algebra, then B is isomorphic with the coordinate ring of an affine variety. In fact, let $x_1, \ldots, x_n$ be generators of B over $\bar{k}$. The map $X_j \to x_j$ gives a $\bar{k}$ homomorphism of $\bar{k}[X_1, \ldots, X_n]$ onto B. Since B is an integral domain, the kernel is a prime ideal I. Then V_I is the required affine variety.

Now let us consider the function field K of Theorem 11.44. Let x be a member of K not in $\bar{k}$. We form the polynomial ring $\bar{k}[x]$. Then K is a finite extension of the quotient field $\bar{k}(x)$. Let B be the integral closure of $\bar{k}[x]$ in K. When the construction of the previous paragraph is applied to this B, the resulting affine variety turns out to be a nonsingular affine curve.

At this stage we bring in C_K. Let R be a discrete valuation ring of K over $\bar{k}$, and choose x to be in R (but not in $\bar{k}$). Then $\bar{k}[x] \subseteq R$. One can show that R is integrally closed in K, and it follows that $B \subseteq R$. The intersection $N = M_R \cap B$ is a maximal ideal of B and thus corresponds to a unique point of the nonsingular affine curve built from B. In other words, R has been attached to a unique point of the curve corresponding to B.

To prove Theorem 11.44, one has to take the product of the projective closures of finitely many such affine curves, map C_K into this product diagonally, and take the image of C_K as the desired curve C. We omit the details.

For our application to $X_0(N)$ it will be important to know conditions under which the curve C in Theorem 11.44 is defined over the subfield k_0. We assume now that $\bar{k} = \mathbb{C}$. Taking a cue from Theorem 11.36, suppose $K = \mathbb{C}(f,g)$ and $K_0 = k_0(f,g)$. Given R in the argument above, we can always choose x to be either f or $1/f$, so that x is in K_0. So we need to deal only with two B's. We are allowed to use more B's, but we insist that they come from x's in K_0. The heart of the matter is to get the affine variety determined by B to be defined over k_0. This step we address in the following lemma.

Lemma 11.45. Let $K = \mathbb{C}(f,g)$ be a function field of dimension 1 over $\mathbb{C}$, let k_0 be a subfield of $\mathbb{C}$, put $K_0 = k_0(f,g)$, and suppose that $\mathbb{C} \cap K_0 = k_0$. Let x be in K_0 but not k_0, let B be the integral closure of $\mathbb{C}[x]$ in K, and let V be a nonsingular affine curve determined by B. Then V is canonically defined over k_0. Under this definition, $k_0(V) \cong K_0$ and $\mathbb{C}(V) \cong K$.

PROOF. Without loss of generality, f is transcendental over $\mathbb{C}$, and g is algebraic over $\mathbb{C}(f)$. Taking into account (c) $\Rightarrow$ (a) in Theorem 11.36, we see that f is transcendental over k_0 and g is algebraic over $k_0(f)$. Therefore K_0 has transcendence degree 1 over k_0.

Since $\mathbb{C} \cap K_0 = k_0$, the given element x of K_0 is not in $\mathbb{C}$, hence is transcendental over k_0. Consequently $\{x\}$ is a transcendence basis of K_0 over k_0. It follows that f and g are algebraic over $k_0(x)$ and that K_0 is a finite algebraic extension of $k_0(x)$. By the Theorem of the Primitive Element, there exists y in K_0 such that $K_0 = k_0(x,y)$. Then $K = \mathbb{C}(x,y)$.

Condition (c) in Theorem 11.36 is independent of the two generators and hence still holds when we replace (f,g) by (x,y). By (c) $\Rightarrow$ (a) in the theorem, we see that

$$[K_0 : k_0(x)] = [K : \mathbb{C}(x)]. \tag{11.73}$$

Let n be the common value of the two sides of (11.73).

Since $\mathbb{C}[x]$ is a principal ideal domain, there exists a basis $\{x_1, \ldots, x_n\}$ of K over $\mathbb{C}(x)$ consisting of members of B such that

$$B = \sum_{j=1}^{n} \mathbb{C}[x]\, x_j. \tag{11.74}$$

We take $x_1, \ldots, x_n, x$ as generators of B over $\mathbb{C}$. Then the map

$$\varphi_B : \mathbb{C}[X_1, \ldots, X_{n+1}] \longrightarrow B$$

given by $X_j \to x_j$ for $j \leq n$ and $X_{n+1} \to x$ exhibits B as the affine coordinate ring of the curve V defined by the ideal $I_B = \ker \varphi_B$. We are to prove that I_B is finitely generated as an ideal by elements of $k_0[X_1, \ldots, X_{n+1}]$, and we are to identify the function fields.

8. CANONICAL MODEL OF $X_0(N)$

Let B_0 be the integral closure of $k_0[x]$ in K_0. As above, there exists a basis $\{y_1, \ldots, y_n\}$ of K_0 over $k_0(x)$ consisting of members of B_0 such that

$$B_0 = \sum_{j=1}^{n} k_0[x] \, y_j. \tag{11.75}$$

We form the map

$$\varphi_{B_0} : k_0[X_1, \ldots, X_{n+1}] \longrightarrow B_0$$

given by $X_j \to y_j$ for $j \leq n$ and $X_{n+1} \to x$. Let $I_{B_0} = \ker \varphi_{B_0}$. By the Hilbert Basis Theorem, I_{B_0} is finitely generated, say by members $P_1, \ldots, P_N$ of $k_0[X_1, \ldots, X_{n+1}]$. We complexify the exact sequence

$$0 \longrightarrow I_{B_0} \longrightarrow k_0[X_1, \ldots, X_{n+1}] \longrightarrow B_0 \longrightarrow 0$$

to a sequence

$$0 \longrightarrow I_{B_0} \otimes_{k_0} \mathbf{C} \longrightarrow \mathbf{C}[X_1, \ldots, X_{n+1}] \longrightarrow B_0 \otimes_{k_0} \mathbf{C} \longrightarrow 0.$$

The result is exact since the operation $(\cdot) \otimes_{k_0} \mathbf{C}$ is an exact functor on k_0 vector spaces. Here $I_{B_0} \otimes_{k_0} \mathbf{C}$ is generated by $P_1, \ldots, P_N$, and

$$B_0 \otimes_{k_0} \mathbf{C} = \sum_{j=1}^{n} \mathbf{C}[x] \, y_j.$$

Let $\varphi_{B_0}^{\mathbf{C}}$ denote the complexified version of φ_{B_0}.

The elements y_j are linearly independent over $\mathbf{C}(x)$ by (a) of Theorem 11.36. Hence both $\{x_j\}$ and $\{y_j\}$ are bases of K over $\mathbf{C}(x)$. Let ψ be the $\mathbf{C}(x)$ linear map of K into itself given by $x_j \to y_j$ for $1 \leq j \leq n$. Then it is clear that

$$\psi \circ \varphi_B = \varphi_{B_0}^{\mathbf{C}}$$

as $\mathbf{C}$ linear maps. Therefore their kernels are equal. But ψ is one-one. Hence

$$I_B = I_{B_0} \otimes_{k_0} \mathbf{C}. \tag{11.76}$$

The right side is generated by $P_1, \ldots, P_N$, which are in $k_0[X_1, \ldots, X_{n+1}]$. Thus I_B is finitely generated as an ideal by elements of $k_0[X_1, \ldots, X_{n+1}]$.

The affine coordinate ring of V is B, and the function field is the field of quotients of B, which is K by (11.74). In the case of V/k_0, we determine the affine coordinate ring by first determining $I(V/k_0)$, which is given by

$$\begin{aligned} I(V/k_0) &= I(V) \cap k_0[X_1,\ldots,X_{n+1}] \\ &= I_B \cap k_0[X_1,\ldots,X_{n+1}] \\ &= (I_{B_0} \otimes_{k_0} \mathbb{C}) \cap k_0[X_1,\ldots,X_{n+1}] \qquad \text{by (11.76)} \\ &= I_{B_0}. \end{aligned}$$

By definition $k_0[V] = k_0[X_1,\ldots,X_{n+1}]/I_{B_0}$, and this is B_0. The function field is the field of quotients of B_0, which is K_0 by (11.75).

Theorem 11.46. Let k_0 be a subfield of $\mathbb{C}$. If C/k_0 is a nonsingular projective curve, then there exist elements f and g in $k_0(C)$ such that $k_0(C) = k_0(f,g)$ and $\mathbb{C}(C) = \mathbb{C}(f,g)$. Moreover,

$$\mathbb{C} \cap k_0(C) = k_0.$$

Conversely let $K = \mathbb{C}(f,g))$ be a function field of dimension 1 over $\mathbb{C}$, let k_0 be a subfield of $\mathbb{C}$, put $K_0 = k_0(f,g)$, and suppose that $\mathbb{C} \cap K_0 = k_0$. Then there exists a nonsingular projective curve C/k_0 such that $k_0(C) \cong K_0$ and $\mathbb{C}(C) \cong K$.

The question of uniqueness of C in Theorem 11.46 is conveniently addressed in the context of a correspondence between certain maps of function fields and certain morphisms between curves.

Recall from the end of §7 that a dominant morphism $F : C_1 \to C_2$ between nonsingular projective curves (over $\bar{k}$) induces a one-one $\bar{k}$ algebra homomorphism $F^* : \bar{k}(C_2) \to \bar{k}(C_1)$ by composition: $F^*(f) = f \circ F$.

Lemma 11.47. A morphism $F : C_1 \to C_2$ between nonsingular projective curves over $\bar{k}$ either is onto or is a constant map.

Theorem 11.48. Let C_1/k_0 and C_2/k_0 be nonsingular projective curves, and let $F : C_1 \to C_2$ be a dominant morphism defined over k_0. Then F^* carries $k_0(C_2)$ into $k_0(C_1)$, and $k_0(C_1)$ is a finite extension of $F^*(k_0(C_2))$. Conversely if $\mathcal{F} : k_0(C_2) \to k_0(C_1)$ is a one-one k_0 algebra homomorphism (carrying 1 to 1), then there exists a unique dominant morphism $F : C_1 \to C_2$ defined over k_0 such that $F^* = \mathcal{F}$.

In particular the nonsingular projective curve given by Theorem 11.46 is unique up to isomorphism. Finally we can put into place the canonical $\mathbb{Q}$ structure on the compact Riemann surface $X_0(N)$.

8. CANONICAL MODEL OF $X_0(N)$

Corollary 11.49. There exist a nonsingular projective curve $C/\mathbf{Q}$ and a biholomorphic mapping $\varphi : X_0(N) \to C(\mathbf{C})$ such that

$$\varphi^*(\mathbf{C}(C)) = K(X_0(N)) = \mathbf{C}(j, j_N)$$

and

$$\varphi^*(\mathbf{Q}(C)) = \mathbf{Q}(j, j_N).$$

The curve $C/\mathbf{Q}$ is unique up to isomorphism defined over $\mathbf{Q}$, and φ is uniquely determined by the isomorphism of $\mathbf{Q}(C)$ with $\mathbf{Q}(j, j_N)$.

REMARK. One says that (φ, C) is a **model** for $X_0(N)$ over $\mathbf{Q}$. Our practice is to identify $C(\mathbf{C})$ with $X_0(N)$ and refer to $X_0(N)/\mathbf{Q}$ as a **Q structure** on $X_0(N)$.

PROOF. Theorem 11.33 gives $K(X_0(N)) = \mathbf{C}(j, j_N)$. Theorem 11.32 shows that $\mathbf{C}(j, j_N)$ and $\mathbf{Q}(j, j_N)$ satisfy (a) of Theorem 11.36 and hence the other equivalent conditions. Theorem 11.46 produces the desired curve $C/\mathbf{Q}$, and Theorem 11.21 produces φ. Finally Theorem 11.48 proves the uniqueness of C and φ.

Our occasional observations in §7 about $\operatorname{Aut}_{k_0}(\bar{k})$ apply with $k_0 = \mathbf{Q}$ and $\bar{k} = \mathbf{C}$ since the subfield of $\mathbf{C}$ fixed by $\operatorname{Aut}_{\mathbf{Q}}(\mathbf{C})$ is $\mathbf{Q}$. This fact allows us to give a reasonable way of deciding what members of $K(X_0(N))$ are in $\mathbf{Q}(j, j_N)$.

Corollary 11.50. A member of $\mathbf{C}(j, j_N)$ is in $\mathbf{Q}(j, j_N)$ if and only if its q expansion at ∞ has all coefficients in $\mathbf{Q}$.

PROOF. Any element φ of $\operatorname{Aut}_{\mathbf{Q}}(\mathbf{C})$ acts on $\mathbf{C}(j, j_N)$ and fixes $\mathbf{Q}(j, j_N)$ because these fields are associated with varieties. Let

$$f(\tau) = \sum_{n=-M}^{\infty} c_n q^n$$

be in $\mathbf{C}(j, j_N)$. If f is written as a rational function in j and j_N, then f^φ is the same rational function but with each complex coefficient a replaced by $\varphi(a)$. The claim is that

$$f^\varphi(\tau) = \sum_{n=-M}^{\infty} \varphi(c_n) q^n, \qquad (11.77)$$

and then the corollary will follow because Corollary 11.49 ensures that the subfield of $\mathbf{C}(j, j_N)$ fixed by $\operatorname{Aut}_{\mathbf{Q}}(\mathbf{C})$ is just $\mathbf{Q}(j, j_N)$.

By Theorem 11.32,
$$C(j, j_N) = \sum_{l=0}^{\psi(N)-1} C(j) j_N^l.$$

Since the coefficients of the q expansion of j_N are integers, it is enough to prove (11.77) for f in $C(j) = C(j^{-1})$. Since the coefficients of the q expansion of j are integers, it is enough to handle
$$f = \frac{1}{1 - P(j^{-1})},$$
where P is a polynomial without constant term. We have
$$f = 1 + P(j^{-1}) + P(j^{-1})^2 + \ldots,$$
and the coefficients of $1, q, q^2, \ldots, q^M$ are the same for f as for
$$1 + P(j^{-1}) + \cdots + P(j^{-1})^M.$$
Similarly they are the same for f^φ as for
$$1 + P^\varphi(j^{-1}) + \cdots + P^\varphi(j^{-1})^M.$$
Then (11.77) follows.

EXAMPLE. If M divides N, then j_M is in $K(X_0(N)) = C(j, j_N)$. Since the q expansion of j_M has rational coefficients, j_M is in $\mathbb{Q}(j, j_N)$. The inclusion $\Gamma_0(N) \subseteq \Gamma_0(N/M)$ induces a holomorphic mapping F of $X_0(N)$ onto $X_0(N/M)$ whose pullback on $K(X_0(N/M))$ is given by $F^*(f)(\tau) = f(M\tau)$. Thus $F^*(j) = j_M$ and $F^*(j_{N/M}) = j_N$, and
$$F^*(\mathbb{Q}(j, j_{N/M})) \subseteq \mathbb{Q}(j, j_N). \tag{11.78}$$
Comparison of Theorem 11.21 and Corollary 11.49 shows that the holomorphic mapping F can be regarded as a morphism between complex varieties. Bringing in (11.78) and looking at Corollary 11.49 again, we see that F is defined over $\mathbb{Q}$.

We mention one further corollary of our results. The proof is in the same spirit as the above results and will be omitted.

Corollary 11.51. Let k_0 be a subfield of C, and let C/k_0 be a nonsingular projective curve. If K_1 is a subfield of $k_0(C)$ of finite index containing k_0, then there exist a nonsingular projective curve C'/k_0 and a dominant morphism $F : C \to C'$ defined over k_0 such that $F^*(k_0(C')) = K_1$. The curve C' is unique up to isomorphism defined over k_0, and F is uniquely determined by the isomorphism of $k_0(C')$ with K_1.

9. Abstract Elliptic Curves and Isogenies

The Eichler-Shimura theory does not work directly with nonsingular Weierstrass equations. Instead it works with a certain kind of projective curve that turns out to be isomorphic to a curve defined by a nonsingular Weierstrass equation. In this section we shall expand our definition of elliptic curve to include this wider kind of projective curve. We omit proofs of general results about curves but usually give sketches of proofs about elliptic curves. Let $k_0 \subseteq \bar{k}$ be fields with the same assumptions as in §7.

Let C_1/k_0 and C_2/k_0 be nonsingular projective curves, and let $F : C_1 \to C_2$ be a morphism. If $F = 0$, we define the **degree** of F by $\deg F = 0$. Otherwise F is onto, by Lemma 11.47, and $F^*(k_0(C_2))$ has finite index in $k_0(C_1)$, by Theorem 11.48. In this case we define

$$\deg F = [k_0(C_1) : F^*(k_0(C_2))]. \tag{11.79}$$

This degree is the same if computed over $\bar{k}$.

For the case of characteristic $\neq 0$, one must pay attention to separability of field extensions in the present context. Any algebraic extension can be obtained in two steps, a separable one followed by a purely inseparable one. For the field extension indicated in (11.79), we let $\deg_s F$ be the degree of the separable part and $\deg_i F$ be the degree of the purely inseparable part. Then $\deg F$ is the product of $\deg_s F$ and $\deg_i F$. We say that F is **separable** or **purely inseparable** if $\deg F = \deg_s F$ or $\deg F = \deg_i F$, respectively.

Recall that the quotient map $\mathcal{H} \to \Gamma_0(N)\backslash\mathcal{H}$ of Riemann surfaces fails to be a covering map at elliptic fixed points of $\Gamma_0(N)$ in $\mathcal{H}$. We shall define the counterpart of this kind of ramification for varieties. For F as above but nonconstant, let x be in C_1 and let $t_{F(x)}$ be a uniformizing parameter at $F(x)$. We define $e_F(x) = \mathrm{ord}_x(F^*(t_{F(x)}))$. The morphism F is **unramified** at x if $e_F(x) = 1$, **ramified** if $e_F(x) > 1$. It is **unramified** if it is unramified at every point. One can prove for every y in C_2 that

$$\sum_{x \in F^{-1}(y)} e_F(x) = \deg F; \tag{11.80}$$

also for all but finitely many y in C_2,

$$\#F^{-1}(y) = \deg_s F. \tag{11.81}$$

It follows immediately from (11.80) that F is unramified if and only if $\#F^{-1}(y) = \deg F$ for all y in C_2.

Suppose C is defined over $\mathbf{Z}_p$ with $I(C)$ generated by some homogeneous polynomials f in $\mathbf{Z}_p[X_1,\ldots,X_n]$. The **Frobenius morphism** ϕ of C is given by

$$\phi[(x_0,\ldots,x_n)] = [(x_0^p,\ldots,x_n^p)]. \tag{11.82}$$

Since $f(x_0^p,\ldots,x_n^p) = \bigl(f(x_0,\ldots,x_n)\bigr)^p$, ϕ really does map C to C and is defined over $\mathbf{Z}_p$.

More generally let $\text{char}(k_0) = p$ and $q = p^r$. If C is defined over k_0 with $I(C)$ generated by some homogeneous polynomials f in $k_0[X_0,\ldots,X_n]$, we define $C^{(q)}$ by the ideal generated by all $f^{(q)}$, where $f^{(q)}$ is f with all coefficients raised to the qth power. The qth power **Frobenius morphism** ϕ carries C to $C^{(q)}$ and is given by (11.82) with p replaced by q. The situation in the previous paragraph is a special case because $C^{(p)} = C$ when C is defined over $\mathbf{Z}_p$. One can prove that

(a) $\phi^*(k_0(C^{(q)})) = \{f^q \mid f \in C\}$
(b) ϕ is purely inseparable
(c) $\deg \phi = q$.

In characteristic p, our original nonconstant morphism $F : C_1 \to C_2$ admits a factorization $F = F_s \circ \phi$, where $\phi : C_1 \to C_1^{(q)}$ is the qth power Frobenius morphism with $q = \deg_i F$ and where $F_s : C_1^{(q)} \to C_2$ is separable.

We can define the abelian group $\text{Div}(C)$ of **divisors** of a nonsingular projective curve C just as we did for a compact Riemann surface in §4. In the case of $f \neq 0$ in $\bar{k}(C)$, we say f has a **zero** at x if $\text{ord}_x(f) > 0$ or a **pole** if $\text{ord}_x(f) < 0$. The divisor $[f]$ of f makes sense because of the following theorem.

Theorem 11.52. If C is a nonsingular projective curve and f is in $\bar{k}(C)$, then f has only finitely many zeros and finitely many poles.

The divisor of a function is called a **principal divisor**. The **degree** of a divisor $\sum_x \text{ord}_x(D)x$ is the integer $\sum_x \text{ord}_x(D)$.

Theorem 11.53. Every principal divisor on a nonsingular projective curve has degree 0.

If $F : C_1 \to C_2$ is a dominant morphism between nonsingular projective curves, then F induces maps on divisors in both directions by

$$F_* : \text{Div}(C_1) \to \text{Div}(C_2) \qquad F^* : \text{Div}(C_2) \to \text{Div}(C_1)$$
$$x \to F(x) \qquad\qquad y \to \sum_{x \in F^{-1}(y)} e_F(x) x. \tag{11.83}$$

9. ABSTRACT ELLIPTIC CURVES AND ISOGENIES

The idea of the map F_* was already implicit in our discussion centered about (11.29).

The theory of divisors on C proceeds as in the case of compact Riemann surfaces (§4). The set of principal divisors is denoted $\text{Div}_0(C)$, and the quotient $\text{Pic}(C) = \text{Div}(C)/\text{Div}_0(C)$ is called the **divisor class group**. Let $\text{Pic}^0(C)$ be the quotient of the divisors of degree 0 by the principal divisors. Our curve C has an analog of meromorphic differentials on Riemann surfaces. These always exist, and the divisor class of such differentials is called the **canonical class**. Linear systems $L(D)$ for divisors are defined just as in §4. If $D < 0$, then $L(D) = 0$ as a consequence of Theorem 11.53. By an analog of Proposition 11.7, $L(0)$ is 1-dimensional.

If $\bar{k} = \mathbb{C}$, then the $\mathbb{C}$ points of C form a compact Riemann surface, and **genus** has the customary topological meaning. One can give an abstract definition of genus even in characteristic $\neq 0$, but for our purposes we may as well define the genus g by its value from the Riemann-Roch Theorem with $D = 0$: $g = \dim L(W)$ for any $W \neq 0$ in the canonical class.

Theorem 11.54 (Riemann-Roch Theorem). Let C be a nonsingular projective curve of genus g, let D be a divisor, and let W be in the canonical class. Then

$$\dim L(D) = \deg D + \dim L(W - D) - g + 1.$$

Corollary 11.55. Let C be a nonsingular projective curve of genus g. If D is a divisor with $\deg D > 2g - 2$, then

$$\dim L(D) = \deg D - g + 1.$$

The **Jacobian variety** $\text{Pic}^0(C)$ will be identified as a variety for the genus 1 case in Corollary 11.60, and for the higher genus case in §10.

If C is defined over k_0, the group $\text{Aut}_{k_0}(\bar{k})$ acts on points of C, fixing exactly those in $C(k_0)$. The action on points induces an action on divisors. We say that a divisor D is **defined over** k_0 if it is fixed by $\text{Aut}_{k_0}(\bar{k})$. (It is not necessary for each of the points with nonzero coefficient in D to be fixed by $\text{Aut}_{k_0}(\bar{k})$.) The following result is fundamental.

Proposition 11.56. Let C/k_0 be a nonsingular projective curve and let D be a divisor defined over k_0. Then the $\bar{k}$ vector space $L(D)$ has a basis defined over k_0 (i.e., consisting of members of $k_0(C)$).

We come to our expanded definition of elliptic curve. An **elliptic curve** is a pair (E, O), where E is a nonsingular projective curve of genus 1 and O is a point in E. The elliptic curve is **defined over** k_0 if E is defined over k_0 and O is in $E(k_0)$.

If a curve E is given by a nonsingular Weierstrass equation (3.23) over k_0 and if O is taken as the usual point at ∞, then (E, O) is an elliptic curve in the new sense, and it is defined over k_0. To see this, we need only compute the genus. The expression (11.21) is a member of the canonical class with neither zeros nor poles, hence with divisor W of degree 0. Then Theorem 11.54 with $D = W$ gives

$$g = \dim L(W) = \deg W + \dim L(0) - g + 1 = 0 + 1 - g + 1.$$

Hence $g = 1$. The converse is as follows.

Theorem 11.57. If (E, O) is an elliptic curve defined over k_0, then there exist functions x and y in $k_0(E)$ such that the map $F : E \to P_2(\bar{k})$ with

$$F = [(x, y, 1)]$$

has $F(O) = [(0, 1, 0)]$ and is an isomorphism defined over k_0 onto a nonsingular curve given by a Weierstrass equation (3.23) whose coefficients are in k_0.

Some comments about the proof may be helpful. For $n \geq 1$, $L(n(O))$ has dimension n, by Corollary 11.55, and we may choose a basis for it from $k_0(E)$ by Proposition 11.56. We take any x so that $\{1, x\}$ is a basis over k_0 of $L(2(O))$ and then any y so that $\{1, x, y\}$ is a basis over k_0 of $L(3(O))$. The seven functions $1, x, y, x^2, xy, y^2, x^3$ are in $L(6(O))$, which has dimension 6, and must be linearly dependent. A nontrivial linear relation among the seven functions can be adjusted to give the desired Weierstrass equation. (This argument can be implemented for specific equations. For example, the change of variables that leads from (3.18) to (3.19) and (3.20) can be deduced from it after (3.18) is desingularized at ∞.)

To show that the Weierstrass equation obtained in the proof of Theorem 11.57 is nonsingular, one proves first that x and y generate $\bar{k}(E)$ over $\bar{k}$. Then it follows that F has degree 1. If the image is singular, Proposition 3.11 allows us to compose F with a certain map to $P_1(\bar{k})$ such that the composition has degree 1. Since E and $P_1(\bar{k})$ are both nonsingular, Theorem 11.48 shows they are isomorphic. But they have different genus, contradiction.

The same kind of argument classifies isomorphisms between elliptic curves that are given by Weierstrass equations.

9. ABSTRACT ELLIPTIC CURVES AND ISOGENIES

Proposition 11.58. If two elliptic curves (E, O) and (E', O') defined over k_0 are given by Weierstrass equations over k_0 (with O and O' corresponding to ∞), and if there exists an isomorphism $F : E \to E'$ defined over k_0 with $F(\infty) = \infty$, then E and E' are related by an admissible change of variables (3.43) with coefficients in k_0.

The idea of the proof is as follows: If $E \leftrightarrow (x, y)$ and $E' \leftrightarrow (x', y')$, then $\{1, x\}$ and $\{1, x'\}$ are bases of $L(2(O))$ and $L(2(O'))$, while $\{1, x, y\}$ and $\{1, x', y'\}$ are bases of $L(3(O))$ and $L(3(O'))$.

As a consequence of the above results, we can transfer the group operation to a general elliptic curve from any of its associated Weierstrass equations. We need to know that this group operation fits in with the kinds of mappings we have been discussing, and we need to know how to recognize the group operation without passing back and forth to a Weierstrass equation.

Proposition 11.59.

(a) For an elliptic curve over k_0 given by a Weierstrass equation over k_0, addition and negative are morphisms defined over k_0.

(b) For an elliptic curve (E, O), a definition of addition that is consistent with the definition in Weierstrass form is as follows: If D is any divisor of degree 0, there exists a unique point z on E such that $D - (z - (O))$ is a principal divisor. The sum of x and y is the point corresponding to the divisor $x + y - 2(O)$, the identity is (O), and the negative of x is the point corresponding to the divisor $(O) - x$.

Part (a) is tedious to check. Part (b) follows by repeated use of Theorem 11.54 and little else.

Corollary 11.60. For an elliptic curve (E, O), the mapping $x \to x - x(O)$ of E into $\text{Div}(E)$ yields a group isomorphism of E onto $\text{Pic}^0(E)$.

The corollary is just a reworded version of Proposition 11.59b. The Riemann surface analog is Corollary 11.16a. From now on, we shall typically say, "Let E be an elliptic curve," dropping reference to the group identity. The identity will be denoted O and for Weierstrass equations will be the point at ∞.

If E_1 and E_2 are elliptic curves over $\bar{k}$, an **isogeny** between them is a morphism $F : E_1 \to E_2$ with $F(O) = O$. All nonzero isogenies are onto, by Lemma 11.47. When $\bar{k} = \mathbb{C}$, we saw in Chapter VI that isogenies are automatically homomorphisms. The same thing is true in general.

Proposition 11.61. An isogeny is a group homomorphism.

SKETCH OF PROOF. We may assume that the isogeny $F : E_1 \to E_2$ is not zero. Referring to (11.83) and Corollary 11.60, we write F as the composition of three homomorphisms $E_1 \cong \text{Pic}^0(E_1)$, $F_* : \text{Pic}^0(E_1) \to \text{Pic}^0(E_2)$, and $\text{Pic}^0(E_2) \cong E_2$.

EXAMPLES.

1) If E is any elliptic curve, then multiplication by the integer m, denoted $[m]$, is an isogeny from E to E.

2) In §1 we wrote down an explicit isogeny $F : E' \to E$ between two elliptic curves defined over $\mathbf{Q}$ such that $j(E') = -16^3/11$ and $j(E) = -2^{12}31^3/11^5$.

3) If E is an elliptic curve over $\mathbf{Z}_p$, then the Frobenius map $\phi : E \to E$ given in (11.82) is an isogeny.

If E_1 and E_2 are elliptic curves defined over $\bar{k}$, we let $\text{Hom}(E_1, E_2)$ be the set of all isogenies from E_1 to E_2. The pointwise sum $F + G$ of two members of $\text{Hom}(E_1, E_2)$ is again in $\text{Hom}(E_1, E_2)$, being the composition of

$$\text{diagonal} : E_1 \to E_1 \times E_1,$$
$$F \times G : E_1 \times E_1 \to E_2 \times E_2, \text{ and}$$
$$+ : E_2 \times E_2 \to E_2.$$

Then $\text{Hom}(E_1, E_2)$ becomes an abelian group. The subset of isogenies defined over k_0 is a subgroup.

We write $\text{End}(E)$ in place of $\text{Hom}(E, E)$ when the two curves are the same. The abelian group $\text{End}(E)$ has a multiplication given by composition. The distributive law $F(G + H) = FG + FH$ follows immediately by applying Proposition 11.61 to F.

Let E be an elliptic curve. For x in E, we define a translation map $T_x : E \to E$ by $T_x(y) = x + y$. Then T_x^* is an automorphism of $\bar{k}(E)$; if E is defined over k_0 and x is in $E(k_0)$, then T_x^* is an automorphism of $k_0(E)$.

Proposition 11.62. If $F : E_1 \to E_2$ is a nonconstant isogeny between elliptic curves defined over $\bar{k}$, then $\ker F$ is finite, and the map $x \to T_x^*$ induces an isomorphism

$$\ker F \cong \text{Aut}_{F^*(\bar{k}(E_2))}(\bar{k}(E_1)). \tag{11.84}$$

If F is separable, then F is unramified and

$$\deg F = \#\ker(F); \tag{11.85}$$

moreover, $\bar{k}(E_1)$ is a Galois extension of $F^*(\bar{k}(E_2))$.

The proof uses (11.80) and (11.81), as well as a little Galois theory. We omit the details.

For an elliptic curve over $\mathbb{Z}_p$ with $\bar{k} = \bar{\mathbb{Z}}_p$, Proposition 11.62 gives a way of calculating $\#E(\mathbb{Z}_p)$. In fact, the Frobenius morphism ϕ of (11.82) fixes exactly those points of E that are in $E(\mathbb{Z}_p)$. Thus

$$x \in E(\mathbb{Z}_p) \iff \phi(x) = x \iff ([1] - \phi)(x) = O \iff x \in \ker([1] - \phi)$$

and

$$\#E(\mathbb{Z}_p) = \#\ker([1] - \phi). \tag{11.86}$$

A fundamental fact, which we shall not prove, is that

$$1 - \phi \quad \text{is a separable isogeny.} \tag{11.87}$$

Combining this fact with (11.85) and (11.86), we obtain

$$\#E(\mathbb{Z}_p) = \deg([1] - \phi). \tag{11.88}$$

Lemma 11.63. Let $F : E_1 \to E_2$ and $G : E_1 \to E_3$ be nonconstant isogenies of elliptic curves over $\bar{k}$. If F is separable and $\ker F \subseteq \ker G$, then there exists a unique isogeny $H : E_2 \to E_3$ such that $G = H \circ F$. If F and G are defined over k_0, so is H.

PROOF. The inclusion of kernels and (11.84) give natural isomorphisms

$$\operatorname{Aut}_{F^*(\bar{k}(E_2))}(\bar{k}(E_1)) \cong \ker F \subseteq \ker G \cong \operatorname{Aut}_{G^*(\bar{k}(E_3))}(\bar{k}(E_1)).$$

Hence every automorphism fixing $F^*(\bar{k}(E_2))$ fixes $G^*(\bar{k}(E_3))$. By Proposition 11.62, $\bar{k}(E_1)$ is Galois over $F^*(\bar{k}(E_2))$. Hence $G^*(\bar{k}(E_3)) \subseteq F^*(\bar{k}(E_2))$. Then Theorem 11.48 gives us a corresponding morphism $H : E_2 \to E_3$ with $F^* \circ H^* = G^*$. By uniqueness in Theorem 11.48, $G = H \circ F$. Then $O = G(O) = H(F(O)) = H(O)$ shows that H is an isogeny. If F and G are defined over k_0, we can apply Theorem 11.48 to $G^*(k_0(E_3)) \subseteq F^*(k_0(E_2))$ to construct H as defined over k_0.

Theorem 11.64. Let $F : E_1 \to E_2$ be a nonconstant isogeny of degree m between elliptic curves defined over $\bar{k}$. Then there exists a unique isogeny $\widehat{F} : E_2 \to E_1$ such that $\widehat{F} \circ F = [m]$. If F is defined over k_0, so is $\widehat{F}$.

REMARKS. $\widehat{F}$ is called the **dual isogeny** to F. We say that E_1 is **isogenous** to E_2 if there is a nonconstant isogeny from E_1 to E_2. As a consequence of this theorem, "is isogenous to" is an equivalence relation. If we restrict attention to elliptic curves defined over k_0, it is still an equivalence relation.

SKETCH OF PROOF. As a morphism, F factors into $F = F_s \circ \phi$, where F_s is separable and ϕ is a qth power Frobenius. Thus it is enough to treat F_s and ϕ individually. We treat only F_s, changing notation and writing F for it. By (11.85), $\#\ker(F) = m$. Thus the map $[m] : E_1 \to E_1$ annihilates $\ker F$, and $\ker F \subseteq \ker[m]$. Application of Lemma 11.63 completes the proof.

Dual isogenies have the following properties, whose proofs we omit.

Proposition 11.65. Let $F : E_1 \to E_2$ be an isogeny of degree m.
(a) $\widehat{F} \circ F = [m]$ on E_1, and $F \circ \widehat{F} = [m]$ on E_2.
(b) $\deg \widehat{F} = \deg F$ and $(\widehat{F})\widehat{} = F$.
(c) $[m]\widehat{} = [m]$ and $\deg[m] = m^2$.
(d) If $G : E_2 \to E_3$ is an isogeny, then $(G \circ F)\widehat{} = \widehat{F} \circ \widehat{G}$.
(e) If $H : E_1 \to E_2$ is an isogeny, then $(F + H)\widehat{} = \widehat{F} + \widehat{H}$.

The next result is a converse to Proposition 11.62.

Theorem 11.66. Let E be an elliptic curve over $\bar{k}$, and let S be a finite subgroup of E. Then there exist an elliptic curve E' and a separable isogeny $F : E \to E'$ such that $\ker F = S$. The curve E' is unique up to isomorphism. If E is defined over k_0 and if S is stable under $\text{Aut}_{k_0}(\bar{k})$, then E' is defined over k_0 and may be taken to be defined over k_0.

IDEA OF PROOF OVER $\bar{k}$ IF $\text{char}(\bar{k}) = 0$. The set $\{T_x^* \mid x \in S\}$ is a finite group of automorphisms of $\bar{k}(E)$. We apply Lemma 11.35 to determine a subfield, Theorem 11.44 to define a nonsingular projective curve C, and Theorem 11.48 to define a dominant morphism $F : E \to C$. One has to prove that F is unramified and then that the genus of C is one.

Let us return to elliptic curves over $\mathbf{Q}$. The L function of such a curve was defined in (10.9).

Theorem 11.67. Let E and E' be elliptic curves over $\mathbf{Q}$ that are isogenous over $\mathbf{Q}$. Then $L(s, E) = L(s, E')$.

SKETCH OF PROOF. The two L functions will be equal factor by factor. We treat the pth factor only for primes p not dividing Δ or Δ', omitting the others. If $F : E \to E'$ is a nonconstant isogeny defined over $\mathbf{Q}$, one has to check that F induces a nonconstant isogeny $F_p : E_p \to E'_p$

defined over $\mathbf{Z}_p$. Taking this for granted, let ϕ_1 and ϕ_2 be the Frobenius morphisms of E_p and E'_p defined in (11.82). We can write $F_p = F_s \phi_1^r$ with $F_s : E_p \to E'_p$ separable and defined over $\mathbf{Z}_p$. So we may as well assume from the outset that $F_p : E_p \to E'_p$ is a separable isogeny defined over $\mathbf{Z}_p$.

If we write out the morphism F_p as $[(F_0, F_1, F_2)]$, we have

$$F_p(\phi_1(x)) = [(F_0(x^p), F_1(x^p), F_2(x^p))]$$
$$= [(F_0(x)^p, F_1(x)^p, F_2(x)^p)] = \phi_2(F_p(x)).$$

Thus
$$F_p \circ \phi_1 = \phi_2 \circ F_p. \tag{11.89}$$

If y is in E'_p, then $y = F_p(x)$ for $\deg F_p$ values of x, by (11.85) since ϕ_1 is onto. For such an x,

$$y \in E'_p(\mathbf{Z}_p) \iff \phi_2(y) = y$$
$$\iff \phi_2 F_p(x) = F_p(x) \iff x \in \ker(([1] - \phi_2)F_p).$$

Thus

$$\# E'_p(\mathbf{Z}_p) = \frac{\# \ker(([1] - \phi_2)F_p)}{\deg F_p}$$
$$= \frac{\# \ker(F_p([1] - \phi_1))}{\deg F_p} \quad \text{by (11.89)}$$
$$= \frac{\deg(F_p([1] - \phi_1))}{\deg F_p} \quad \text{by (11.87) and (11.85)}$$
$$= \frac{\deg(F_p) \deg([1] - \phi_1)}{\deg F_p}$$
$$= \deg([1] - \phi_1)$$
$$= \# E_p(\mathbf{Z}_p).$$

Hence the pth factors of $L(s, E)$ and $L(s, E')$ match.

10. Abelian Varieties and Jacobian Variety

In this section our algebraically closed field $\bar{k}$ will be $\mathbf{C}$, and the subfield k_0 will be $\mathbf{Q}$. An **abelian variety** A is a nonsingular projective variety over $\mathbf{C}$ with a distinguished point O and with an abelian group structure such that O is the identity and the operations of addition and negative are morphisms. The abelian variety is said to be **defined over** $\mathbf{Q}$ if A is defined over $\mathbf{Q}$, O is in $A(\mathbf{Q})$, and addition and negative are defined over $\mathbf{Q}$. Any elliptic curve (over $\mathbf{C}$ or $\mathbf{Q}$) is an example, by Proposition 11.59a. Conversely an abelian variety of dimension 1 is an elliptic curve. (We have only to see that the genus is 1. We can work with $A(\mathbf{C})$, which is a compact Riemann surface. Translations are fixed-point free automorphisms homotopic to the identity, and the Lefschetz Fixed-Point Theorem allows these only in genus 1.)

The goal of this section is to introduce the Jacobian variety $J(C)$ of a nonsingular projective curve C defined over $\mathbf{Q}$. $J(C)$ is to be an abelian variety defined over $\mathbf{Q}$ and is to play the same role in geometry over $\mathbf{Q}$ that the Jacobian variety of §4 plays in the theory of compact Riemann surfaces.

Let A and B be abelian varieties. A **homomorphism** $F : A \to B$ is a morphism of varieties that is also a homomorphism of groups. We say that F is an **isomorphism** if it is an isomorphism of varieties and an isomorphism of groups. If A and B are defined over $\mathbf{Q}$, we say F is **defined over** $\mathbf{Q}$ if it is defined over $\mathbf{Q}$ as a morphism.

The set $\mathrm{Hom}(A, B)$ of homomorphisms from A to B is an abelian group under pointwise addition. For the special case that $A = B$, we write $\mathrm{End}\, A$ in place of $\mathrm{Hom}(A, A)$. This is a ring with composition as multiplication.

If A is an abelian variety and C is a **subvariety** (obtained by enlarging the defining homogeneous ideal of polynomials), then the inclusion map is automatically a morphism. If C has a group structure compatible with that of A, then inclusion and addition give morphisms $C \times C \to A \times A \to A$ with image in C that show that addition on C is a morphism. Similarly negative on C is a morphism. We say that C is an **abelian subvariety** of A. The kernel of a homomorphism between abelian varieties is an example if it is connected in the complex manifold topology. We shall make use of the following two propositions.

Proposition 11.68. If $F : B \to A$ is a homomorphism between abelian varieties, then $C = \mathrm{image}\, F$ is an abelian subvariety of A. If A, B, and F are defined over $\mathbf{Q}$, then C may be taken to be defined over $\mathbf{Q}$.

Proposition 11.69. Let A be an abelian variety, and let C be an abelian subvariety. Then A/C may be given the structure of an abelian variety in the following sense: There exists a pair (A', F) such that A' is an abelian variety, $F : A \to A'$ is a homomorphism onto, $\ker F =$

C, and any homomorphism $F'' : A \to A''$ of abelian varieties with $\ker F'' \supseteq C$ factors through F, i.e., $F'' = F' \circ F$ for a homomorphism $F' : A' \to A''$ of abelian varieties. The pair (A', F) is unique up to canonical isomorphism. If A and C are defined over $\mathbf{Q}$, then A' and F will be defined over $\mathbf{Q}$; in this case, when A'' and F'' in the universal property are defined over $\mathbf{Q}$, F' may be taken to be defined over $\mathbf{Q}$.

The next proposition gives some properties of abelian varieties and homomorphisms in order to shed some light on the definitions. Although the proposition helps to motivate parts of §11 and Chapter XII, it will not be used explicitly.

Proposition 11.70. Let A and B be abelian varieties.

(a) A homomorphism $F : A \to B$ is onto if and only if its kernel is finite. (In this case, A and B are said to be **isogenous**.)

(b) Isogeny is symmetric and hence is an equivalence relation.

(c) If V is a nonsingular projective variety and $F : V \to A$ is a rational map, then F is a morphism.

(d) Any morphism $F : A \to B$ is the composition of a homomorphism and a translation.

(e) If B is an abelian subvariety of A, then there exists an abelian subvariety C of A such that $B \cap C$ is a finite group and $A = B + C$ as groups. If A and B are defined over $\mathbf{Q}$, then C may be taken to be defined over $\mathbf{Q}$.

(f) A is isogenous to a product of abelian varieties that are simple (in the sense of not having any proper nonzero abelian subvarieties), and the factors of the decomposition are unique up to isogeny.

For X a compact Riemann surface of genus g, we defined the Jacobian variety $J(X) = \mathbf{C}^g/\Lambda(X)$ in §4. With a base point x_0 fixed in X, there is a canonical holomorphic mapping $\Phi : X \to J(X)$ with the following universal property. Whenever $F : X \to T$ is a holomorphic mapping into a complex torus, then F factors through the Jacobian variety:

$$F = f \circ \Phi + F(x_0)$$

for some holomorphic homomorphism $f : J(X) \to T$.

When the base point in X is changed, Φ gets composed with a translation. The universal property makes $J(X)$ unique up to a biholomorphic group isomorphism. If $J(X)$ is fixed, Φ is unique up to composition with a translation and a biholomorphic automorphism of $J(X)$.

If C is a nonsingular projective curve over $\mathbf{C}$, then $C(\mathbf{C})$ has an underlying complex manifold structure, hence is a compact Riemann surface. Thus C has a Jacobian variety in the above sense. The following theorem solves the problem of putting on the Jacobian variety the structure of a nonsingular projective variety.

Theorem 11.71. Let C be a nonsingular projective curve over $\mathbf{C}$, let $J(C)$ be the Jacobian variety of the underlying compact Riemann surface, and let $\Phi : C \to J(C)$ be the canonical holomorphic mapping with base point x_0. Then $J(C)$ admits the structure of a nonsingular projective variety in such a way that

  (a) its group structure makes it into an abelian variety,
  (b) Φ is a morphism, and
  (c) whenever $F : C \to A$ is a morphism into an abelian variety, then F factors through $J(C)$ as $F = f \circ \Phi + F(x_0)$ for some homomorphism $f : J(C) \to A$ of abelian varieties.

Moreover, if C is defined over $\mathbf{Q}$, then $J(C)$ can be defined over $\mathbf{Q}$ in such a way that (a), (b), and (c) are valid with structures defined over $\mathbf{Q}$.

There is a uniqueness statement for Theorem 11.71, just as in the setting of compact Riemann surfaces, and it is again easy. Existence, however, is a deep question. Theorem 11.71 is due to Lefschetz for $\mathbf{C}$ and to Weil and Chow for $\mathbf{Q}$. The following result will be handy in detecting morphisms that involve $J(C)$.

Theorem 11.72 (Chow). If V_1 and V_2 are nonsingular projective varieties over $\mathbf{C}$ and if $F : V_1 \to V_2$ is a holomorphic mapping between their underlying complex manifolds, then F is rational over $\mathbf{C}$.

We shall apply Theorem 11.71 to $C = X_0(N)$, which is defined over $\mathbf{Q}$. The ring of holomorphic homomorphisms of $J(X_0(N))$ into itself is the same as the ring $\mathrm{End}(J(X_0(N)))$ of abelian variety homomorphisms defined over $\mathbf{C}$, as a consequence of Theorem 11.72. Two key steps in the Eichler-Shimura theory are to transform the Hecke operators into members of $\mathrm{End}(J(X_0(N)))$ and to show that the resulting homomorphisms are defined over $\mathbf{Q}$. We carry out the first of these steps now.

For this purpose we fix notation as follows: Let $X_0(N)$ have genus g, and let $\omega_1, \ldots, \omega_g$ be the basis of $\Omega_{\mathrm{hol}}(X_0(N))$ used to define $J = J(X_0(N))$ concretely as $\mathbf{C}^g / \Lambda(X_0(N))$. (Recall that $\mathbf{C}^g$ and $\Lambda(X_0(N))$ are not affected by changing the $\mathbf{Z}$ basis of cycles.) Let $\pi : \mathcal{H}^* \to \Gamma_0(N) \backslash \mathcal{H}^*$ be the quotient map and $\tilde{\Phi} : \mathcal{H}^* \to J$ be the composition

$\tilde{\Phi} = \Phi \circ \pi$. If we put $f_j(\tau)\,d\tau = \pi^*(\omega_j)$, then $f_1,\ldots,f_g$ is a basis for $S_2(\Gamma_0(N))$ and $\tilde{\Phi}$ is given by

$$\tilde{\Phi}(\tau) = \left\{ \int_{\tau_0}^{\tau} f_j(\zeta)\,d\zeta \right\}_{j=1}^{g} \tag{11.90}$$

for any base point $\tau \in \pi^{-1}(x_0)$.

Since $J = \mathbf{C}^g/\Lambda(X_0(N))$, we may identify the tangent space $\mathfrak{j}$ to J at the group identity O as the same $\mathbf{C}^g$. Its standard basis will be denoted $e_1,\ldots,e_g$. The group J is a Lie group, and its Lie algebra is $\mathfrak{j} = \mathbf{C}^g$, column-vector space. Any real analytic homomorphism f of J into itself induces (by passage to the differential df at O) an $\mathbf{R}$ linear map of $\mathbf{C}^g$ into itself, and distinct homomorphisms yield distinct linear maps. The members of $\mathrm{End}(J)$ are holomorphic, and their differentials are consequently $\mathbf{C}$ linear maps of $\mathfrak{j} = \mathbf{C}^g$ into itself. In this way we get a one-one ring homomorphism of $\mathrm{End}(J)$ into the algebra of all g-by-g complex matrices:

$$\mathrm{End}(J) \longrightarrow M(g, \mathbf{C}). \tag{11.91}$$

We can use $z_1,\ldots,z_g$ as coordinates on J, and the space $\Omega_{\mathrm{hol}}(J)$ of holomorphic 1-forms is the $\mathbf{C}$ linear span of $dz_1,\ldots,dz_g$. The spaces $\Omega_{\mathrm{hol}}(J)$ and $\mathfrak{j}$ are in natural duality, the pairing being given by

$$\langle dz_i, e_j \rangle = \delta_{ij}.$$

If we identify $\mathfrak{j}$ with the space of invariant vector fields on J, we can regard this formula also as a valid pairing at any point of J. For f in $\mathrm{End}(J)$, we define an endomorphism δf of $\Omega_{\mathrm{hol}}(J)$ by

$$\langle (\delta f)u, v \rangle = \langle u, (df)v \rangle \qquad \text{for } u \in \Omega_{\mathrm{hol}}(J) \text{ and } v \in \mathfrak{j}. \tag{11.92}$$

Let us check that
$$\tilde{\Phi}^*(dz_j) = f_j(\tau)\,d\tau. \tag{11.93}$$

In fact, at any point τ_1, differentiation of (11.90) at τ_1 gives

$$\langle \tilde{\Phi}^*(dz_j), \frac{d}{d\tau}\Big|_{\tau_1} \rangle = \langle dz_j, d\tilde{\Phi}\left(\frac{d}{d\tau}\Big|_{\tau_1}\right) \rangle = \langle dz_j, \begin{pmatrix} f_1(\tau_1) \\ \vdots \\ f_g(\tau_1) \end{pmatrix} \rangle = f_j(\tau_1).$$

Since $\langle d\tau, \frac{d}{d\tau}\Big|_{\tau_1} \rangle = 1$, we obtain (11.93). Since $\tilde{\Phi}^*$ maps basis to basis, $\tilde{\Phi}^*$ is a vector space isomorphism. Therefore it makes sense to define $\mu : S_2(\Gamma_0(N)) \to \Omega_{\mathrm{hol}}(J)$ by

$$\tilde{\Phi}^*(\mu(f)) = f(\tau)\,d\tau \qquad \text{for } f \in S_2(\Gamma_0(N)). \tag{11.94}$$

Let $M(n,N) = \bigcup_{i=1}^{K} \Gamma_0(N)\alpha_i$ as in (11.28). In (11.29) and (11.32) we defined consistently a Hecke operator $T(n) : X_0(N) \to \text{Div}(X_0(N))$ by

$$T(n)(\pi(\tau)) = \sum_{i=1}^{K} (\pi(\alpha_i \tau)). \tag{11.95}$$

The mapping $\Phi : X_0(N) \to J$ extends by additivity to a map

$$\Phi^\# : \text{Div}(X_0(N)) \to J,$$

and then we can form the composition $T^\#(n) = \Phi^\# \circ T(n)$ given by

$$T^\#(n)(\pi(\tau)) = \sum_{i=1}^{K} \Phi(\pi(\alpha_i \tau)) = \begin{pmatrix} \sum_i \int_{\tau_0}^{\alpha_i \tau} f_1(\zeta)\,d\zeta \\ \vdots \\ \sum_i \int_{\tau_0}^{\alpha_i \tau} f_g(\zeta)\,d\zeta \end{pmatrix}. \tag{11.96}$$

Equation (11.96) exhibits $T^\#(n)$ as holomorphic; by Theorem 11.72 it is a morphism of varieties over $\mathbf{C}$. In §11 one of the steps will be to prove that $T^\#(n)$ is defined over $\mathbf{Q}$.

Applying the universal mapping property to $T^\#(n) : X_0(N) \to J$ (with $A = J$), we obtain a member $t(n)$ of End J such that

$$T^\#(n)(\pi(\tau)) = t(n)(\Phi(\tau)) + T^\#(n)(\pi(\tau_0))$$

for all τ. What this equation says is

$$t(n)\begin{pmatrix} \int_{\tau_0}^{\tau} f_1(\zeta)\,d\zeta \\ \vdots \\ \int_{\tau_0}^{\tau} f_g(\zeta)\,d\zeta \end{pmatrix} = \begin{pmatrix} \sum_i \int_{\tau_0}^{\alpha_i \tau} f_1(\zeta)\,d\zeta \\ \vdots \\ \sum_i \int_{\tau_0}^{\alpha_i \tau} f_g(\zeta)\,d\zeta \end{pmatrix}. \tag{11.97}$$

Since $t(n)$ is additive, we can evaluate (11.97) at τ_1 and subtract the result from (11.97) to obtain

$$t(n)\begin{pmatrix} \int_{\tau_1}^{\tau} f_1(\zeta)\,d\zeta \\ \vdots \\ \int_{\tau_1}^{\tau} f_g(\zeta)\,d\zeta \end{pmatrix} = \begin{pmatrix} \sum_i \int_{\tau_1}^{\alpha_i \tau} f_1(\zeta)\,d\zeta \\ \vdots \\ \sum_i \int_{\tau_1}^{\alpha_i \tau} f_g(\zeta)\,d\zeta \end{pmatrix}. \tag{11.98}$$

To compute the differential $dt(n)$, we can differentiate the formula for $t(n)$ along any curve extending from the O vector. Differentiating (11.98)

10. ABELIAN VARIETIES AND JACOBIAN VARIETY

at $\tau = \tau_1$ and then writing τ for τ_1 (since τ_1 is arbitrary), we find

$$dt(n)\begin{pmatrix} f_1(\tau) \\ \vdots \\ f_g(\tau) \end{pmatrix} = \begin{pmatrix} \sum_i f_1(\alpha_i(\tau)) \frac{d(\alpha_i \tau)}{d\tau} \\ \vdots \\ \sum_i f_g(\alpha_i(\tau)) \frac{d(\alpha_i \tau)}{d\tau} \end{pmatrix}$$

$$= \begin{pmatrix} \sum_i f_1 \circ [\alpha_i]_2(\tau) \\ \vdots \\ \sum_i f_g \circ [\alpha_i]_2(\tau) \end{pmatrix} = \begin{pmatrix} T_2(n) f_1(\tau) \\ \vdots \\ T_2(n) f_g(\tau) \end{pmatrix} \quad (11.99)$$

for all τ.

Consequently the matrix of $dt(n)$ is $A(T_2(n))$, in the notation of (11.47). We shall use the reformulation below of this result.

Proposition 11.73 (Shimura-Taniyama). For f in $S_2(\Gamma_0(N))$,

$$(\delta t(n))(\mu(f)) = \mu(T_2(n)f). \tag{11.100}$$

PROOF. Write $f = \sum c_k f_k$, and let e_l be a standard basis vector of j. Then we have

$$\langle (\delta t(n))(\mu(f)), e_l \rangle = \langle \mu(f), dt(n)e_l \rangle \quad \text{by (11.92)}$$

$$= \sum_k c_k \langle dz_k, dt(n)(e_l) \rangle$$

$$= \sum_k c_k (dt(n)e_l)_k$$

$$= \sum_k c_k A(T_2(n))_{kl}$$

and

$$\langle \mu(T_2(n))f, e_l \rangle = \sum_k c_k \langle \mu(T_2(n)f_k), e_l \rangle$$

$$= \sum_{k,m} c_k \langle \mu(A(T_2(n))_{km} f_m), e_l \rangle$$

$$= \sum_{k,m} c_k A(T_2(n))_{km} \langle dz_m, e_l \rangle$$

$$= \sum_k c_k A(T_2(n))_{kl}.$$

The above equations show that the two sides of (11.100) are paired equally with any member of j and therefore must be equal.

11. Elliptic Curves Constructed from $S_2(\Gamma_0(N))$

We come to the main theorem of the Eichler-Shimura theory. We continue with the notation $X_0(N)$, g, $\Lambda(X_0(N))$, J, $\tilde{\Phi}$, $\{f_1, \ldots, f_g\}$, τ_0, j, End J, $\Omega_{\text{hol}}(J)$, d and δ, μ, $T(n)$, $T^\#(n)$, $t(n)$, $dt(n)$, and $\delta t(n)$ as in the latter part of §10.

In order to handle one step in the proof of the main theorem, we shall assume that the basis $\{f_1, \ldots, f_g\}$ of $S_2(\Gamma_0(N))$ over $\mathbb{C}$ has been constructed as in Theorem 11.25, so that the matrices of the Hecke operators $T_2(n)$ have integer entries. By (11.99) these integer matrices are the matrices of $dt(n)$.

We define $\text{End}_\mathbb{Q}(J) = \text{End}(J) \otimes_\mathbb{Z} \mathbb{Q}$. By (11.91), $\text{End}_\mathbb{Q}(J)$ can be regarded as an algebra over $\mathbb{Q}$ of g-by-g complex matrices.

Theorem 11.74 (Eichler-Shimura). Let $f(\tau) = \sum_{n=1}^\infty c_n e^{2\pi i n \tau}$ be a newform in $S_2(\Gamma_0(N))$ normalized to have $c_1 = 1$, and suppose that all c_n are in $\mathbb{Z}$. Then there exists a pair (E, ν) such that

(a) E is an elliptic curve defined over $\mathbb{Q}$, and (E, ν) is a quotient of J by an abelian subvariety of J defined over $\mathbb{Q}$,

(b) the members $t(n)$ of $\text{End}(J)$ leave A stable and act on the quotient E as multiplication by the integers c_n,

(c) $\mu(f)$ is a nonzero multiple of $\nu^*(\omega)$, where ω is the invariant differential (11.21) of E,

(d) if
$$\Lambda_f = \left\{ \Phi_f(\gamma) = \int_{\tau_0}^{\gamma(\tau_0)} f(\zeta)\, d\zeta \;\Big|\; \gamma \in \Gamma_0(N) \right\},$$
then Λ_f is a lattice in $\mathbb{C}$, and E is isomorphic to $\mathbb{C}/\Lambda_f$ over $\mathbb{C}$,

(e) the L functions of E and f coincide as Euler products except possibly at finitely many primes.

Moreover, properties (a) and (b) characterize A uniquely and therefore determine (E, ν) up to isomorphism defined over $\mathbb{Q}$.

REMARKS. Composing ν in (a) with $\Phi : X_0(N) \to J$, we obtain a morphism $\nu \circ \Phi : X_0(N) \to E$ defined over $\mathbb{Q}$. The proof of the theorem does not use that f is a newform, only that f is an eigenfunction of all the Hecke operators $T_2(n)$. Work of Igusa shows that the Eichler-Shimura argument for (e) is applicable to all primes not dividing N. If f is a newform, then Theorem 12.8 will note that $L(E, s)$ and $L(f, s)$ match exactly.

11. ELLIPTIC CURVES CONSTRUCTED FROM $S_2(\Gamma_0(N))$

The proof has a characteristic 0 part and a characteristic p part, the latter to handle (e). We shall give the full characteristic 0 part in this section and shall discuss the characteristic p part in §12. We need the following extension of the standard Wedderburn theorem on semisimple associative algebras.

Lemma 11.75 (Wedderburn). Let T be a finite-dimensional associative algebra with identity defined over a field k, and let $\mathcal{R}$ be its **nilradical** (largest two-sided ideal). Then there exists a semisimple subalgebra $\mathcal{S}$ of T such that $T = \mathcal{S} \oplus \mathcal{R}$ as vector spaces. Moreover, $\mathcal{S}$ is a direct sum of ideals, each of which is a simple algebra isomorphic to a full matrix algebra over a division algebra over k.

REMARK. We need the lemma only in the case that T is commutative, and then the simple algebras in $\mathcal{S}$ are isomorphic to fields that are finite algebraic extensions of k.

In order to get to the body of the proof quickly, we postpone to the end of this section the proof of the next lemma.

Lemma 11.76. The members $t(n)$ of $\text{End}(J)$ are defined over $\mathbf{Q}$.

PROOF OF THEOREM 11.74 EXCEPT PART (e). Let T be the commutative $\mathbf{Q}$ subalgebra of $\text{End}_\mathbf{Q}(J)$ generated by all the $t(n)$ in $\text{End}(J)$. Since each $t(n)$ may be identified with the g-by-g matrix of its differential $dt(n)$, which has integer entries, T is isomorphic to a subalgebra of $M(g, \mathbf{Q})$. Therefore T is finite-dimensional over $\mathbf{Q}$.

We apply Lemma 11.75 to the algebra T and the field $\mathbf{Q}$, obtaining

$$T = \mathcal{S} \oplus \mathcal{R} = (k_1 \oplus \cdots \oplus k_r) \oplus \mathcal{R},$$

where $\mathcal{R}$ is the nilradical and each k_j is an ideal of $\mathcal{S}$ isomorphic to a finite algebraic extension of $\mathbf{Q}$. By Proposition 11.73, we have

$$(\delta t(n))(\mu(f)) = c_n \mu(f).$$

Hence there exists a well defined $\mathbf{Q}$ algebra homomorphism ρ of T into $\mathbf{Q}$ given by the formula

$$\rho(t(n)) = c_n.$$

It is clear that $\rho(\mathcal{R}) = 0$. Changing the order of the factors $k_1, \ldots, k_r$ if necessary, we may assume that $\rho(k_1) = \mathbf{Q}$. Since k_1 is a field, $k_1 \cong \mathbf{Q}$. Let $\rho' : \mathbf{Q} \to k_1$ be the inverse of ρ to k_1, and define an ideal $\mathcal{U}$ of T by

$$\mathcal{U} = (k_2 \oplus \cdots \oplus k_r) \oplus \mathcal{R}.$$

The abelian subvariety A will be the sum of all $\alpha(\mathcal{T})$ for α in $\mathcal{U} \cap \operatorname{End}(J)$. Let us see that A is indeed a subvariety and is defined over $\mathbf{Q}$. The members of $\operatorname{End}(J)$ may be viewed via (11.91) as members of $M(g, \mathbf{C})$ that carry $\Lambda(X_0(N))$ into itself. With this identification the members of $\operatorname{End}_{\mathbf{Q}}(J)$ are $\mathbf{Q}$ linear combinations of these matrices. Let $\alpha \in \mathcal{U} \cap \operatorname{End}(J)$ be given. Being in $\mathcal{T}$, α is a polynomial in the $t(n)$'s with rational coefficients. Hence there is a nonzero integer m such that $m\alpha$ is an integer polynomial combination of the $t(n)$'s. According to Lemma 11.76, each $t(n)$ is defined over $\mathbf{Q}$. Therefore $m\alpha$ is defined over $\mathbf{Q}$. Since α and $m\alpha$ have the same image in J, Proposition 11.68 shows that image(α) is an abelian subvariety of J defined over $\mathbf{Q}$. Now suppose we have two abelian subvarieties A_1 and A_2 of J defined over $\mathbf{Q}$. Their sum $A_1 + A_2$ is the image of $A_1 \times A_2 \subseteq J \times J \to J$ under addition. Since $A_1 \times A_2$ and addition are defined over $\mathbf{Q}$, another application of Proposition 11.68 shows that $A_1 + A_2$ is defined over $\mathbf{Q}$. We now iterate this construction. With each iteration, either the dimension goes up (of the abelian subvariety as a connected complex Lie group) or nothing new happens. We conclude that A, as we have defined it, is an abelian subvariety of J defined over $\mathbf{Q}$.

By Proposition 11.69 we can form the quotient (E, ν) of J by A, and it can be taken to be defined over $\mathbf{Q}$. We shall prove shortly that E has dimension 1, and then (a) will be proved. In the meantime, each $\beta \in \mathcal{T} \cap \operatorname{End}(J)$ maps J into A. In fact, if a is in A, write $a = \sum \alpha_k(z_k)$ with $\alpha_k \in \mathcal{U} \cap \operatorname{End}(J)$ and $z_k \in J$. Since $\mathcal{U}$ is an ideal, each $\beta\alpha_k$ is in $\mathcal{U} \cap \operatorname{End}(J)$, and we have

$$\beta(a) = \sum \beta\alpha_k(z_k) \in \sum \beta\alpha_k(J) \subseteq A.$$

Consequently each $\beta \in \mathcal{T} \cap \operatorname{End}(J)$ maps J into A. Applying this conclusion to $t(n)$, we obtain the first conclusion of (b). Since $t(n)(A) \subseteq A$, we have $(\ker \nu \circ T(n)) \subseteq \ker \nu$. By the universal mapping property of (E, ν) in Proposition 11.69, there exists $\bar{t}(n) \in \operatorname{End}(E)$ with

$$\bar{t}(n) \circ \nu = \nu \circ t(n). \tag{11.101}$$

In other words $t(n)$ acts on E as $\bar{t}(n)$. To calculate what this action is, we observe that $t(n) - \rho'(c_n)$ is in $\mathcal{U}$ by construction and $\rho'(c_n) - [c_n]$ is in $\mathcal{U}$ also, where $[c_n]$ denotes multiplication by the integer c_n. Hence $t(n) - [c_n]$ is in $\mathcal{U} \cap \operatorname{End}(J)$. In other words, $t(n) - [c_n]$ passes to the quotient and acts as 0. Thus $\bar{t}(n) = [c_n]$. This proves (b).

To prove that $\dim E > 0$, we are to prove that $A \neq J$. Let $m \geq 0$ be the integer for which $k_1 \mathcal{R}^m \neq 0$ and $k_1 \mathcal{R}^{m+1} = 0$, and let β be a

11. ELLIPTIC CURVES CONSTRUCTED FROM $S_2(\Gamma_0(N))$

nonzero member of $k_1 \mathcal{R}^m$. Possibly multiplying β by a nonzero integer (as in an argument earlier in the proof), we may assume β is in $\text{End}(J)$, not just $\text{End}_\mathbb{Q}(J)$. If a is in $\mathcal{U}$, then $\beta a = 0$ (because $k_1 k_j = 0$ for $j > 1$ and $\mathcal{R}^m \mathcal{R} = 0$). If a is in A, we can write $a = \sum \alpha_k(z_k)$ with $\alpha_k \in \mathcal{U} \cap \text{End}(J)$ and $z_k \in J$. Since $\beta \alpha_k = 0$, we have $\beta(a) = 0$. Therefore β is a nonzero member of $\text{End}(J)$ that annihilates A. Hence $A \neq J$.

We shall now prove (c) and prove that $\dim E \leq 1$, thereby completing the proof of (a). Let ω' be a nonzero member of $\Omega_{\text{hol}}(E)$ (existence by previous paragraph), and let ν^* be the pullback mapping $\nu^* : \Omega_{\text{hol}}(E) \to \Omega_{\text{hol}}(J)$ induced by ν. Since ν on the complex manifold level is just a homomorphism of one complex torus onto another, ν^* is one-one. Applying $(\cdot)^*$ to (11.101), we have

$$\nu^* \circ \delta \bar{t}(n) = \delta t(n) \circ \nu^*.$$

Now $\bar{t}(n) = [c_n]$ implies that $\delta \bar{t}(n) = c_n \cdot 1$. Hence

$$\delta t(n)(\nu^*(\omega')) = c_n \nu^*(\omega').$$

If we define $f' = \mu^{-1}(\nu^*(\omega'))$, then $\mu(f') = \nu^*(\omega')$, and Proposition 11.73 gives

$$\mu(T_2(n)f') = c_n \mu(f')$$

and hence

$$T_2(n)f' = c_n f'. \tag{11.102}$$

If $\dim E > 1$, then there exist linearly independent ω' and ω'' in $\Omega_{\text{hol}}(E)$, and $\nu^*(\omega')$ and $\nu^*(\omega'')$ will be independent. If we put $f'' = \mu^{-1}(\nu^*(\omega''))$, then the above argument shows that

$$T_2(n)f'' = c_n f''. \tag{11.103}$$

Since f' and f'' are independent, (11.102) and (11.103) together contradict Proposition 9.20b. Thus $\dim E = 1$, and the proof of (a) is complete. We can take ω' to be the invariant differential (11.21) of E in the above argument, and (11.102) and Proposition 9.20b force f' and f to be linearly dependent. This proves (c).

Let us prove uniqueness. Suppose A' and (E', ν') satisfy conclusions (a) and (b). Let ω' and ω be the invariant differentials (11.21) of E' and E, respectively. Then the argument in the previous paragraph shows that $\nu'^*(\omega')$ is a multiple of $\mu(f)$. Since we already know that $\nu^*(\omega)$ is a multiple of $\mu(f)$, $\nu'^*(\omega')$ and $\nu^*(\omega)$ are multiples of each other. Hence

$\nu'^*(\omega')$ and $\nu^*(\omega)$ annihilate the same members of j. The respective annihilators are the tangent spaces in j of $\ker \nu' = A'$ and $\ker \nu = A$. Since A' and A are connected Lie subgroups of J with the same Lie subalgebras, we conclude that $A' = A$. Then (E, ν) is isomorphic to (E', ν') over $\mathbf{Q}$ as a consequence of Proposition 11.69.

Let us prove (d). The natural pairing of $\Omega_{\text{hol}}(J)$ with $j \cong \mathbf{C}^g$ makes $\mu(f)$ acts as a linear functional on j. We calculate the effect of $\mu(f)$ on $\Lambda(X_0(N)) = \ker(j \to J)$. The lattice $\Lambda(X_0(N))$ has generators

$$u_k = \begin{pmatrix} \int_{c_k} f_1(\zeta) \, d\zeta \\ \vdots \\ \int_{c_k} f_g(\zeta) \, d\zeta \end{pmatrix},$$

where $c_1, \ldots, c_{2g}$ represent a $\mathbf{Z}$ basis of $H_1(X_0(N), \mathbf{Z})$. Write $f = \sum r_j f_j$. Then

$$\mu(f)(u_k) = \langle \mu(f), u_k \rangle = \langle \sum_j r_j \mu(f_j), u_k \rangle$$
$$= \sum_j r_j \langle dz_j, u_k \rangle = \sum_j r_j \int_{c_k} f_j(\zeta) \, d\zeta = \int_{c_k} f(\zeta) \, d\zeta.$$

By (11.37) and Proposition 11.22, we conclude that

$$\mu(f)(\Lambda(X_0(N))) = \sum_k \mathbf{Z} \int_{c_k} f(\zeta) \, d\zeta = \Lambda_f. \tag{11.104}$$

Let $\mathfrak{a} \subseteq \mathfrak{j}$ be the Lie algebra of A (the tangent space at the identity). We shall verify that

$$\ker \mu(f) = \mathfrak{a}. \tag{11.105}$$

In fact, $\mu(f)$ is a nonzero multiple of $\nu^*(\omega)$, and thus

$$\ker \mu(f) = \{u \in \mathfrak{j} \mid \langle \nu^*(\omega), u \rangle = 0\}$$
$$= \{u \in \mathfrak{j} \mid \langle \omega, (d\nu)(u) \rangle = 0\}$$
$$= \{u \in \mathfrak{j} \mid d\nu(u) = 0\} \qquad \text{since } \omega \text{ spans } \Omega_{\text{hol}}(E)$$
$$= \ker(d\nu) = (\text{Lie algebra of } A) = \mathfrak{a}.$$

This proves (11.105).

The homomorphism $\mathfrak{j} \to J$ with kernel $\Lambda(X_0(N))$ is really the exponential map of Lie theory and thus must implement $\mathfrak{a} \to A$. Hence the kernel of $\mathfrak{a} \to A$ is $\mathfrak{a} \cap \Lambda(X_0(N))$. Since A is compact, $\mathfrak{a} \cap \Lambda(X_0(N))$ is a

11. ELLIPTIC CURVES CONSTRUCTED FROM $S_2(\Gamma_0(N))$

lattice in $\mathfrak{a}$, evidently of rank $2g - 2$. Let $x_1, \ldots, x_{2g-2}$ be a $\mathbf{Z}$ basis for it, and adjoin x_{2g-1} and x_{2g} in $\Lambda(X_0(N))$ so that $\Lambda' = \sum_{j=1}^{2g} \mathbf{Z} x_j$ has rank $2g$. Then Λ' has finite index in $\Lambda(X_0(N))$, say m, and it follows that $\Lambda(X_0(N))$ is contained in $\frac{1}{m}\Lambda'$. Now

$$\mathbf{C} = \mu(f)(\mathfrak{j}) = \mu(f)(\sum \mathbf{R} x_j) = \mu(f)(\mathbf{R} x_{2g-1} + \mathbf{R} x_{2g})$$

shows that

$$\mu(f)(x_{2g-1}) \text{ and } \mu(f)(x_{2g}) \text{ are linearly independent over } \mathbf{R}. \quad (11.106)$$

On the other hand,

$$\mu(f)(\mathbf{Z} x_{2g-1} + \mathbf{Z} x_{2g}) = \mu(f)(\sum_j \mathbf{Z} x_j) = \mu(f)(\Lambda')$$
$$\subseteq \mu(f)(\Lambda(X_0(N))) \subseteq \mu(f)(m^{-1}\Lambda')$$
$$= \mu(f)(\sum_j m^{-1} \mathbf{Z} x_j)$$
$$= \mu(f)(m^{-1} \mathbf{Z} x_{2g-1} + m^{-1} \mathbf{Z} x_{2g}). \quad (11.107)$$

Combining (11.104) with (11.106) and (11.107), we see that $\Lambda_f = \mu(f)(\Lambda(X_0(N)))$ is a free abelian subgroup of $\mathbf{C}$ of rank 2 that spans $\mathbf{C}$ over $\mathbf{R}$. Consequently Λ_f is a lattice.

By Theorem 6.14, $E' = \mathbf{C}/\Lambda_f$ is an elliptic curve over $\mathbf{C}$. Let $\eta : \mathbf{C} \to \mathbf{C}/\Lambda_f$ be the quotient homomorphism. The composition $\eta \circ \mu(f) : \mathfrak{j} \to E'$ is given by

$$\mathfrak{j} \longrightarrow \mathfrak{j}/\mathfrak{a} \cong \mathbf{C} \longrightarrow \mathbf{C}/\Lambda_f = E'$$

and has kernel $\mu(f)^{-1}(\Lambda_f)$, which equals $\mathfrak{a} + \Lambda(X_0(N))$ by (11.104) and (11.105). Since $\mathfrak{j} \to J$ is a covering homomorphism with kernel $\Lambda(X_0(N))$, $\eta \circ \mu(f)$ factors through $\mathfrak{j} \to J$. Let us say

$$\eta \circ \mu(f) = \varepsilon \circ (\mathfrak{j} \to J)$$

with $\varepsilon : J \to E'$ a holomorphic homomorphism with kernel the image of $\mathfrak{a} + \Lambda(X_0(N))$ under $\mathfrak{j} \to J$, namely A. By Theorem 11.72, ε is a morphism defined over $\mathbf{C}$. Since $\ker \varepsilon = A$, the universal mapping property of (E, ν) in Proposition 11.69 implies that $\varepsilon = \varepsilon' \circ \nu$ for a morphism $\varepsilon' : E \to E'$ defined over $\mathbf{C}$. The kernel of ε' is trivial. Combining (11.85) with our results on dual isogenies, we see that ε' is an isomorphism over $\mathbf{C}$. This completes the proof of (d) and all of the characteristic 0 part of Theorem 11.74 except for Lemma 11.76.

PROOF OF LEMMA 11.76. Because of the universal mapping property of J, it is enough to prove that $T^\#(n)$ is defined over $\mathbb{Q}$. Here $T^\#(n)$ is given by

$$T^\#(n)(\tau) = \sum_{i=1}^K \int_{\tau_0}^{\alpha_i \tau} \mathbf{f}(\zeta)\, d\zeta, \qquad \tau \in X_0(N), \qquad (11.108)$$

where $\mathbf{f} = \begin{pmatrix} f_1 \\ \vdots \\ f_g \end{pmatrix}$. (To make sense of the $\alpha_i \tau$'s in this formula, we must choose a representative of τ in $\mathcal{H}^*$, compute each $\alpha_i \tau$, and project back to $X_0(N)$. We always assume this has been done.)

Let $G = \operatorname{Aut}_{\mathbb{Q}}(\mathbb{C})$. This group operates on various things in our setting. When a member φ of G acts on a polynomial P to give P^φ, P^φ is obtained by having φ act on just the coefficients. When the polynomial gets regarded as a member of a local coordinate ring of a variety, for example, φ acts on the points of the variety and the relationship is

$$P^\varphi(x) = \varphi(P(\varphi^{-1} x)). \qquad (11.109)$$

We shall use this notion in order to localize the proof of rationality. If $(F_0, \ldots, F_m)$ gives the imbedding of J as a nonsingular variety in a projective space, we may take $F_0, \ldots, F_m$ to be defined over $\mathbb{Q}$.* Then $T^\#(n)$ is certainly defined over $\mathbb{Q}$ if we show that

$$\tau \longrightarrow H(\tau) = \left(\sum_i \int_{\tau_0}^{\alpha_i \tau} \mathbf{f}(\zeta)\, d\zeta \right) \quad \text{for } F = F_0, \ldots, F_m \qquad (11.110)$$

is in $\mathbb{Q}(X_0(N))$, i.e., that $H^\varphi = H$ for all $\varphi \in G$. In view of (11.109) it is enough to show for all $\tau \in X_0(N)$, $\varphi \in G$, and $F \in \mathbb{Q}(J)$ that

$$\varphi\left(F\left(\sum_i \int_{\tau_0}^{\alpha_i \varphi^{-1} \tau} \mathbf{f}(\zeta)\, d\zeta \right) \right) = F\left(\sum_i \int_{\tau_0}^{\alpha_i \tau} \mathbf{f}(\zeta)\, d\zeta \right) \qquad (11.111)$$

whenever F is defined at all points of the relevant finite set.

---

*As in Theorem 6.14, one version of $F_0, \ldots, F_m$ may not handle all points of J simultaneously. We may have to normalize $(F_0, \ldots, F_m)$ by one or more members of $\mathbb{Q}(J)$ to handle all points. This is not a serious matter for the present proof, and we shall ignore it.

11. ELLIPTIC CURVES CONSTRUCTED FROM $S_2(\Gamma_0(N))$

Since F is defined over $\mathbf{Q}$ and addition within J is defined over $\mathbf{Q}$, we can move φ past F and the summation sign on the left side of (11.111), and it is enough to prove that

$$\sum_i \varphi\left(\int_{\tau_0}^{\alpha_i \varphi^{-1}\tau} \mathbf{f}(\zeta)\,d\zeta\right) = \sum_i \int_{\tau_0}^{\alpha_i \tau} \mathbf{f}(\zeta)\,d\zeta \qquad (11.112)$$

for all τ and φ. Since $\Phi : X_0(N) \to J$ is defined over $\mathbf{Q}$,

$$\varphi\left(\int_{\tau_0}^{\tau'} \mathbf{f}(\zeta)\,d\zeta\right) = \int_{\tau_0}^{\varphi(\tau')} \mathbf{f}(\zeta)\,d\zeta \qquad (11.113)$$

for all τ'. Thus (11.112) comes down to proving that

$$\sum_i \int_{\tau_0}^{\varphi(\alpha_i \varphi^{-1}\tau)} \mathbf{f}(\zeta)\,d\zeta = \sum_i \int_{\tau_0}^{\alpha_i \tau} \mathbf{f}(\zeta)\,d\zeta$$

or equivalently (under the change of variables $\tau \to \varphi(\tau)$)

$$\sum_i \int_{\tau_0}^{\varphi(\alpha_i \tau)} \mathbf{f}(\zeta)\,d\zeta = \sum_i \int_{\tau_0}^{\alpha_i \varphi(\tau)} \mathbf{f}(\zeta)\,d\zeta. \qquad (11.114)$$

Fix $\tau \in X_0(N)$ and $\varphi \in G$ (together with representatives of τ and $\varphi(\tau)$ in $\mathcal{H}^*$, called by the same names). We shall prove for a suitable permutation $i \to l(i)$ that

$$\varphi(\alpha_i \tau) = \alpha_{l(i)} \varphi(\tau) \qquad \text{for all } i, \qquad (11.115)$$

and this will prove (11.114) and the lemma.

Let h be an arbitrary member of $\mathbf{Q}(j, j_N)$ that is defined at every point of a finite set of points of interest in $X_0(N)$ (including all points in (11.115)). We form the polynomial

$$\prod_{i=1}^K (X - h \circ \alpha_i) = X^K + h_{K-1} X^{K-1} + \cdots + h_0. \qquad (11.116)$$

If γ is in $\Gamma_0(N)$, write $\alpha_i \gamma = \gamma_i \alpha_{j(i)}$ for a permutation $i \to j(i)$. Then

$$\prod_{i=1}^K (X - h \circ \alpha_i \circ \gamma) = \prod_{i=1}^K (X - h \circ \alpha_{j(i)}) = \prod_{i=1}^K (X - h \circ \alpha_i),$$

from which it follows that $h_{K-1},\ldots,h_0$ in (11.116) are invariant under $\Gamma_0(N)$. By the same argument as in Proposition 9.5, all of $h_{K-1},\ldots,h_0$ are in $A_0(\Gamma_0(N))$. From §6 we know we can view these functions as in $K(X_0(N))$, which Theorem 11.33 identifies as $\mathbf{C}(j,j_N)$. We shall prove they are in $\mathbf{Q}(j,j_N)$.

Let $h(\tau) = \sum_{n=-M}^{\infty} c_n q^n$ be the q expansion of h at ∞, with $q = e^{2\pi i \tau}$. Since h is in $\mathbf{Q}(j,j_n)$, the c_n's are in $\mathbf{Q}$. Let $\alpha_i = \begin{pmatrix} a & b \\ 0 & d \end{pmatrix}$, and put $q_d = e^{2\pi i \tau/d}$ and $\zeta_d = e^{2\pi i/d}$. Then

$$h \circ \alpha_i(\tau) = h\left(\frac{a\tau + b}{d}\right) = \sum_{n=-M}^{\infty} c_n \zeta_d^b q_d^q.$$

Expanding out the left side of (11.116), we see that the coefficients in the q expansions of $h_{K-1},\ldots,h_0$ are in $\mathbf{Q}(\zeta_N)$. Now we can argue as in the last paragraph of the proof of Theorem 11.32 to see that the coefficients for $h_{K-1},\ldots,h_0$ are in fact in $\mathbf{Q}$. Applying Corollary 11.50, we conclude that $h_{K-1},\ldots,h_0$ are in $\mathbf{Q}(j,j_N)$.

Let us evaluate (11.116) at τ:

$$\prod_{i=1}^{K}(X - h(\alpha_i \tau)) = X^K + h_{K-1}(\tau)X^{K-1} + \cdots + h_0(\tau).$$

Since h is in $\mathbf{Q}(j,j_N)$, we have $h^{\varphi^{-1}} = h$. Thus $h(\tau') = \varphi^{-1}(h(\varphi(\tau')))$, and

$$\prod_{i=1}^{K}(X - \varphi^{-1}(h(\varphi(\alpha_i \tau)))) = X^K + h_{K-1}(\tau)X^{K-1} + \cdots + h_0(\tau).$$

At this stage we are dealing with a polynomial over $\mathbf{C}$, and it makes sense to apply φ to both sides (by applying it to the coefficients). We obtain

$$\prod_{i=1}^{K}(X - h(\varphi(\alpha_i \tau))) = X^K + \varphi(h_{K-1}(\tau))X^{K-1} + \cdots + \varphi(h_0(\tau))$$

$$= X^K + h_{K-1}(\varphi(\tau))X^{K-1} + \cdots + h_0(\varphi(\tau)),$$

since $h_{K-1},\ldots,h_0$ are in $\mathbf{Q}(j,j_N)$ and hence are fixed by φ. The above expression is

$$= \prod_{i=1}^{K}(X - h(\alpha_i \varphi(\tau)))$$

by (11.116). Consequently

$$h(\varphi(\alpha_i \tau)) = h(\alpha_{l(i)} \varphi(\tau)) \tag{11.117}$$

for a permutation $i \to l(i)$ depending on h (as well as τ and φ). We need to obtain (11.117) with a permutation that is independent of h.

For the algebra of h's in $\mathbb{Q}(j, j_N)$ under study, let

$$S_h = \{\text{permutations } i \to l(i) \mid (11.117) \text{ holds for } h\}.$$

If $\{h_1, \ldots, h_m\}$ is a finite set of such h's, then for each $c \in \mathbb{Q}$ there is a permutation that works for

$$h_1 + ch_2 + c^2 h_3 + \cdots + c^{m-1} h_m.$$

Since $\mathbb{Q}$ is infinite, there are m values of c, say $c_1, \ldots, c_m$, that work with some common permutation $l(i)$. Then we have

$$\begin{pmatrix} 1 & c_1 & \cdots & c_1^{m-1} \\ 1 & c_2 & \cdots & c_2^{m-1} \\ \vdots & & & \\ 1 & c_m & \cdots & c_m^{m-1} \end{pmatrix} \begin{pmatrix} h_1(\varphi(\alpha_i(\tau))) - h_1(\alpha_{l(i)}\varphi(\tau)) \\ h_2(\varphi(\alpha_i(\tau))) - h_2(\alpha_{l(i)}\varphi(\tau)) \\ \vdots \\ h_m(\varphi(\alpha_i(\tau))) - h_m(\alpha_{l(i)}\varphi(\tau)) \end{pmatrix} = \begin{pmatrix} 0 \\ 0 \\ \vdots \\ 0 \end{pmatrix}.$$

Since the Vandermonde matrix on the left is nonsingular, we see that $l(i)$ is in $S_{h_1} \cap \cdots \cap S_{h_m}$.

Thus for every finite set $\{h_1, \ldots, h_m\}$, $S_{h_1} \cap \cdots \cap S_{h_m}$ is nonempty. Choose a finite set $\{h_1, \ldots, h_M\}$ for which $S_{h_1} \cap \cdots \cap S_{h_M}$ has the smallest possible cardinality. Then we have

$$\bigcap_{\text{all } h} S_h = S_{h_1} \cap \cdots \cap S_{h_M} \neq \emptyset.$$

For any member of the left side, (11.115) is valid for all h. This completes the proof of the lemma.

12. Match of L Functions

The proof of Theorem 11.74 is now complete except for part (e), which says that the L functions of E and f coincide as Euler products except possibly at finitely many primes. This is the characteristic p part of the proof. Carrying out the details would involve too much further preparation in algebraic geometry, and we shall settle for some discussion.

If p is one of the primes for which (e) is asserted, the idea is to prove an identity of the type
$$\tilde{T}_p = \phi + \hat{\phi} \tag{11.118}$$
in $X_0(N)$, where $\tilde{T}_p$ is a reduction modulo p of the operator $T(p)$ of (11.95) and where ϕ is the Frobenius map.

Before trying to make sense of (11.118) in general, let us consider the special case where $X_0(N)$ has genus 1. (The first such example is when $N = 11$.) Then we can identify $X_0(N)$ with the elliptic curve E produced by Theorem 11.74. From §V.2 it is meaningful to speak of E_p, the reduction of E modulo p. In E_p, (11.118) becomes an identity within $\mathrm{End}(E_p)$, with $\tilde{T}_p$ acting as $[c_p]$ in accordance with (b) of the theorem. Also within $\mathrm{End}(E_p)$ we have

$$\begin{aligned}[][\#E(\mathbb{Z}_p)] &= [\deg([1] - \phi)] &&\text{by (11.88)} \\ &= ([1] - \hat{\phi}) \circ ([1] - \phi) &&\text{by Proposition 11.65} \\ &= [1] - (\phi + \hat{\phi}) + \hat{\phi} \circ \phi \\ &= [1] - (\phi + \hat{\phi}) + [p], \end{aligned}$$

the last equality holding by item (c) after (11.82) and by Proposition 11.65 again. Thus
$$\phi + \hat{\phi} = [p + 1 - \#E(\mathbb{Z}_p)] = [a_p],$$
in the notation of (10.7).

In other words, (11.118) as an identity within $\mathrm{End}(E_p)$ says that $[c_p] = [a_p]$. It can be shown that $\mathrm{End}(E_p)$ has no zero divisors, and hence $c_p = a_p$. This equality says that the pth factor of $L(s, f)$ coincides with the pth factor of $L(s, E)$.

It is not hard to imagine a similar interpretation of (11.118) within a reduction modulo p of J in the case of general $X_0(N)$. The identity passes to E_p, and we get the same conclusion.

In fact, reduction modulo p of J was too technically difficult to be the method that was used. Instead (11.118) was proved in $X_0(N)$, and then some consequences were transferred to J and E; the identity itself was not transferred. We shall not pursue this transfer of consequences.

The meaning that is customarily attached to (11.118) is as an identity of "correspondences" on $X_0(N)_p$, the reduction modulo p of $X_0(N)$

12. MATCH OF L FUNCTIONS

(which we have not defined in genus > 1). Intuitively a correspondence is like the graph of a function, except that the function need not be single-valued, and one must be careful in counting multiplicities. More precisely but still not exactly, the building blocks of correspondences in $X_0(N)_p$ are irreducible subvarieties of dimension 1 in $X_0(N)_p \times X_0(N)_p$ (curves, in other words), and a **correspondence** is a divisor-like **Z** combination of such irreducible curves. Both $\tilde{T}_p$ and ϕ are to be regarded as correspondences, and $\hat{\phi}$ refers to the correspondence ϕ with the two coordinates interchanged. The sum on the right side of (11.118) is the sum in the sense of divisors. We shall not go into the proof of (11.118) but shall end our discussion of the characteristic p part of the proof at this point.

CHAPTER XII

TANIYAMA-WEIL CONJECTURE

1. Relationships among Conjectures

Throughout this chapter, E will denote an ellliptic curve defined over $\mathbf{Q}$. By Corollary 10.6 the L function $L(s,E)$ converges for $\operatorname{Re} s > \frac{3}{2}$ and is given there by an absolutely convergent Dirichlet series. It is expected that deep arithmetic information is encoded in the behavior of $L(s,E)$ beyond the region of convergence. For example, the Birch and Swinnerton-Dyer Conjecture (Conjecture 1.10) predicts that the rank of $E(\mathbf{Q})$ is the order of vanishing of $L(s,E)$ at $s=1$.

To make sense of such conjectures, we must expect that $L(s,E)$ has an analytic continuation. In view of the behavior of Dirichlet series that are better understood, it is natural to expect also that $L(s,E)$ will satisfy a functional equation. Here is a reasonably precise formulation.

Conjecture 12.1. The function $L(s,E)$ extends to be entire, and for a suitable sign ε and a suitable positive integer N, $L(s,E)$ satisfies a functional equation given in terms of

$$\Lambda(s,E) = N^{s/2}(2\pi)^{-s}\Gamma(s)L(s,E) \tag{12.1}$$

by

$$\Lambda(s,E) = -\varepsilon\Lambda(2-s,E). \tag{12.2}$$

Hasse had the further idea that $L(s,E)$ should have as good transformation properties as any other reasonable Dirichlet series. In the case of $L(s,f)$ for a cusp form $f \in S_k(\Gamma_0(N))$ in one of the eigenspaces $S_k^\varepsilon(\Gamma_0(N))$ of w_N, there is more that can be said beyond Theorem 9.8. For one thing, by examining (9.27), (9.36), and (9.37), we see that the entire function

$$\Lambda(s,f) = N^{s/2}(2\pi)^{-s}\Gamma(s)L(s,f) \tag{12.3}$$

is actually bounded in vertical strips. For another thing, let us consider some modifications of $L(s,f)$. Write

$$L(s,f) = \sum_{n=1}^{\infty} \frac{c_n}{n^s}, \tag{12.4a}$$

1. RELATIONSHIPS AMONG CONJECTURES

let m be an integer relatively prime to N, let χ be a primitive Dirichlet character modulo m, and let

$$L(s,f,\chi) = \sum_{n=1}^{\infty} \frac{c_n \chi(n)}{n^s}. \tag{12.4b}$$

Define

$$\Lambda(s,f,\chi) = (m^2 N)^{s/2}(2\pi)^{-s}\Gamma(s)L(s,f,\chi). \tag{12.4c}$$

Recall from (7.46) the **Gauss sum**

$$c(m,\chi) = \sum_{l=0}^{m-1} e^{2\pi i l/m}\chi(l). \tag{12.5}$$

The following theorem is proved by combining the techniques for Theorem 7.19 and Theorem 9.8.

Theorem 12.2. Let f be in $S_k^\varepsilon(\Gamma_0(N))$, let $\mathrm{GCD}(m,N) = 1$, let χ be a primitive Dirichlet character modulo m, and let $L(s,f,\chi)$ and $\Lambda(s,f,\chi)$ be defined by (12.4). Then $L(s,f,\chi)$ is initially defined for $\mathrm{Re}\, s > \frac{k}{2}+1$ and extends to be entire in s. Moreover, $\Lambda(s,f,\chi)$ is entire and bounded in vertical strips, and it satisfies the functional equation

$$\Lambda(s,f,\chi) = \varepsilon(-1)^{k/2} \frac{c(m,\chi)\chi(-N)}{c(m,\bar\chi)} \Lambda(k-s,f,\bar\chi),$$

where $c(m,\chi)$ is the Gauss sum (12.5).

The Hasse-Weil Conjecture expects $L(s,E)$ to share all these properties. It is given in terms of $L(s,E) = \sum_{n=1}^{\infty} \frac{c_n}{n^s}$ and its modifications $L(s,E,\chi) = \sum_{n=1}^{\infty} \frac{c_n \chi(n)}{n^s}$.

Conjecture 12.3 (Hasse-Weil). The function $L(s,E)$ extends to be entire and, for a certain positive integer N, so does $L(s,E,\chi)$ for every Dirichlet character whose conductor m is prime to N. Moreover, the modified functions

$$\Lambda(s,E) = N^{s/2}(2\pi)^{-s}\Gamma(s)L(s,E)$$
$$\Lambda(s,E,\chi) = (m^2 N)^{s/2}(2\pi)^{-s}\Gamma(s)L(s,E,\chi)$$

extend to be entire and, for a suitable sign ε, satisfy the functional equations

$$\Lambda(s, E) = -\varepsilon\Lambda(2 - s, E)$$

$$\Lambda(s, E, \chi) = -\frac{\varepsilon c(m,\chi)\chi(-N)}{c(m,\bar{\chi})}\Lambda(2 - s, E, \bar{\chi}). \quad (12.6)$$

Weil addressed the question how close $L(s, E)$ is to coming from a modular form if it satisfies functional equations as in Conjecture 12.1 or 12.3. Suppose that $L(s, E)$, adjusted by factors as in (12.1), is really the Mellin transform of $f(i\sigma)$, where f is an analytic function on the upper half plane $\mathcal{H}$ transforming under a group Γ of linear fractional transformations. The fact that we are obtaining $L(s, E)$ from a Mellin transform incorporates a transformation law under the translation corresponding to $T = \begin{pmatrix} 1 & 1 \\ 0 & 1 \end{pmatrix}$. The functional equation (12.2) incorporates a transformation law under some inversion element, such as the one from $\alpha_N = \begin{pmatrix} 0 & -1 \\ N & 0 \end{pmatrix}$. Conjecture 12.1 does not give any reason for thinking that Γ is any larger than the group Γ' generated by the transformations corresponding to T and α_N. Unfortunately $\Gamma'\backslash\mathcal{H}^*$ is noncompact and not close to compact. So Γ' does not give us much control over an analytic function on $\mathcal{H}$.

Weil found, however, that the additional functional equations (12.6) are enough to force matters to be controlled by $\Gamma_0(N)$. The statement of the Weil Converse Theorem is as follows. We shall not give the proof.

Theorem 12.4 (Weil). Let $L(s) = \sum_{n=1}^{\infty}\frac{c_n}{n^s}$ be a Dirichlet series with $c_n = O(n^c)$ for some $c > 0$. Fix a positive integer N, an even positive integer k, and a sign ε. Suppose that

(a) the function

$$\Lambda(s) = N^{s/2}(2\pi)^{-s}\Gamma(s)L(s)$$

is entire, is bounded in every vertical strip, and satisfies

$$\Lambda(s) = \varepsilon(-1)^{k/2}\Lambda(k - s),$$

(b) for every integer m with $\mathrm{GCD}(m, N) = 1$ and every primitive Dirichlet character χ modulo m, the modified function

$$L_\chi(s) = \sum_{n=1}^{\infty}\frac{c_m\chi(n)}{n^s}$$

1. RELATIONSHIPS AMONG CONJECTURES

is such that the function

$$\Lambda_\chi(s) = (m^2 N)^{s/2} (2\pi)^{-s} \Gamma(s) L_\chi(s)$$

is entire, is bounded in every vertical strip, and satisfies

$$\Lambda_\chi(s) = \varepsilon(-1)^{k/2} \frac{c(m,\chi)\chi(-N)}{c(m,\bar\chi)} \Lambda_{\bar\chi}(k-s),$$

(c) the series defining $L(s)$ converges absolutely at $s = k - \delta$ for some $\delta > 0$.

Then

$$f(\tau) = \sum_{n=1}^{\infty} c_n e^{2\pi i n \tau}$$

is a cusp form in $S_k(\Gamma_0(N))$.

In fact, it is not necessary to assume (b) for quite so many m's, but this improvement will not concern us. Theorems 12.2 and 12.4 show that the Hasse-Weil Conjecture is completely equivalent with the following fundamental conjecture.

Conjecture 12.5. If E is given, then there exist a positive integer N and a sign ε such that the function $L(s, E)$ equals $L(s, f)$ for some eigenform f in $S_2^\varepsilon(\Gamma_0(N))$.

The notion of newform was not known at the time of Weil's work, but we can freely substitute the word "newform" for "eigenform" in Conjecture 12.5. In fact, if f is an eigenform for $S_2^\varepsilon(\Gamma_0(N))$, it comes from a newform for some $S_2^\varepsilon(\Gamma_0(N/M))$ with M dividing N. If $M \neq 1$, then Theorem 9.8 gives two contradictory functional equations that f must satisfy. Thus $M = 1$, and f is a newform.

Conjecture 12.5 is consistent with an important theme that has been discussed on more than one occasion earlier in this book: the identification of geometric L functions as automorphic L functions. The conjecture raises two immediate questions: What are N and ε? Is there a deeper relationship between E and f?

Eichler-Shimura theory gives part of the answer to the second question. Starting from any normalized newform f in $S_2(\Gamma_0(N))$ whose Fourier coefficients are integers, the theory provides us with a mapping from $X_0(N)$ to a canonical elliptic curve over $\mathbb{Q}$. Even before this theory, Taniyama had suggested looking for such a mapping in connection with an arbitrary elliptic curve over $\mathbb{Q}$, but Taniyama's suggestion was apparently initially overlooked. Weil made a similar suggestion as the Eichler-Shimura theory unfolded.

Conjecture 12.6 (Taniyama-Weil). If E is given, then there exists an integer N for which there is a nonconstant morphism $F : X_0(N) \to E$.

We say that such an E has a **modular parametrization** of level N, and, following, custom, we call E a **Weil curve**. If E is given by the usual equation (3.23), then it is equivalent to say that there are functions $\xi(\tau)$ and $\eta(\tau)$ in $A_0(\Gamma_0(N))$ whose q expansions at ∞ have rational coefficients and are such that $(x, y) = (\xi(\tau), \eta(\tau))$ satisfies (3.23b) for all τ where neither ξ nor η has a singularity. The passage from F to (ξ, η) is just a matter of unwinding the definitions; the passage from (ξ, η) to F requires Corollary 11.50.

The modular parametrization in the Eichler-Shimura theory has additional properties. From a newform f of level N, Theorem 11.74 produces a morphism from $X_0(N)$ to an elliptic curve E over $\mathbf{Q}$ such that the pullback of the invariant differential is a multiple of $f(\tau)\,d\tau$ and the period lattice of $\int f(\tau)\,d\tau$ is equivalent with the period lattice of E. If E' is isogenous to E over $\mathbf{Q}$, one can compose the Eichler-Shimura map with the isogeny to get a modular parametrization of E', but the match of the period lattices may be lost.

We saw in an example following Corollary 11.50 that there is a morphism from $X_0(MN)$ to $X_0(N)$. The Eichler-Shimura mapping to a curve E can be composed with such maps to produce modular parametrizations of E of nonminimal level.

Theorem 12.7. If E has a modular parametrization of level N but of no level M for $M < N$, then E comes via the Eichler-Shimura theory from the map defined by a normalized newform in $S_2(\Gamma_0(N))$, followed by an isogeny defined over $\mathbf{Q}$.

In constructing an elliptic curve E from a normalized newform $f \in S_2(\Gamma_0(N))$, the Eichler-Shimura theory shows that the pth L factor of E agrees with the pth L factor of f for all but finitely many p, and Igusa's work shows that any exceptional p must divide N. There are two loose ends—to identify N and to handle the exceptional p's.

There was an early suspicion of what N was. The **conductor** N of E is an isogeny invariant of E with a cohomological definition, but it can be described almost completely in a simple way as follows. First let us adjust E so that it is given by a global minimal Weierstrass equation (§X.1). Then let Δ be the discriminant of E. The conductor N divides Δ and has the same prime factors as Δ. The power of a prime dividing N is 1 if and only if E_p has a node. If $p > 3$, then the power of p dividing N is 2 if and only if E_p has a cusp. For the case of a cusp with $p = 2$ or

1. RELATIONSHIPS AMONG CONJECTURES

3, the conductor can be computed by an algorithm due to Tate. Those curves in Tables 3.1 and 3.2 whose conductor is not obviously $|\Delta|$ are listed in Table 12.1.

Elliptic Curve	Δ	N
$y^2 + y = x^3$	-27	27
$y^2 + xy - y = x^3$	-28	14
$y^2 = x^3 - x^2 + x$	-48	24
$y^2 = x^3 + x^2 + x$	-48	48
$y^2 + 3xy - y = x^3$	-54	54
$y^2 + xy = x^3 + x$	-63	21
$y^2 = x^3 + x$	-64	64
$y^2 = x^3 - x$	64	32
$y^2 = x^3 - 2x + 1$	80	40
$y^2 = x^3 - 2x - 1$	80	80
$y^2 = x^3 + x^2 - x$	80	20
$y^2 = x^3 - x^2 - x$	80	80
$y^2 + 5xy + y = x^3$	98	14

TABLE 12.1. Conductors of some elliptic curves

The following theorem ties up both loose ends in the Eichler-Shimura theory.

Theorem 12.8 (Carayol). Let $f \in S_2(\Gamma_0(N))$ be a normalized newform whose Fourier coefficients are integers, and let E be the elliptic curve over $\mathbb{Q}$ associated to f by Eichler-Shimura theory. Then $L(s, E) = L(s, f)$, and N is the conductor of E.

It follows that the Taniyama-Weil Conjecture implies the Hasse-Weil Conjecture. In fact, if E is given, we choose N to be the smallest positive integer such that E has a modular parametrization of level N. Theorem 12.7 says that E comes from the map defined by a normalized newform f in $S_2(\Gamma_0(N))$, possibly followed by an isogeny over $\mathbb{Q}$. Since isogeny over $\mathbb{Q}$ does not affect the L function (Theorem 11.67), Theorem 12.8 says that $L(s, E) = L(s, f)$. This is the conclusion of Corollary 12.5, which we have seen is equivalent with the Hasse-Weil Conjecture.

We can come close to proving that the Hasse-Weil Conjecture implies the Taniyama-Weil Conjecture. Namely let E be given with conductor N. By the Hasse-Weil Conjecture in the form of Conjecture 12.5, there is a normalized newform f in $S_2(\Gamma_0(N'))$ such that $L(s,E) = L(s,f)$. By the Eichler-Shimura construction and Theorem 12.8, f leads to an elliptic curve E' of conductor N' with $L(s,E') = L(s,f)$. Since $L(s,E') = L(s,E)$, a theorem of Serre allows us to conclude that E and E' are isogenous over $\mathbb{Q}$ provided $j(E)$ is not an integer. In this case the modular parametrization for E yields one for E' by composition.

2. Strong Weil Curves and Twists

Let E and E' be Weil curves, and let $F : X_0(N) \to E$ and $F' : X_0(N) \to E'$ be modular parametrizations, with N minimal in both cases. Without loss of generality, we suppose that ∞ of $X_0(N)$ maps to the group identity of E and E'. We say that (E, F) **dominates** (E', F') if F' factors through F, i.e., if $F' = \varphi \circ F$ for an isogeny φ. A Weil curve E is a **strong Weil curve** if (E, F) is maximal in this ordering for some F. (The F yielding maximality will not be unique, as it can always be composed with -1 in E.)

Proposition 12.9. In every isogeny class of Weil curves, there is a strong Weil curve and it is unique up to isomorphism. The strong Weil curve and map (E, F) are characterized within the isogeny class by either of the following conditions:

(a) the induced map F_* on homology $H_1(X_0(N), \mathbb{Z}) \to H_1(E, \mathbb{Z})$ is onto.

(b) when F is factored through the Jacobian variety as $F_1 \circ \Phi$, the kernel of F_1 is a (connected) variety.

EXAMPLE. In §XI.1, $X_0(11)$ led to three Weil curves, of which the first two were
$$E: \quad y^2 + y = x^3 - x^2 - 10x - 20 \qquad (12.7)$$
and
$$E': \quad y^2 + y = x^3 - x, \qquad (12.8)$$
defined by respective lattices Λ and Λ' in $\mathbb{C}$. The lattice Λ was defined as the image of the cycles of $X_0(N)$. Hence $H_1(X_0(N), \mathbb{Z})$ maps onto $H_1(E, \mathbb{Z})$, and (12.7) is a strong Weil curve. We identified E' by using the sublattice Λ', which does not include the image of the cycles of $X_0(N)$, and then used a dual isogeny to map E to E'. It would have amounted to the same thing if we had used $\frac{1}{5}\Lambda'$, which properly contains Λ, to define E'. In this case the cycles of $X_0(N)$ do have image in $\frac{1}{5}\Lambda'$, but the homology map is not onto. Thus E' in (12.8) is not a strong Weil curve.

2. STRONG WEIL CURVES AND TWISTS

The Eichler-Shimura construction uses the full image of cycles of $X_0(N)$ as the set of periods to define an elliptic curve, by Theorem 11.74d. Thus it always leads directly to a strong Weil curve.

What we see computationally from the Eichler-Shimura construction is a lattice in $\mathbf{C}$, from which we can compute the j-invariant approximately, as in §XI.1. Even if we assume that our approximation to the j-invariant is an approximation to a particular rational number, we have not completely solved the problem of identifying the strong Weil curve as a particular curve over $\mathbf{Q}$. There remains the problem of distinguishing a curve over $\mathbf{Q}$ from its twists, as we saw by example in §XI.1. We now take up the question of twists more generally.

Two elliptic curves E and E' over $\mathbf{Q}$ are called **twists** of each other if $j(E) = j(E')$. We shall skip over the cases that the common value of j is 0 or 1728, which require special treatment. We may assume that E and E' are in global minimal form (3.23b) with respective coefficients $a_1, \ldots, a_6$ and $a_1', \ldots, a_6'$ as usual. Since j is not 0 or 1728, none of c_4, c_4', c_6, c_6' is equal to 0. Put $v = \Delta/\Delta' \in \mathbf{Q}$. Since $j(E) = j(E')$ and

$$j(E) = \frac{c_4^3}{\Delta} = 1728 + \frac{c_6^2}{\Delta},$$

we see that

$$v = \frac{\Delta}{\Delta'} = \frac{c_4^3}{c_4'^3} = \frac{c_6^2}{c_6'^2}.$$

It follows that $v = u^6$ for some $u = \dfrac{r}{s}$ in $\mathbf{Q}$ with $\mathrm{GCD}(r, s) = 1$. Choosing s as > 0 and the sign of r to match the sign of c_6/c_6', we have

$$(s^2 c_4, s^3 c_6, s^6 \Delta) = (r^2 c_4', r^3 c_6', r^6 \Delta'). \tag{12.9}$$

By Lemma 10.1 and Theorem 10.3, neither r nor s has any prime square factor p^2 with $p > 3$. Conversely if E and E' are related as in (12.9), then $j(E) = j(E')$.

Proposition 12.10. Let E and E' be twists of each other, related as in (12.9), and let their L functions have respective coefficients a_p and a_p'. If p is a prime not dividing $6\Delta\Delta'$, then

$$a_p' = \left(\frac{rs}{p}\right) a_p,$$

where $\left(\dfrac{rs}{p}\right)$ is the Legendre symbol.

PROOF. Let N_p and N_p' be the respective numbers of *affine* solutions modulo p. In view of (10.7), we are to show that $N_p = N_p'$ if $\left(\dfrac{rs}{p}\right) = 1$ and that $N_p + N_p' = 2p$ if $\left(\dfrac{rs}{p}\right) = -1$. Since p is prime to 6, we may take the two equations to be

$$y^2 = x^3 - 27c_4 x - 54c_6 \qquad (12.10a)$$
$$y^2 = x^3 - 27c_4' x - 54c_6'. \qquad (12.10b)$$

Suppose $\left(\dfrac{rs}{p}\right) = 1$. Choose $a \in \mathbb{Z}$ with $ra^2 \equiv s \bmod p$. If (x_0, y_0) solves (12.10a) modulo p, then

$$y_0^2 = x_0^3 - 27\left(\frac{r}{s}\right)^2 c_4' x_0 - 54\left(\frac{r}{s}\right)^3 c_6'$$

shows that

$$(sar^{-1} y_0)^2 = (sr^{-1} x_0)^3 - 27c_4'(sr^{-1} x_0) - 54c_6',$$

i.e., that $(sr^{-1} x_0, sar^{-1} y_0)$ solves (12.10b) modulo p. Thus $N_p \leq N_p'$, and similarly $N_p' \leq N_p$.

Now suppose that $\left(\dfrac{rs}{p}\right) = -1$. In the same way, we check, for each fixed x_0, that the sum of the number of solutions (x_0, y) of (12.10a) and the number of solutions $(sr^{-1} x_0, y)$ of (12.10b) is 2. Summing on x_0, we obtain $N_p + N_p' = 2p$.

By quadratic reciprocity, Proposition 12.10 says that $L(s, E')$ in almost all factors equals $L(s, E, \chi)$ for a quadratic Dirichlet character modulo $4rs$. Taking Theorems 12.2 and 12.4 into account, we see that Proposition 12.10 almost proves that the question whether E is a Weil curve depends only on $j(E)$.

3. Computation of Equations of Weil Curves

How can we check definitively whether a particular elliptic curve is a Weil curve? And how can we write down equations for the Weil curves that arise from $X_0(N)$? Let us first address these questions for strong Weil curves.

3. COMPUTATION OF EQUATIONS OF WEIL CURVES

The classical method of proof is to exhibit $\xi(\tau)$ and $\eta(\tau)$ in $A_0(\Gamma_0(N))$ such that $(\xi(\tau), \eta(\tau))$ parametrizes the curve. This method requires constructing some cusp forms, especially of weight 2, and manipulating identities to get the right results. Constructing cusp forms of weight 2 is a problem in its own right that we address below. In any event, the classical method has been carried out in genus 1 and for some higher cases, particularly in genus 2. It becomes really tedious as the genus increases.

g	N
0	$1 \leq N \leq 10$; 12, 13, 16, 18, 25
1	11, 14, 15, 17, 19, 20, 21, 24, 27, 32, 36, 49
2	22, 23, 26, 28, 29, 31, 37, 50

TABLE 12.2. Curves $X_0(N)$ of low genus

In the case that the conductor is of the form $N = 2^a 3^b$, one knows all possible elliptic curves by classifying all relevant integer solutions of $1728\Delta = c_4^3 - c_6^2$. All such elliptic curves are Weil curves. To prove this, we start with an elliptic curve of conductor N, we compute many coefficients a_p from (10.7), and we check that they correspond to the first part of the q expansion of a newform in $S_2(\Gamma_0(N))$ (see below). Then using the method of §XI.1, we can compute to several decimal places the lattice in $\mathbb{C}$ generated by images of cycles of $X_0(N)$. Next we can compute the period lattice of the given elliptic curve as in Chapter VI, using the Gauss arithmetic-geometric mean, and we can use some $\mathbb{C}^\times$ multiple of it to organize the information about the lattice in $\mathbb{C}$.

Once we know a basis for the images of cycles of $X_0(N)$, we can compute g_2, g_3, and j as in §XI.1. If the computed j does not match j for the elliptic curve we have in mind, then the elliptic curve that we have in mind is not a strong Weil curve. But it may still be an ordinary Weil curve, and a method for approaching that question will be discussed later.

The difficulty now from the point of view of proof is in identifying the computed j as an exact rational number. We have it only to several decimal places. But when $N = 2^a 3^b$, the classification of curves of conductor N has bounded the denominator of j, which is Δ. Thus it tells us the value of j as a rational number, provided we have computed enough decimal places.

In the case of conductor $N = 2^a 3^b$ we can therefore tell exactly whether our computed j matches j for the given elliptic curve. If so, the curves are twists of one another. Since the discriminant can have only 2 and 3 as prime factors, the set of possible twists is quite limited. Using Proposition 12.10, we can check for the correct Fourier coefficients.

Recent progress allows use of a similar technique in the case of prime conductor, and it should be expected that further progress will expand the method. The following theorem bounds the denominator of j in the case of prime conductor and allows us to proceed as above.

Theorem 12.11 (Mestre-Oesterlé). If a Weil curve E is in global minimal form and its conductor N is a prime, then the discriminant Δ of E divides N^5.

The problem of writing down all ordinary Weil curves coming from $X_0(N)$, with no side information, is a hard one. The problem is that it is easy to miss a member of an isogeny class. Instead let us address the question of proving that a particular curve E of conductor N is a Weil curve. We start as above by computing many coefficients a_p from (10.7) and checking that they come from an initial segment of a newform in $S_2(\Gamma_0(N))$. We form the lattice of E and the lattice from $X_0(N)$. Even if they do not match (after some multiplication by $\mathbb{C}^\times$), it may happen that the lattice from E corresponds to a sublattice of the images of cycles of $X_0(N)$. Let n be the index of the lattice from E in the lattice from $X_0(N)$, and let E' be the strong Weil curve. The goal is to find an isogeny of degree n from E to E' or from E' to E. We shall not discuss the details of how to proceed, but there are formulas that help in the effort.

Success in proving that particular elliptic curves are Weil curves depends on being able to produce initial segments of q expansions of newforms. Although the methods of §IX.3 can be a little helpful, ultimately one has to come to grips with an actual construction of newforms. For low conductor the standard technique is to compute the action of the Hecke operators on $H_1(X_0(N), \mathbb{Z})$ and then on $H_1^+(X_0(N), \mathbb{C})$. (See §XI.5.) Proposition 11.24b yields matrices representing $T_2(n)$ in some basis. Computation of a basis of $H_1(X_0(N), \mathbb{Z})$ takes some time but is manageable.

A second technique is to produce initial segments of q expansions of eigenforms by using trace formulas for the Hecke operators. Once the trace is known for $T_2(n)$, one can compute the characteristic polynomial and then the eigenvalues; from the eigenvalues, one can piece together the q expansions of the eigenforms.

To obtain the characteristic polynomial of $T_2(n)$, we use an observation from linear algebra. If L is a linear transformation on an r dimensional complex vector space with characteristic polynomial

$$\det(XI - L) = \prod_{j=1}^{r}(X - \lambda_j) = X^r - c_{r-1}X^{r-1} + \cdots + (-1)^r c_0, \quad (12.11)$$

then c_i is the ith elementary symmetric polynomial in $\lambda_1, \ldots, \lambda_r$ and is a linear combination of the symmetric polynomials

$$\lambda_1^j + \lambda_2^j + \cdots + \lambda_r^j = \text{Tr}(L^j), \quad 1 \leq j \leq r.$$

For $L = T_2(n)$, Theorem 9.17 allows us to express $T_2(n)^j$ as a linear combination of the operators $T_2(m)$ for which $m \mid n$. Hence (12.11) can be computed from the traces of the Hecke operators.

Once we have the eigenvalues, we have to correlate the effects of the operators $T_2(n)$ as n varies. For example, if the space is 2-dimensional and $T_2(3)$ has a and b as eigenvalues and $T_2(5)$ has c and d as eigenvalues, then we can look at $T_2(10)$ to see what the simultaneous eigenvalues are. If $T_2(10)$ has eigenvalues ac and bd, then the simultaneous eigenvalues are $\{a,c\}$ and $\{b,d\}$; otherwise they are $\{a,d\}$ and $\{b,c\}$.

Everything is easier if we have a candidate for an initial segment of an eigenform, such as the corresponding coefficients of an elliptic curve. Then we just have to check that the coefficients solve (12.11) and satisfy recursions consistent with Theorem 9.17.

4. Connection with Fermat's Last Theorem

Suppose $\alpha^l + \beta^l = \gamma^l$ is a relatively prime counterexample to Fermat's Last Theorem, with l prime and ≥ 5. The elliptic curve E over $\mathbb{Q}$ given by

$$E: \quad y^2 = x(x - \alpha^l)(x - \gamma^l) \quad (12.12)$$

has

$$\Delta = 16\alpha^{2l}\beta^{2l}\gamma^{2l}$$
$$c_4 = 16(a^{2l} - \alpha^l\gamma^l + \gamma^{2l}). \quad (12.13)$$

Let us put E in global minimal from and compute its conductor N. Any odd prime p dividing Δ divides $\alpha\beta\gamma$. Since α, β, γ are relatively prime in pairs, we see that p does not divide c_4. By Lemma 10.1, E is minimal at p. If $p \mid \alpha\gamma$, we readily check that E_p has a node at $(0,0)$, and if $p \mid \beta$, we readily check that E_p has a node at $(\alpha^l, 0)$. Thus

$$N = 2^{\text{power}} \prod_{\substack{p \mid \alpha\beta\gamma, \\ p \text{ odd}}} p. \tag{12.14}$$

We still have to check $p = 2$. Since $2^4 \parallel c_4$, at most one reduction is possible. We may assume that γ is even, so that

$$\gamma^l \equiv 0 \mod 32. \tag{12.15a}$$

Since $\beta^l \equiv -\alpha^l$ mod 4 and l is odd, we may assume, possibly by interchanging α and β, that

$$\alpha^l \equiv 1 \mod 4. \tag{12.15b}$$

Putting $x = 4X$ and $y = 8Y + 4X$, we are led to

$$Y^2 + XY = X^3 + \tfrac{1}{4}(1 - \alpha^l - \gamma^l)X^2 + \tfrac{1}{16}\alpha^l\gamma^l X. \tag{12.16}$$

The congruences (12.15) show that the coefficients here are integers (so that (12.16) is a global minimal form) and that the equation modulo 2 is

$$Y^2 + XY = \begin{cases} X^3 & \text{or} \\ X^3 + X^2. \end{cases}$$

The singularity is at $(X, Y) = (0,0)$; since neither $Y^2 + XY$ nor $Y^2 + XY + X^2$ is a square, the singularity is a node.

Taking (12.14) into account, we see that the conductor is given by

$$N = \prod_{p \mid \alpha\beta\gamma} p. \tag{12.17}$$

We shall use the following theorem.

Theorem 12.12 (Ribet). Suppose E is an elliptic curve over $\mathbf{Q}$ given in global minimal form and having discriminant $\Delta = \prod_{p \mid \Delta} p^{\delta_p}$ and conductor $N = \prod_{p \mid \Delta} p^{f_p}$. Suppose further that E is a Weil curve with a modular parametrization of level N given via a normalized newform $f(\tau) = q + \sum_{n=2}^{\infty} c_n q^n$ in $S_2(\Gamma_0(N))$. Fix a prime l, and let

$$N_1 = \frac{N}{\prod_{\substack{p \text{ with} \\ f_p = 1 \\ \text{and } l \mid \delta_p}} p}. \tag{12.18}$$

Then there exists f_1 in $S_2(\Gamma_0(N_1))$ such that $f_1(\tau) = \sum_{n=1}^{\infty} d_n q^n$ with integral coefficients and with $c_n \equiv d_n$ mod l for $1 \leq n < \infty$.

4. CONNECTION WITH FERMAT'S LAST THEOREM

Corollary 12.13 (Frey-Serre-Ribet). The Taniyama-Weil Conjecture implies Fermat's Last Theorem.

PROOF. Assuming that Fermat's Last Theorem is false, we apply Theorem 12.12 to the curve E in (12.12). Comparison of (12.17) and (12.18) shows that $N_1 = 2$. The theorem produces f_1 in $S_2(\Gamma_0(2))$, which is the 0 space. Thus $d_n = 0$ for all n. But $c_1 = 1$. So $c_n \equiv d_n \bmod l$ fails for $n = 1$, and we have a contradiction.

NOTES

Chapter I

The organization of Chapter I is based on Zagier [1988] and is influenced also by the Introduction of Husemöller [1987]. References to papers before 1940, among others, may be found in Shimura [1971a] and Serre [1973a]. For Theorem 1.1, see Borevich and Shafarevich [1966]. Cubic counterexamples to the Hasse Principle were studied by Selmer [1951]. Theorem 1.4 is discussed in Chapter III.

What we have called the "Weierstrass form" of a cubic is really Tate's modification of the Weierstrass form. The first ten chapters of the book work in the context of plane projective geometry, and our definition of "elliptic curve" (as a nonsingular plane cubic curve in Weierstrass form) is appropriate for that context. In Chapters XI and XII, and usually in algebraic geometry, an elliptic curve is a nonsingular projective curve of genus one, and one proves that all such curves can be realized in Weierstrass form. See §XI.9.

Theorem 1.5 is the subject of Chapter IV, and Theorem 1.6 is the main result of Chapter V. Theorem 1.7 appears in Mazur [1977] and [1978]. For Theorem 1.8, see Chapter X. Conjectures 1.11 and 1.12, as well as Theorem 1.13, are discussed in Chapter XII.

Conjecture 1.10 appeared originally in Birch and Swinnerton-Dyer [1963-65]. For more recent numerical evidence, see Brumer and McGuinness [1990]. There are several expositions of Conjecture 1.10 in the literature. See Swinnerton-Dyer and Birch [1975], Zagier [1984], Appendix C of Silverman [1986], Chapter 17 of Husemöller [1987], and Chapter 20 of Ireland and Rosen [1990].

Chapters II-V

The standard book on elliptic curves from an arithmetic standpoint is Silverman [1986]. Three other books on this subject are Koblitz [1984], Husemöller [1987], and Charlap and Robbins [1988]. Koblitz concentrates on elliptic curves with complex multiplication, especially the curves $y^2 = x^3 - n^2 x$ that arise in connection with congruent numbers. Some books on curves are Brieskorn and Knörrer [1986], Fulton [1989], and Walker [1950]. Chapter II is based chiefly on Husemöller [1987] and Walker [1950].

The examples at the beginning of Chapter III are from Zagier [1988]. The standard notation is as in Tate [1974] and Silverman [1986]. Tables

3.1 and 3.2 are the result of a simple home-computer calculation. The idea of the proof of Theorem 3.8, as we give it, is a standard one and may be found, for example, in Husemöller [1987]. Usually the proof in this form is not carried through to completion. The discussion of singular points is based on Silverman [1986].

Most of Chapter IV is from Zagier [1988]. The examples were in Zagier's lectures. Proofs of most of the results in §9 may be found in Ireland and Rosen [1990]. Theorem 4.11 goes also under the name **Mordell-Weil Theorem**. Mordell [1922] proved the theorem as given here, and Weil [1930] extended the result to elliptic curves defined over number fields.

In Chapter V, Theorem 5.1 is due in a different form to Nagell [1935] and in essentially the form here to Elisabeth Lutz [1937]. Our treatment is similar to that of Lutz, as modified in Chapter 5 of Husemöller [1987]. The examples are based on Silverman [1986] and Husemöller [1987]. Theorems 5.2 and 5.3 are due to Nagell [1935] and Fueter [1930], respectively. Our attention was brought to these results by Lichtenbaum [1960], and we give Lichtenbaum's proofs. The material of §5 is based on Husemöller [1987]; Husemöller credits H. P. Kraft, F. Knopp, G. Menzel, and E. Senn. See also Exercise 8.12 of Silverman [1986].

Chapter VI

The material in §§1-3 is fairly standard and can be found in many places. For example, there are individual chapters in Lang [1987], Silverman [1986], and Husemöller [1987] on this topic. Sections 5-8 are based on Siegel [1969]. The material in §9 is from Zagier [1988]. Gauss's use of the **arithmetic-geometric mean** extended to complex numbers and is applicable in evaluating period integrals when the roots of the cubic are not all real. See Cox [1984] for an exposition.

Chapter VII

The definitive book about the Riemann zeta function is Titchmarsh [1951]. Dirichlet series are discussed in many analysis books. For Euler products, see pp. 217-220 of Husemöller [1987]. The material on Dirichlet's Theorem is taken from Serre [1973a]. Further motivation in terms of the zeta function of a number field may be found in Hecke [1981].

For the definition and properties of the gamma function $\Gamma(s)$, see Ahlfors [1966]. For the elementary properties of the Fourier transform on the line, see Stein and Weiss [1971]. The results of §5 have several points of contact with §3.6 of Shimura [1971a]. The passage from the transformation laws of a function like $\theta(\tau)$ to the functional equation of an associated Dirichlet series is one of the main themes of Ogg [1969a].

Chapters VIII-IX

Modular forms appear in many books. The definitive book is Shimura [1971a]. Other full-length books on the subject in the spirit presented here are Gunning [1962], Ogg [1969a], Lang [1976], and Miyake [1989]. In addition, there are chapters on the subject in Apostol [1976], Koblitz [1984], and Serre [1973a], and there are summaries in Gelbart [1975], Silverman [1986], and Husemöller [1987].

Our presentation of much of the material in Chapter VIII is based on Serre [1973a]. For §5 and §8, see also Ogg [1969a]. Our proof of Corollary 8.9 follows Siegel [1954]; for a different proof that gives different extra information, see Serre [1973a].

Some authors define weight differently. Also Hecke operators have more than one standard definition. Hecke operators, when defined as operators given by one double coset, as in Shimura [1971a], are slightly different from Hecke operators as we have defined them.

The subgroup $\Gamma_0(N)$ is treated in only a few places. The term "Hecke subgroup" follows Weil [1971b] and Gelbart [1975]. Our presentation is a reinterpretation of Ogg [1969a], Lang [1976], and Atkin and Lehner [1970]. The expositions by Ogg [1973] and Zagier [1990] are helpful also.

For more detail about Example 4 in §IX.3, see Serre [1973] and Ogg [1973]. Example 5 is discussed in Ogg [1973], and a generalization is in Newman [1959]. In connection with §VIII.5 and §IX.4, see Ogg [1969a].

The details of the proof of Theorem 9.9 were omitted. For the Riemann surface structure, see Chapter 1 of Shimura [1971a] or Chapter 1 of Miyake [1989]. For the comparison of zeros and poles of meromorphic functions, see Proposition II.5.4 of Farkas and Kra [1980]. The discussion of Theorem 9.10 mentions the Riemann-Hurwitz formula, which is given as Theorem I.2.7 of Farkas and Kra [1980], and the fact that the space of holomorphic differentials has dimension g, which is given as Proposition III.2.7. For Theorem 9.10, see §2.6 of Shimura [1971a].

The material of §IX.7 is taken from Atkin and Lehner [1970]. The proof of Theorem 9.22 may be found in that paper.

Chapter X

The material in §1 is due to Néron [1964]; see §VII.1 and §VIII.8 of Silverman [1986]. Our discussion in §2 of zeta functions and L functions is abbreviated. Silverman [1986] includes more detail in his Chapter V. The L function of an elliptic curve often has the name **Hasse-Weil** attached to it, and its shape is a motivating example for the **Weil conjectures** about varieties over finite fields. The final Weil conjecture generalizes Theorem 10.5, which is due to Hasse [1936]. We have given a nonstandard elementary proof of this theorem, following Manin [1956]; Birch discovered an error in Manin's proof, and the error is corrected in Cassels [1957] and here. The standard proof, given in §§V.1 and V.2 of Silverman [1986], is longer but has the advantage that its methods are applicable to other situations. For an exposition and history of Hasse's Theorem and the Weil conjectures and their proofs, see Katz [1976].

Chapter XI

Eichler-Shimura theory refers both to the kinds of results in §§11-12 and to a theory that seeks to generalize the material in §1 from $S_2(\Gamma_0(N))$ to $S_k(\Gamma_0(N))$. The main references for the theme that leads to §§11-12 are Eichler [1954] and Shimura [1958], [1971a], and [1973b]. The article Swinnerton-Dyer and Birch [1975] includes a very helpful overview. For the generalization of §1, the references are Eichler [1957] and Shimura [1959]; Chapters V and VI of Lang [1976] give an exposition for $SL(2, \mathbf{Z})$.

In connection with §1, economical generators of $X_0(N)$ are known if N is prime. This result is due to Rademacher [1930] and is quoted by Apostol [1990], p. 78. For the equations of E, E', and E'' in §1, as well as the isogeny, see Vélu [1971a]. See also Langlands [1990].

For proofs of the results in §2, see §1.5 of Shimura [1971a]. The material in §§3-4 about general Riemann surfaces is taken from Farkas and Kra [1980], and detailed proofs may be found there: For our Proposition 11.4, see Theorem II.5.1 of that book. For our Proposition 11.5, see Theorem III.2.7. Proposition 11.7 through Corollary 11.12 are in §III.4 of that book, Proposition 11.13 through Theorem 11.19 are in §III.6, and Theorems 11.20 and 11.21 are in §IV.11.

The invariant differential of an elliptic curve is treated in detail in Silverman [1986], pp. 52-53 and §III.5.

The idea of the construction in the first part of §5 is in Swinnerton-Dyer and Birch [1975], and the proof has been assembled from arguments scattered throughout Shimura [1971a]. For example, our Proposition 11.22 is implicit in the proof of Shimura's Theorem 2.20, and many of the steps in the proof of Proposition 11.23 appear in §8.3 of Shimura's book. Shimura gives two proofs that the Hecke operators are given by integer matrices and thus have algebraic integers as eigenvalues. One proof uses the generalization of our §1 to weight k and appears on p. 84 and in Chapter 8 of Shimura [1971a]; the other uses what we have called Proposition 11.73 and is on pp. 171-172 of that book. Theorem 11.27 is a modification of Shimura's Theorem 3.51.

The material in §6 through Theorem 11.33 and its lemmas is classical. Our treatment is based on §§3.3 and 5.2 of Lang [1987] and lectures of

H. Matumoto. Theorem 11.36 is implicit in §§6.7 and 6.8 of Shimura [1971a].

Our treatment of general varieties and curves in §§7-8 is a merger of Chapter I of Hartshorne [1977] and Chapters I and II of Silverman [1986], and the reader is referred to those sources for proofs or references to proofs of results in algebraic geometry. In places where neither a proof nor a reference appears for such a result in either book, we have tried to give some indication of proof.

For quoted results in commutative algebra, the reference is Zariski and Samuel [1958] and [1960]. The Hilbert Basis Theorem is on p. 201 of volume 1, and the Hilbert Nullstellensatz is on p. 164 of volume 2. Integral closure is discussed in pp. 254-265 of volume 1, and the result quoted to obtain (11.74) is on p. 265.

Corollary 11.50 is implicit in Chapters 6 and 7 of Shimura [1971a]. The material in §9 is abstracted from Silverman [1986], largely Chapters II and III. For more information about isogenies, see Vélu [1971b].

Lang [1983] treats abelian varieties in his §§I.1 and II.1, and he introduces the Jacobian variety in §II.2. Proofs of Propositions 11.68 to 11.70 and of some of the assertions about Jacobian varieties may be found there. Historically Weil [1946] and [1948, reprinted in 1971a] reworked the foundations of algebraic geometry in part to deal rigorously with the Jacobian variety. Page 88 of Weil [1971a] remarks on the understanding of the Italian school of the relationship between correspondences and the Jacobian variety; using correspondences involves counting intersections properly, and this subject had never been pinned down. Lefschetz [1921] had constructed the Jacobian variety as a projective variety over C, and Weil's work constructed Jacobian varieties of curves defined over Q as "abstract varieties." Chow [1954] realized the Jacobian variety of a curve defined over Q as a projective variety over Q, and his paper contains the results we quote as Theorem 11.71. For Theorem 11.72, see Theorem VII of Chow [1949]. For Proposition 11.73, see §2.9, Proposition 9, of Shimura and Taniyama [1961].

For the theorem of Wedderburn given as Lemma 11.75, see pp. 63, 66, and 116 of Jacobson [1943].

Concerning Theorem 11.74, Eichler [1954] discovered (11.118) and used it to relate the L function of $X_0(N)$ to the action of the Hecke operators. Shimura [1971a] adapted this algebraic argument and added the idea of looking at a subvariety of J to get a version of Theorem 11.74; see his Theorems 7.14 and 7.15. The theorem as we have given it is a sharpened version that appeared in Shimura [1973b]. Much of Shimura's proof, including the rationality of the Hecke operators, is couched in the language of adeles in order to gain generality; insight into what Shimura is doing may be obtained from Chapter 7 of Lang [1987]. Shimura's proof of the equality of L functions uses his own general theory (Shimura [1955]) of reduction of varieties modulo p and is valid for all but finitely many primes. Igusa [1959] proves results about the reduction modulo p of $X_0(N)$ implying that Shimura's argument is valid for all primes not dividing N.

Chapter XII

The presentation in this chapter, especially in §1, owes a great deal to Zagier [1988] and Swinnerton-Dyer and Birch [1975]. For Theorem 12.2, see §3.6 of Shimura [1971a]. Theorem 12.4 originally appeared in Weil [1967]; Chapter V of Ogg [1969a] gives an exposition. For the Taniyama-Weil Conjecture, see Taniyama [1957] and Weil [1967]. Theorem 12.7 is Theorem 4 of Swinnerton-Dyer and Birch [1975].

The notion of conductor dates to Artin and became manageable as a result of Ogg [1967]. An actual algorithm for computing it is the subject of Tate [1975].

Swinnerton-Dyer, Stephens, et al. [1975] assembled extensive tables of information about elliptic curves over $\mathbb{Q}$ with low conductor. Vélu [1976] reported a small number of errors that had been discovered in the tables, including the omission of a page listing curves of conductors 121 to 124. Our Table 12.1 relies partly on those tables and partly on computation using Tate's algorithm. Miyake [1989] includes long tables of information about a few particular curves.

Theorem 12.8 is due to Carayol [1986], p. 411. The arguments relating the Hasse-Weil Conjecture and the Taniyama-Weil Conjecture are in Swinnerton-Dyer and Birch [1975]. The end of our §1 alludes to Serre's Isogeny Theorem, which is on p. IV-14 of Serre [1968].

In §2, Proposition 12.9 is noted by Mazur [1973]. Twists appear in Atkin and Lehner [1970], Shimura [1973b], and Atkin and Li [1978]. See also §X.5 of Silverman [1986].

Ligozat [1975] works extensively with $X_0(N)$ of genus 1, and the classical method of proving that curves are Weil curves can be seen there. Mazur and Swinnerton-Dyer [1974] show how $N = 37$ (of genus 2) can be handled, and Birch [1973] shows how $N = 50$ (of genus 2) can be handled. We have taken Table 12.2 from Mazur [1973], but it can readily be computed directly.

Conductor $2^a 3^b$ was treated in the unpublished Manchester thesis of F. B. Coghlan, and the numerical results are in Table 4 of Swinnerton-Dyer, Stephens, et al. [1975]. Theorem 12.11 is due to Mestre and Oesterlé [1989]. In this connection, see also Theorem 1 of Coates [1970] and its interpretation in Theorem 8 of Tate [1974].

A subject that this book has not treated is elliptic curves over $\mathbf{Q}$ with complex multiplication. All such curves are Weil curves, by a theorem of Deuring [1953-57]. The j-invariants of curves over $\mathbf{Q}$ with complex multiplication lie in a finite set.

Trace formulas for Hecke operators are the subject of Chapter 6 of Miyake [1989]. For original papers, see Eichler [1957], Selberg [1957], Eichler [1973], and Hijikata [1974]. The argument centered about (12.11) is taken from Miyake's book.

Frey [1986] studied an equation more general than (12.12) and noted its relevance to Fermat's Last Theorem. Serre [1987b] studied this question at length and gave a conjecture whose truth would yield Corollary 12.13. Ribet [1990] proved enough of Serre's conjecture to obtain Corollary 12.13. We have followed Serre's paper in computing the conductor of (12.12).

Connection with Representation Theory

Langlands [1990] pulls together L functions, class field theory, elliptic curves, modular forms, infinite-dimensional representations of the general linear group, adeles, and automorphic representations, all in the space of 30 pages. His paper is a good one to start with, and a good one to return to.

The idea that modular forms are connected with the representations of $GL(2)$ seems to have originated with Gelfand and Fomin in the 1950's. Some expositions of the connection are in parts of Gelfand, Graev, and Pyatetskii-Shapiro [1969], Weil [1971b], Deligne [1973], Gelbart [1975], and Piatetski-Shapiro [1979]. In part, the connection is that one can identify cusp forms for $\Gamma_0(N)$ in two stages with functions on groups. In the first stage the identification is with certain functions on $SL(2,\mathbf{R})$ transforming on one side by $\Gamma_0(N)$ and on the other side by the rotation subgroup. In the second stage it is with certain functions on $GL(2,\mathbf{A})$, where $\mathbf{A}$ is the group of adeles of $\mathbf{Q}$. The cusp forms are then intimately connected with the decomposition of the representation of $GL(2,\mathbf{A})$ on L^2 of the quotient space $GL(2,\mathbf{Q})Z_\mathbf{A}\backslash GL(2,\mathbf{A})$, where $Z_\mathbf{A}$ is the center of $GL(2,\mathbf{A})$. Under this correspondence the Hecke operators end up with a surprisingly simple interpretation.

Once this construction is complete, the objects of interest are pairs consisting of a certain kind of infinite-dimensional irreducible unitary representation of $GL(2,\mathbf{A})$ and a finite-dimensional holomorphic representation of $GL(2,\mathbf{C})$ (initially taken to be the standard representation). To such a pair the theory attaches an L function. To construct the L function, one recognizes the infinite-dimensional representation as a kind of infinite tensor product of representations of $GL(2,\mathbf{R})$ and

$GL(2, p\text{-adics})$ for all p (see Flath [1979]), and one constructs a factor of the L function for each factor of the representation. The intention is to be guided by Hecke's classical theory, to obtain an analytic continuation and functional equation for the resulting L function, and to see connections with L functions in number theory. This program for $GL(2)$ is carried out in detail in Jacquet and Langlands [1970]. For an exposition, see Robert [1973]. In retrospect, Tate's 1950 thesis (Tate [1967]) did the same thing for $GL(1)$.

This construction has become in part a way of generating automorphic L functions, and it lends itself to generalization. For generalization to $GL(n)$, see Godement and Jacquet [1972]. For generalization to other kinds of groups, several things have to be brought to bear simultaneously. One can see them all at once in a preliminary form in Langlands [1970], and all at once in a later form in Langlands [1979]. For more detail one can consult Borel [1979], Borel and Jacquet [1979], and Tate [1979]. General sources for representation theory of reductive groups are Knapp [1986] for real groups and Silberger [1979] for p-adic groups. For a status report on construction of L functions, see Shahidi [1990].

The theory does more, however, than just produce L functions. It provides tools for working with them and bringing together the fields of mathematics in which L functions play a role. Three papers that explain such relationships are Gelbart [1984], Gelbart and Shahidi [1988], and Clozel [1990]. And, once again, one should look at Langlands [1990].

REFERENCES

Conferences

"Antwerp-Bonn": *Modular Forms in One Variable, I-VI*, Lecture Notes in Math. 320, 349, 350, 476, 601, 627 (1973-77), Springer-Verlag, Heidelberg Berlin New York.

"Corvallis": *Automorphic Forms, Representations, and L-functions*, Proc. Symposia in Pure Math. XXXIII, parts 1-2 (1979), American Mathematical Society.

"Ann Arbor": Clozel, L., and J. S. Milne, *Automorphic Forms, Shimura Varieties, and L-functions I-II: Proceedings of a Conference Held at the Univesity of Michigan, Ann Arbor, July 6-16, 1988*, Perspectives in Mathematics 10-11 (1990), Academic Press Inc., San Diego, Academic Press, Boston.

"Motives": Proc. Symposia in Pure Math., American Mathematical Society, to appear.

Books and Papers

Ahlfors, L. V., *Complex Analysis*, McGraw Hill Inc., New York, 2nd edition, 1966.

Apostol, T. M., *Modular Functions and Dirichlet Series in Number Theory*, Springer-Verlag, New York Berlin Heidelberg, 2nd edition, 1990 (1st edition, 1976).

Atkin, A. O. L., and J. Lehner, Hecke operators on $\Gamma_0(m)$, *Math. Annalen* 185 (1970), 134-160.

Atkin, A. O. L., and W. W. Li, Twists of newforms and pseudo-eigenvalues of W-operators, *Invent. Math.* 48 (1978), 221-243.

Birch, B., Some calculations of modular relations, *Modular Functions of One Variable I*, Lecture Notes in Math. 320 (1973), Springer-Verlag, Berlin Heidelberg New York, 175-186.

Birch, B. J., and H. P. F. Swinnerton-Dyer, Notes on elliptic curves I and II, *J. Reine Angew. Math.* 212 (1963), 7-25; 218 (1965), 79-108.

Borel, A., Automorphic L-functions, *Automorphic Forms, Representations, and L-functions*, Proc. Symposia in Pure Math. XXXIII, part 2 (1979), American Mathematical Society, 27-61.

Borel, A., and H. Jacquet, Automorphic forms and automorphic representations, *Automorphic Forms, Representations, and L-functions*, Proc. Symposia in Pure Math. XXXIII, part 1 (1979), American Mathematical Society, 189-202.

Borevich, Z. I., and I. R. Shafarevich, *Number Theory*, Academic Press, New York, 1966.

Brieskorn, E., and H. Knörrer, *Plane Algebraic Curves*, Birkhäuser Verlag, Basel Boston Stuttgart, 1986.

Brumer, A., and O. McGuinness, The behavior of the Mordell-Weil group of elliptic curves, *Bull. Amer. Math. Soc.* 23 (1990), 375-382.

Carayol, H., Sur les représentations l-adiques associées aux formes modulaires de Hilbert, *Ann. Scient. Ecole Norm. Sup.* 19 (1986), 409-468.

Cassels, J. W. S., Review of Yu I. Manin's, "On cubic congruences to a prime modulus," *Math. Reviews* 18 (1957), 380-381.

Coates, J., An effective p-adic analogue of a theorem of Thue II, *Acta Arithmetica* 16 (1970), 399-412.

Charlap, L. S., and D. P. Robbins, An elementary introduction to elliptic curves, preprint, 1988.

Chow, W., On compact complex analytic varieties, *Amer. J. Math.* 71 (1949), 893-914.

Chow, W., The Jacobian variety of an algebraic curve, *Amer. J. Math.* 76 (1954), 453-476.

Clozel, L., Motifs et formes automorphes: applications du principe de fonctorialité, *Automorphic Forms, Shimura Varieties, and L-functions I: Proceedings of a Conference Held at the Univesity of Michigan, Ann Arbor, July 6-16, 1988,* , Perspectives in Mathematics 10 (1990), Academic Press Inc., San Diego, 77-159.

Cox, D. A., The arithmetic-geometric mean of Gauss, *L'Enseignement Math.* 30 (1984), 275-330.

Deligne, P., Formes modulaires et représentations de GL(2), *Modular Functions of One Variable II*, Lecture Notes in Math. 349 (1973), Springer-Verlag, Berlin Heidelberg New York, 55-105.

Deligne, P., and M. Rapoport, Les schémas de modules de courbes elliptiques, *Modular Functions of One Variable II*, Lecture Notes in Math. 349 (1973), Springer-Verlag, Berlin Heidelberg New York, 143-316.

Deuring, M., Die Zetafunktion einer algebraischen Kurve vom Geschechte Eins I, II, III, IV, *Nachr. Akad. Wiss. Göttingen* (1953) 85-94, (1955) 13-42, (1956) 37-76, (1957) 55-80.

Eichler, M., Quaternäre quadratische Formen und die Riemannsche Vermutung für die Kongruenzzetafunktionen, *Arch. der Math.* 5 (1954), 355-366.

Eichler, M., Eine Verallgemeinerung der Abelschen Integrale, *Math. Zeitschr.* 67 (1957), 267-298.

Eichler, M., The basis problem for modular forms and the traces of the Hecke operators, *Modular Functions of One Variable I*, Lecture Notes in Math. 320 (1973), Springer-Verlag, Berlin Heidelberg New York, 75-151.

Farkas, H. M., and I. Kra, *Riemann Surfaces*, Springer-Verlag, New York Heidelberg Berlin, 1980.

Flath, D., Decomposition of representations into tensor products, *Automorphic Forms, Representations, and L-functions*, Proc. Symposia in Pure Math. XXXIII, part 1 (1979), American Mathematical Society 179-183.

Frey, G., Links between stable elliptic curves and certain Diophantine equations, *Annales Universitatis Saraviensis*, Series Mathematicae, 1 (1986), 1-40.

Fueter, R., Ueber kubische diophantische Gleichungen, *Comm. Math. Helv.* 2 (1930), 69-89.

Fulton, W., *Algebraic Curves*, Addison-Wesley, Redwood City, California, 1989.

Gelbart, S. S., *Automorphic Forms on Adele Groups*, Annals of Math. Studies 83, Princeton University Press, Princeton, 1975.

Gelbart, S. S., An elementary introduction to the Langlands program, *Bull. Amer. Math Soc.* 10 (1984), 177-219.

Gelbart, S., and F. Shahidi, *Analytic Properties of Automorphic L-Functions*, Perspectives in Mathematics 6 (1988), Academic Press Inc., San Diego.

Gelfand, I. M., M. I. Graev, and I. I. Pyatetskii-Shapiro, *Representation Theory and Automorphic Functions*, W. B. Saunders Co., Philadelphia, 1969.

Godement, R., and H. Jacquet, *Zeta Functions of Simple Algebras*, Lecture Notes in Math. 260 (1972), Springer-Verlag, Berlin Heidelberg New York.

Gunning, R. C., *Lectures on Modular Forms*, Annals of Math. Studies 48, Princeton University Press, Princeton, 1962.

Hartshorne, R., *Algebraic Geometry*, Springer-Verlag, New York Heidelberg Berlin, 1977.

Hasse, H., Zur Theorie der abstrakten elliptischen Funktionkörper I, II, III, *J. Reine Angew.. Math.* 175 (1936), 55-62, 69-88, 193-208.

Hecke, E., *Lectures on the Theory of Algebraic Numbers*, Springer-Verlag, New York Heidelberg Berlin, 1981.

Hijikata, H., Explicit formula of the traces of Hecke operators for $\Gamma_0(N)$, *J. Math. Soc. Japan* 26 (1974), 56-82.

Husemöller, D., *Elliptic Curves*, Springer-Verlag, New York Berlin Heidelberg, 1987.

Igusa, J., Kroneckerian models of fields of elliptic modular functions, *Amer. J. Math.* 81 (1959), 561-577.

Ireland, K., and M. Rosen, *A Classical Introduction to Modern Number Theory*, Springer-Verlag, New York Berlin Heidelberg, 2nd edition, 1990.

Jacobson, N., *The Theory of Rings*, Mathematical Surveys II (1943), American Mathematical Society.

Jacquet, H., and R. P. Langlands, *Automorphic Forms on $GL(2)$*, Lecture Notes in Math. 114 (1970), Springer-Verlag, Berlin Heidelberg New York.

Katz, N. M., An overview of Deligne's proof of the Riemann hypothesis for varieties over finite fields, *Mathematical Developments Arising from Hilbert Problems*, Proc. Symposia in Pure Math. XXVIII (1976), American Mathematical Society, 275-305.

Knapp, A. W., *Representation Theory of Semisimple Groups: An Overview Based on Examples*, Princeton University Press, Princeton, 1986.

Koblitz, N., *Introduction to Elliptic Curves and Modular Forms*, Springer-Verlag, New York Berlin Heidelberg, 1984.

Lang, S., *Introduction to Modular Forms*, Springer-Verlag, Berlin Heidelberg New York, 1976.

Lang, S., *Abelian Varieties*, Springer-Verlag, New York Berlin Heidelberg, 1983.

Lang, S., *Elliptic Functions*, Springer-Verlag, New York Berlin Heidelberg, 2nd edition, 1987.

Lang, S., Old and new conjectured Diophantine inequalities, *Bull. Amer. Math. Soc.* 23 (1990), 37-75.

Langlands, R. P., Problems in the theory of automorphic forms, *Lectures in Modern Analysis and Applications III*, Lecture Notes in Math. 170 (1970), Springer-Verlag, Berlin Heidelberg New York, 18-61.

Langlands, R. P., Modular forms and l-adic representations, *Modular Functions of One Variable II*, Lecture Notes in Math. 349 (1973), Springer-Verlag, Berlin Heidelberg New York, 361-500.

Langlands, R. P., Automorphic representations, Shimura varieties, and motives. Ein Märchen, *Automorphic Forms, Representations, and L-functions*, Proc. Symposia in Pure Math. XXXIII, part 2 (1979), American Mathematical Society, 205-246.

Langlands, R. P., Representation theory: its rise and its role in number theory, *Proceedings of the Gibbs Symposium, Yale University, May 15-17, 1989*, American Mathematical Society, 1990, 181-210.

Lefschetz, S., On certain numerical invariants of algebraic varieties with application to Abelian varieties, *Trans. Amer. Math. Soc.* 22 (1921), 327-482.

Li, W. W., Newforms and functional equations, *Math. Annalen* 212 (1975), 285-315.

Lichtenbaum, S., Torsion Subgroups of Elliptic Curves, Undergraduate Honors Thesis, Harvard University, Cambridge, Massachusetts, 1960.

Ligozat, G., Courbes modulaires de genre 1, *Bull. Soc. Math. France, Supplément*, Mémoire 43 (1975).

Lutz, E., Sur l'equation $y^2 = x^3 - Ax - B$ dans les corps $\mathfrak{p}$-adiques, *J. Reine Angew. Math.* 177 (1937), 238-247.

Manin, Yu I., On cubic congruences to a prime modulus, *Izv. Akad. Nauk SSSR, Ser. Mat.* 20 (1956), 673-678 (Russian), *Amer. Math. Soc. Transl.* (2) 13 (1960), 1-7.

Mazur, B., Courbes elliptiques et symboles modulaires, *Séminaire Bourbaki, vol. 1971/72, Exposées 400-417*, Lecture Notes in Math. 317 (1973), Springer-Verlag, Berlin Heidelberg New York, 277-294.

Mazur, B., Modular curves and the Eisenstein ideal, *I.H.E.S. Publ. Math.* 47 (1977), 33-186.

Mazur, B., Rational isogenies of prime degree, *Invent. Math.* 44 (1978), 129-162.

Mazur, B., and P. Swinnerton-Dyer, Arithmetic of Weil Curves, *Invent. Math.* 25 (1974), 1-61.

Mestre, J.-F., and J. Oesterlé, Courbes de Weil semi-stables de discriminant une puissance m-ième, *J. Reine Angew. Math.* 400 (1989), 173-184.

Miyake, T., *Modular Forms*, Springer-Verlag, Berlin Heidelberg New York, 1989.

Mordell, L. J., On the rational solutions of the indeterminate equations of the third and fourth degrees, *Proc. Cambridge Philos. Soc.* 21 (1922-23), 179-192.

Nagell, T., Solution de quelques problèmes dans la théorie arithmétique des cubiques planes du premier genre, *Skrifter Norske Videnskaps-Akademi i Oslo*, 1935, No. 1, pp. 1-25.

Néron, A., Modèles minimaux des variétés abéliennes sur les corps locaux et globaux, *I.H.E.S. Publ. Math.* 21 (1964), 5-128.

Newman, M., Construction and application of a class of modular functions II, *Proc. London Math. Soc.* 9 (1959), 373-387.

Ogg, A. P., Elliptic curves and wild ramification, *Amer. J. Math.* 89 (1967), 1-21.

Ogg, A., *Modular Forms and Dirichlet Series*, W. A. Benjamin Inc., New York, 1969a.

Ogg, A., On the eigenvalues of Hecke operators, *Math. Annalen*, 179 (1969b), 101-108.

Ogg, A., Survey of modular functions of one variable, *Modular Functions of One Variable I*, Lecture Notes in Math. 320 (1973), Springer-Verlag, Berlin Heidelberg New York, 1-35.

Patterson, S. J., Automorphic forms and number theory, Ecole Européenne de Théorie des Groupes, C.I.R.M. Marseille-Luminy, preprint, 1991.

Piatetski-Shapiro, I., Classical and adelic automorphic forms. An introduction, *Automorphic Forms, Representations, and L-functions*, Proc. Symposia in Pure Math. XXXIII, part 1 (1979), American Mathematical Society, 185-188.

Rademacher, H., Über die Erzeugenden von Kongruenzuntergruppen der Modulgruppe, *Abhandlungen Math. Seminar Hamburg* 7 (1930), 134-148.

Ribet, K. A., On modular representations of $\text{Gal}(\bar{\mathbf{Q}}/\mathbf{Q})$ arising from modular forms, *Invent. Math.* 100 (1990), 431-476.

Robert, A., Formes automorphes sur GL_2, *Séminaire Bourbaki, vol. 1971/72, Exposées 400-417*, Lecture Notes in Math. 317 (1973), Springer-Verlag, Berlin Heidelberg New York, 295-318.

Selberg, A., Automorphic functions and integral operators, *Seminars on analytic functions*, Institute for Advanced Study, Princeton, N. J., and U. S. Air Force Office of Scientific Research, 2 (1957), 152-161.

Selmer, E. S., The diophantine equation $ax^3 + by^3 + cz^3 = 0$, *Acta Math.* 85 (1951), 203-362, and 92 (1954), 191-197.

Serre, J.-P., *Abelian l-adic Representations and Elliptic Curves*, W. A. Benjamin Inc., New York, 1968.

Serre, J.-P., *A course in Arithmetic*, Springer-Verlag, New York Heidelberg Berlin, 1973a.

Serre, J.-P., Congruences et formes modulaires, *Séminaire Bourbaki, vol. 1971/72, Exposées 400-417*, Lecture Notes in Math. 317 (1973b), Springer-Verlag, Berlin Heidelberg New York, 319-338.

Serre, J.-P., Lettre à J.-F. Mestre, *Current Trends in Arithmetical Algebraic Geometry*, Contemporary Mathematics Series 67 (1987a), American Mathematical Society, Providence, 263-268.

Serre, J.-P., Sur les représentations modulaires de degré 2 de $\text{Gal}(\bar{\mathbf{Q}}/\mathbf{Q})$, *Duke Math. J.* 54 (1987b), 179-230.

Serre, J.-P., *Lectures on the Mordell-Weil Theorem*, Fried. Vieweg & Sohn, Braunschweig, 2nd edition, 1990.

Shahidi, F., Automorphic L-functions: a survey, *Automorphic Forms, Shimura Varieties, and L-functions I: Proceedings of a Conference Held at the Univesity of Michigan, Ann Arbor, July 6-16, 1988,* , Perspectives in Mathematics 10 (1990), Academic Press Inc., San Diego, 415-437.

Shimura, G., Reduction of algebraic varieties with respect to a discrete valuation of the basic field, *Amer. J. Math.* 77 (1955), 134-176.

Shimura, G., Correspondances modulaires et les fonctions ζ de courbes algébriques, *J. Math. Soc. Japan* 10 (1958), 1-28.

Shimura, G., Sur les intégrales attachées aux formes automorphes, *J. Math. Soc. Japan* 11 (1959), 291-311.

Shimura, G., *Introduction to the Arithmetic Theory of Automorphic Functions*, Princeton University Press, Princeton, 1971a.

Shimura, G., On elliptic curves with complex multiplication as factors of the jacobians of modular function fields, *Nagoya Math. J.* 43 (1971b), 199-208.

Shimura, G., Complex multiplication, *Modular Functions of One Variable I*, Lecture Notes in Math. 320 (1973a), Springer-Verlag, Berlin Heidelberg New York, 37-56.

Shimura, G., On the factors of the jacobian variety of a modular function field, *J. Math. Soc. Japan* 25 (1973b), 523-544.

Shimura, G., and Y. Taniyama, *Complex Multiplication of Abelian Varieties and Its Applications to Number Theory*, Publ. Math. Soc. Japan, no. 6, 1961, Kenkyusha Printing Co., Tokyo.

Siegel, C. L., A simple proof of $\eta(-1/\tau) = \eta(\tau)\sqrt{\tau/i}$, *Mathematika* 1 (1954), p. 4.

Siegel, C. L., *Topics in Complex Function Theory*, vol. 1, John Wiley & Sons, New York, 1969.

Silberger, A. J., *Introduction to Harmonic Analysis on Reductive p-Adic Groups*, Princeton University Press, Princeton, 1979.

Silverman, J. H., *The Arithmetic of Elliptic Curves*, Springer-Verlag, New York Berlin Heidelberg, 1986.

Stein, E. M., and G. Weiss, *Introduction to Fourier Analysis on Euclidean Spaces*, Princeton University Press, Princeton, 1971.

Swinnerton-Dyer, H. P. F., and B. J. Birch, Elliptic curves and modular functions, *Modular Functions of One Variable IV*, Lecture Notes in Math. 476 (1975), Springer-Verlag, Berlin Heidelberg New York, 2-32.

Swinnerton-Dyer, H. P. F., N. M. Stephens, J. Davenport, J. Vélu, F. B. Coghlan, A. O. L. Atkin,, D. J. Tingley, Numerical tables on elliptic curves, *Modular Functions of One Variable IV*, Lecture

Notes in Math. 476 (1975), Springer-Verlag, Berlin Heidelberg New York, 74-144.

Taniyama, Y., *L*-functions of number fields and zeta functions of abelian varieties, *J. Math. Soc. Japan* 9 (1957), 330-366.

Tate, J. T., Fourier analysis in number fields and Hecke's zeta functions, in J. W. S. Cassels and A. Fröhlich, *Algebraic Number Theory*, Academic Press, London, 1967, 305-347.

Tate, J. T., The arithmetic of elliptic curves, *Invent. Math.* 23 (1974), 179-206.

Tate, J., Algorithm for determining the type of a singular fiber in an elliptic pencil, *Modular Functions of One Variable IV*, Lecture Notes in Math. 476 (1975), Springer-Verlag, Berlin Heidelberg New York, 33-52.

Tate, J., Number theoretic background, *Automorphic Forms, Representations, and L-functions*, Proc. Symposia in Pure Math. XXXIII, part 2 (1979), American Mathematical Society, 3-26.

Titchmarsh, E. C., *The Theory of the Riemann Zeta-Function*, Oxford University Press, Oxford, 1951.

Vélu, J., Courbes elliptiques sur **Q** ayant bonne réduction en dehors de $\{11\}$, *C. R. Acad. Sci. Paris*, Ser. A 273 (1971a), 73-75.

Vélu, J., Isogénies entre courbes elliptiques, *C. R. Acad. Sci. Paris*, Ser. A 273 (1971b), 238-241.

Vélu, J., Review of "Numerical tables on elliptic curves," *Math. Reviews* 52 (1976), #10557.

Walker, R. J., *Algebraic Curves*, Dover Publications Inc., New York, 1950.

Weil, A., Sur un théorème de Mordell, *Bull. Sci. Math.* 54 (1930), 182-191.

Weil, A., *Foundations of Algebraic Geometry*, Amer. Math. Soc. Colloquium Publications XXIX, American Mathematical Society, 1946; 2nd edition, 1962.

Weil, A., Über die Bestimmung Dirichletscher Reihen durch Funktionalgleichungen, *Math. Annalen* 168 (1967), 149-156.

Weil, A., *Courbes Algébriques et Variétés Abéliennes*, Hermann, Paris, 1971a. (Reprint of volumes 1041 and 1064 of *Actualités Scientifiques et Industrielles*, 1948.)

Weil, A., *Dirichlet Series and Automorphic Forms*, Lecture Notes in Math. 189 (1971b), Springer-Verlag, Berlin Heidelberg New York.

Zagier, D. B., *L*-series of elliptic curves, the Birch-Swinnerton-Dyer conjecture, and the class number problem of Gauss, *Notices Amer. Math. Soc.* 31 (1984), 739-743.

Zagier, D. B., Modular parametrizations of elliptic curves, *Canad. Math. Bull.* 28 (1985), 372-384.

Zagier, D., Elliptic Curves, lecture series, Tata Institute for Fundamental Research, Bombay, January 1988.

Zagier, D., Introduction to modular forms, preprint, 1990.

Zariski, O., and P. Samuel, *Commutative Algebra*, vol. I, D. Van Nostrand Company Inc., Princeton, 1958.

Zariski, O., and P. Samuel, *Commutative Algebra*, vol. II, D. Van Nostrand Company Inc., Princeton, 1960.

INDEX OF NOTATION

See also the list of Standard Notation on page xv. In the list below, Latin, German, and script letters appear together and are followed by Greek symbols and non-letters.

a_1, a_2, a_3, a_4, a_6, 42, 56, 57
a_p, 294, 295
$A(T)$, 331
$A_k(r)$, 283
$A_k(\Gamma_0(N))$, 333
b_2, b_4, b_6, b_8, 57
c_1, 329
c_4, c_6, 57
c_n, 225
$c(m, \chi)$, 214, 387
C, 273
$C(x, y, w)$, 37
C_K, 353
d, 59
deg, 317, 359
$\deg_s, \deg_i$, 359
df, 371
$dt(n)$, 372
$\mathrm{Div}(X)$, 316, 360
$\mathrm{Div}_0(X)$, 316, 361
e, 170
$e_F(x)$, 359
$E(k)$, 25
$E(\mathbb{Q})$, 15, 80
$E(\mathbb{Q})_{\mathrm{tors}}$, 130
$E^{(n)}(\mathbb{Q})$, 137, 138
$\mathrm{End}(E)$, 364, 368
$\mathrm{End}_\mathbb{Q}(J)$, 374
E_p, 130, 135
$\widehat{f}$, 210
$\tilde{f}$, 243
$f_1(x, y)$, 26
f_H, 345

f_r, 212
f_ω, 314
$\mathbf{f}(\tau)$, 330
$\widehat{F}$, 365
F^*, 349, 360
F_*, 360, 392
$F(k)$, 25
$F(\tau)$, 302
F_p, 135
F^Φ, 28
g, 272, 317, 361
$g_2, g_2(\Lambda)$, 158, 222
$g_3, g_3(\Lambda)$, 158, 222
$g_2(\tau), g_3(\tau), G_{2k}(\tau)$, 223
$G_2(\tau)$, 266
$G_2^{(2)}(\tau)$, 266
$G_{2k}, G_{2k}(\Lambda)$, 158, 222
h, 97, 263
h_0, 95
H, 30
$H_1(\mathbb{R}), H_1(\mathbb{C})$, 326
$H_1^\pm(\mathbb{R}), H_1^\pm(\mathbb{C})$, 327
$H_1(X_0(N), \mathbb{Z})$, 323
$\mathrm{Hom}(E_1, E_2)$, 364, 368
$\mathcal{H}$, 223
$\mathcal{H}^*$, 311
$i(P, L, F)$, 32
I, 341
$I(V)$, 342
$I(V/k_0)$, 342
$I(W)$, 344
$I(W \cap \bar{k}^n)$, 345
j, 65, 333

419

INDEX OF NOTATION

$j(\Lambda)$, 222
$j(\tau)$, 223, 333
j_N, 335
$\mathfrak{j}$, 371
J, 371
$J(X)$, 318, 369
$J(X_0(N))$, 310
$k[x,y,w]_d$, 24
$k_0[V]$, $\bar{k}[V]$, 342
$k_0(V)$, $\bar{k}(V)$, 342
$k_0(W)$, $\bar{k}(W)$, 346
$\bar{k}[V]_x$, 343
$\bar{k}[W]$, 346
$\bar{k}[W]_x$, 346
$K(X)$, 312
L, 19
$L(k)$, 19
$L(s)$, 17
$L(s,f)$, 238, 267, 386
$L(s,f,\chi)$, 387
$L(s,E)$, 295
$L(s,\chi)$, 201
$L(D)$, 316, 361
$L(T)$, 329
L^Φ, 20
m_x, 343
$[m]$, 365
$M(a,b)$, 186
$M(n)$, 244
$M(n,N)$, 276, 320
$M^*(n)$, 256
M_k, 234
$M_k(\Gamma_0(N))$, 265
M_x, 344
M_R, 351
$n(C)$, 170
n_1, n_2, 108
N, 390
$N(p)$, 16
$\text{ord}_x(f)$, 316, 350

$\text{ord}_x(D)$, 316, 360
O, 11, 67, 74, 362, 368
$\wp$, 151, 153
$-P$, 12, 74
$P_2(k)$, 19
$P_n(k)$, 344
P_Λ, 273
$\text{Pic}(X)$, 316, 360
$\text{Pic}^0(C)$, 361
$P+Q$, 11, 67
$P \cdot Q$, 10, 43
PQ, 10, 43
q, 225
q_h, 263
q_N, 263
Q, 285
r, 102
r_p, 130, 135
R, 227
$\tilde{R}$, 251
$R(f,g)$, 45
$[R(f,g)]$, 45
$R(n)$, 248, 278
$R_{E/\mathbf{Q}}$, 106
R_N, 260
$\mathcal{R}$, 170, 271, 305, 375
$\mathcal{R}^*$, 172, 271, 305
S, 228
S_k, 234
$S_k(\Gamma_0(N))$, 265
$S_k^\varepsilon(\Gamma_0(N))$, 270
$S_k^{\text{old}}(\Gamma_0(N))$, 283
$S_k^{\text{new}}(\Gamma_0(N))$, 283
$\mathcal{S}$, 375
$\mathcal{S}(\mathbf{R}^1)$, 211
$t(n)$, 372
T, 228, 305
$T(n)$, 244, 275, 320, 321, 323
$T^\#(n)$, 372
$T_k(n)$, 244, 245, 275, 277

INDEX OF NOTATION

$\tilde{T}_p$, 384
$\mathcal{T}$, 375
u, 63, 290, 293
$\mathcal{U}$, 375
v, 351
v_∞, v_i, v_ρ, v_τ, 231
V, 342
V/k_0, 342
$V(k_0)$, 342
V_4, V_6, 305
V_I, 341
$w(C)$, 174
$w(Q)$, 285
w_N, 269
w_Q, 285
W, 344
$W(k_0)$, 344
W_I, 344
$X_0(11)$, 305
$X_0(N)$, 310, 311
$\alpha^\#$, 253, 261
α_i, 244, 277, 320, 372
α_N, 269
$[\alpha]_k$, 245, 261
$[\gamma]$, 323
$\Gamma(N)$, 250, 256
$\Gamma_0(N)$, 256
$\Gamma_0(r,p)$, 287
Γ_1, Γ_2, 170
$\tilde{\Gamma}_1$, $\tilde{\Gamma}_2$, 177
Γ_a, Γ_b, Γ_c, 170
Γ^{ab}, 321
Γ_α, Γ_β, Γ_γ, 177
$\delta(g,\tau)$, 229, 261
δf, 371
$\delta t(n)$, 373
Δ, 58
$\Delta(\Lambda)$, 222
$\Delta(\tau)$, 223
$\zeta(s)$, 189

$\eta(\tau)$, 235
$\theta(\tau)$, 208
$\theta(\tau,\chi)$, 216
Λ, 153, 223, 371
$\Lambda(s)$, 18, 209
$\Lambda(s,f)$, 240, 270, 386
$\Lambda(s,f,\chi)$, 387
$\Lambda(s,E)$, 386, 387
$\Lambda(s,E,\chi)$, 387
$\Lambda(s,\chi)$, 216
Λ_f, 374
Λ_τ, 223
(Λ,C), 273
μ, 272, 371
μ_∞, μ_2, μ_3, 272
ν, 374
π, 370
Π, 153
ρ, 208, 375
σ, 208
$\sigma_l(n)$, 225
$\tau = \rho + i\sigma$, 208, 224
τ_0, 302
ϕ, 360, 384
$\hat{\phi}$, 384
φ, 272, 357, 380
$\varphi(\tau)$, 239, 268
φ_α, 90
Φ, 20, 318, 345, 369
$\tilde{\Phi}$, 370
$\Phi^\#$, 372
$\Phi_f(\gamma)$, 303
$\Phi_N(X)$, 335
χ, 201
$\chi^\#$, 213
χ_0, 201
ω, 314, 370, 374
ω_1, ω_2, 153, 223, 274
$\Omega(X)$, 312
$\Omega_{\text{hol}}(X)$, 313, 370

… INDEX OF NOTATION

∞, 79, 231, 261
$\square$, 80
$\|$, 90
$(\cdot)\widehat{}$, 210, 365
$(\cdot)^*$, 328
$\langle \cdot, \cdot \rangle$, 242, 280, 326, 371

$[\cdot]$, 323, 365
$[\cdot]_k$, 245, 261
$|\cdot|_p$, 134
$\left(\dfrac{a}{p}\right)$, 272, 298, 393
$\left|\dfrac{i}{j}\right|_\infty$, 83

INDEX

Abel's Theorem, 318
abelian
  subvariety, 368
  variety, 367
addition, 162, 363
  elliptic curve, 11
  formula, 76
adeles, 407
admissible change of variables, 63, 363
affine
  algebraic set, 342
  coordinate ring, 342
  curve, 25, 343
  local coordinates, 21, 345
  problem, 3
  variety, 342
algebraic integer, 122, 123
algebraic numbers, 122
algebraic set
  affine, 342
  projective, 344
analytic continuation, 166, 167
arithmetic-geometric mean, 185, 402
associate Dirichlet characters, 213
associativity, 12, 67
Atkin-Lehner Theorem, 283, 289
automorphic function, 333
automorphic L function, xi, 208

bad prime, 108
Bezout's Theorem, 27, 47
birational map 348
Birch and Swinnerton-Dyer Conjecture, 17, 208, 386

canonical class, 316, 361
canonical height, 95, 97
canonical $\mathbb{Q}$ structure, 333, 341, 349, 357
Carayol, 391
change of variables, admissible, 63, 362

character, 200
chord case, 75
chord-tangent composition rule, 11, 43
class group
  divisor, 316, 361
  ideal, 127
class number, 126
complex multiplication, 164, 407
complex torus, 160, 183, 370
conductor, 213, 390
congruence subgroup, 250, 256
congruent numbers, 52, 110, 112, 115
conic, 25
continuation, analytic, 166, 167
convergent product, 195
coordinate ring, 342
coordinates, affine local, 345
correspondence, 385
cubic, 25, 40, 56
  nonsingular, 58
  singular, 58
curve
  affine, 25, 343
  elliptic, 42
  nonsingular, 27
  projective, 345
  projective plane, 24
  same, 24
  singular, 27
  smooth, 27
cusp, 77, 262
cusp form, 225, 261

defined at, 344, 346, 348
defined over, 342, 344, 347, 361, 362, 368
degenerate, 68
degree, 243, 249, 273, 317, 359, 360
descent, 80
differential
  holomorphic, 312
  invariant, 314
  meromorphic, 312

dimension, 343, 346
dimension formula, 235, 272
Diophantus, 3, 6, 7, 10, 50
Diophantus method, 6, 10, 115, 119
Dirichlet character, 201
  associate, 213
  conductor of, 213
  extension, 213
  primitive, 213
  principal, 201
Dirichlet L function, xii, 201
Dirichlet series, 192
Dirichlet Unit Theorem, 125
Dirichlet's Theorem, xii, 148, 189
discrete valuation ring, 351
discriminant, 15, 58, 59, 125, 226
divisor, 316, 360
divisor class group, 316, 361
divisor, principal, 316, 360
dominant morphism, 349
dominate, 392
Double Series Theorem, 158
doubling formula, 76
doubly periodic, 152
dual isogeny, 309, 365

Eichler-Shimura theory, xii, 302, 374, 389, 390, 404
eigenform, 280
elliptic curve, 13, 42, 160, 183, 362
  modular, 221
elliptic element, 303
elliptic function, 152
elliptic integral, 174
elliptic regulator, 106
equivalent eigenform, 280
equivalent ideals, 126
Euler product, 196, 202, 255, 282, 289, 295
  first degree, 197
  second degree, 199
even total winding number, 170
extension of Dirichlet character, 213

fairly bad prime, 108

Fermat, 52, 110, 112
Fermat's Last Theorem, xii, 3, 18, 50, 54, 58, 81, 397, 399
Fermat's method of descent, 80
Fermat's problem for Mersenne, 55, 119
Fibonacci, 52
first degree Euler product, 197
flex, 35
Fourier inversion formula, 200, 210
Fourier transform, 200, 210
Frey-Serre-Ribet, xii, 18, 399
Frobenius morphism, 360, 384
  q^{th} power, 360
function element, 166
function field, 312, 319, 342, 346
  of dimension 1, 349
functional equation, 209, 216, 240, 270, 289, 386, 387, 388
fundamental domain, 228, 260
fundamental parallelogram, 152

Gauss, 185, 402
Gauss sum, 214, 387
genus, 272, 317, 361, 395
global minimal, 290
good prime, 108

Hasse Principle, 5, 15, 16
Hasse's Theorem, 16, 296
Hasse-Minkowski Theorem, 5
Hasse-Weil Conjecture, 18, 387
Hasse-Weil L function, 403
Hecke operator, 244, 275, 320, 321, 323, 372
Hecke subgroup, 256
Hecke's Theorem, 249, 279
Hecke-Petersson Theorem, 255, 282
height, 95, 97
Hessian matrix, 30
Hilbert Basis Theorem, 342
Hilbert Nullstellensatz, 342
holomorphic at ∞, 225
holomorphic at the cusp, 261, 263
holomorphic differential, 312
homogeneous, 243, 273
homogeneous ideal, 344

homogeneous polynomial, 22
homology, 317, 321, 392
homomorphism, 368

ideal class group, 127
ideal of variety, 342, 344
identity, elliptic curve, 11
Igusa, 374, 390
inflection point, 35
integral, elliptic, 174
intersection multiplicity, 32
invariant differential, 314
inversion, 269, 285
inversion problem, 165
irreducible, 342, 345
isogenous, 365, 369
isogeny, 164, 309, 363, 369
    dual, 309, 369
isomorphic elliptic curves, 63
isomorphic over, 349
isomorphic varieties, 349
isomorphism, 368

j-invariant, 65, 226
Jacobi Inversion Theorem, 319
Jacobian variety, 310, 318, 361, 369, 370

L function, xi, 17, 201, 238, 267, 295
    automorphic, xi, 208
    Dirichlet, xii, 201
    motivic, xi, 207
Langlands program, xiii
lattice, 243, 318
Legendre symbol, 272, 298, 393
Legendre's Theorem, 5
level, 261, 333, 390
Lie group, 376
line, 19, 25
    at infinity, 19
    same, 19
    tangent, 27
linear system, 316, 361
Liouville Theorems, 152, 153
local coordinates, affine, 21, 345
local L factor, 294
local ring, 343, 346

Lutz-Nagell Theorem, xi, 15, 130, 144

Mazur's Theorem, 15
Mellin transform, 238
meromorphic differential, 312
Mestre-Oesterlé, 396
method of descent, 80
method of Diophantus, 6, 10, 115, 119
minimal Weierstrass equation, 290
    at prime p, 290
    global, 290
Minkowski, 5, 104
model, 333, 341, 349, 357
modular elliptic curve, 221
modular form, 224, 225, 261
    unrestricted, 224, 261
modular pair, 273
modular parametrization, 310, 390
modular polynomial, 335
Mordell's Theorem, xi, 14, 80, 95, 402
Mordell-Weil Theorem, xi, 80, 95, 402
morphism, 348, 356
    dominant, 349
    Frobenius, 360, 384
motivic L function, xi, 207
multiplicative, 196
    strictly, 197
Multiplicity One Theorem, 283
multiplicity, intersection, 32

Nagell, xi, 15, 130, 144
naive height, 95
negative, 12
newform, 283
Newton, 10
node, 77
Noetherian, 342
nondegenerate, 68
nonsingular, 25, 27, 58, 161, 343, 347
nonsplit case, 79
norm, 123
Nullstellensatz, 342
number field, 122

oldform, 283

order, 153

p-adic filtration, 138
p-adic norm, 134
p-integral, 135
p-reduced, 135
parabolic element, 303
parallelogram
  fundamental, 152
  period, 152
period, 151, 184
period lattice, 153
period parallelogram, 152
Petersson inner product, 242, 252, 280
plane curve, 24
Poincaré, 13, 67
point
  nonsingular, 25, 343, 347
  of inflection, 35
  singular, 25, 77
points, 19, 25, 342, 344
  at infinity, 19
Poisson Summation Formula, 211
pole, 360
prescribed torsion, 145
primitive, 213
Principal Axis Theorem, 37
principal Dirichlet character, 201
principal congruence subgroup, 250, 256
principal divisor, 316, 360
product
  convergent, 195
  of ideals, 126
  of varieties, 347
projective
  algebraic set, 344
  closure, 345
  curve, 345
  n-space, 344
  plane, 19
  plane curve, 24
  problem, 3
  transformation, 20, 345
  variety, 345
purely inseparable, 359
Pythagorean triple, 7

q-expansion, 225, 263, 265
quadratic residue symbol, 272, 298, 393

ramified, 359
rank, 15, 102, 107
rational map, 347
rational points, 19, 25, 342, 344
reciprocal roots, 198
reduction modulo p, 130, 134, 384
regular, 344, 346, 348
regulator, 106
representation theory, xiii, 407
result of rearranging, 167
resultant, 44
Ribet, xii, 18, 398, 399
Riemann-Roch Theorem, 317, 361
Riemann surface, 170, 311, 312, 316
Riemann zeta function 189, 194

same curve, 24
same line, 19
Schwartz function, 211
second degree Euler product, 199
Segre embedding, 347
Selmer's example, 8
separable, 359
Serre, xii, 18, 392, 399
Shimura, xii, 302, 374, 389, 390, 390, 404
Shimura-Taniyama, 373
singular, 12, 25, 27, 58, 77
smooth curve, 27
split case, 79
square root, 169
strictly multiplicative, 197
strong Weil curve, 392
subvariety, 368
summation by parts, 192
Swinnerton-Dyer, 17, 208, 386

tangent case, 75
tangent line, 27
Taniyama, 373
Taniyama-Weil Conjecture, xii, 18, 208, 221, 390, 399

INDEX

Tate normal form, 147
theta function, 209, 216
torsion subgroup, 15, 130
   prescribed, 145
torus, complex, 160, 183, 370
total winding number, 170
trace, 123, 396
translation, 364
twist, 308, 393

ultrametric inequality, 134
uniformizer, 350
unique factorization, 92, 127
unit, 92, 124
unramified, 359
unrestricted modular form, 224, 261

valuation, 351
vanishes at the cusp, 263
variety
   abelian, 367
   affine, 342
   projective, 345

very bad prime, 108

Wedderburn's Theorem, 375
Weierstrass
   Double Series Theorem, 158
   equation, 362
      minimal, 290
   form, 13, 42, 56, 57, 401
   $\wp$ function, 153
weight, 63, 224, 261, 333
Weil, xii, 18, 208, 221, 387, 390, 399
Weil conjectures, 403
Weil Converse Theorem, 388
Weil curve, 390
   strong, 392
width, 263
winding number, total, 170

zero, 360
zeta function, xi, 189, 194, 295